T0073663

RELATIVITY PRINCIPLES AND THEORIES
FROM GALILEO TO EINSTEIN

Relativity Principles and Theories from Galileo to Einstein

OLIVIER DARRIGOL

Research Director, Centre National de la Recherche Scientifique: SPHere

OXFORD

UNIVERSITY PRESS

OXFORD
UNIVERSITY PRESS

Great Clarendon Street, Oxford, OX2 6DP,
United Kingdom

Oxford University Press is a department of the University of Oxford.
It furthers the University's objective of excellence in research, scholarship,
and education by publishing worldwide. Oxford is a registered trade mark of
Oxford University Press in the UK and in certain other countries

Published in the United States of America by Oxford University Press
198 Madison Avenue, New York, NY 10016, United States of America

British Library Cataloguing in Publication Data

Data available

Library of Congress Control Number: 2021938605

ISBN 978-0-19-284953-3(Hbk)

DOI: 10.1093/oso/9780192849533.001.0001

Printed and bound by
CPI Group (UK) Ltd, Croydon, CR0 4YY

CONTENTS

PREFACE

> [The theory of relativity] did not at all originate in a revolutionary act but as a natural development of a line that can be traced through centuries.[1] (Albert Einstein, 1921)

The relativity theories of the early twentieth century deeply altered our most basic notions of space, time, and motion. They made time depend on the observer, they entwined space and time, and they reduced gravitation to inertial motion in a curved space–time. Historians have tried to explain the radicalness of these changes in various ways: through Einstein's peculiar genius, through contemporary developments in philosophy or technology, and through a major crisis in physics around 1900. These three explanations correspond to three alleged preconditions of radical change: singular minds, external causation, and internal crisis. The internal crisis approach is perhaps the most convincing, because philosophical and technological considerations may have been internalized before they served the founders of relativity, and because the singularity of these founders may boil down to their ability to detect and solve crises. But this approach can only be a first approximation, and a fuller history should combine internal, external, and biographical elements.

Whether or not they come close to this ideal, the received histories of relativity are short-term, fragmented histories. Special and general relativity are usually treated separately, and the maximal time span is about a century long (in the internal crisis approach), with a focus on the late nineteenth and early twentieth century. There are exceptions: Jürgen Renn has compared the emergence of relativity theory with the emergence of early modern mechanics; Jean Eisenstaedt has developed analogies between eighteenth-century Newtonian optics and general relativity; several philosophers, including Roberto Torretti and Robert DiSalle, have explored the various concepts of space and time in the history of physics from Aristotle to relativity; and one physicist, Julian Barbour, has probed the long-term history of dynamics from a Machian viewpoint. But their main intention was not to demonstrate genuine, long-term historical connections: it was to instruct on the variety of possible concepts, to reflect on their similarities and differences, or (in Renn's case) to illustrate a generic theory of conceptual development.[2]

In contrast, this book presents a long-term, multi-approach history of relativity from Galileo to Einstein, in the various contexts of mechanics, mathematics, philosophy, astronomy, optics, and electrodynamics. One might fear that any attempt to connect events three

---

[1] Einstein [1921], p. 431.

[2] Torretti 1978; Renn 1993; Barbour 2001; Eisenstaedt 2005b; DiSalle 2006. Also worth mentioning are Marie-Antoinette Tonnelat's *Histoire du principe de relativité* (Tonnelat 1971) for an analysis of the successive meanings of relativity from antiquity to general relativity; and Raffaella Toncelli's dissertation on the role of principles in the construction of relativity theory (Toncelli 2010), which also has a very large scope.

centuries apart would be superficial and artificial, and this was my own opinion until I realized, after a few partial pre-histories, that genuine historical connections existed between the various uses of relativity principles across three centuries. I observed two kinds of connectedness. First, at least until the early twentieth century physicists had a much longer memory than they have nowadays. They read and exploited literature written centuries earlier, whereas today scientists tend to disregard anything older than a few years. Second, there were indirect historical connections through chains of successive borrowings. In both cases it is essential, in a properly historical study, to take into account the change of context in the borrowing process, even when the borrower is not aware of this change. Otherwise we would misunderstand the successive systems of thought, and we would misrepresent the transfer of knowledge from one system to another.[3]

When developed with sufficient care, this long-term view has several advantages. It shows that questions about the nature of space, time, and inertia traversed the history of physics from the early modern period to the relativity theories, although the answers given to these questions varied considerably. It explains how the relativity principle emerged as a true, constructive principle as early as the seventeenth century, how it came to be named so in the mid-nineteenth century to mark its constructive power in mechanics, and how this name and concept reached the young Einstein. It demonstrates deep analogies between some authors' dismissal of Newton's absolute space and Einstein's later dismissal of the electromagnetic ether. It establishes an indirect historical link between a corollary of Newton's *Principia* and Einstein's equivalence principle, through a French principle of accelerative relativity. It situates both principles in a space–time–inertia tangle that originated in Galileo's and Newton's theories of motion and evolved into Einstein's final reduction of gravitation to inertia, through intermediate steps in the eighteenth and nineteenth century. Also, there are advantages in treating the histories of special and general relativity in the same volume: this highlights the continuity of Einstein's endeavors, and shows the importance of Minkowski's and Laue's versions of the special theory in Einstein's quest for a relativistic field theory of gravitation.

Any long-term history faces difficulties in naming basic concepts and principles across time. Today, we usually define the relativity principle through the complete equivalence of all inertial reference frames. It is impossible or at least dangerous to apply this definition uniformly from Galileo to Einstein. The name "reference frame" first appeared in 1884, under James Thomson's pen. The concept earlier existed under different names, guises, and extensions: a metonymical boat for Galileo, a "space" for Newton, and a "system of axes" for Laplace. Statements of relativity had a variable status: an empirical fact for Galileo, a theorem for Newton, and a genuine principle for Huygens, Laplace, and later French authors. There were different theorems and principles of relativity (hence the plural in my title) according to the implied class of equivalent frames: inertial frames and accelerated frames for Newton, Laplace, and Bélanger; just inertial frames for Huygens, Poincaré, and the early Einstein; and any accelerated frame and even a framing "mollusk" for the later Einstein. Until the late nineteenth century, relativity principles and theorems were

---

[3] By context I here mean intellectual and experimental context, although the changes in this narrower context usually imply the broader socio-cultural context.

most frequently formulated in the active form, as the absence of effects of a commonly impressed motion on the relative motions of a system of bodies under given forces, and more
rarely in the passive form, as the lack of effects of a change of reference frame. All these
ambiguities should be kept in mind when, for the convenience of the reader, I will sometimes use "reference frame" instead of the true historical term. That said, there is enough
kinship between the various kinds of relativity to justify a global historical account of their
meaning and purpose.

   Chapter 1 describes the new theories of motion of the seventeenth century and the emergence of relativity facts, theorems, and principles in conjunction with new concepts of
mechanical inertia. Chapter 2 recounts how, following Huygens's pioneering derivation
of the laws of free fall, a number of French authors used a relativity principle to derive
Newton's second law of motion, how the name "principle of relative motions" appeared in
this context, how Poincaré modified this name, and how he passed it to Einstein. Chapter 3
discusses difficulties in the correlative definitions of space, time, and inertia from Newton
to late-nineteenth-century critics of the foundations of mechanics, the repeated rejections
of Newton's absolute space, and Streintz's final suggestion of an intrinsic Galilean space–
time. At this point, the reader will have viewed the emergence of a mechanical relativity
principle with constructive and representational functions. Chapter 4 is about the relativity that emerged in corpuscular optics and survived wave optics despite the existence of the
ether as a privileged frame. As is explained in Chapter 5, a similar tension between ether
and relativity existed in the electrodynamics of moving bodies, and Lorentz alleviated it
without suppressing it in his electron theory of the 1890s. Chapter 6 shows how Poincaré
raised optical and electrodynamic relativity to a principle on par with the mechanical relativity principle, and how he transformed Lorentz's theory into a first version of special
relativity. Chapter 7 describes the elaboration of more radical and more potent forms of
special relativity in the hands of Einstein, Minkowski, and Laue. Chapter 8 shows how
Riemann and his followers invented the mathematics Einstein employed in general relativity, also how the emergence of non-Euclidean geometry inspired Poincaré and Minkowski.
Chapter 9 is a medium-sized account of the long and twisted genesis of general relativity.
Chapter 10 covers the difficulties Einstein's early readers encountered in trying to make
sense of this theory, and the great clarification brought by Weyl, Eddington, and others.
The epilogue brings out the interconnections of the developments described in the previous
chapters. Two of these chapters, the eighth and the ninth, are formally and mathematically
more demanding than the others. Readers who do not wish to enter technicalities may
content themselves with the summaries given in their concluding sections.

   There is plenty of secondary literature for most of the topics addressed in this book.
The most important studies are indicated at the beginning of each chapter, and the rest
in footnotes. For the moment, it will be sufficient to signal the sources I found most
useful for my project. On the ether-theoretical background to special relativity, there is
Edmund Whittaker's old and rich *A History of Aether and Electricity*, Tetu Hirosige on
Lorentz and electrodynamic origins, and Russell McCormmach on the electromagnetic
worldview. Arthur Miller's *Albert Einstein's Special Relativity* remains a standard reference
of great value. More recent contributions to the history and contexts of special relativity include Michel Janssen's dissertation on the relation between Lorentz's theory and

special relativity in the light of the Trouton–Noble experiments, Scott Walter's dissertation on Minkowski and the mathematical reception of relativity, Peter Galison's *Einstein's Clocks, Poincaré's Maps*, Richard Staley's *Einstein's Generation*, Jean-Pierre Provost and Christian Bracco's studies of Poincaré's relativity theory, Marco Giovanelli's thorough study of principles-based vs. constructive approaches to relativistic dynamics, and Galina Weinstein's *Einstein's Pathway*. On the history of general relativity, much value can be found in Abraham Pais's *Subtle Is the Lord*, in John Stachel's contemporary discussions of the rigid-body problem, in John Norton's groundbreaking study of Einstein's Zürich notebook, in Jean Eisenstaedt's history of the Schwarzschild singularity, in Weinstein's *General Relativity Conflict and Rivalries*, and in the multi-volume *Genesis* directed by Jürgen Renn. This last, collective enterprise groups highly competent and thorough studies by John Norton, John Stachel, Jürgen Renn, Tilman Sauer, Michel Janssen, and a few other scholars. Useful studies of pre-relativistic considerations on the relativity of motion include Christiane Vilain's *La mécanique de Christiaan Huygens*, Giulio Maltese's study of Euler's relativity considerations, Marius Stan's articles on Huygens, Kant, and relativity, Jean Eisenstaedt's *Avant Einstein*, and Alberto Martínez's *Kinematics*. Lastly, a few historico-philosophical studies helped me clarify conceptual issues encountered in the primary sources: Lawrence Sklar's two *Spacetime* books, Roberto Torretti's *Philosophy of Geometry*, Michael Friedman's *Foundations of Spacetime Theories* and his *Kant's Construction of Nature*, John Earman's *World Enough*, Michel Paty's *Einstein Philosophe*, Julian Barbour's *The Discovery of Dynamics*, Harvey Brown's *Physical Relativity*, Robert DiSalle's *Understanding Spacetime*, Thomas Ryckman's *The Reign of Relativity* and his recent *Einstein*.[4]

This book is the end result of an old interest of mine in the origins of relativity theory. I fell in love with this theory in my teen years, while reading Landau and Lifshitz on a Cap Ferret beach on the western shores of France. In the 1980s, I contributed to the French Einstein project under the direction of Françoise Balibar, benefiting from John Stachel's expertise for the volumes on relativity. In the 1990s, I studied the electrodynamic origins of the special theory as the natural conclusion of a history of electrodynamics. More recently I worked on the optics of moving bodies, on Riemann's geometry and consequences, on the early reception of general relativity, and on the genesis of this theory. My return to general relativity and my idea to write a history of relativity from Galileo to Einstein derived from my philosophical interest in the coordination between theory and experience, and from my decision, in the light of Ryckman's *Reign of Relativity*, to focus on the most basic and oldest problem of coordination, the spatial and temporal ordering of events. Ryckman having explored this problem through the philosophical reception of general relativity, I wanted to retrace its evolution from early modern mechanics to general relativity.

---

[4]Whittaker 1951; Hirosige 1966, 1969, 1976; McCormmach 1970; Sklar 1974, 1985; Torretti 1978; Stachel 1980, 1989a, 1989b; Miller 1981a; Eisenstaedt 1982, 2005; Pais 1982; Friedman 1983, 2013; Earman 1989; Paty 1993; Janssen 1995; Vilain 1996; Walter 1996a; Maltese 2000; Barbour 2002; Galison 2003; Brown 2005; Ryckman 2005, 2017; Provost and Bracco 2006, 2009; DiSalle 2006; Staley 2008; Martínez 2009; Stan 2009, 2016; Weinstein 2015a, 2015b; Giovanelli 2020; Renn et al. *GGR*. Also worth mentioning is Christopher Ray's *The evolution of relativity* (Ray 1987) for its analysis of the foundations of general relativity from Einstein to Hawking, with Mach's philosophy of motion as a backdrop.

During all these years, I benefited from the lively and pleasant environment of the SPHere research team at Centre national de la recherche scientifique (CNRS) in Paris, with unabated support from its successive directors Michel Paty, Karine Chemla, Pascal Crozet, David Rabouin, and Sabine Rommevaux. I had numerous, instructive conversations with my closest colleagues in this team: Nadine de Courtenay, Jan Lacki, Sara Franceschelli, and Martha-Cecilia Bustamante. My research was greatly eased and enriched by my affiliation to UC-Berkeley's renowned Office for History of Science and Technology under the wings of John Heilbron, Cathryn Carson, and Massimo Massotti. While researching the history and prehistory of relativity theory, I profited from enlightening exchanges with John Heilbron, Jed Buchwald, John Norton, Peter Galison, Harvey Brown, Tom Ryckman, and Thibault Damour. Back on the very beach on which I first encountered relativity theory, I fondly remember all these colleagues and friends and I warmly thank them for easing, guiding, and illuminating my intellectual voyage through the mysteries of space and time coordination.

CONVENTIONS AND NOTATIONS

- For the sake of the physicist reader, I use modernized and standardized notation whenever the original notation did not essentially condition the historical developments. For example, I use the modern vector notation in older mechanics and electrodynamics although its propagation began only in the late nineteenth century. For general relativity, I use the modern tensor notation, which departs from Einstein's original notation in ways indicated at the beginning of Chapter 9. For the history of these notations, I invite the reader to read the classical studies by Michael Crowe and Karin Reich. While the historical significance of a change of notation may sometimes be high, its conceptual and practical importance has often been exaggerated. For instance, in older mechanics, the geometrical representation of vector quantities permitted an intrinsic conception of vectors without the vector notation; in older electrodynamics, the practice of writing only one of the three components of a Cartesian-coordinate equation (the two other components being implicitly given by circular permutation) compensated for the lack of an intrinsic notation.[5]

- Citations are in the author–date format and refer to the appended bibliography (I use "cf." in the French way, to refer to a source in the secondary literature). Square brackets around a date indicate a manuscript source. Abbreviations are listed on p. 416 before the bibliography.

- Translations from Latin, German, and French are mine unless the cited source already is a translation.

---

[5] Crowe 1967; Reich 1994. For the historical representation of mechanical quantities, cf. Martínez 2009, chap. 4. On the use of four-vectors in Minkowski space, cf. Walter 2007b.

1

RETHINKING MOTION IN THE SEVENTEENTH CENTURY

> I know that you, Simplicio, have gone from Padua by boat many times, and, if you will admit the truth of the matter, you have never felt within yourself your participation in that motion except when the boat has been stopped by running aground or by striking some obstacle, when you and the other passengers, taken by surprise, have stumbled perilously.[1] (Galileo Galilei, 1632)

The Scientific Revolution brought a drastic rethinking of motion, now subjected to laws never dreamt of by Aristotelian philosophers. A few brave supporters of the Copernican cosmos, Kepler and Galileo foremost, measured the consequences of changing the reference of planetary motion from the earth to the sun. By erasing the ancient difference between the sublunary and celestial worlds, the new system required common concepts of motion in the two worlds. In addition, Galileo and Descartes promoted a fully mathematical, geometric understanding of the physical world, based on the simplest and least speculative concept of change, that is, locomotion. Beeckman, Galileo, and Newton appealed to the motion of atoms; Descartes and Huygens to the relative motions of the particles of a plenum. The purpose of this chapter is to introduce the relevant theories of motion, with special attention to the relativity issue.[2]

The motion of a body is always defined in reference to something regarded as immobile. As we will see in this chapter, the concept of this "something" varied considerably in seventeenth-century mechanical philosophy. For Descartes (as for Aristotle), it was the medium in which the body is immersed. For Galileo, it was a distinguished material system, for instance the sun and fixed stars, the earth, or a boat; for Newton, it was space regarded as an immaterial, rigid, immobile, penetrable substratum of all things. To each kind of reference corresponded a different concept of inertia. Galileo, Beeckman, Descartes, Huygens, and Newton all assumed the persistence of the motion of a body left to itself for a variety of reasons: response to the need to harmonize celestial and terrestrial motions, continuity with the medieval concept of impetus, logical necessity in a neo-atomist world picture, God's predilection for permanence. For Galileo, inertia could be circular, and it covered both the motions of planets around the sun and the motion of a ball on a horizontal table on earth; for Descartes, inertial motion was rectilinear and referred to

---

[1] Galilei 1632, p. 255.

[2] On the Scientific Revolution, cf. Cohen 2015, and further references therein.

Relativity Principles and Theories from Galileo to Einstein. Olivier Darrigol, Oxford University Press.
© Olivier Darrigol (2022). DOI: 10.1093/oso/9780192849533.003.0001

the surrounding medium; for Newton, it was rectilinear in absolute space. These different concepts of inertia implied different laws of motion and different takes on the relativity of motion.

For Galileo, the motion of a system of bodies obeyed the same laws whether it was referred to the earth or to a boat uniformly moving on the sea. For Huygens, the similarity extended to any reference frame in which the law of uniform, rectilinear inertia is valid. For Newton, this similarity applied to any reference frame moving uniformly and rectilinearly in absolute space; it also applied to a system of bodies subjected to a common acceleration, as is the case when the bodies are all subjected to the attraction of a remote heavy mass. For Descartes, the laws of motion could depend only on the relative position of the particles of the world. In modern parlance, Descartes anticipated Mach's elimination of any immaterial reference of motion, while Galileo, Huygens, and Newton detected a basic invariance of the laws of motion within a privileged class of reference frames. Newton included uniformly and rectilinearly accelerated frames in this class, thus partially anticipating Einstein's equivalence principle.

Another important distinction concerns the architectonic role of the invariance asserted in the relativity statements of Galileo, Newton, and Huygens. As we will see, Galileo was mostly concerned with a law of relativity, directly inferred from observations (on a boat) or indirectly from the empirical laws of free fall. In contrast, Newton's relativity statements were theorems deduced from his three laws of motion (which he believed he had deduced from experience). Huygens inaugurated a relativity principle that comes before the mechanical laws and contributes to their derivation. This was a crucial step in regard to the later role of the principle in justifying or constructing theories of motion.

It would be artificial to limit this chapter to the consideration of a few extracts of the major text in which relativity statements appeared in the seventeenth century. As was just indicated, such considerations cannot be separated from the theories of motion and the philosophies of nature to which they belong. Also, these theories and philosophies form part of the background in which later recourse to relativity principles occurred. With this in mind, we will successively consider the decisive contributions of Galileo, Beeckman, Descartes, Newton, and Huygens.

1.1 Galileo's science of motion

Inertia and relativity

In 1632, after many years of self-restraint, Galileo Galilei published his lengthy, witty, and risky *Dialogo sopra i due massimi sistemi del mondo*, a powerful defense of the Copernican system through a dialogue between three fictional characters: the brilliant Copernican Salviati, the obtuse scholiast Simplicio, and the inquisitive Sagredo. An important part of their second day is devoted to a then frequent objection to the rotation of the earth: that it would imply never observed alterations of the way bodies move on earth. In particular, a stone, when falling from a vertical tower, would not fall at the foot of the tower because during the fall the ground and the tower would be carried eastward by the rotating earth. To fortify this point by experiment, the objector might add that on a moving ship, a stone released from the top of a mast should fall behind the mast as a result of the progression of the ship.[3]

Salviati denies both the point and the experiment. The Aristotelians, he explains, have erred in ignoring the horizontal velocity of the stone at the beginning of its fall from the top of a tower. This velocity, being part of the natural circular motion of all terrestrial bodies, has to remain constant during the fall so that stone and tower constantly share the same horizontal motion and the stone therefore falls along the mast. The persistency of horizontal motion was indeed alien to Aristotle's physics. In this doctrine, natural motion in the sublunary world is directed to or from the center of the universe, which coincides with the center of the earth. Horizontal motion is "violent" or forced motion of external origin, and it cannot persist without a sustaining force. In the first day of the dialogue, Salviati has purposely redefined natural motion as motion that does not disturb the global arrangement of the world. Accordingly, circular motion is equally natural for the rotating earth and for the circulation of the earth and the planets around the sun, whereas for Aristotle circular motion is natural only for celestial bodies moving around the earth.[4]

Besides this cosmological consideration, Salviati asserts that for the moving ship too, the falling stone accompanies the ship in its horizontal motion. If Aristotle's followers had truly performed the experiment, Salviati tells Simplicio, they would have seen that the ship's motion has no effect on the fall observed by the sailors. Salviati bases his conviction on two empirical facts: the constant velocity of a ball rolling on a perfectly smooth horizontal table (the resistance of the air being negligible), and the independence between vertical and horizontal motion during the rolling of the ball on an inclined plane. Most vividly, he argues that common experience confirms the absence of effects of a ship's uniform motion on the passengers' internal observations:[5]

> Shut yourself up with some friend in the main cabin below decks on some large ship, and have with you there some flies, butterflies, and other small flying animals. Have a large bowl of water with some fish in it; hang up a bottle that empties drop by drop into a wide vessel beneath it. With the ship standing still, observe carefully how the little animals fly with equal speed to all sides of the cabin. The fish swim indifferently in all directions; the drops fall into the vessel beneath; and, in throwing something to your friend, you need throw it no more strongly in one direction than another, the distances being equal; jumping with your feet together, you pass equal spaces in every direction. When you have observed all these things carefully (though there is no doubt that when the ship is standing still everything must happen in this way), have the ship proceed with any speed you like, so long as the motion is uniform and not fluctuating this way and that. You will discover not the least change in all the effects named, nor could you tell from any of them whether the ship was moving or

[3] Galilei 1632, p. 126; also Galilei to Ingoli 1624, in Finocchiaro 1989, pp. 182–188. Cf. Drake 1978; Vilain 1996, chap. 2; Heilbron 2010, pp. 262–263, 270–271; Swerdlow 2013.

[4] Galilei 1632, pp. 138–143 (stone and tower), 31–32 (perpetual circular motion).

[5] Galilei 1632, pp. 145–149, 186–188 (citation). As Galileo understood it, the inertial motion from the top of the tower slightly exceeds the horizontal motion of the lower parts of the tower (since they are closer to the axis of the rotating earth), so that the falling body should experience a slight deviation to the east; cf. Burstyn 1965. On anticipations of Galileo's ship-cabin by Giordano Bruno and others, cf. Capecchi 2014, pp. 158–159.

standing still. In jumping, you will pass on the floor the same spaces as before, nor will you make larger jumps toward the stern than toward the prow even though the ship is moving quite rapidly, despite the fact that during the time that you are in the air the floor under you will be going in a direction opposite to your jump. In throwing something to your companion, you will need no more force to get it to him whether he is in the direction of the bow or the stern, with yourself situated opposite. The droplets will fall as before into the vessel beneath without dropping toward the stern, although while the drops are in the air the ship runs many spans. The fish in their water will swim toward the front of their bowl with no more effort than toward the back, and will go with equal ease to bait placed anywhere around the edges of the bowl. Finally, the butterflies and flies will continue their flights indifferently toward every side, nor will it ever happen that they are concentrated toward the stern, as if tired out from keeping up with the course of the ship, from which they will have been separated during long intervals by keeping themselves in the air. And if smoke is made by burning some incense, it will be seen going up in the form of a little cloud, remaining still and moving no more toward one side than the other. The cause of all these correspondences of effects is the fact that the ship's motion is common to all the things contained in it, and to the air also. That is why I said you should be below decks; for if this took place above in the open air, which would not follow the course of the ship, more or less noticeable differences would be seen in some of the effects noted. No doubt the smoke would fall as much behind as the air itself. The flies likewise, and the butterflies, held back by the air, would be unable to follow the ship's motion if they were separated from it by a perceptible distance. But keeping themselves near it, they would follow it without effort or hindrance; for the ship, being an unbroken structure, carries with it a part of the nearby air. For a similar reason we sometimes, when riding horseback, see persistent flies and horseflies following our horses, flying now to one part of their bodies and now to another. But the difference would be small as regards the falling drops, and as to the jumping and the throwing it would be quite imperceptible.

It is easy to misunderstand the kind of relativity Galileo had in mind here. In Salviati's words, "the cause of all these correspondences of effects is that the ship's motion is common to all the things contained in it." This seems to be echoing his earlier remark that the common reference for the motion of a system of bodies can be arbitrarily changed without altering the relative motions of these bodies:

> It is obvious . . . that motion which is common to many moving things is idle and inconsequential to the relation of these movables among themselves, nothing being changed among them, and that it is operative only in the relation that they have with other bodies lacking that motion, among which their location is changed.

The context of this statement is what we would now call the full observational equivalence between Copernic's heliocentric system and Tycho's system in which the same relative motions are described in an earth-bound frame (to put it in modern terms). The common motion need not be uniform in this case, whereas Salviati requires his ship to move uniformly. As he later makes clear, nonuniformity would disturb the similarity of the phenomena observed in the moving ship and in the anchored ship:

> I know that you, Simplicio, have gone from Padua by boat many times, and, if you will admit the truth of the matter, you have never felt within yourself your participation in that motion except when the boat has been stopped by running aground or by striking some obstacle, when you and the other passengers, taken by surprise, have stumbled perilously.

The point is that a sudden acceleration of the ship is not shared by its passengers, so that the motion (or rest) of the passengers with respect to the ship is not conserved. A modern physicist here recognizes the effect of inertia: Galileo implicitly selects the reference frame so that the principle of inertia holds in it. Namely, he wants the rolling of a ball on a table or the horizontal motion of a falling stone to be still uniform when the ship is moving.[6]

The Galilean invariance of the other phenomena observed in a moving ship's cabin is an experimental fact. In addition, the Galilean invariance of free fall derives from the constancy of horizontal motion and the independence of vertical and horizontal motions in the earth-based frame. So there is no doubt that Galileo perceives the intimate connection between the persistence of unimpeded motion and his relativity principle. Still, one should think twice before identifying his concept of persisting motion with the modern concept of inertia. In the case of a stone falling from a tower, he regards the horizontal motion as a natural motion shared with the rotating earth. This motion is circular and not rectilinear as modern inertia would have it. In the case of a stone falling from the top of a mast in a moving ship, the horizontal motion is no longer natural but its persistence remains empirically true. As Galileo extrapolates from a ball rolling on a curved horizontal surface, he regards the persisting motion as circular in this case too.[7]

Yet, when Salviati later discusses the effects of the earth's rotation on the motion of a bullet shot by a cannon, he assumes the bullet's (absolute) motion to be rectilinear and uniform as long as the effects of gravity and air resistance can be ignored. He is here nearing the modern concept of inertia, even more so when he espouses the medieval concept of an inherent tendency to motion, the "impetus," and denies the orthodox Aristotelian view that the medium is responsible for the persistence of motion started by human force. In Galileo's view, the medium can only impede the motion and the bullet's motion is conserved as long as the effects of the medium can be ignored.[8]

Salviati concedes to Simplicio that the kind of relativity he has in mind is not easily conceived, at least for anyone with a scholastic background: "You are not the first to feel a great repugnance toward recognizing this nonoperative quality of motion among the things which share it in common." This is why Salviati spends so much time arguing that the conservation of horizontal velocity and the independence of the vertical and horizontal velocity in free fall imply the fall of a stone along the mast of the moving ship. Sagredo then notes that the trajectory of the falling stone, as seen from the shore, would be analogous to the trajectory of a bullet shot from a horizontal cannon. Accordingly, the bullet should take the same time to reach the ground (which is supposed to be perfectly level) for any

[6] Galilei 1632, pp. 116, 255.

[7] Galilei 1632, pp. 142 (stone and tower), 148 (stone and mast).

[8] Galilei 1632, pp. 178 (cannon), 149–151(medium). Giambattista Benedetti anticipated rectilinear inertia in 1585: cf. Drake and Drabkin 1969, p. 156. On impetus, cf. Drake 1975; Wallace 1981; Capecchi 2014, pp. 63–78.

value of the initial velocity. At first glance, this looks like a proto-application of Galilean relativity to deriving a law of motion. But this is not quite so because, in order to prove the fall of the stone along the mast, Salviati has earlier relied on the independence of the vertical and horizontal motions, which by itself implies Sagredo's law. All in all, Galileo's relativity statement is more a law than a principle.[9]

The laws of free fall

In his response to a classical objection to a rotating earth, Salviati is naturally led to the consideration of free fall and he repeatedly relies on results earlier established by his friend "the Academician," that is, Galileo. For example, Salviati relies on the independence of the vertical and horizontal components of motion, on the slower fall of a ball rolling on an inclined plane, on the symmetry of the ascent and descent of a ball shot vertically upward, and even on the proportionality between the descent and the square of the time of fall from rest (when Sagredo asks him what would be the trajectory of a body falling with an initial horizontal velocity).[10]

Although Galileo obtained these results early in the century, he published the evidence very late, in the *Discorsi* of 1638. In the received scholastic view, a falling body is suddenly accelerated until it reaches a well-defined velocity. The velocity of fall is proportional to the weight of the body, and inversely proportional to the density of the medium. Galileo refutes all these claims by a mixture of reasoning and experimentation. Experiments with inclined planes and pendulums have convinced him that the velocity and impetus acquired during the fall depend only on the vertical distance traveled, and that this impetus would bring the falling body back to its original height if the motion were redirected upward (as occurs for a pendulum when the bob passes its lowest point). For slightly inclined planes, the fall is much slower than in free vertical fall, because the traveled distance is much longer. Galileo was able to measure the time of this fall with a water clock and found it to be proportional to the square root of the descent. As he understood, the effect of the inclined plane is to diminish the effective propelling force in a constant proportion (the sine of the inclination), and the free vertical fall therefore obeys a similar law.[11]

Galileo then proves that this law implies a constant acceleration of the falling body, that is, a velocity linearly increasing with the time of descent. For a body launched with a horizontal velocity, he derives the parabolic shape of the trajectory by composing a uniform horizontal motion with a constantly accelerated fall. For a body thrown upward or moving against the slope of an inclined plane, he understands that gravity implies a decrease of the velocity at a constant rate until reversal at the point of zero velocity.[12]

Galileo's success in unveiling these simple laws crucially depends on his focus on the ideal case in which the resistance of the air can be ignored because the density of the falling

[9] Galilei 1632, pp. 171 (citation), 154–155 (canon).

[10] Galilei 1632, pp. 30–31, 145–149, 163–164, 221–222.

[11] Galilei 1638, pp. 105–106 (scholastics), 205 (inclined plane), 206–207 (pendulum), 212–213 (experiment).

[12] Galilei 1638, pp. 208–210 (acceleration law), 268–308 (parabola), 200–201 (upward free motion), 243–245 (upward motion on inclined plane).

body is sufficiently large and the time of fall is sufficiently small. Also, he refrains from speculating on the cause of the acceleration and rather computes and measures its effects. He knows only that some "mutual cooperation of the parts" or some "concordant instinct and natural tendency" is responsible for the rotundity of the earth, the moon, the sun, and the planets, for the weight and fall of objects near their surface, and perhaps even for the velocity of the earth and planets around the sun (if this velocity results from a fall from a fixed solar distance).[13]

As for the effect of the weight of bodies on their fall, Galileo claims that sufficiently dense bodies all fall with the same velocity, irrespective of their weight. Whether or not he dropped objects from the top of the Tower of Pisa, he could easily verify this law for constrained fall on an inclined plane or with a pendulum. In the *Discorsi*, he also reasons that the contrary assumption of a faster fall of heavier bodies would lead to a logical contradiction: on the one hand, the compound body formed by attaching a heavy body to a light body should fall at an intermediate speed because the light body slows down the heavy one; on the other hand, as the weight of the compound body is greater than the weight of its two components, it should fall faster than both of them. This reasoning has sometimes been regarded as a piece of sophistry, because it would for instance imply that all electric charges should move at the same speed in a given uniform electric field. The objection is invalid because Galileo assumes the speed of fall depends on weight only, whereas the acceleration of an electric charge depends on two parameters, charge and mass. That two different parameters, inertial mass and gravitational mass, might also apply to free fall was not an option in Galileo's context of reasoning: even though he rejected the Aristotelian proportionality between speed of fall and weight, he conserved the weaker assumption that the speed of fall depends on weight only. Quite often, the singularity of Galileo's reasoning in our eyes depends on his being at an intermediate stage between Aristotle and Newton.[14]

The definition of motion

What is now called "motion" truly was *moto locale* or *movimiento locale*, in order to avoid confusion with the broader Aristotelian concept of motion, which includes any change of state. In *Il Saggiatore* (1623), Galileo made clear that he regarded the kind of motion implied in change of place as more definite and less subjective than changes in other perceived properties of bodies. He even flirted with the atomist idea of reducing the latter properties (secondary qualities) to the shape, size, and motion of atoms. At any rate, he was mostly concerned with astronomy and mechanics in which the narrower concept of motion is dominant.[15]

---

[13] Galilei 1638, pp. 275–277 (idealization), p. 202 (cause unknown); Galilei 1632, pp. 33–34 (universal attraction), pp. 20–21, 29 (velocity of planets).

[14] Galilei 1638, pp. 128–129 (pendulums), 107–108 (compound body). Galileo restricts his reasoning to bodies made of the same material, presumably because otherwise the Archimedean push would not be a constant fraction of the weight. But he believes the result to extend to all bodies (ibid. pp. 116–117). On the alleged sophistry, see, for example, Hacyan 2015.

[15] Galilei 1623, pp. 275–276 [Drake transl.]. Cf. Drake and O'Malley 1960, pp. 310–311.

In the *Dialogo*, Galileo made clear that for him place and motion were defined with re-spect to a concrete body or system of bodies considered at rest: it could be a ship, the earth, or the sun together with the fixed stars. For free motion referred to a spherical body like the earth, he assumed constant acceleration toward the center as well as uni-form circular motion around the center. For free motion referred to distant bodies (the sun and fixed stars), he assumed uniform rectilinear motion. He thus subverted the distinc-tion between natural and violent motion, and he denied the essential role of the medium in the Aristotelian understanding of these two kinds of motion. That said, his concept of free motion remained tied to the natural distribution of matter in the world, and he did not give a complete theory of motion that could serve as the foundation of a new physics.[16]

1.2 Beeckman and Descartes on free fall

As stated in the previous paragraph, for Galileo the place of an object is defined with respect to a concrete reference body. For Aristotle, the place (τόπος) of an object is the touching surface of the medium in which it is immersed; and change of place, or locomotion, depends on the properties of this medium. For both thinkers, there is no concept of empty space as a container in which objects are placed and can move. The word "space" (χῶρος, διάστημα, *spazio*), whenever it occurs, refers to room or interval between concrete bodies.[17]

The modern concept of empty space as an immovable receptacle of objects has roots in ancient Greek atomism and also in peripatetic criticism of Aristotle's concept of mo-tion (e.g., by Aristotle's definition, the spinning of a globe does not seem to involve any motion). In the late Renaissance, this new concept of space penetrated the writings of Fran-ciscus Patricius, Giordano Bruno, and Tommaso Campanella. In the seventeenth century, it prospered in the neo-atomist hands of Isaac Beeckman in the Netherlands and Pierre Gassendi in France. For an atomist philosopher, the world is made of atoms traveling through emptiness, and it becomes natural to define space as the immovable container of matter.[18]

Space, being empty, is also homogeneous and nonsubstantial. Consequently, an atom originally at rest must remain at rest and an atom originally in motion must keep moving in the same direction with the same velocity until a collision with another atom occurs. Both Beeckman and Gassendi asserted the persistence of motion in empty space. As Beeckman put it in 1614: "The mind conceives very easily that in a vacuum motion never turns to rest because there is no cause to alter the motion: indeed, nothing can be changed without some cause of change." By a similar argument, the direction of motion in a vacuum can-not change because nothing can cause this change. Beeckman and Gassendi nonetheless

---

[16] Galilei 1632, pp. 115–117. As Edmund Halley asserted in 1718 from an unreliable comparison between ancient Greek and Tycho's observations, and as Jacques Cassini definitely proved in 1738, the so-called fixed stars actually move (Halley 1718; Cassini 1738; cf. Verbunt and van der Sluys 2019).

[17] Cf. Jammer 1954, chap. 1.

[18] Jammer 1954, chaps. 2–3; De Risi 2015.

maintained the naturalness of circular motion (around the sun and around the center of the earth) and thus did not quite reach the modern concept of rectilinear inertia.[19]

Beeckman also had an interest in free fall, well before Galileo published his views on this topic. Beeckman believed (rectilinear) inertia played a fundamental role in this process. In June 1618, he entered the following remark in his diary:

> Here is the reason why two motions are compounded [for a falling stone]: firstly, there is the natural motion downwards; secondly, the stone, once set in motion, persists in its motion, and the natural motion again adds to this motion.

This idea is similar to Galileo's composition of horizontal inertial motion and vertical accelerated motion in the case of a horizontally shot bullet. Later in the same year, Beeckman asked young René Descartes, then in Breda as a mercenary for the Dutch army, to determine the law of free fall based on this idea. A portion of Descartes's reply is recorded in a letter to Marin Mersenne of November 1629:

> Firstly, I assume that the motion impressed on a body goes on forever unless it is removed by some other cause. That is to say, once [a body] has been set in motion, it will keep moving forever with equal velocity. Now suppose that a weight at A is pulled by its gravity toward C. As soon as it has begun to move, if its gravity is suppressed, it will nevertheless persist in the same motion until it reaches C, and it will descend from A to B at the same speed as from B to C [A, B, and C are three vertically aligned points, with AB = BC]. Since in reality gravity keeps acting and adds new downward forces at each successive instant, the body must travel much faster in BC than in AB because while moving in BC it retains all the impetus of its motion in AB and in addition it keeps acquiring new impetus by gravity.

Descartes goes on with a false deduction of the time ratios for the successive intervals AB and BC. What matters to us is the recourse to a principle of inertia as well as the superposition of previously acquired and newly impressed motion. Note that for Beeckman and Descartes the superposition concerns parallel, vertical velocities, whereas for Galileo a horizontal inertial motion is combined with a vertical fall.[20]

A month later, Descartes communicated to Mersenne his proof of the linear increase of velocity of a falling body:

> In the first moment, the velocity one is impressed by gravity; then again the velocity one in the second moment, and so forth. One in the first moment and one in the second moment makes two; and one in the third moment makes three, so that the velocity increases in arithmetic proportion. I thought I had sufficiently proved this from the fact that gravity acts permanently. . . . Suppose, for example, that a plumb mass falls by

[19] Beeckman 1939–1953, Vol. 1, pp. 24–25. On Beeckman, cf. Arthur 2007; van Berkel and Ultee 2013. On Gassendi, cf. Pav 1966; Fischer 2014.

[20] Beeckman 1939, Vol. 1, p. 174; Descartes to Mersenne, November 13, 1629, in Adam and Tannery 1897, pp. 72 (citation, mostly in Latin and probably taken from a manuscript of 1619), 75 (Tannery on the Beeckman connection). Cf. Bouasse 1895, pp. 103–105; Jouguet 1908, Vol. 1, pp. 81–82; Damerow et al. 1992; Richard 2007. As noted by Damerow et al., Jean Buridan anticipated Beeckman's approach.

the force of gravity and that, after the first moment of fall, God withholds all gravity from the plumb. Then this mass keeps going down in a vacuum because it has been set in motion and because no reason can be given why its velocity would increase; but it cannot decrease either (remember that I suppose that once [a body] has been set in motion, it will move forever in a vacuum; and I will prove it in my treatise). If after some time God restitutes gravity to the plumb for an equal moment, and then again withholds gravity, the force of gravity will pull the plumb as much as it did during the first moment, so that the velocity of the motion is doubled.

Again, the proof probably dates from 1618, when Descartes was under the influence of Beeckman's atomism. Descartes was then assuming vacuum, inertial motion in a vacuum, and the mutual independence of the impressed and inertial motions.[21]

In 1631, Descartes told Mersenne that he no longer held the third of these assumptions:

I not only assumed a vacuum, but I also assumed that the moving force . . . was acting in an always equal manner, which openly conflicts with the laws of Nature: indeed all natural powers act to a larger or smaller extent according as the object is more or less disposed to receive their action; and it is certain that a stone is not equally disposed to receive a new motion, or a velocity increase, when it is already moving very rapidly and when it is moving very slowly.

When, a year later, Descartes became aware of Galileo's *Dialogo*, he disagreed with the universality of free fall as well as with Galileo's treatment of fall with an initial horizontal velocity. He never published his early rational deduction of the law of constant acceleration, since he had ceased to believe in its truth. As we are about to see, in his mature philosophy the accelerating effect of a force on a body had to depend on its initial velocity.[22]

1.3 Descartes's world

While Descartes was reporting to Mersenne his old reply to Beeckman's query on free fall, he was also writing an ambitious treatise, *Le monde*, propounding a new mechanical philosophy to supplant Aristotle's system. From Beeckman he retained the corpuscles, their inertia, and their collisions, but he strongly rejected the concept of empty space at the heart of neo-atomism. In Descartes' world, any spatial extension is matter, and vice versa. The infinite world is made of contiguous rigid figures of various shape: the gross particles of ordinary matter (third element), contiguous balls between the former particles (second element), and dust filling the remaining interstices (first element). These corpuscles or rigid figures possess no quality other than their shape and extension. They move with respect to each other in a perpetual rearrangement implying mutual collisions. There is no absolute, empty space. Motion, in common parlance, being referred to a large rigid body (most frequently the earth), is ill-defined since the world contains many rigid assemblies

---

[21] Descartes to Mersenne, December 18, 1629, in Adam and Tannery 1897, pp. 88–90 (in Latin).

[22] Descartes to Mersenne, October or November 1631, in Adam and Tannery 1897, p. 230; November or December 1632, ibid. p. 261; 14 Aug. 1634, ibid. pp. 304–305.

of corpuscles that could serve as reference bodies. This is why Descartes defines the true motion of a body as motion with respect to adjacent bodies:[23]

> If, instead of settling on a notion founded on ordinary usage only, we wish to know what is motion in truth, we shall say, for the sake of determinateness, that motion is the transport of a portion of matter or of a body from the vicinity of those that touch it immediately.

From God's immutability, Descartes derives the global conservation of this motion as well as the tendency of individual particles to preserve their rest or their motion. This leads him to three "laws of nature":[24]

(1) Every thing remains in its state of being as long as nothing changes it.

(2) Every moving body tends to continue its motion on a straight line.

(3) If a moving body encounters a stronger body, it does not lose any of its motion; if a moving body encounters a weaker one that it may move, the former body loses as much motion as it gives to the latter.

A modern reader of Descartes may be tempted to conflate the two first laws with our principle of inertia, and the third with the conservation of momentum or of kinetic energy in a collision. These guesses are deeply incompatible with Descartes's true meaning. First, Descartes's statement of the persistence of rest and motion should not be confused with the modern principle of inertia, because in his system the rest and motion of a body are defined with respect to the adjacent corpuscles, not with respect to an abstract immovable space. In particular, Descartes explains the rigidity of a macroscopic body by persistence of the mutual rest of its particles, and fluidity by the constant, erratic motion of the particles of the fluid. Second, the proposition that a larger particle remains at rest when impacted by a smaller particle contradicts our conservation of momentum (yet for Descartes it is compatible with the contrary behavior of macroscopic bodies, because the medium contributes to this behavior). Third, Descartes's idea that in a collision one particle loses as much motion as the other gains does not agree with any modern conservation law, because he measures the motion of a particle by volume times velocity modulus.

Descartes's original definition of true motion has important consequences. First, the earth and the planets do not have any true motion, because they are at rest with respect to the subtle matter in which they bathe:

> Properly speaking, [motion] is only the transport of a body from the vicinity of those that touch it. . . . In earth and the other planets no motion in the proper sense can be found, because they are not transported from the vicinity of the parts of the sky that touch them. . . . If we seem to attribute some motion to the earth, we should think that we do so improperly, in the same way as we sometimes say that those who sleep and rest in a ship nonetheless travel from Calais to Dover, owing to the ship that carry them.

---

[23] Descartes [1633]; 1644, p. 140.

[24] Descartes 1644, pp. 150–156. Cf. Garber 1992; Vilain 1996, pp. 60–67.

In Descartes' cosmology, the sun is surrounded by a huge vortex of subtle matter that drags the earth and the planets circularly around it. Within this vortex, he imagines smaller vortices around the earth and the planets. The latter vortices are responsible for the spinning of the planets and for the rotation of their moons around them. There is no relative motion between a planet and the adjacent parts of these vortices.[25]

With this mechanism, Descartes could accommodate the relative motion (in the ordinary sense) assumed by Copernicus or Tycho for celestial bodies, and still pretend that the earth did not truly move, in conformity with the teachings of the Church. In addition, his system immediately explained why a rotation of the earth with respect to the fixed stars does not have any effect on the motion of bodies near the surface of the earth: the earth is constantly at rest with respect to the neighboring fluid matter (subtle or not) on which this motion depends. Descartes thus satisfied a Mach relativity in which the relative motion of a system of bodies does not depend on a common motion imparted to these bodies and to the surrounding fluids. Indeed, at the fundamental level of his elements, corpuscular inertia and collisions depend only on relative distances and motions.[26]

Accordingly, the motion of a projectile with respect to the earth should not depend on the motion of the earth. In contrast this motion depends on the projectile's progress through the medium (air and subtle matter). In the absence of gravity and air resistance, Descartes expects any preexisting motion of the projectile to be continued on a straight line with a constant velocity. Indeed, his first and second laws imply rectilinear uniform motion with respect to the medium. At the same time, he expects the accelerating effect of gravity to depend on the initial velocity of the projectile, perhaps because according to his collision rules the momentum acquired by impact (of the corpuscles of the medium on the corpuscles of the projectile) depends on the initial velocity of the impacted body. This may explain why he abandoned his early derivation of the law of constant acceleration in free fall and why he questioned Galileo's superposition of a constant horizontal motion and a vertical acceleration for a horizontally shot projectile.

However, Descartes does not truly explain gravity through corpuscular collisions. In his system, what causes gravity is the disparity of the centrifugal forces on ordinary matter and on subtle matter. A weighing body rotates together with the earth and is therefore subjected to an upward centrifugal force. The subtle matter permeating and surrounding this body is subjected to a stronger centrifugal force because in order to keep the earth rotating the celestial vortex must rotate faster than the earth. While the subtle matter moves upward under the effect of its higher centrifugal force, the ordinary matter must move downward to fill the resulting vacuum.[27]

---

[25] Descartes 1644, pp. 194, 195, 197–198. In effect, Descartes defines the true motion of a body as motion with respect to the average position of contiguous bodies. This precision is necessary in the case of planets because the celestial corpuscles must be in constant motion for the sky to be a liquid.

[26] For Mach's kind of relativity, see pp. 68–70, this volume.

[27] Descartes 1644, pp. 345–347.

There are many problems with this fanciful mechanism. First, the faster rotation of subtle matter seems to contradict Descartes's earlier statement that the earth does not have any true motion.[28] Second, centrifugal force is incompatible with Descartes's own concept of inertia, which concerns true motion only (not motion with respect to a distant body).[29] Third, it is hard to imagine how centrifugal force could have spherically symmetric effects.[30] Still, we may try to imagine what Descartes would have replied to Simplicio when asked about the stone falling from a tower on a rotating earth: on a non-rotating earth there would be no gravity, and therefore the stone would not even fall; on a rotating earth, the stone should fall vertically because centrifugal force is directed from the center and because, when convenient, Descartes ignores the differential rotation he assumes between the earth and the surrounding vortex.

Descartes easily impressed some of his contemporaries through the beautifully simple basis of his mechanical philosophy: collisions between the corpuscles of a fragmented extension, under rules derived from God's immutability. Yet, in order to fill the gap between this corpuscular foundation and observed phenomena, he frequently appealed to concrete analogies that were not necessarily compatible with the foundations. For instance, he justified centrifugal force in the vortex around the earth by analogy with a stone in a rotating catapult.[31] He overlooked an essential difference between the two cases: the stone moves through the surrounding medium, whereas corpuscles in the vortex have no true motion. According to his own principle of inertia, there should be a centrifugal force in the first case but not in the second. Such inconsistencies make it difficult to judge how much relativity there is in Descartes's system. On the one hand, his foundations imply the absence of effect of any common motion on the relative motions in a system. On the other hand, his cosmic centrifugal force implies an effect of common rotation.

1.4 Newton's laws of motion

Collisions

For the neo-atomists and for Descartes, collisions between corpuscles were in principle responsible for every evolution of the world. The laws of collision therefore had basic import. Beeckman regarded his atoms as completely inelastic hard bodies, in which case two colliding atoms move in tandem after collision. In addition, he assumed the product of velocity and corporeity (*corporeitas*) to be conserved when a moving body collides with a body at rest. For instance, a moving body colliding with an equal body at rest communicates half its velocity to the double body formed by collision (in analogy with a lever: the same force is needed to raise a given weight at a given speed as is needed to raise double the weight at half this speed). In a frontal collision in which two equal bodies meet with equal and

---

[28] Descartes 1644, p. 131, Descartes suggests that the orthoradial motions of the various corpuscles of the medium compensate for each other.

[29] A similar criticism is found in Newton [c. 1668?].

[30] This objection is found, for example, in a letter to Descartes written in 1649 by Henry More (More 1712, pp. 97–98).

[31] Descartes 1644, pp. 221–222. On Descartes's analogies, cf. Galison 1984.

opposed velocities, the two motions cancel each other out. Beeckman was thus groping toward the relation

$$m_1 \bar{v}_1 + m_2 \bar{v}_2 = (m_1 + m_2)\bar{v}' \qquad (1.1)$$

for momentum conservation, where m_1 and m_2 are the masses of the two colliding inelastic bodies, $\bar{v}_1$ and $\bar{v}_2$ their (algebraic) initial velocities, and $\bar{v}'$ their final common velocity. This relation implies a constant loss of motion in the universe, which Beeckman's benevolent God compensated for by providing the early universe with an infinite stock of motion.[32]

Descartes's collision rules were more numerous and harder to grasp. They were meant to comply with his special concept of inertia and also to preserve the global amount of motion (measured by spatial extension times velocity modulus) that God gave to this world. They were purely rational and not meant to apply to macroscopic bodies, whereas Beeckman's rules applied to binding collisions between concrete bodies. Like Descartes but in an atomist framework, Gassendi distinguished between conservative micro-laws for the collision and combination of atoms, and more complex laws for macroscopic bodies. Early mechanical philosophy easily ignored the "analogy of nature" later promoted by Newton between the micro- and macro-worlds.[33]

The young Isaac Newton appreciated the neo-atomism of Gassendi's British follower Walter Charleton and found much to criticize in Descartes's grand system. He adopted the concept of matter as made of immutable and indivisible particles moving in absolute empty space, and he rejected Descartes's identification of matter with extension as contradicting the immateriality and omnipresence of God. He knew and read the Cambridge neo-Platonist theologian Henry More, who, while he promoted Descartes's mechanical philosophy, alienated it with the "Spirit of Nature," an omnipresent, infinitely extended and "indescerpible" (indivisible) emanation of God responsible for the immortality of the soul and also for gravity. While More generally approved of the Cartesian endeavor to explain physical and celestial phenomena by matter and God-given motion, he denied that the fall of a stone could be explained by mechanical power alone: an immaterial entity or force had to be implied.[34]

Between neo-atomism and spiritualized Cartesianism, Newton saw the importance of knowing the laws of collision between bodies. He gave his own theory of impact in notes written around 1665. The two first axioms in the relevant section of his "Wastebook" read:

1) If a quantity once move it will never rest unlesse hindered by some externall caus{e.}

2) A quantity will always move on in the same streight line (not changing the determination {nor} celerity of its motion) unlesse some externall cause divert it.

---

[32] Beeckman 1939, Vol. 1, pp. 265–266 (November–December 1918). Cf. Arthur 2007. On pre-Newtonian collision rules and analogy with the lever, cf. Bertoloni Meli 2006, chap. 5.

[33] Gassendi's atoms had an inherent velocity, whose modulus remained the same after collision (cf. Pav 1966, pp. 31–32).

[34] Newton [1668?]; More 1659, preface (Spirit), pp. 449–458 (gravity). Cf. Westfall 1962; Henry 2016.

Although these axioms seem reminiscent of Descartes's first two laws of motion, their meaning is different because the motion is now referred to "Extension," regarded as an abstract empty space: "When a body Quantity {passeth} from one parte of Extension to another it is saide to mo{ve}."[35]

Newton also differs from Descartes by assuming a concept of force or pressure as the cause of motion. For the French philosopher, the only direct cause of motion is impact between absolutely rigid bodies, and "force" refers to the tendency of a body to persist in its state of rest or motion (Newton's later *vis insita*); for Newton, (impulsive) force mediates between the two colliding bodies. Force is defined as what causes a body originally at rest to assume a given amount of motion, or (with sign reversed) what causes a body to stop when initially in motion: "There is exactly required so much & noe more force to reduce a body to rest as there was {to} put it {in a give}n motion." Like Descartes and Beeckman, Newton measures the motion by the product of size and velocity (matter, at its most fundamental, has uniform, constant density), and therefore measures a force by the quantity of motion it can produce (or suppress). In his view, force and quantity of motion are directional (they have a "determination") and their relationship is the geometric counterpart of the modern vector formula $\mathbf{I} = \Delta(m\mathbf{v})$ relating the impulsive force $\mathbf{I}$ to the variation of the momentum $m\mathbf{v}$ of the body to which this force is applied. In addition, Newton assumes the equality of action and reaction; that is to say, the force $\mathbf{I}_{12}$ acting from body 1 to body 2 is equal and opposed to the force $\mathbf{I}_{21}$ acting from body 2 to body 1:

> If two quantitys (a & b) move towards one another & meete in o, Then the difference of theire motion shall not bee lost nor loose its determination. For at their occursion they presse equally uppon one another, & (p) therefore one must loose noe more motion than the other doth; soe that the difference of their motions cannot be destroyed.

In symbols, this statement reads

$$m_1\mathbf{v}_1 + m_2\mathbf{v}_2 = m_1\mathbf{v}_1' + m_1\mathbf{v}_2' \text{ since } m_1\mathbf{v}_1' - m_1\mathbf{v}_1 = \mathbf{I}_{21} = -\mathbf{I}_{12} = m_2\mathbf{v}_2 - m_2\mathbf{v}_2'. \qquad (1.2)$$

We do not know how Newton arrived at this reasoning. He may have regarded the equality of action and reaction as an obvious consequence of the impossibility of perpetual motion (as he later did in the *Principia*), or he may have derived it from the observed conservation of the vector quantity of motion in real collisions.[36]

In order to determine the final velocities of frontally colliding elastic spheres, Newton further assumes the reversal of the relative velocity during the collision (Axioms 10–11). This is obvious in the case of a symmetric collision between two equal spheres moving toward each other with opposite velocities, because, as Newton explains, the local spring-like compression beginning when the spheres first meet must be followed by a symmetric dilation until the spheres separate. Newton extends this result to any asymmetric collision (with different masses and different velocity moduli), without any explanation (I am not

---

[35] Newton [c. 1665], f. 10v. Cf. Herivel 1965; Westfall 1980, pp. 145–148. On the origins of Newton's concept of space, cf. De Risi 2015.

[36] Newton [c. 1665], ff. 10v, 11r.

aware of any simple intuitive justification). He also treats the case of arbitrary shaped and rotating bodies.[37]

Principia Mathematica

Most importantly, Newton already has in hand the three laws of motion he will state as follows in the *Principia* of 1687:

1) Every body perseveres in its state of rest, or of uniform motion in a right line, unless it is compelled to change that state by forces impress'd thereon.

2) The alteration of motion is ever proportional to the motive force impress'd; and is made in the direction of the right line in which that force is impress'd.

3) To every action there is always an opposed and an equal reaction: or the mutual actions of two bodies upon each other are always equal, and directed to contrary parts.

As is clear from Newton's comments and from the way he applies these laws, the implied "forces" truly are impulses or integrals of time-dependent forces over time (the impressed force is proportional to the added momentum "whether that force is impressed all at once, or gradually and successively"). When Newton, later in the *Principia*, investigates the motion of a planet continuously attracted by the sun, he replaces this attraction with a series of impulses and analyzes the motion of the planet as a rapid alternation of inertial motions and momentum jumps caused by the impulses. Note that if the second law were directly enunciated as the acceleration law $\mathbf{f} = m\ddot{\mathbf{r}}$, then the first law would be superfluous: it would be a special case of the second law for $\mathbf{f} = \mathbf{0}$. Newton's formulation thus reflects the discrete, collision-like character of basic interactions he remotely inherited from Beeckman.[38]

In deliberate opposition to Descartes's and More's rationalism, Newton insists that the true foundation of his laws of motion is empirical: they are inferred from the empirical laws of collision and from Galileo's empirical laws of free fall. At the same time, he clearly sees that these laws presuppose a few definitions and conventions, which he expounds at the very beginning of his treatise, in a series of definitions regarding matter, motion, force, and their measures, followed by a scholium on space, time, and motion. Most fundamentally, in order that the first law make any sense, the reference of the motion must be specified. This reference is what Newton now calls absolute space: an imagined, unbounded, rigid body, through which material bodies freely move, and regarded as immobile. In Newton's parlance, "Absolute Space, in its own nature, without regard to any thing external, remains always similar and immoveable." Unlike relative spaces bound to rigid material bodies, absolute space is not detectable by the senses. We may nonetheless detect absolute rotation through centrifugal forces curving the surface of water in a bucket or stretching a string between two globes. Absolute uniform translation is not detectable, as Newton expresses in the fifth corollary to his laws of motion:

---

[37] Newton [c. 1665], f. 11r.

[38] Newton 1687, pp. 12–13; 1729, Book I, pp. 19–20 (citation). On the *Principia*, cf. Smeenk and Schliesser 2013.

The motions of bodies included in a given space are the same among themselves [*inter se*] whether that space is at rest, or moves uniformly forwards in a right line without any circular motion.

This is a direct consequence of the laws of motion, granted that the forces are the same in the two compared spaces. As Newton notes in a probable allusion to Galileo, the motions observed on a moving ship confirm this invariance.[39]

Newton's next corollary, the sixth, reads:

If bodies moved in any manner among themselves, are urged in the direction of parallel lines by equal accelerative forces, they will all continue to move among themselves, after the same manner as if they had not been urged by those forces.

In other words, if a force like uniform gravity equally accelerates all the bodies of a system, and if other forces simultaneously act on these bodies, their relative motion will be the same as if the former accelerative force did not exist. This is an obvious consequence of the second law: if, in anachronistic notation, $\mathbf{g}$ denotes the common acceleration (owing to gravity), m_i the mass of the body i, $\mathbf{f}_i$ the additional force acting on it, we have

$$\mathbf{f}_i + m_i\mathbf{g} = m_i\ddot{\mathbf{r}}_i \;\Rightarrow\; \ddot{\mathbf{r}}_i - \ddot{\mathbf{r}}_j = \frac{\mathbf{f}_i}{m_i} - \frac{\mathbf{f}_j}{m_j}, \tag{1.3}$$

so that the relative motion does not depend on the uniform gravity $\mathbf{g}$. The modern reader here recognizes Einstein's free-falling elevator, in which mechanical experiments yield the same result as they would if the elevator were at rest in zero gravity—as long as the inertial and gravitational masses are confounded.[40]

Why would Newton introduce this relativity with respect to common acceleration? He uses it in his proof of Proposition 3 of Book I:

Every body, that by a radius drawn to the centre of another body, however moved, describes areas about that centre proportional to the time, is urged by a force compounded of the centripetal force tending to that other body, and of all the accelerative force by which that other body is impelled.

Using the sixth corollary, the acceleration of the second body can be subtracted from both bodies without altering the relative motion and its constant areal velocity. The second body is then at rest (or in uniform rectilinear motion) and Newton can rely on his previous proposition, which states the central character of a force for which the area law is satisfied when the center is fixed in absolute space. In Book III about the "System of the World," Newton relies on Proposition 3 of Book I to derive the central character of the action of Jupiter on its moons from the areal property of the observed motion of these moons. He

---

[39] Newton 1687, pp. 6–7, 9–11, 19; 1729, Book I, pp. 9 (citation), 15–18, 30 (citation). Cf. Westfall 1980, pp. 417–420. On Newton's views on space, time, and motion, cf. Rynasiewicz 2014.

[40] Newton 1687, p. 20; 1729, Book I, p. 31. Cf. Stein 1977, p. 19; Torretti 1983, p. 19; Barbour 2001, pp. 577–578; Saunders 2013, pp. 34–39; Smeenk and Schliesser 2013, pp. 124–125.

needs this proposition because Jupiter and its moons are all attracted by the sun and thus subjected to a common acceleration. Of course Newton understands that the equality of all the accelerations caused by the sun depends on the equality of inertial and gravitational mass. He notes that even a tiny difference between these two masses would visibly distort the orbits of Jupiter's satellites.[41]

Newton's two relativity statements were potential threats to his idea of absolute space. Formally, if the laws of motion are true in a given space, they will be so in any space moving rectilinearly and uniformly with respect to the former space. They will even be true in a rectilinearly accelerated space, if only a very remote huge mass is imagined as the cause of this acceleration. In practice, Newton imagined two means of determining the absolute space: through the absence of centrifugal forces, and through the motion of the center of mass of a closed system. As a consequence of his three laws of motion, the center of mass of a system of mutually interacting bodies (ignoring interactions with any external body) should move with constant velocity. By a natural "hypothesis," Newton deems absolute the space in which the center of mass of the solar system (better: of the entire universe) is at rest and in which the stars are seen in fixed directions. Yet he is aware that according to his sixth corollary a global rectilinear acceleration of the solar system would have no observable effect:[42]

> It may be alleged that the sun and planets are impelled by some other force equally and in the direction of parallel lines, but by such a force (by Cor. vi of the Laws of Motion) no change would happen in the situation of the planets one to another, nor any sensible effect follow: but our business is with the causes of sensible effects. Let us, therefore, neglect every such force as imaginary and precarious, and of no use in the phenomena of the heavens.

Absolute time is another mathematical abstraction: "By its own nature [it] flows equably without regard to anything external." The way it differs from the time empirically given by the periods of celestial bodies must be determined by applying the laws of motion to these bodies.[43]

Force is the third basic concept appearing in the laws of motion. Newton defines the force appearing in his second law as follows:

> An impressed force is an action exerted upon a body, in order to change its state, either of rest, or of moving uniformly forward in a right line.

When measured by the quantity of motion (momentum) it may produce or destroy, an impressed force (such as gravity) is called motive force. With this definition, the second law almost sounds like a tautology. Yet it is not, since force is defined as a cause of motion to

---

[41] Newton 1687, pp. 39 (Prop. 3), 402 (Hyp. 5), 405 (Prop. 1); Newton 1729, Book I, p. 62 (citation); Book III, pp. 206 (Prop. 1), 213 (Phen. 1), 221–223 (equivalence).

[42] Newton 1687, pp. 402, 417; 1729, Book III, p. 232; citation from 1728 draft of Book III, in Cajori 1962, Vol. 2, p. 558.

[43] Newton 1687, pp. 5, 7; 1729, Book I, pp. 9 (citation), 11.

be found in external circumstances, and since the second law further implies the stability and consistency of the measurement of the motive force through the induced momentum variation. That is to say, the momentum acquired by a test body under a given force should not depend on its mass or on its initial velocity.[44]

For a continually impressed force, induced momentum is not the only way Newton knows of measuring the motive force. Another way is to balance the force statically with another force that is a multiple of a given unit force. The principle of such static measurement is (a special case of) the parallelogram of forces, according to which two forces acting conjointly on the same point are equivalent to a single force given by their vector sum. This is why Newton's two first corollaries to his laws of motion are devoted to the parallelogram of force and the derived laws of statics (then called mechanics). The first corollary reads:

> A body by two forces conjoined will describe the diagonal of a parallelogram, in the same time that it would describe the sides, by those forces apart.

Here again the forces are meant to be impulsive. Newton implicitly assumes that the two conjoined impulses have the same effect whether they are applied simultaneously or with a slight delay. Then, by the second law the momentum induced by the second impulse adds vectorially to the momentum induced by the first impulse, as required by the parallelogram law. We thus see that in the context of impulsive forces or of forces reducible to a rapid succession of impulses, Newton's second law is intimately related to the parallelogram of forces.[45]

Newton's definitions of space, time, and force will prove important in later attempts to justify or derive his laws of motion. For the moment, it is sufficient to remember that together with the laws of motion they imply two relativity theorems: a first theorem according to which the relative motions of a system of bodies obey the same laws whether they are referred to absolute space or to a uniformly and rectilinearly moving space; and another theorem according to which the relative motions of a system of bodies under given forces are unaffected when a uniform gravity acts on the bodies in addition to these forces. The first theorem justifies Galileo's empirical law or relativity. The second theorem is entirely new and replaces Galileo's and others' notion of circular inertia in explaining why, for example, the circulation of the moon around the earth is not affected by the circulation of the earth–moon system around the sun.

1.5 Huygens's mechanics

Collisions

Let us return to the 1660s. Newton was not alone in reflecting about collisions in those years. The newly founded Royal Society pursued the topic from 1666, and in 1668 its

---

[44] Newton 1687, pp. 2, 4; 1729, Book I, 3 (citation), 7–9. Cf. Jammer 1957, chap. 7.

[45] Newton 1687, pp. 3–4; 1729, Book I, p. 7 ("The weight [as motive force] is ever known by the quantity of a force equal and contrary to it, that is just sufficient to hinder the descent of the body"), p. 21 (citation).

curator Robert Hooke invited three renowned geometers, Christiaan Huygens, Christopher Wren, and John Wallis, to communicate their collision rules. Wren dealt with strictly hard bodies (i.e., soft from the modern point of view) and obtained the correct laws through momentum conservation (which he guessed by analogy with the principle of virtual velocities, just as Beeckman had privately done some forty years earlier). Wallis obtained the correct laws for elastic collisions by semi-empirical reasoning. The Royal Society failed to publish Huygens's submission, which contained the same laws. Huygens had privately deduced them in 1652 from three principles: rectilinear inertia, Galilean relativity, and the impossibility of perpetual motion (together with Galileo's laws of free fall). In 1661 he had performed and explained a few simple cases of collisions between pendulum bobs in front of a London audience including Wallis and Wren.[46]

In his approach, Huygens starts with a restricted class of collisions in which the rules are evident enough, and extends this class by making the restricted collisions happen on a moving boat and considering them from the shore. He does so in a vivid manner, imagining a boat moving along the bank of a canal, one gentleman in the boat holding two pendulums and communicating contrary velocities to their bobs, the other gentleman gently accompanying the motion of his friend's hands with his own hands (Figure 1.1). Huygens's account is here modernized for the convenience of today's readers.[47]

A first obvious case is the symmetric collision between two equal bodies (elastic balls). Seen from the boat, the collision amounts to a sign change of the two equal velocities. Seen from the shore, it provides the rule for any frontal collision between equal bodies, since any values of the two initial velocities can be obtained by properly choosing the boat's velocity and the invariant measure of the relative velocity. More generally, Huygens observes that

Fig. 1.1.   Pendulum collisions induced and observed by two gentlemen, one on a river bank, the other on a boat. From Huygens 1703, p. 369.

[46]Wallis 1668; Wren 1668. Cf. Jouguet 1908, Vol. 1, chap. 2; Volgraff 1929a.

[47]Huygens 1669 (for the results), 1703 (for the proof). Cf. Gabbey 1980; Vilain 1996, chap. 4.

for any collision between two bodies of initial velocities $\mathbf{v}_1$ and $\mathbf{v}_2$, there is a velocity $\mathbf{u}$ of the boat for which the relative velocity of the first body simply changes sign:

$$\mathbf{v}_1' - \mathbf{u} = -(\mathbf{v}_1 - \mathbf{u}), \quad \text{if } \mathbf{u} = \frac{1}{2}(\mathbf{v}_1 + \mathbf{v}_1'). \tag{1.4}$$

In this case, the intuition of an elastic collision implies that the other body should also experience a mere sign change of its velocity:

$$\mathbf{v}_2' - \mathbf{u} = -(\mathbf{v}_2 - \mathbf{u}). \tag{1.5}$$

Consequently, the relative velocity of the two bodies simply changes sign:

$$\mathbf{v}_2' - \mathbf{v}_1' = -(\mathbf{v}_2 - \mathbf{v}_1). \tag{1.6}$$

This is the rule privately enunciated by Newton in 1665 without any apparent justification.

In order to determine the final velocities $\mathbf{v}_1'$ and $\mathbf{v}_2'$, Huygens needs another relation between initial and final velocities. He relies on what we would now call the conservation of kinetic energy, obtained as follows.

By the principle of inertia, the two colliding bodies originally move on a horizontal table with the constant velocities $\mathbf{v}_1$ and $\mathbf{v}_2$. By Galileo's laws of free fall, these velocities can be generated by having the two bodies fall from the heights $v_1^2/2g$ and $v_2^2/2g$, respectively, and then bounce elastically on oblique planes. A subsequent collision between the two bodies produces the horizontal velocities $\mathbf{v}_1'$ and $\mathbf{v}_2'$, which can then be converted into permanent elevations by the reverse of the former procedure. Globally, the elevation of the center of mass of the two bodies has thus been changed reversibly from $m_1 v_1^2/4g + m_2 v_2^2/4g$ to $m_1 v_1'^2/4g + m_2 v_2'^2/4g$. This process or the inverse process would allow the fabrication of a perpetual motion, unless we require

$$m_1 v_1^2 + m_2 v_2^2 = m_1 v_1'^2 + m_2 v_2'^2. \tag{1.7}$$

Huygens combines this relation with the reversal of the relative velocity to determine the final velocities as a function of the initial velocities. The same combination leads to the equation

$$m_1 \mathbf{v}_1 + m_2 \mathbf{v}_2 = m_1 \mathbf{v}_1' + m_1 \mathbf{v}_2' \tag{1.8}$$

of momentum conservation, which Huygens has indeed enunciated in a brief communication of 1669. There is no written evidence that Huygens (also) reached this relation by applying the conservation of the potential elevation on a boat moving at the velocity $\mathbf{u}$. Ignaz Schütz gave the first known derivation of this kind in 1897.[48]

---

[48] Huygens 1669. On Schütz's derivation, see pp. 49–50, this volume.

The Cartesian geometer Frans van Schooten discouraged his past student Huygens from publishing a theory of collisions that contradicted Descartes's rules. This theory, completed in the mid-1650s, is found in a manuscript posthumously published in 1703. Besides Galilean recourse to a nautical thought experiment, Huygens there offers a formal statement of his relativity principle:

> The motion of bodies and their equal and unequal speeds are to be understood in relation to other bodies considered to be at rest, even if both the former and the latter bodies happen to be involved in some additional common motion. As a result, when two bodies collide, even if both are further subject to an additional uniform motion, they will push each other with respect to a body that is carried by the same common motion no differently than if this additional motion were absent.

Like Galileo's, Huygens's statement concerns the relative motions within a closed system of bodies and the absence of effect of a common uniform, rectilinear motion on the relative motions; it is not stated in terms of a change of reference frame or change of "space" as Newton puts it. That said, what was an empirical generalization for Galileo and a theorem for Newton, becomes a constructive principle for Huygens. This is a crucial innovation.[49]

Free fall

Even before his first thoughts on collisions, Huygens had a vivid interest in free fall. In 1646, at age seventeen, he defended the truth of Galileo's laws in face of Mersenne's skepticism. He may already have had the proof of constant acceleration found in a manuscript of 1659 and published in 1673 in his influential *Horologium Oscillatorium*. In this treatise, he bases his analysis of free fall on the three following principles:

1) If there were no gravity and if the air did not impede the motion of bodies, any body, once set into motion, would go on moving with an equal velocity on a straight line.

2) The action of gravity, whatever be its origin, is to make bodies assume a motion composed of the uniform motion they have in such and such direction and of the downward motion impressed by gravity.

3) These two motions can be considered independently of each other, and do not interfere with each other.

In the first principle we recognize the principle of inertia (which Huygens traces to Descartes's laws of nature). The second and third principles allow one to determine the motion of a body falling with an initial velocity from the motion of a body falling from rest simply by superposing an inertial motion at the initial velocity to the latter motion. As

---

[49] Huygens 1703, p. 369. Newton's "space" very nearly corresponds to our "reference frames." More on this in Chapter 2, p. 42 (note 37).

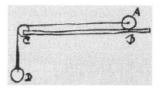

Fig. 1.2.   Huygens' contraption for the horizontal sliding of a body
under constant force. From Huygens [1659], p. 126.

we saw, Galileo introduced this rule in the case in which the initial velocity is horizontal, and Beeckman and Descartes in the case in which the initial velocity is vertical.[50]

In the manuscript of 1659, Huygens relates this rule to Galilean relativity and he uses it to derive the laws of free fall. In order to benefit from Galileo's nautical imagery, he conceives the device of Figure 1.2, in which the body A slides without friction on a horizontal plane and is pulled with constant force by the descending weight D. Let the body A move from rest during the time τ and thus acquire the velocity $\mathbf{u} = \mathbf{v}(\tau)$. During the next time interval of the same duration, it moves under the same force but with the initial velocity $\mathbf{u}$. Judged from a boat moving at the constant velocity $\mathbf{u}$, the initial velocity of A is still zero and the acquired velocity should therefore be the same as in the first time interval. Judged from the shore, the final velocity is the sum of the velocity $\mathbf{u}$ of the boat and the velocity $\mathbf{v}(\tau)$ acquired during the first interval. Hence we have $\mathbf{v}(2\tau) = 2\mathbf{v}(\tau)$, and more generally $\mathbf{v}(n\tau) = n\mathbf{v}(\tau)$ for any integer n and for any time τ. Consequently, $\mathbf{v}(t)$ increases linearly with the time t, the acceleration is a constant, and the space traveled is proportional to the squared time in conformity with Galileo's laws.[51]

Huygens thus derived the laws of free fall from the principle of relativity. Or he almost did: he implicitly assumed that the motion of a falling body depended only on its original position and velocity, and not on any other feature of its anterior motion. Without this assumption, any law of motion of the form

$$\mathbf{f} = \Phi(\dot{\mathbf{r}}, \ddot{\mathbf{r}}, ...)\ddot{\mathbf{r}} \tag{1.9}$$

would comply with both the principle of inertia and the principle of relativity. What matters most, however, is that Huygens regarded the relativity principle as a major constructive principle both in collisions and in free fall. Since the law of constant acceleration in free fall can be regarded as a special case of Newton's second law, some of Huygens's readers were tempted to extend his way of reasoning to a proof of the latter law. This happened repeatedly in the eighteenth century, as we will see in the next chapter.

Conclusions

In this chapter, we walked through the chief mechanical philosophies that subverted the inherited scholastics in the seventeenth century. We observed the deep interrelatedness of

---

[50] Huygens to Mersenne, October 26, 1646, in Huygens 1888–1950, Vol. 1, pp. 24–27; Huygens [1659]; 1673, p. 21. There is no evidence that Huygens was aware of Beeckman's and Descartes's private considerations on free fall.

[51] Huygens [1659], pp. 125–128. I have somewhat modernized Huygens's reasoning.

three components of these philosophies: definition of motion, concept of inertia, and relativity principles or theorems. The different mechanical philosophies may be characterized by their diverging takes on this conceptual triplet, with consequences for the theory of the motions naturally occurring in the absence of human intervention.

Motion, inertia, and relativity

Time necessarily enters any definition of motion. Although the concepts of motion varied considerably from one author to another, there was not much debate on the nature of time. From antiquity to the early modern period, time was defined so that some distinguished motions would be uniform (the motion of celestial bodies for the ancients; inertial motion for Galileo and the neo-atomists). While Newton similarly posited the uniformity of inertial motion, he vaguely and metaphysically characterized time as that which "by its own nature flows equably," and more precisely as the temporal parameter with respect to which his laws of motion hold true. In the modern view, the uniformity of celestial motions could only be an approximation, and the ancient astronomical definitions of time needed correction. The definition of motion also implied Euclidean geometry to measure the changes of location, a point no one believed worth noting. In contrast, there was much variety of disagreement on the proper definition of the reference of motion.

For Galileo, motion always has a concrete, macroscopic reference; for instance, a ship, the earth, or the sun and fixed stars. Inertial motion, namely motion persisting without impressed force (weight or mechanical pressure), is uniform and it is circular or rectilinear. Mechanical phenomena display the same regularities whether they are referred to the stellar frame, to the earth, or to a uniformly progressing ship. This is partly a consequence of combining inertial motion with impressed motion and partly an inductive generalization.

For the neo-atomists and for Newton, (true) motion is referred to absolute space, an imagined immobile and immaterial container of all things. Although logically the homogeneity and isotropy of space thus conceived should imply rectilinear inertia, most neo-atomists admit circular inertia to explain the motion of celestial bodies. Newton does not and instead traces the circular motion of planets to the combined effect of rectilinear inertia and centripetal gravitational force. His laws of motion imply a relativity theorem according to which the relative motions of a system of bodies are the same in absolute space and in a space in absolute uniform translation. For Copernican neo-atomists, absolute space is empirically determined as the space in which free bodies have the expected inertial motion. In Newton's system, this criterion is not sufficient because his rectilinear inertia and the resulting relativity theorem make it impossible to discriminate between spaces in uniform translation with respect to each other. This is why Newton complements his laws of motion with the hypothesis that the center of mass of the universe is at rest. Nowadays, we would say that definite laws of motion require only a definite space–time structure and we would find this structure in the equivalence class of inertial frames. This is not an option for Newton: in his mind definite laws of motion require a concept of absolute motion. He indeed makes the determination of this absolute or true motion the chief aim of the *Principia*:[52]

---

[52]Newton 1687, p. 11; 1729, Vol. 1, p. 18.

How we are to collect the true motions from their causes, effects, and apparent differences; and vice versa, how from the motions, either true or apparent, we may come
to the knowledge of their causes and effects, shall be explain'd more at large in the
following Tract. For to this end it was that I compos'd it.

For Descartes, the true motion of a body is motion with respect to adjacent bodies in
a plenum. His concept of inertia is peculiar, for it is defined as the persistence of rest or
motion in this local sense. It implies, together with the associated collision rules, that the
relative motions within a portion of the universe are not affected by any global motion of
this portion even if this motion departs from uniformity or rectilinearity. At the same time,
Descartes assumes that a corpuscle rotating around the sun within a vortex tends at every
instant to take the tangent even though the corpuscle does not have any true motion. This
inconsistency results from misleading analogy with a stone rotating in a catapult through a
stationary medium (and thus truly moving). At any rate, no major disciple of Descartes followed him in his peculiar definition of true motion and inertia. What seduced his followers
was the general endeavor to reduce all physics to matter and motion, without mysterious
action at a distance or immanent forces.

Huygens, being some sort of Cartesian, agrees with Descartes that actions should be
mechanically mediated and he requires motion to be referred to material bodies. Being
more concerned than Descartes with the mechanics of macroscopic bodies, he wants this
reference to be concretely realizable (Descartes's subtle medium cannot serve for this purpose) and he wants it to be compatible with simple laws of motion at our scale. In practice
he picks the earth, a moving boat, or the sun and fixed stars, just as Galileo has done before
him. In Chapter 3 we will see how, late in his life, he theoretically constructed the inertial
frames by means of free particles mutually at rest.

Natural motions

These diverse concepts of motion, inertia, and relativity imply diverse approaches to free
fall and celestial motion. Galileo treats planetary motion as a case of circular inertia
(around the sun) and free fall as a combination of inertial motion and uniformly accelerated vertical motion. In the latter case, the inertial motion can be circular (around the
earth) or rectilinear (with respect to sun and fixed stars) according to the problem treated;
the accelerated vertical motion results from a tendency of masses to come together; and
this tendency and inertia are both regarded as natural. In addition, Galileo speculates that
the circular velocity of the planets may have resulted from their fall from a common large
height (measured from the sun). This is universal gravitation *avant la lettre*. As any of
his contemporaries, Galileo measures quantity of matter through weight, even when this
quantity enters the definition of impetus as the product of velocity and quantity of matter.
Weight thus being the only parameter of free fall, all bodies have to fall at the same speed
irrespective of their size and density as long as the resistance of the medium can be ignored.

For Descartes, the circular motion of planets results from their being dragged by a vortex
centered on the sun. Similarly, the rotation of the earth is caused by a surrounding sub-
vortex. The rotation of the subtle matter in the propelling vortex being faster than the
rotation of the planet, the centrifugal force density on a body near the planet's surface is

smaller than the centrifugal force density on the surrounding subtle matter. This is how Descartes explains weight. According to this view, the active role of the medium makes it plausible that the acceleration of a free-falling body would depend on its weight and on its initial velocity, against Galileo's laws.

In contrast, for Beeckman and for the young Descartes free fall is to be understood as the superposition, at every instant, of inertial motion at the acquired velocity and acceleration independent of this acquired velocity. This mechanism implies the constancy of the acceleration, in conformity with Galileo's empirical laws. In essence, Newton assumes the same mechanism, since his second law makes the momentum increase at a given time of the fall independent of the acquired momentum. Evidently, this mechanism implies that the law of constant vertical acceleration in free fall remains true on Galileo's moving ship. In Huygens's eyes, Galilean relativity is more basic than Newton's second law, and it is the deeper reason why the acceleration of a falling body does not depend on its previously acquired motion.

Whereas Beeckman imagines gravity to be caused by atomic impacts and Descartes traces it to ethereal vortices, Galileo and Newton refrain from any specific mechanism. Instead they both assume some mutual attraction of the particles of matter, and they both assume a relationship between free fall and planetary motion. What permitted Newton's final breakthrough was his belief in rectilinear inertia, which demands an active force to keep the planets on their curved trajectory, together with Hooke's assumption that this force is proportional to the inverse squared distance from the sun. Newton henceforth assumed the universal attraction we would now write as Gmm'/r^2 between two point-masses m and m' separated by the distance r.

Galileo's observations, his and Newton's experiments with pendulums, and his interpretation of celestial motions require the mass m occurring in this formula to be the same as the inertial mass in Newton's second law. Newton argues for this identity in the third book of the *Principia*, thereby indicating that it agrees with the assumption that all matter is made of particles of the same homogeneous substance dispersed in a vacuum. Indeed, under this assumption the inertial mass and the gravitational mass of a body are both proportional to the total volume of the primitive substance. As Newton and his followers usually assumed this much in their atomistic speculations, the mass independence of motion under gravity was natural to them. More puzzling in their eyes was the existence of forces, such as magnetic forces, for which this independence does not hold.[53]

The existence of "accelerating forces," such as gravity, under which the motion of a body does not depend on its mass, has a singular consequence that did not escape Newton's attention. In a portion of space in which the accelerating field is homogeneous, all bodies of a system are subjected to the same acceleration. Therefore, as Newton puts it, the relative motion of these bodies under additional forces is the same as if the accelerating field did not exist. As Albert Einstein would later put it, the effects of a homogeneous gravitational field can be eliminated by referring the motions to a frame moving with the acceleration induced by this field. This relativity theorem was important to Newton, for it enabled him to ignore the gravitational effect of a distant body when analyzing the motion of a binary

---

[53] Newton 1687, pp. 408–411; 1713, p. 368 (Corollaries 3–4); 1729, Vol. 2, pp. 224–225.

system. For instance, the attraction of the sun does not affect the relative motion of the moon and the earth.

Altogether, the mechanics of Galileo, Huygens, and Newton produced two kinds of relativity: one concerning Galilean frames, the other concerning rectilinearly accelerated frames. As we will see in the next chapter, both played a role in the later history of mechanics, not only as theorems but also as principles.

2

DERIVING NEWTON'S SECOND LAW FROM RELATIVITY PRINCIPLES

> The principle of relative motions leads, with as much rigor as ease, to the theorem of the proportionality of forces to the accelerations they produce on the same body, and to the theorem for the composition of force.[1] (Jean-Baptiste Bélanger, 1847)

In the previous chapter, we saw that Huygens used the Galilean relativity principle to derive several laws of mechanics for collisions and free fall. He did not attempt a derivation of Newton's second law beyond the limited case of constant force. As for Newton, he regarded his laws of motion as inductively drawn from the observed behavior of falling, colliding, and gravitating bodies. We will now see that in the two following centuries, there were numerous attempts to justify the law of acceleration through relativity principles. We will also see that the motivations for such reasoning varied. The relativity-based deductions could be, in the eighteenth century, part of an endeavor to give mechanics a purely rational foundation. Or, after Laplace's pivotal intervention, they were usually meant to simplify the basis of mechanics, without denying its empirical origins. In both cases, the deductions had a similar structure, and much of their appeal depended on the constructive power of the relativity principles.

My plural in "relativity principles" alludes to an interesting aspect of this story: it involves not only the Galilean relativity principle more or less as we know it, but also an *accelerative* relativity principle according to which a rectilinear motion, uniform *or not*, commonly impressed on a system of bodies, does not affect the relative motions of these bodies under given forces. We saw in Chapter 1 that Newton derived this accelerative relativity from his laws of motion. We will now see that Jean-Baptiste Bélanger and a number of French authors after him turned this law into a constructive principle from which the law of acceleration could be derived. We will also see that the term "relativity principle" can be traced to Bélanger's "principle of relative motions."

Sections 2.1 and 2.2 of this chapter follow the historical intricacies of eighteenth-century and nineteenth-century derivations of the second law. A more synthetic, less chronological, and more easily readable account is given in Section 2.3. The conclusion contains ahistorical remarks on the respectability of these derivations, as well as an assessment of their historical importance in a suggested new approach to theory construction.

---

[1] Bélanger 1847, p. III.

Relativity Principles and Theories from Galileo to Einstein. Olivier Darrigol, Oxford University Press.
© Olivier Darrigol (2022). DOI: 10.1093/oso/9780192849533.003.0002

2.1 Rational mechanics in the eighteenth century

Newton's mechanics and his theory of gravitation gradually diffused through Europe at the turn of the eighteenth century. British and Dutch Newtonians remained close to the spirit and style of Newton's *Principia*: they reasoned geometrically and inductively, they accepted absolute space and time, and they maintained Newton's statement of the laws of motion. The only exception worth mentioning is the third edition of Willem Jacob 's Gravesande's widely read *Physices elementa mathematica*, which offered the following justification of Newton's second law for impulsively generated change of momentum:

> We derive this law from the phenomena: indeed, in a ship an impelled body undergoes the same motion whether the ship rests or moves forward uniformly with some velocity. This proves that *two motions do not interfere with each other*, as is also true for a larger number of motions.

The remark seems reminiscent of Huygens's derivation of the laws of free fall, except that Newton's impulses now replace the continuous action imagined by Huygens. It represents, in a sketchy form, a first derivation of Newton's second law based on a relativity principle.[2]

Newtonian mechanics took its modern mathematical form in the hands of a few Swiss and French Newtonians in the first third of the eighteenth century. Leonhard Euler, Jean le Rond d'Alembert, and Joseph Lagrange later extended it to continua and connected systems. Mixed mathematics gradually absorbed all physics, under a mechanical umbrella. At the close of the century, Pierre Simon de Laplace's ambition was to reduce all physics to Newtonian interactions between point-like atoms. This century-long transformation naturally included reflections on the foundations of mechanics. The new builders, especially Euler, d'Alembert, and Laplace, wanted to consolidate the basic laws of mechanics. Euler and d'Alembert argued for their rational necessity; Laplace their empirical necessity. They all employed Galilean relativity in this endeavor.

Euler's Mechanica

Galileo and Newton asserted the empirical origin of the fundamental laws of physics, and they denounced Aristotle's and Descartes's belief in the rational necessity of these laws. As a nondogmatic Cartesian, Huygens wanted to derive the most important laws from commonly acceptable principles (inertia, relativity, impossibility of perpetual motion), without committing himself on the rational or empirical origin of these principles. In contrast, the two most influential geometers of the eighteenth century, Euler and d'Alembert, firmly asserted the purely rational character of the laws of mechanics.

In his *Mechanica* of 1736, Euler follows Newton in defining absolute motion as change of place in the unlimited space of the entire world. He derives the law of inertia from the principle of sufficient reason: a particle at rest in the infinite empty space remains at rest because it has no more reason to move in one direction than in any other direction; if

---

[2] 's Gravesande 1742, p. 94. The earlier editions (from 1720 on) did not contain this remark.

the particle is originally moving, its velocity has no more reason to decrease than it has to increase, and its path has no more reason to deviate one way than any other. Whereas for Newton inertia served to define absolute space, for Euler absolute space implies inertia. Euler further believes inertia to require "a true, essential cause in the body," echoing Newton's *vis insita*.[3]

Euler's reasoning sounds incorrect because one could very well admit that the velocity **v** of a particle in empty space varies like $\mathbf{v} = \mathbf{v}_0 e^{-t/\tau}$ without contradicting the homogeneity and isotropy of space. It could be saved by appealing to the isotropy of time. Or one could use Galilean relativity to derive the permanence of motion from the permanence of rest. Instead Euler notes that the law of inertia remains valid for motion in a relative space moving uniformly and rectilinearly with respect to the absolute space. When observing a free particle, Euler comments, there are no means to determine how much of its motion is absolute and how much is relative. Unlike Newton, Euler does not try to remove this underdetermination by requiring the center of mass of the universe to be at rest.[4]

For Newton and Euler, a force is what causes departure from inertial motion. More specifically, Euler defines an "absolute force" (*potentia absoluta*) as a continuous force whose action on a material point (body) does not depend on the motion of this point. Like Huygens, he means that for a given value of the absolute force **f**, the motion of a body initially moving at the velocity **u** can be obtained by composing the motion of a body initially at rest with a uniform motion at the velocity **u**. Considering the motion **r**(*t*) in a time τ so small that the variation of the force **f** is negligible and slightly modernizing Euler's reasoning, this implies

$$\dot{\mathbf{r}}(t + \tau) = \dot{\mathbf{r}}(t) + \mathbf{\Phi}[\mathbf{f}(t)]\tau, \quad \text{or } \ddot{\mathbf{r}} = \mathbf{\Phi}(\mathbf{f}). \tag{2.1}$$

That is to say, there is a one-to-one correspondence between the force **f** and the acceleration $\ddot{\mathbf{r}}$ of the body at any instant. Euler thus generalizes Huygens's deduction of the laws of free fall from Galilean relativity. He then determines the correspondence $\mathbf{\Phi}$ in two steps. First, he argues that the force on a body must be proportional to its mass, or number of points, because a mass M subjected to the force F can be replaced by n masses M/n each subjected to the force F/n. Second, through a peculiar thought experiment he argues that the parallelogram of forces implies the proportionality of force and acceleration.[5]

Unfortunately, this thought experiment implicitly assumes the nonevident fact that the motion of the center of mass of two particles subjected to the forces $\mathbf{f}_1$ and $\mathbf{f}_2$ is the same as if this center were a single particle subjected to the resultant $\mathbf{f}_1 + \mathbf{f}_2$. This flaw probably explains why later authors ignored Euler's deduction of the law $\mathbf{f} = m\ddot{\mathbf{r}}$. Still, they may have been seduced by his reliance on Galilean relativity in the first part of his proof.[6]

There is an important difference between Euler's and Newton's concepts of force. Like Huygens, Euler treats force as continuous, and he measures it statically, through balances.

[3] Euler 1736, Section 1–8 (motion and space), Section 56–76 (inertia). Cf. Jammer 1957, pp. 215–217; Pulte 1989, pp. 108, 115; Maltese 2000, pp. 321–322.

[4] Euler 1736, Section 77–82.

[5] Euler 1736, Section 99 (force), 111 (absolute), 118–135 ($\ddot{\mathbf{r}} = \mathbf{\Phi}(\mathbf{f})$), 136 ($\mathbf{f} \propto m$), 146 ($\mathbf{f} \propto \ddot{\mathbf{r}}$).

[6] Cf. Darrigol 2014a, pp. 26–27.

In contrast, Newton relies on impulsive forces and he measures force dynamically, through the product of mass and acceleration. Newton uses his second law to derive the parallelogram of forces and other laws of statics, whereas Euler regards these laws as previously given. Euler must have been aware of earlier rational deductions of the parallelogram of forces (for instance Daniel Bernoulli's) and of other laws of statics. These differences are worth noting here, because they will inform our discussion of later attempts to derive Newton's second law from relativity principles.[7]

D'Alembert's dynamics

In his *Traité de dynamique* of 1743, d'Alembert embraces the Cartesian endeavor to reduce mechanics to the contact or collision between strictly impenetrable and rigid bodies, with no obscure forces between the bodies. He nonetheless departs from Descartes in several ways. First, he defines motion in Newton's and Euler's manner, as change of place in an unlimited empty space. Second, he assumes the destruction of motion during collisions between particles of the primitive hard matter. Third, despite his rejection of forces as quantities defined and measured independently of their effects, he takes into account noncollisional causes of motion such as the pressure of a spring, the pull of a weight, or gravitation. He does this by defining the force acting on a body through the product of the mass and acceleration of the body. There are difficulties with this subterfuge: it is not clear that the force thus defined is in a one-to-one correspondence with its physical cause, and it is not obvious that it complies with the parallelogram of forces.[8]

D'Alembert derives the principle of inertia in Euler's manner, by appealing to the principle of sufficient reason. There is an interesting difference: for d'Alembert the permanence of motion (when the body is left to itself) is a mere corollary of the permanence of rest:

> *Law*: A body at rest will remain at rest, unless an extraneous cause comes into play. For a body cannot by itself come to move.

> *Corollary*: Whence follows that if a body receives motion from any cause whatsoever, this body will not be able to accelerate or retard this motion by itself.

The parallelism of these two statements suggests an implicit recourse to Galilean relativity: a body that has been set to move at a given velocity in absolute space is a body at rest with respect to a frame moving at this velocity. Persistency of rest in the moving frame is equivalent to persistency of motion in absolute space.[9]

D'Alembert then derives his second principle of the "composition of motions"—our parallelogram of forces—by having two impulses I_1 and I_2 act simultaneously on the same body at rest and comparing the induced velocity v_{12} with the velocities v_1 and v_2 produced by each impulse separately. For this purpose, he imagines the body to be attached to a massless rigid plane that can slide in a direction parallel to the first impulse through coulisses

---

[7] Euler 1736, Section 103–110.

[8] D'Alembert 1743. Cf. Hankins 1970; Firode 2001; Darrigol 2014, pp. 31–39.

[9] D'Alembert 1743, pp. 3–8.

that can slide in a direction parallel to the second impulse. Suppose this plane to be then set into motion by applying the impulse $-\mathbf{I}_1$ directly to it and the impulse $-\mathbf{I}_2$ to its coulisses. The body is thus subjected to two pairs of contradictory impulses and remains at absolute rest. With respect to the sliding plane, it moves at the velocity produced by $\mathbf{I}_1$ and $\mathbf{I}_2$ acting together, while the plane evidently moves at the velocity $-(\mathbf{v}_1 + \mathbf{v}_2)$. Therefore, the velocity produced by $\mathbf{I}_1$ and $\mathbf{I}_2$ acting together is $\mathbf{v}_1 + \mathbf{v}_2$, in agreement with the parallelogram of forces.[10]

This intricate kinematic reasoning amounts to a concrete application of Galilean relativity. Relativity indeed implies that the velocity jump under a given impulse cannot depend on the initial velocity, in agreement with Newton's second law. Then the parallelogram of force is obtained by adding the velocity jumps produced by the two impulses one after the other, as Newton did in the first corollary to his laws of motion. Truly, d'Alembert's second principle is Newton's second law in disguise. As long as forces can be replaced by a dense series of impulses, it warrants that the "accelerating force" $m\ddot{r}$ depends only on the external circumstances of the motion and agrees with the static measure of force.

D'Alembert's third principle is the "principle of equilibrium" according to which two colliding hard bodies come to rest if and only if their momenta are equal and opposite. He also calls this principle a theorem, because he thinks he can derive it from the parallelogram of forces and the principle of sufficient reason. In the case of two bodies of equal mass, the latter principle evidently implies rest after collision. If the mass of the first body is twice the second, then the initial velocity of the second body is twice the initial velocity of the first. According to the parallelogram of forces, the impact of this body should have the same effect as the impact of two bodies of the same mass and half the velocity. We are thus brought back to the symmetric case, and rest must ensue after the collision. Similar reasoning applies to the case of two bodies with commensurable masses, and the general case is obtained as a limit of this case.[11]

The former reasoning shows only that the balance of momenta is a sufficient condition of equilibrium. D'Alembert proves the necessity of this condition in the second edition of his treatise. There he imagines the collision to occur on a sliding plane, moving uniformly at the original velocity of the center of mass of the two bodies. With respect to the plane, the momenta of the two bodies are balanced and their collision brings them to rest according to the former reasoning. They therefore have a nonvanishing absolute motion after collision, unless the absolute velocity of the center of mass vanishes. The latter condition is equivalent to the balance of momenta. Again, d'Alembert is here applying the relativity principle in a concrete guise reminiscent of Galileo and Huygens.[12]

D'Alembert's third principle is a counterpart to Newton's third law, since from Newton's point of view momentum conservation during the collision between two bodies (hard or not) is a consequence of the third law (together with the second law) applied to the impulses acting on the two bodies during the collision. The main difference is that for d'Alembert

---

[10] D'Alembert 1743, pp. 22–25.

[11] D'Alembert 1743, pp. 37–40. D'Alembert calls his three fundamental laws *principes* in his foreword; in the main text, he calls the first a *loi*, and the others *théorèmes*.

[12] D'Alembert 1758, pp. 53–54.

an impulse is not a well-defined physico-mathematical entity. What is well defined is the momentum change of the body that is acted upon during the impulse. The impenetrability of primitive matter and the concomitant destruction of motion in collisions, not any obscure concept of force, are the rational sources of the laws of motion.

Consequently, for d'Alembert statics is just a special case of dynamics. Equilibrium is only the perfect destruction of the motions externally impressed on a mechanical contraption. In this view, the collision between two hard bodies of opposite momentum is the simplest case of equilibrium. The generalization to nonequilibrium is smooth: d'Alembert need only assume that for an arbitrary collision in which the final momenta of the two colliding bodies no longer vanish, the momentum changes of the two bodies still balance each other:

$$(m_1\mathbf{v}_1 - m_1\mathbf{v}_1') + (m_2\mathbf{v}_2 - m_2\mathbf{v}_2') = \mathbf{0} \quad \text{(with } \mathbf{v}_1' = \mathbf{v}_2'). \tag{2.2}$$

In d'Alembert's words, the motion destroyed in one body balances the motion destroyed in the other. This is a special case of d'Alembert's general principle of dynamics, which stipulates the balance of destroyed motions in any connected system moving under impressed forces. Possibly d'Alembert obtained the collision subcase by eliminating the parenthesis in Newton's following statement:

> If a body impinge upon another, and by its force change the motion of the other; that body . . . (because of the equality of the mutual pressure) will undergo an equal change, in its own motion, towards the contrary parts.

Alternatively, d'Alembert could have reasoned that the balance of momenta in a space sliding with the original velocity of the center of mass implies the balance of destroyed momenta in absolute space.[13]

Laplace's derivation of the second law

Even though d'Alembert's principle remains the most general principle of the dynamics of connected systems to this day, no one ever followed d'Alembert's rational deduction of all the laws of equilibrium and motion from the impenetrability of bodies. Later authors usually accepted Newton's concept of force as a cause of motion and his view of the second law as a quantitative relation between force and change of motion. This is, for instance, the case of the Marquis de Laplace, the highest authority on mechanical matters at the turn of the nineteenth century.[14]

For Laplace, motion is again defined as change of place in absolute space. The law of inertia is the simplest law we can imagine for free bodies, and it is confirmed by experience. Force is the cause of departures from inertial motion. Like Newton and d'Alembert, Laplace regards impulsive force as most basic, and his discussion of the relation between force and motion concerns impulses. A priori, the impulse could be any function of the

---

[13] D'Alembert 1743, pp. 49–51, 138; Newton 1687, p. 13; 1729, p. 20.

[14] Cf. Grattan-Guinness 2005.

velocity it impresses on a body. In order to determine this function, Laplace appeals to a kind of relativity established by experiment:

> It is observed upon the Earth that a body solicited by any force, moves in the same manner, whatever be the angle which the direction of this force makes with the direction of the motion which is common to the body and to the part of the terrestrial surface to which it corresponds. The same thing takes place in a vessel whose motion is uniform: a moveable body submitted to the action of a spring, or of gravity, or any other force, moves relatively to the parts of the ship in the same manner whatever be the velocity and direction of the vessel. It may then be established as a general law of terrestrial motions, that if in a system of bodies carried on by a common motion, any force be impressed on one of them, its apparent or relative motion will be the same, whatever be the general motion of the system, and the angle which its direction makes with the impelling force.

Here we meet again Galileo's ship, with a dubious transposition to the moving earth. As Laplace must have known (since his theory of tides took inertial forces properly into account), the absence of effects of the earth's motion is true only to a first approximation. However, for the impulses considered by Laplace in his investigation of the relation between force and velocity, his relativity statement is exactly true because inertial forces do not act during the infinitesimal time lapse of an impulse.[15]

This kind of relativity being granted, when two impulses act on the same body, the acquired velocity can be computed by having the second impulse act in a space moving at the velocity imparted by the first impulse. It must therefore be equal to the sum of the velocities that the two impulses would separately impart on the body. More generally, the impulse must be proportional to the acquired velocity, in conformity with Newton's second law.

For this law to be generally established, the reasoning requires the relativity principle to be verified for any velocity of the moving space and for any velocities of the moving bodies. Laplace considers the best available verification, by means of bodies moving at relatively small velocities with respect to the moving earth. A body of unit mass on earth is dragged by the earth with the absolute velocity $\mathbf{u}$. This velocity is related to the impulse $\mathbf{I}$ that would produce it through the relation

$$\mathbf{u} = \mathbf{I}\Phi(\mathbf{I}^2). \tag{2.3}$$

Let a terrestrial impulse $\mathbf{I}'$ act on this body (Laplace imagines a horizontally moving cylindrical cane whose end hits a ball on a horizontal plane). Granted the parallelogram of forces, the resulting absolute velocity $\mathbf{v}$ is given by

$$\mathbf{v} = (\mathbf{I} + \mathbf{I}')\Phi[(\mathbf{I} + \mathbf{I}')^2]. \tag{2.4}$$

---

[15] Laplace 1796, pp. 246–252; 1809, p. 300 (citation).

From mechanical experiments on earth, we know that the relative velocity $\mathbf{v} - \mathbf{u}$ is always found to be parallel to the impulse $\mathbf{I}'$. Using

$$\mathbf{v} - \mathbf{u} \approx \mathbf{I}'\Phi[(\mathbf{I} + \mathbf{I}')^2] + 2\mathbf{I}(\mathbf{I} \cdot \mathbf{I}')\Phi'(\mathbf{I}^2), \qquad (2.5)$$

this implies $\Phi'(\mathbf{I}^2) = 0$. Although we can test this relation only for a limited number of values of the impulse $\mathbf{I}$ (corresponding to various values of the absolute velocity of the earth), Laplace anticipates that the derivative of the function Φ vanishes for every value of its argument, in which case impulse and imparted velocity become proportional, as was to be proved. Laplace adds that the excellent verification of the consequences of this law for the motion of celestial bodies further justifies it.[16]

Conversely, if the force acting on a material point is proportional to its acceleration, then a common, not necessarily uniform motion of a system of material points does not affect their relative motions under given additional forces. Indeed, if $\mathbf{u}(t)$ denotes the variable common velocity of the material points, their absolute velocities $\mathbf{v}_i$ vary according to

$$m_i d\mathbf{v}_i = \mathbf{f}_i dt + m_i d\mathbf{u} \qquad (2.6)$$

when the forces $\mathbf{f}_i$ act on them during the time dt, so that the relative velocity variations $d\mathbf{v}_i - d\mathbf{v}_j$ do not depend on the common motion $\mathbf{u}$. This is so, Laplace tells us, because the same rule of vector addition (the parallelogram rule) applies to the composition of forces and to the composition of induced velocity increments. He concludes:

> The relative motions of a system of bodies, animated by any forces whatever, are the same, whatever be their common motion. . . . It is therefore impossible to judge of the absolute motion of a system of which we make a part, by any appearances that can be observed.

This is similar to Newton's sixth corollary to the laws of motions, which states the lack of effect of a common accelerative force on the relative motions of a system of bodies. The only difference is that Laplace, unlike Newton, relates this property to the empirical (in)determination of absolute motion. He thus gets closer to something like Einstein's free-falling elevator. That said, his derivation of Newton's second law requires the weaker relativity only, and the stronger relativity is of no use to him.[17]

Laplace judges the laws of mechanics to be of empirical origin, thereby siding with Newton against Descartes, Euler, and d'Alembert:

> Hence we have two laws of motion, that is to say, inertia, and that of the force proportional to the velocity, given by observation. They are the most simple and most natural that can be imagined, and without doubt they derive from the very nature of matter; but this nature being to us unknown, these laws are to us only observed facts. They are the only ones that the science of mechanics borrows from experience.

Although, for Laplace, the relativity principle can be used to justify Newton's second law, this law remains an empirical law because the relativity principle itself is an empirical

---

[16]Laplace 1796, pp. 250–251; 1799, pp. 15–18 for a sharper version.

[17]Laplace 1796, p. 249; 1809, p. 301 (citation); 1799, p. 18. As was previously mentioned (p. 18), Newton anticipated Laplace's remark in an unpublished draft of Book III of his *Principia*.

generalization. Laplace's allusion to a possible derivation of these laws "from the very nature of matter" plausibly refers to d'Alembert's assertion that the laws of dynamics are
"those that result from the existence of matter and motion."[18]

2.2 Nineteenth-century French textbooks

Nineteenth-century authors of texts on mechanics shared Laplace's and Newton's conviction regarding the empirical origin of mechanical laws (with one exception found in the
second edition of Poisson's treatise). Most of them were happy with directly assuming Newton's laws or variants of them as the result of a broad induction from experience. Others,
mostly in France, tried to derive these laws from simpler empirical principles, including the
two relativities assumed by Laplace.

The authors of British and German texts on mechanics generally ignored Laplace's
derivation of Newton's second law. They most commonly postulated this law, with some
empirical justification, through Galileo's inclined plane or through Atwood's machine.
Others, like Gustav Kirchhoff and Ernst Mach, simply defined force as the product of
mass and acceleration in the differential equations of motion, in which case the second law
is no longer needed (or so it seems).[19]

In France too, there were physicists who preferred to postulate the second law, and there
were also those who, following Adhémar Barré de Saint-Venant, based mechanics on the
mutual acceleration of pairs of point-masses. However, a fair number of authors of physics
and mechanics texts relied on Laplace's derivation of the second law: Louis Benjamin
Francœur in 1804, Eugène Péclet in 1823, Auguste Pinaud in 1846, Pierre-Adolphe Daguin
in 1861, and Augustin Privat-Deschanel in 1869. As we saw, this derivation concerns the
original version of Newton's second law, concerning the momentum change induced by an
impulsive force. In the nineteenth century, continuous forces gradually replaced Newton's
impulses and Newton's second law became the proportionality between force and acceleration. In this context, the constancy of the acceleration of a body under constant force
can still be derived in Huygens's manner: so did, for instance, Gaspard Coriolis in 1844,
Jean-Marie Duhamel in 1845, and Ossian Bonnet in 1858. However, Galilean relativity is
no longer sufficient to derive the proportionality of acceleration to a variable force. We will
now see how the authors of three of the most influential French expositions of mechanics
introduced Newton's second law.[20]

Poisson

In 1811 Laplace's closest disciple, Siméon Denis Poisson, published the first edition of
a widely used treatise on mechanics. Strangely, Poisson does not reproduce his mentor's
derivation of the second law of motion. Instead he remarks that the static measure of force
(through concrete equality and concrete addition) does not imply any relation between this

---

[18] Laplace 1796, p. 252; 1809, pp. 302–303 (citation); d'Alembert 1758, p. xxix.

[19] Kirchhoff 1876; Mach 1883. For exceptions, see "Laplace's System of the World," *Edinburgh Review*, 15:
396–417, on 401–402; Whewell 1819, pp. 270–272.

[20] Saint Venant 1851; Francœur 1804; Péclet 1823, 1838; Pinaud 1846; Daguin 1861; Privat-Deschanel 1869;
Coriolis 1844; Duhamel 1845; Bonnet 1858.

measure and the velocity impressed on a material point in a given time, and that the best one can do is to pick the simplest possible relation, proportionality, and verify its conformity with experiments. In particular, the fall of a body on an inclined plane, by reducing the effect of gravity to its oblique component, establishes the proportionality of force and acceleration in the case of constant force,and the laws of free fall establish the independence of the acceleration from the initial velocity. In the case of a variable force, Poisson considers a time interval so small that the variation of the force can be neglected, and applies the former relation within this interval.[21]

In the second edition of his treatise, published in 1833, Poisson changed his mind:

> It is customary to present [the second law] as a hypothesis. We give it here as a necessary consequence of the fact that the velocities impressed by arbitrary forces in infinitely small time intervals are always infinitely small, and that at the same time the displacements of the mobiles are also infinitely small.

In order to prove this, Poisson has two forces f and f' act on a material point during the infinitesimal time τ on a given axis. Call u the velocity variation induced by f alone, and u' the velocity variation induced by f' alone. The only way the action of f' could depend on the simultaneous action of f, Poisson argues, would be through a modification of the initial velocity and position of the material point during the time τ. This modification being infinitesimal, the resulting modification of the induced velocity u' would be a second-order infinitesimal. We therefore have $f + f' \propto u + u'$ to first order. More generally, f must be proportional to u.[22]

In this fallacious reasoning, Poisson implicitly assumes that the velocity increment caused by a given force does not depend on the initial velocity, and he arbitrarily excludes any effect of the velocity variation caused by f during τ on the action of f'. Evidently, the conditions expressed in the former citation are met by any equation of motion of the kind

$$\mathbf{f} = m(\ddot{\mathbf{r}}, \dddot{\mathbf{r}}, \ldots)\ddot{\mathbf{r}} \quad \text{with } m > 0. \tag{2.7}$$

Even an analyst as powerful as Poisson could err in intuitively manipulating infinitesimals of various orders. His mathematical proof of the second law was quickly and justly forgotten.

Bélanger's "principle of relative motions"

Jean-Baptiste Bélanger was a Polytechnique-trained engineer, and he taught mechanics for several years both at the École des Ponts et Chaussées and at the École Centrale. His *Cours de mécanique*, published in 1847, stands out for its clear conceptual architecture and for its central reliance on a relativity principle. Bélanger praised his masters Gaspard Coriolis and Jean-Victor Poncelet for placing dynamics before statics and for giving prominence to the concept of work. He followed André Marie Ampère's advice to expound pure kinematics

---

[21] Poisson 1811, pp. 278, 285–286.

[22] Poisson 1833, p. 215.

(the study of motion independently of its causes) before dynamics proper (the study of motion under given causes). He criticized those who, like d'Alembert, pretended to eliminate the concept of force from mechanics and he emphasized that forces were causes that could be measured since concrete operations existed for the comparison and addition of forces. He joined Euler and Coriolis in rejecting impulses in favor of fundamentally continuous forces.[23]

Most innovatively, Bélanger based dynamics on the *principe général des mouvements relatifs*:

> When an arbitrary system of material points undergoes a rectilinear varied motion, if another point having at a given instant the dragging velocity of the system, receives, from this instant on, besides the force necessary to have it participate in the acceleration of the system, the action of one or several forces, it takes, relatively to the system, the same motion that these forces would induce if the velocities and forces of the common motion did not exist.

We recognize Laplace's statement that "the relative motions of a system of bodies, animated by any forces whatever, are the same, whatever be their common motion," here stated more precisely and turned into a principle. Bélanger excludes a direct verification of this principle, and bases its truth on the truth of its consequences, especially in astronomy.[24]

Bélanger first notes that the weaker version of this principle, in which the common motion is uniform, implies the uniformly accelerated character of motion under a constant force. This is Huygens's old argument, frequently duplicated in earlier French texts on mechanics. Next, Bélanger proves that his principle, together with the concrete additivity of parallel forces, implies the proportionality between force and acceleration. One just has to imagine two identical particles moving under the same force, and then have an additional force act on one of them with the same intensity and in the same direction. The complete motion of this particle being, according to Bélanger's principle, composed of the motion of the other particle and the relative motion, and the net force acting on this particle being twice the force acting on the other particle, a double force induces a double acceleration, and so forth until proportionality is proved. Since the motions considered can be infinitesimal, the proportionality extends to variable forces. This is how Bélanger turns Newton's second law into a "theorem."[25]

Further, Bélanger derives the parallelogram of forces. If two identical particles are subjected to the same force $\mathbf{F}$ and if one of them is subjected to the additional force $\mathbf{F}'$, according to Bélanger's principle the acceleration caused by the two forces acting jointly must be the vector sum of the accelerations they separately induce. Taking into account the already proven proportionality of force and acceleration, the compound force must be the vector sum $\mathbf{F} + \mathbf{F}'$, in conformity with the parallelogram of forces. Bélanger here departs from the dominant tradition of deriving this rule within statics. His reasoning may be seen

[23] Bélanger 1847, I–II (Poncelet), 38n–39n (forces). Cf. Chatzis 1994; 1995, pp. 238–241. On French kinematics, cf. Martínez 2009, chap. 3.

[24] Bélanger 1847, p. 75.

[25] Bélanger 1847, pp. 73, 75–76.

as a continuous version of Newton's own derivation based on successive impulses acting according to the second law.[26]

Bélanger's innovations were influential, the more so because in 1851 he was in charge of reforming the teaching of mechanics at the École Polytechnique.[27] Many authors adopted his "principle of relative motions" and gave variants of his derivation of the law of acceleration: Nicolas Deguin in 1853, Charles-Eugène Delaunay in 1856, Ossian Bonnet in 1858, Charles Sturm in 1861, Aimé Henry Resal in 1862, Éleuthère Mascart (replacing Galileo's ship with a train) in 1866, Jules Violle in 1883, and Paul Appel in 1993.[28]

Violle

Jules Violle's *Cours de physique*, published in 1883, is worth special attention. Violle then was a physics professor at the University of Lyon and about to become Maître de Conférences at his *alma mater*, the École Normale Supérieure. His *Cours* was widely appreciated and used. Unlike most textbook writers, Violle provided an abundant bibliography, in which he included the mechanical treatises by Laplace, Duhamel, Bélanger, Delaunay, and Resal.[29] Like Bélanger, Violle based dynamics on the principle of inertia and on the "principle of relative motions":

> Consider a system animated by a motion of translation, and consider, in this system, a body singly subjected to the action of a force: this body undergoes a relative motion in the system. The principle states that *the relative motion of the body is independent of the translation of the system and is therefore the same as if the system were at rest.* This principle, whose first notion belongs to Galileo, is clearly indicated by experience only in the case when the translatory motion is rectilinear and uniform, as is for instance the motion of a boat calmly descending a river. We will nonetheless admit it in a general manner, under the sole condition that the motion of the system be one *of translation.*

Here is again the stronger relativity of Newton and Laplace, with Bélanger's name for it. From his two principles, Violle derives the uniform acceleration of a material point under constant force, and proves by similar reasoning that two forces acting jointly produce the vector sum of the accelerations that each force would produce individually; force is consequently proportional to acceleration. He does not detail the reasoning and does not spell out implicit assumptions about the combination of forces, which he has already defined in his statics. He is as floppy as his forerunners when generalizing to variable forces.[30]

In 1894, the official program for the *baccalauréat ès sciences* included the following paragraph:

---

[26] Bélanger 1847, pp. 122–124.

[27] Cf. Chatzis 1995, p. 237.

[28] Deguin 1853; Delaunay 1856; Jamin 1858; Bonnet 1858; Sturm 1861; Resal 1862; Mascart 1866; Violle 1883; Appel 1893.

[29] He did so in the second edition, Violle 1883, p. viii.

[30] Violle 1883, pp. 96–97 (citation), 98, 100.

Law of inertia and Law of relative motion. Whence it will be deduced that a constant force acting on a material point starting from rest or animated with an initial velocity parallel to the force impresses a uniformly accelerated motion. Reciprocal proposition. Two constant forces are proportional to the accelerations that they produce by acting separately on the same material point starting from rest or animated with an initial velocity parallel to the force.

I do not know when the program started to include this topic. I only know that the program for the year 1870 did not have it. In an address of 1900, Poincaré mentioned that "there had long been traces of this demonstration [of the second law] in the programs of the *baccalauréat*."[31]

The past tense in Poincaré's cited remark suggests that by 1900 the *baccalauréat* programs no longer included the relativity-based derivation of the second law. Appel gave it up in the 1902 edition of his treatise under the influence of the Nancy-based physicist René Blondlot. At the international congress of philosophy in Paris in 1900, Blondlot had praised Mach's *Mechanik* and recommended that the science of motion be founded on the mutual acceleration of two material points, in Saint-Venant's and Mach's manner. The Academician Léon Lecornu adopted a similar presentation in his *Cours de mécanique* of 1914 for the École Polytechnique. In his view, the formula $\mathbf{f} = m\ddot{\mathbf{r}}$ was there only to "abbreviate the discourse." The winds had turned against the Laplace–Bélanger approach to the laws of mechanics and in favor of Kirchhoff's or Mach's approach.[32]

Poincaré's "relativity principle"

Talking after Blondlot at the international congress of philosophy of 1900, Poincaré mentioned past attempts to derive the second law of motion from "the principle of relative motion," which he restricted to Galilean relativity:

> The attempt has sometimes been made to attach the law of acceleration to a more general principle. The motion of any system must obey the same laws, whether it be referred to fixed axes, or to movable axes carried along in a rectilinear and uniform motion. This is the principle of relative motion, which forces itself upon us for two reasons: first, the commonest experience confirms it, and second, the contrary hypothesis is singularly repugnant to the mind.

In conformity with his conventionalist analysis of the foundations of mechanics, Poincaré declared that all attempts to derive the second law from first principles were blocked by the impossibility of defining force independently of mass and acceleration.[33]

As we will see in Chapter 6, since 1895 Poincaré had repeatedly expressed his conviction that optical (and electromagnetic) phenomena should depend only on the motion of matter

---

[31] *Programme de l'examen du baccalauréat ès sciences* (Paris: Nony, 1894), p. 23. Poincaré 1901a, p. 477. I have also seen the programs for the years 1851, 1859, and 1870.

[32] Appell 1902; Lecornu 1914, p. 211. Kirchhoff's views also impregnated two major British treatises: Love 1897 and Whittaker 1904.

[33] Poincaré 1901a, p. 477.

with respect to matter, and not on the motion of matter with respect to the ether. For the first time in 1900, he expressed this condition as the validity of the "principle of relative motion" when applied to matter alone. More exactly, the principle of relative motion had to hold in the following sense: the results of optical and electromagnetic experiments should be the same in any inertial frame to which the experimental setup is attached. Poincaré called this principle *le principe de relativité* for the first time at the end of *La science et l'hypothèse*, published in 1902. In his Saint-Louis lecture of 1904 he enunciated "the principle of relativity, according to which the laws of physical phenomena should be the same, whether for an observer fixed, or for an observer carried along in a uniform movement of translation; so that we have not and could not have any means of discerning whether or not we are carried along in such a motion."[34]

To summarize, Poincaré first used the phrase "the principle of relative motion" in connection with the French tradition of deriving Newton's second law from Galilean or accelerative relativity. Evidently, this mechanical principle applies to electrodynamics as long as it is considered as the theory of a mechanical ether (a change of reference frame thereby affects both the motion of matter and the motion of the ether). In this perspective, the ether and its hidden mechanical parts are to be placed on the same footing as the parts of ordinary mechanical devices. Therefore, motion with respect to the ether is principally detectable. Poincaré rejected this received view. As we will see in Chapter 6, he denied the ether any materiality and postulated the validity of Galilean relativity when applied to ordinary matter alone. "The principle of relative motion" or "the principle of relativity" thus acquired a new meaning and scope, even before the ether had truly disappeared from physical theory.

Einstein's relativities

When Albert Einstein, in 1905, wrote about the *Prinzip der Relativität*, he was probably borrowing this phrase from the German translation of Poincaré's *La science et l'hypothèse*. No one had earlier used it for relativity with respect to inertial frames. To summarize, Poincaré encountered "the principle of relative motions" in a derivation of Newton's second law that had long been popular in France. He rejected the derivation but kept the principle and its name, while he extended it to electromagnetic and optical phenomena. This is a first, terminological link between the French relativity-based derivations of Newton's second law and Einstein's relativity theory.[35]

Another possible connection is through Violle's course. There is direct evidence that in 1895 Einstein read the section in which Violle derived the second law from the "principle of relative motions" (*Princip der relativen Bewegungen* in the German). As has already been noted by Einstein scholars, this section was an eye-catching example of the constructive use of a relativity principle. Yet no historian seems to have noticed that Violle's "principle of relative motions" was not the Galilean relativity principle but Bélanger's principle of

---

[34] Poincaré 1900a, p. 271; 1902a, p. 281; 1904a, pp. 306 (citation), 310–312. Cf. Darrigol 1995a.

[35] Einstein 1905a, p. 891; Poincaré 1904b, p. 243. The phrase "principle of relativity" (of knowledge) had been abundantly used in Spencerian philosophy, and John Stallo had used it to refer to the general idea that all motion is relative (Stallo 1882, p. 200; cf. Martínez 2009, pp. 98–101).

accelerative relativity. The latter principle, being akin to the imperceptibility of gravitation in a free-falling frame, may have inspired Einstein when in 1907 he started to make the equivalence principle the basis for a new theory of gravitation.[36]

2.3 Principles and deductions

There being some difficulty in following the meanderings of the history of quasi-rational derivations of the second law, it may help to give a synthetic, less chronological account of these derivations. Let us first list and clarify the basic concepts and principles used in the (quasi-) rational derivations of Newton's second law.

Space and time

All authors of mechanical texts in the Newtonian tradition assumed the existence of space and time as basic concepts necessary for the formulation of the laws of mechanics. Remember that space, in Newton's sense, is an imaginary, rigid, indefinitely extended, and perfectly penetrable body in which material bodies have a definite place. Time is a continuous indicator of temporal succession. Among all possible spaces and times, there is, according to Newton, an absolute space and an absolute time in which the laws of mechanics hold. "Space" in the old Newtonian sense is congruent with the "system of axes" and the "system of reference" introduced in nineteenth-century mechanical texts. Implicitly, all our authors regarded absolute space as homogeneous and isotropic, and time as homogeneous.[37]

The principle of inertia

The first commonly accepted principle is the principle of inertia, first stated in its modern form by Huygens and Newton. According to this principle, the motion of a free material point in absolute space is rectilinear and uniform. Most authors since Newton have regarded this principle as the result of an empirical generalization from common observations such as the frictionless motion of a ball on a horizontal table. As we saw earlier in this chapter, a few authors of the eighteenth century, especially Euler and d'Alembert, regarded it as a consequence of the principle of sufficient reason. Later authors usually ignored Euler's and d'Alembert's rational deductions of the principle of inertia and treated it as an empirical law.[38]

---

[36]Violle 1892, pp. 90–91. Cf. Stachel et al. 1989, pp. 258–259. Einstein had used Violle's text to prepare for the ETH entrance examination in 1895. A marginal annotation on p. 94 of Einstein's copy of this book proves that he read the derivation of the second law based on the principle of relative motions. On the equivalence principle, see pp. 274–275, this volume.

[37]*Système d'axes* is found in Coriolis 1844; "reference frame" in Thomson 1884a, 1884b, and Tait 1884 "reference system" in Love 1897; *Bezugssystem* in Boltzmann 1897. On the history of reference frames, cf. Bertoloni Meli 1993; Maltese 2000.

[38]See pp. 16 (Newton), 29–30 (Euler), 31–32 (d'Alembert), 33 (Laplace), this volume.

Relativity principles

As we saw in Chapter 1, Galileo asserted that mechanical behavior was unaffected by a common uniform motion of all bodies concerned. Huygens later employed Galilean relativity (properly adapted to rectilinear inertia) as a general principle fit to determine other laws of mechanics. Newton proved that his laws of motion implied Galilean relativity but, unlike Huygens, he regarded this result as a corollary or a theorem, and not as an organizing principle. Newton also derived a more general kind of relativity, which we may call *accelerative relativity*: for a system of bodies freely moving in a uniform gravitational field, the relative motion of these bodies under additional forces is the same as if the gravitational field did not exist. Again, Newton did not regard this theorem as a constructive principle.

Until the late nineteenth century, relativity principles were rarely stated in terms of a change of reference frame, or of a change of "space" in Newton's parlance (with Euler's exception). Instead these principles were phrased as the lack of effect of a common rectilinear motion of the bodies of a system on their "relative motions" (Newton's *inter se motus*) under given forces. This is why in 1847 Bélanger called his own relativity principle *le principe des mouvements relatifs*. In order to reach the modern phrasing in terms of reference frames, we may consider that one of the bodies is a ship, a train wagon, or an elevator and attach a reference frame to this rigid body. Then the Galilean relativity principle implies that the laws of motion are the same with respect to any inertial frame, and the accelerative relativity principle implies that the laws of motion in a free-falling frame are the same as in an inertial frame.[39]

Consider, in the old relativity conception, a system of bodies (material points) that share the same rectilinear motion at the velocity $\mathbf{u}(t)$. Call $\mathbf{r}_i$ the position of the body i and m_i its mass. According to the acceleration law, the force $m_i\dot{\mathbf{u}}$ is needed to move the body i. Now suppose the additional force $\mathbf{f}_i$ to be acting on the body i. The acceleration law and the vector composition of forces imply

$$m_i\dot{\mathbf{u}} + \mathbf{f}_i = m_i\ddot{\mathbf{r}}_i, \quad \text{whence follows } \ddot{\mathbf{r}}_i - \ddot{\mathbf{r}}_j = \frac{\mathbf{f}_i}{m_i} - \frac{\mathbf{f}_j}{m_j}. \tag{2.8}$$

Consequently, the relative motion is unaffected by the common translation at the velocity $\mathbf{u}$, even if this velocity varies in time. This is a modernized derivation of a result obtained by Newton and Laplace. The question is whether, reciprocally, the principle of relative motions jointly with other principles implies the acceleration law.

The second-order principle

Starting with Huygens, several authors tacitly assumed that the motion of a material point with vanishing initial velocity was completely determined under a given force. In other words, the motion was the same as if the body had been at rest before the initial time. Since this assumption implies that the equation of motion cannot involve derivatives of order higher than two, I call it *the second-order principle*. It was often implicitly included in the

---

[39] Newton 1687, pp. 19–20; Bélanger 1847, p. 75.

Galilean relativity principle, so that derivations allegedly based on the latter principle truly required the second-order principle in addition.

The secular principle

Until the early nineteenth century, mechanical authors usually distinguished between two kinds of force: forces acting instantaneously (or during a very brief time), and forces acting continuously. For the sake of clarity, I call the former forces *impulses*, and the second just forces, although most authors used the same name. From Newton to Laplace, impulses were usually regarded as more fundamental, and continuous forces were assumed to be equivalent, in their observable effects, to a very rapid succession of impulses. This equivalence is what I call *the secular principle*. It did not necessarily imply a commitment to an underlying mechanism in which continuous distance forces such as those of gravitation would result from collisions between invisible particles. This sort of mechanism dwindled together with Cartesian natural philosophy. What favored the impulsive approach was its higher conceptual and mathematical simplicity: an impulse generates a velocity, whereas a force generates an acceleration. In the first third of the nineteenth century, the impulsive approach gradually vanished, presumably because it was associated with the increasingly obsolete appeal to collisions between strictly hard bodies.[40]

The additivity of forces

Since d'Alembert, many teachers of mechanics prefer to *define* force as the product of mass and acceleration (or as the product of mass and velocity in the case of impulsive forces). With this definition, the law of acceleration reduces to a tautology and its derivation from other principles becomes pointless. The move is problematic because it makes force an intrinsic property of a moving material point (since its mass and acceleration are so), whereas the basic intuition of force refers to an external cause for the acceleration.

Instead we may follow Newton and define force as a cause of motion, which, at a given time and a given point of space, is *measured* by the product of the mass and acceleration that a test particle would take if it were placed at this point at this time. Then the acceleration law becomes a statement of consistency of the measurement of force: whatever be the motion of the test particle, the product of its mass and acceleration is the same (for impulsive forces, whatever be the mass and the initial velocity of the particle, its momentum change is the same). In this view, it still makes no sense to prove that the force is the product $m\ddot{\mathbf{r}}$ of mass and acceleration and not any other expression of the kind $\Phi(m\ddot{\mathbf{r}})m\ddot{\mathbf{r}}$ (wherein Φ is any scalar function of a vector). This is simply assumed. Yet it becomes necessary to check that the dynamic measurement of force is compatible with what we already know of forces from statics. This is why Newton derives the parallelogram of forces from his second law of motion.[41]

---

[40] Strictly hard bodies and their collisions had played a fundamental role in eighteenth-century French rational mechanics (d'Alembert, Carnot, Lagrange, even Laplace): cf. Darrigol 2001.

[41] See p. 19, this volume.

There is still another concept of force, the one adopted by most of those who tried to derive the second law from a relativity principle: force is defined as a (point-driving, instantaneous, and directed) cause of motion whose measure is defined by a concrete equality and a concrete addition. Two parallel forces are said to be of equal intensity if and only if they can be balanced by the same force when they act together on the same point at the same time. The concrete addition is defined by assuming that the intensities of two parallel forces add up when they act together on the same point (in the same direction). As is well known, a concrete equality and a concrete addition together define a measure of a physical quantity (as long as the equality is symmetric and transitive, and the addition is commutative and associative, which is obvious in our case). This definition of force is inherently static since it is based on equilibrium under forces simultaneously acting on the same point. As Daniel Bernoulli proved in 1726, it can be used, together with the requirement that the resultant of two forces of equal intensity bisects their angle, to derive a basic law of statics: the *parallelogram of forces* according to which the resultant of two forces is the vector sum of these two forces. However, there is no need to develop a full theory of statics before defining force as we just did. The comparison and the addition of forces are sufficient, and their operational significance is simple and immediate.[42]

Two basic derivations of the second law

We are now equipped to explain the structure of historical derivations of Newton's second law. Two cases must be distinguished: impulsive forces à la Newton and continuous forces à la Huygens. In Laplace's impulsive treatment, the principle of inertia and Galilean relativity together imply that under a given impulse the change of velocity of a material point cannot depend on its initial velocity $\mathbf{u}$. Indeed, this change must be the same in an inertial frame moving at the velocity $\mathbf{u}$ so that the relative initial velocity vanishes. Now, by the secular principle a uniform gravitational field can be replaced by a rapid, periodic succession of equal impulses. Each impulse adds a constant velocity increment to a falling body. In the limit of an infinitesimally small period between two successive impulses, the velocity increases linearly with time, and the acceleration is constant. Galileo's laws of free fall thus derive from Galilean relativity, inertia, and secularity.

Call $\mathbf{I}_1$ and $\mathbf{I}_2$ two impulses that would separately induce the velocities $\mathbf{v}_1$ and $\mathbf{v}_2$ when applied to a material point at rest. In order to evaluate the simultaneous action of these two impulses on a material point initially at rest, we may consider (in harmony with secularity) that the second impulse acts right after the first with a negligible delay. According to Galilean relativity, the velocity induced by the second impulse in a frame moving at the velocity $\mathbf{v}_1$ induced by the first impulse is equal to $\mathbf{v}_2$. Therefore, the final absolute velocity is $\mathbf{v}_1 + \mathbf{v}_2$. According to the additivity principle, the responsible impulse is $\mathbf{I}_1 + \mathbf{I}_2$ for parallel impulses. This property, together with continuity and rotational invariance, implies the proportionality of impulse and induced velocity, in conformity with Newton's second law. In addition, the velocity variation of a material point under a given impulse does not depend on the prior velocity of the material point. According to the secularity

---

[42] Bernoulli 1726. Cf. Darrigol 2014, pp. 23–26.

principle, the action of a variable force can be mimicked by the action of a rapid succession of periodic impulses with a value proportional to the value of the force at the time of each impulse. If the succession is sufficiently rapid, the motion is quasi-continuous and according to the previously established proportionality of impulse and velocity increment, the instantaneous (secular) acceleration must be proportional to the force.

To summarize, inertia, Galilean relativity, and the additivity of impulses together imply Newton's second law in its original form (proportionality of impulse and velocity increment). Together with secularity, this law implies the acceleration law.[43]

Let us move to reasoning based on continuous forces, as first given by Huygens and by Bélanger. First consider, with Huygens, a material point moving under a constant force $\mathbf{f}$. According to Galilean relativity, the velocity acquired by the material point during the time interval $[t,\ t+\tau]$ does not depend on the velocity at time t. According to the second-order principle, this acquired velocity is the same as if the material point were starting from rest. It is therefore a well-defined function $\mathbf{u}(\tau, \mathbf{f})$ of the time lapse τ and of the force $\mathbf{f}$. Altogether, we have

$$\mathbf{v}(t + \tau) = \mathbf{v}(t) + \mathbf{u}(\tau, \mathbf{f}). \tag{2.9}$$

Deriving with respect to τ at $\tau = 0$, this gives

$$\ddot{\mathbf{r}} = \frac{\partial \mathbf{u}}{\partial \tau}(0, \mathbf{f}) = \text{constant vector}. \tag{2.10}$$

Thus, Galileo's laws of free fall result from Galilean relativity enriched with the second-order principle.

According to the latter principle and for a variable force $\mathbf{f}(t)$ acting on a material point, the equation of motion has the generic second-order form

$$\Psi(\mathbf{r}, \dot{\mathbf{r}}, \ddot{\mathbf{r}}, \mathbf{f}) = 0. \tag{2.11}$$

The homogeneity of absolute space implies that $\mathbf{r}(t) + \mathbf{a}$ is a solution if $\mathbf{r}(t)$ is a solution. Therefore, Ψ cannot depend on $\mathbf{r}$. Galilean relativity further implies that $\dot{\mathbf{r}}(t) + \mathbf{b}$ satisfies the equation of motion if $\dot{\mathbf{r}}(t)$ does. Therefore, Ψ can depend only on $\ddot{\mathbf{r}}$ and $\mathbf{f}$. We may then write $\mathbf{f} = \Omega(\ddot{\mathbf{r}})$. We now assume the material point to be subjected, in addition to the force $\mathbf{f}$, to a uniform gravitational field of acceleration $\mathbf{g}$. According to the additivity of forces, the net force acting on the material point is $\mathbf{f} + \Omega(\mathbf{g})$. According to the principle of accelerative relativity in Bélanger's fashion, the acceleration $\ddot{\mathbf{r}}$ of the material point in an accelerated frame of acceleration $\mathbf{g}$ is the same as it would be in absolute space without gravitation. In absolute space, we therefore have

$$\Omega(\mathbf{g} + \ddot{\mathbf{r}}) = \Omega(\mathbf{g}) + \Omega(\ddot{\mathbf{r}}). \tag{2.12}$$

---

[43] I skip considerations of mass dependence.

The function Ω being continuous and rotationally invariant, this relation requires the proportionality of force and acceleration. This proportionality, at the heart of the modern version of Newton's second law, may therefore be regarded as a consequence of the principle of inertia and the principle of accelerative relativity.

Conclusions

From Huygens to Violle, many authors relied on relativity principles to derive Newton's second law. This tradition was abundantly illustrated in French textbooks of the nineteenth century, it was known to Einstein and Poincaré, and it contains the origin of the phrase "principle of relativity." Although most non-French authors and many French ones ignored these derivations, and although they were generally dismissed toward the end of the nineteenth century, they are philosophically significant, and they belong to a historically important turn in the way of founding physical theories. Let us begin with the philosophical significance.

Philosophical lessons

In today's philosophy of science, the most common view on mechanics is that it is based on a purely mathematical scheme for constructing models (in the mathematical sense) based on a few vector and scalar spaces (for forces, positions, and masses in the case of the mechanics of material points) and relations between quantities belonging to these spaces. Physical interpretation is then made by observing some analogy between the models and concrete systems. Questions about the physical definition of force, mass, and reference frames are regarded as remnants of a naive, inductive, empiricist view of physical theory in which every physico-mathematical concept would have a direct empirical counterpart.[44]

In the late nineteenth century, Ludwig Boltzmann came close to this modern semantic view by declaring that a consistently constructed mathematical picture, not a piece-by-piece adaptation to concrete entities, defines a physical theory in general and mechanics in particular. He traced the often deplored obscurities of mechanics to the covering up of its true nature as a hypothetical, mental construct:

> In my opinion, the obscurities in the principles of mechanics come from our inclination not to begin with hypothetical mental pictures and instead to draw on experience. Then one tried more and more to hide the transition to hypotheses and even to fabricate a proof that the entire construct was necessary and assumption-free, only to lapse into obscurity.

Mach, in his analysis of the foundations of mechanics, required a well-defined operational basis, which he found in the motion of interacting material points. Like Kirchhoff, he regarded the concepts of space and time and the derived concepts of velocity and acceleration as sufficient for an empirically significant description of mechanical motion. He defined mass through the ratio of the accelerations of two interacting material points, and

---

[44] See, for example, Suppe 1974; Giere 1988; van Fraassen 1980.

force by the product of mass and acceleration. Poincaré condemned any separate empirical justification of the laws of mechanics as futile, because it was physically impossible to disconnect these laws from each other. He downgraded the basic laws to definitions or conventions introduced for the sake of convenience. Different conventions of measurement would lead to different laws for the same set of mechanical phenomena.[45]

All these views share some holism and some nominalism. They deny the possibility of introducing the concepts of force, mass, and acceleration separately, and they treat forces as mere names introduced for the sake of mathematical representation or just for linguistic convenience. In this case, the relation $\mathbf{f} = m\ddot{\mathbf{r}}$ is just a definition and it does not have any empirical content. Attempts to derive it from the principle of relativity or from any other physical principle are totally meaningless.

This conception of $\mathbf{f} = m\ddot{\mathbf{r}}$ was and still is resisted by many physicists, for good reasons. First, it does not explain why there are simple relations between the forces thus defined and the spatial configuration of a system of bodies. More fundamentally, it does not account for the idea of force as a cause of motion. Historically, d'Alembert, Kirchhoff, and Mach rejected this idea because they believed it to be contaminated with the Cartesian propensity to imagine obscure mechanisms. This need not be so. As Bélanger rightly emphasized, the existence of an external circumstance on which the motion of a body depends can be asserted independently of any hidden mechanism. Moreover, forces can be measured independently of any acceleration law, because the equality of two forces is concretely defined by their being balanced by the same force and because their sum is concretely defined by having them act simultaneously on the same material point. All we need for this purpose is the notion of equilibrium. In other words, there is a static module of mechanics with well-defined physico-mathematical concepts of force and equilibrium. This is not naive operationalism; this is just the recognition that our theories have self-consistent modules through which the implied concepts acquire separate empirical meaning. In mechanics, this modular structure allows us to measure force and acceleration separately, so that their proportionality has genuine empirical meaning.[46]

Then it makes sense to ask whether $\mathbf{f} = m\ddot{\mathbf{r}}$, as empirical law, can be derived from more fundamental principles. As was shown in the previous section, it can be derived from the principle of inertia, the principle of Galilean relativity, and the secular principle according to which a time-dependent force can be replaced by a rapid succession of impulses. What do we gain from this knowledge? Laplace, who first gave this derivation, and his followers did not regard it as purely rational, because in their eyes the principle of relativity was an inductive generalization from experience. There is indeed, within the confines of classical mechanics, no resource other than empirical to justify the peculiar structure of space–time embodied in the principles of inertia and relativity. As emphasized by Harvey Brown, this structure presupposes a strange pre-established harmony between the relative motion of free particles in a portion of space. All we can do, without general relativity, is to admit

---

[45] Boltzmann 1897, Vol. 1, pp. 2; Mach 1883, pp. 202–207; Poincaré 1901a.

[46] On the modular structure of physical theories, cf. Darrigol 2008.

the empirical soundness of this structure. Then all we can say is that the acceleration law is deducible from simpler but not purely rational principles.[47]

There is more to say on these principles, however. Without the inertial structure we would lack basic means to measure time. The most basic way to measure time indeed is to measure the distance traveled by a freely moving particle, and this cannot be done consistently without assuming the Galilean space–time. As for the secular principle, it results from the broader assumption that the motions observed at our scale do not depend on details belonging to a finer timescale. In other words, we assume an effective causality at our scale, irrespective of any sub-mechanics at a finer scale. All these requirements of measurability and causality may be regarded as quantitative expressions of the comprehensibility of nature. Classical mechanics and its laws result from a form of comprehensibility that happens to be warranted at our scale.[48]

A physics of principles

These philosophical reflections show that the now-forgotten derivations of the second law are not mere historical aberrations. They are no less respectable than other principle-based constructions. To be sure, another principle, the impossibility of perpetual motion, played a more important role in historical derivations of mechanical laws, in deriving or justifying energy conservation, and in the thermodynamics of Carnot, Clausius, and Thomson. Later in the century, the latter theory was often regarded as the model of a new sort of theoretical physics, in which general principles replaced detailed mechanical pictures in the construction of theories. This is what Boltzmann called "phenomenology" and Poincaré "the physics of principles." Laplace's and Bélanger's derivations of Newton's second law were in the same spirit.

There are other examples of the constructive role of a relativity principle in nineteenth-century physics. The first that comes to mind is the derivation of electrodynamic laws. As we will see in Chapter 5, Wilhelm Weber designed his fundamental law for the force between two electric particles so as to depend only on the relative motion of these particles, and several authors later derived the electromagnetic induction law in a moving conductor from the induction law for a conductor at rest combined with the principle that electromagnetic induction should depend on the relative motion of moving conductors and magnets only. As was already mentioned and as will be recounted in Chapter 6, later in the century Poincaré extended this relativity of electromagnetic induction to all electromagnetic and optical phenomena and used it to justify new theoretical constructs, and, ultimately, the Lorentz invariance of the fundamental equations of the theory.

Let us return to mechanics. In 1897, Boltzmann's former assistant Ignaz Schütz, then in Göttingen, was aware of the difficulty of a complete, convenient mechanical picture of electrodynamic and thermodynamic phenomena. In his opinion, what was at fault was not so much the idea of a mechanical reduction, but the usual formulation of mechanics based on Newton's three laws. He knew about Georg Helm's and Wilhelm Ostwald's energeticist

---

[47] Brown 2005.

[48] Cf. Darrigol 2014, chaps. 1–2.

program and about the Lübeck meeting of 1895 during which Planck and Boltzmann bran-
dished the impossibility of deriving the laws of mechanics from the energy principle only.
Schütz proposed to base mechanical-energetic reduction on the *principle of the absolute
conservation of energy*, obtained by supplementing the usual energy principle with the condition that "the energy principle holds independently of any constant progressive motion of
our material world with respect to geometrical space." In other words, he assumed both the
energy principle and the Galilean relativity principle and went on to prove that Newton's
three laws derived from their combination.[49]

Schütz's first example is the elastic, frontal collision of two balls of initial velocities v_1
and v_2. Like Galileo and Huygens, he illustrates the progression of "the material world"
with a boat. Energy conservation for the velocity u of the boat in which the collisions occur
reads

$$m_1(v_1 + u)^2/2 + m_2(v_2 + u)^2/2 = m_1(v_1' + u)^2/2 + m_2(v_2' + u)^2/2. \qquad (2.13)$$

The velocity u being arbitrary, this implies the equation

$$m_1v_1 + m_2v_2 = m_1v_1' + m_1v_2' \qquad (2.14)$$

for momentum conservation, usually derived by combining Newton's second and third
laws. For a particle of mass m and velocity v subjected to the force F, Schütz expresses
energy conservation as

$$(F \cdot v)dt = d(mv^2/2). \qquad (2.15)$$

Assuming all natural forces to depend on the relative motion of matter only, energy
conservation in a world moving at any velocity u reads

$$F \cdot (v + u)dt = d[m(v + u)^2/2] \qquad (2.16)$$

and requires Newton's second law $F = mdv/dt$. For a system of particles with forces deriving
from the potential $U(r_1, r_2, ...r_n)$, absolute energy conservation stipulates

$$\frac{d}{dt}\left(U + \sum_{i=1}^{n}\frac{1}{2}m(v_i + u)^2\right) = 0 \text{ for any } u, \qquad (2.17)$$

which implies the equality of action and reaction in the form

$$\sum_{i=1}^{n}F_i = 0, \quad \text{with} \quad F_i = m_i\frac{dv_i}{dt}. \qquad (2.18)$$

As will be seen in Chapter 6, Poincaré presented a similar argument in 1900, with the
intention of showing that the principle of reaction and the relativity principle were intimately related, both in mechanics and in electrodynamics. In his celebrated *Raum und*

---

[49] Schütz 1897, p. 114.

Zeit lecture of September 1908, Hermann Minkowski noted that in relativity theory energy conservation, being the first component of a four-vector relation, implicitly contained the momentum equation. He referred to Schütz's memoir for the classical counterpart of this result. These results were foreshadowing the general relation between symmetries and conservation laws expressed in Emmy Noether's theorem of 1915.[50]

The pre-relativistic history of relativity principles as tools of theory construction has usually been ignored by historians of relativity. Yet it matters for a variety of reasons: it directly inspired some of Poincaré's reflections toward a theory of relativity, including the name he gave to the relativity principle; it indirectly informed Einstein's reflections through some of Poincaré's; it plausibly contributed to the genesis of the equivalence principle; and it belonged to a rising physics of principles in which Einstein inscribed his own efforts.

[50] Minkowski 1909, p. 109. For Poincaré, see pp.166–167. this volume.

3

THE SPACE–TIME–INERTIA TANGLE

> Non est mathematicè difficilis materia, sed physicè aut hyperphysicè.[1] (Christiaan Huygens, c. 1690)

The three most basic concepts of Newton's mechanics, absolute space, absolute time, and inertia, were essentially new and they were difficult to grasp for a variety of reasons: they contradicted the more intuitive ideas of locomotion enshrined in Aristotelian physics; they involved an opaque mix of mathematical, empirical, and theological considerations; and, as we may retrospectively judge, they implied a strange pre-established harmony in the free motion of bodies. One could learn with relative ease how to apply these concepts to concrete problems of celestial or terrestrial motion, but one could not easily penetrate their deeper meaning.

In the two following centuries, a few inquisitive geometers, physicists, and philosophers addressed these difficulties, often criticizing each other's positions (Huygens–Leibniz, Leibniz–Clarke, Leibniz–Euler, Leibniz–Kant, Neumann–Streintz, and Streintz–Mach debates). This happened mostly in the early years of Newtonian mechanics and in the last third of the nineteenth century, with long intervals of dogmatic slumber. There were three attitudes: plainly reject Newton's foundations (Leibniz and Mach), consolidate them (d'Alembert, Euler, Maxwell, and Neumann), or revise them critically (Kant, Thomson and Tait, James Thomson, Streintz, Lange, Calinon, and Poincaré). The motivation varied: it could be the inclusion of mechanics in a grand philosophical scheme (Leibniz, Kant, Mach, and Poincaré); it could be the quest for secure foundations; or it could be basic pedagogy. The chief outcomes were an intimate connection between the definitions of space and time, the law of inertia, and Galilean relativity; a growing awareness of the conventional aspects of mechanical laws; a declining recourse to metaphysics; and an increasing empiricism.[2]

The first, eighteenth-century section of this chapter covers Huygens's rejection of Newton's absolute space and his pioneering construction of a class of strictly equivalent inertial frames, Leibniz's wavering between a purely relational space and a space partially determined by live force, Euler's defense of Newton's absolute space and time, and Kant's metaphysical reconstruction of Newton's mechanics. The second section shows how, in

---

[1] Huygens, on absolute/relative motion, [c. 1690], p. 213: "This is not a mathematically difficult subject, but a physically or hyperphysically [metaphysically] difficult one."

[2] Relevant secondary literature includes: Jammer 1954, 1957, 2006; Sklar 1974, 1985, 2013; Friedman 1983; Torretti 1983; Earman 1989; Barbour 2001; DiSalle 2006.

Relativity Principles and Theories from Galileo to Einstein. Olivier Darrigol, Oxford University Press.
© Olivier Darrigol (2022). DOI: 10.1093/oso/9780192849533.003.0003

the last third of the nineteenth century, conditions became favorable for reconsidering the foundations of Newtonian mechanics in Germany (Neumann, Lange, and Budde), in Austria (Mach and Streintz), in France (Duhamel, Calinon, and Poincaré) and in the UK (Thomson and Tait, James Thomson, and Maxwell). The last and third section is devoted to the rarely and tardily discussed topic of the measurement of time (James Thomson, Streintz, Calinon, and Poincaré, with pioneering remarks by d'Alembert and Poisson). A last preliminary remark: this chapter is not about reformulations of mechanics in general (in which case it would naturally include many more important authors like the Bernoullis, Varignon, Lagrange, Carnot, Hamilton, Kirchhoff, or Hertz). As indicated in the title, it is more narrowly about the definition of space, time, and inertia in Newtonian mechanics, because, as will appear in the following, this definition is intimately related to Galilean relativity.

3.1 From Huygens to Kant

Huygens's constructed frames

As we saw in Chapter 1, Huygens greatly contributed to the foundations of mechanics, with a focus on concrete devices such as the pendulum clock and a distaste for philosophical musing. He nonetheless left behind a few manuscripts, written late in his life in the early 1890s and first published in 1929, in which he criticizes Newton's concept of absolute motion and offers an alternative view.[3]

Huygens rejects any concept of true or absolute motion, as conceived by Newton and those who before him assumed an empty space in which bodies move. First, empty space is a pure abstraction that cannot serve to define motion in a concrete, useful sense. Motion should always be defined with respect to a concrete reference:

> The motion or rest of bodies is always relative. We cannot say or understand the motion or rest of something unless it is about the mutual relation of bodies. Those who imagine immobile, fixed, and infinitely extended spaces are in error because this immobility cannot be conceived without referring to things at rest.

All we can assert is the mutual distances of bodies. Huygens then assumes systems of (point-like) free bodies whose mutual distances are invariable, which he calls "bodies in mutual rest" (*inter se quiescientes corpora*). His formulation of the principle of inertia reads:

> If, with respect to such bodies in mutual rest, an additional body moves freely and without any obstacle, it will travel uniformly on a straight line with respect to them.

In modern terms, there exist reference frames in which the motion of free particles is rectilinear and uniform, and these reference frames can be constructed as systems of free particles in mutual rest. The constructed inertial frame is not unique; it is defined up to a uniform, global translation, in harmony with the relativity principle. The freedom of the

---

[3]Huygens [c. 1690]. Cf. Volgraff 1929b; Mormorino 1994; Stan 2016.

frame-defining particles is essential, because a system of rigidly connected particles could be rotating with respect to the formerly defined frames.[4]

As Huygens insists, Galilean relativity makes it impossible to select a specific frame among the possible inertial frames and thus to define absolute motion. Following Newton's suggestion, we may infer the absolute rotation of two balls connected through a string from the string's tension. But we still cannot infer an absolute motion of the balls. All we can say is that the balls must be moving with respect to a system of free bodies in mutual rest. Their velocities at a given time need not be equal and opposed, because with respect to the reference system they could share a uniform translation in addition to the uniform circulation around the middle of the connecting string.[5]

Huygens thus succeeded in reformulating the principle of inertia without any concept of absolute space, based on principally observable distances and motions. His *inter se quiescentes corpora* determine the reference of motion just as much as is needed to express and apply the other laws of mechanics. From the observed invariability of the mutual distances of fixed stars and the sun, Huygens concludes that they constitute a legitimate inertial frame (to put it in modern terms) and that Newton's laws apply to motion defined in this frame, even though this motion cannot be deemed absolute.[6]

Leibniz's relational space

In 1894, the old Huygens reproached Leibniz with his reluctance to accept the inexistence of "real motion":

> In your notes on des Cartes I have noticed that you believe *absonum esse nullum dari motum realem, sed tantum relativum* [it is incongruous that [in Descartes's philosophy] no real motion but only relative motion is given to us]. Yet I believe this is very much the case, notwithstanding Newton's reasoning in his *Principles of Philosophy*, which I know to be erroneous.

To which Leibniz replied:

> As to the difference between absolute and relative motion, I believe that if the motion—better, the moving force of bodies—is something real, as seems to be the case, then this force should have a *subjectum*. If A and B move toward each other, I admit that all phenomena will be the same whichever be the one in which movement or rest is assumed; would there be 1000 bodies, I still agree that the phenomena (or even the angels) cannot give us an infallible reason for determining the subject of motion or its degree; and that each of them could be taken to be at rest, and I think this is also what you want. But (I think) you will not deny that in truth each of them has a certain degree of motion, or, if you prefer, of force, despite the equivalence of the hypotheses. Indeed, from this remark I conclude that there is in nature something that Geometry cannot determine. And this is not the least of the several reasons I use to prove that besides extension and its variations, which are purely geometrical things, we must admit something superior, which is force.

---

[4] Huygens [c. 1690], pp. 222, 218.

[5] Huygens [c. 1690], pp. 213–216, 220, 223–225.

[6] Huygens [c. 1690], pp. 215–216.

In brief, Leibniz agrees with Descartes and Huygens that motion is merely relative as long as it is defined as a change in geometric relations, but in his philosophy matter is not pure extension, it has an essential activity measured by *vis viva*; this live force being proportional to velocity squared, motion must be more real than mere kinematic analysis would suggest.[7]

In reply, Huygens wrote: "You seem to be close [to my views on motion], except that when several bodies are in relative motion, you want each of them to have a certain degree of motion or true force, in which I do no share your opinion." Leibniz then toned down his claim:[8]

> I hold that all hypotheses [about which body is moving] are equivalent, and when I assign certain motions to certain bodies, I do not and cannot have any other reason than the simplicity of hypothesis, believing that we can hold the simplest (everything considered) for the true one. Thus, having no better criterion, I think we differ only in the manner of speaking, which I try to adjust to common usage as much as I can, *salva veritate.*

In this exchange, it appears that Leibniz, unlike Huygens, was not much concerned with Newton's laws and the needed reference of motion. As expressed in a letter he wrote in 1715 to Abbot Conti, he disagreed with the most basic tenets of Newton's philosophy:

> His philosophy appears to me rather strange and I cannot believe it can be justified. If every body is heavy, it follows necessarily (whatever his supporters may say and however passionately they may deny it) that gravity will be a scholastic occult quality or else the effect of a miracle . . . I do not find the existence of a vacuum proved by the argument of M. Newton or his followers, any more than the universal gravity which they suppose, or the existence of atoms. One can only accept the existence of a vacuum or of atoms, if one has very limited views. . . . Space is the order of co-existents and time is the order of successive existents. They are things true but ideal, like numbers.

Newton read this criticism, and fustigated its author in a letter to Conti. The mildest section reads:[9]

> He [Leibniz] prefers hypotheses to arguments of induction drawn from experiments, accuses me of opinions which are not mine; and instead of proposing questions to be examined by experiments before they are admitted into philosophy, he proposes hypotheses to be admitted and believed before they are examined.

---

[7]Huygens to Leibniz, May 29, 1694, in Gerhardt 1899, p. 731; Leibniz to Huygens, June 12/22, 1694, ibid. pp. 738–739. Leibniz nonetheless agreed with Huygens that Newton's arguments for the existence of absolute rotation were invalid (and he reminded Huygens that in a conversation in Paris, well before Newton's *Principia*, Huygens had himself used centrifugal force to argue for absolute motion). On *vis viva*, cf. Leibniz [1692], 1695, [1698]; Costabel 1960; Duchesneau 1994. On Leibniz's views on motion, cf. Earman 1989, pp. 15, 71–72, 116–22; Roberts 2003; Lodge 2003; Slowik 2006.

[8]Huygens to Leibniz, August 24, 1694, in Gerhardt 1899, p. 745; Leibniz to Huygens, September 4/14, 1894, ibid. p. 751.

[9]Leibniz to Conti, November or December 1715, in Alexander 1956, pp. 184–185; Newton to Conti, February 26, 1716, ibid. p. 187.

In the wake of their dispute over the discovery of infinitesimal calculus, Newton and Leibniz were no longer on speaking terms. The discussion went on between Leibniz and one of Newton's disciples, the theologian Samuel Clarke. One important point of contention was Newton's suggestion, in Query 20 of his *Optice*, that space was like God's *sensorium*:

> *Does it not appear from Phænomena that there is a Being incorporeal, living, intelligent, omnipresent, who in infinite Space, as it were in his Sensory* [tanquam Sensorio suo], *sees the things themselves intimately, and thoroughly perceives them, and comprehends them wholly by their immediate presence to himself: Of which the Images only carried through the Organs of Sense into our little Sensorium, are there seen and beheld by that which in us perceives and thinks.*

Leibniz misread Newton as equipping God with an unnecessary sensory organ, and gave a few other reasons to exclude space as divine attribute. At any rate, Leibniz rejected any concept of empty space as incompatible with his principle of the identity of indiscernibles (there could not be two different points in an empty space since there would be nothing to differentiate them), with God's predilection for a full universe, and with gravitational interaction. In conformity with his earlier rejection of absolute motion, he redefined space as the order of situations of objects at a given time:[10]

> As for my own opinion, I have said more than once, that I hold space to be something merely relative, as time is; that I hold it to be an order of coexistences, as time is an order of successions. For space denotes, in terms of possibility, an order of things which exist at the same time, considered as existing together; without enquiring into their manner of existing.

Yet, as Leibniz had earlier told Huygens, he believed that his concept of live force somehow called for absolute motion. Hence the following concession to Newton and Clarke:

> However, I grant there is a difference between an absolute true motion of a body, and a mere relative change of its situation with respect to another body. For when the immediate cause of the change is in the body, that body is truly in motion; and then the situation of other bodies, with respect to it, will be changed consequently, though the cause of that change be not in them. 'Tis true that, exactly speaking, there is not any one body, that is perfectly and entirely at rest; but we frame an abstract notion of rest, by considering the thing mathematically.

With the "cause . . . in the body," Leibniz was probably alluding to the *vis viva*. It is not clear how much and on what grounds he wanted the *vis viva* to be univocally defined. Was it just to agree with common usage (as he told Huygens), or for mathematical convenience (as he may be suggesting here), or for metaphysical reasons (to comply with the activity of monads)? What is certain is that in his eyes live force, not Newtonian inertia and force, mattered in the definition of motion.[11]

---

[10] Newton 1706, p. 145 (original in italics); Leibniz to Clarke, February 23, 1716, in Alexander 1956, pp. 25–26.

[11] Leibniz, August 18, 1716, in Alexander 1956, p. 74.

Euler's defense of absolute space

In an unusually philosophical essay of 1748, Euler, without naming Leibniz, attacked the "metaphysical view" of space and time that made them a mere order of coexistence and a mere order of succession, respectively. In his opinion, metaphysics should be constrained by the laws of physics, and not vice-versa. In particular, the law of inertia excludes the "metaphysical view" because in the definition of rest or inertial motion, the reference system cannot possibly be a set of concrete bodies. For instance, Euler explains, inertia implies that a material body suspended in a stagnant, frictionless fluid should remain at rest when the fluid is set into motion. As for the option to refer the inertial behavior to the sun and stars, Euler rejects it because remote bodies cannot possibly cause the inertial behavior:

> The conservation of the state [of motion] of bodies is ruled by location, as it is conceived in mathematics, and not at all by their relationship to other bodies. Now, we could not say that this principle of mechanics is founded in something that is a mere figment of our imagination, so that we are forced to conclude that the mathematical idea of location is not imaginary but that something real in the world corresponds to this idea. There is in the world, besides the bodies that constitute it, some reality that we represent through the idea of location.

Moreover, the "metaphysical view" fails to account for the measures of space and time assumed in the law of inertia. Euler knew that in response to a similar objection by Clarke, Leibniz had given the desired measure by counting the monads in a given interval of time or space. Euler flatly rejected this subterfuge. The laws of mechanics, he concluded, compel us to regard space and time as real entities, despite their escaping direct perception:[12]

> As the equality of the times during which a uniformly moving bodies travels through equal space cannot be explained by the order of succession any more than the equality of spaces can be explained by the order of coexistence, and as this equality is an essential component of the principle of motion, we cannot say that bodies rule their motion on something that is a mere figment of our imagination. We are therefore forced to admit, as we were in the case of space, that time is something that subsists beyond our mind, or that time is something real, just as space is.

Kant's philosophical reconstruction of Newton's mechanics

In his earliest writings in the 1840s and 1850s, Kant adopted Leibniz's view of space as the order of coexistence of things, just supplementing it with the idea that this order or mutual relation depends on a force without which extension could not exist at all. Like Leibniz, Kant regarded the principle of identity of indiscernibles as the chief objection to Newton's view:

> I should never use ["motion" and "rest"] in an absolute sense but only relatively. I ought never to say that a body is at rest without adding in respect of what things it is at rest, and never to say that it is in motion without naming the objects in respect of

---

[12] Euler 1748, pp. 329, 349. Cf. Streintz 1883, pp. 35–50. As noted by Streintz, ibid. p. 37, the body-in-fluid argument is invalid, because a force acts on the fluid and not on the body.

which it changes its relation. Even if I thought I could imagine a mathematical space empty of everything as a receptacle for bodies, that would not help me. For how could I distinguish the parts of it and the different places, if they were not occupied with some thing corporeal?

Kant changed his mind in the 1760s, under the force of Euler's arguments. He now agreed with Euler that Leibnizian space could not sustain the laws of mechanics and he accepted Newton's absolute space. Without space as a real entity distinct from the contained things, he further argued, we would have no means to distinguish a right hand from a left hand: the order or mutual relations of their parts would be exactly the same despite the lack of congruence.[13]

Kant reached his final view of space as the pure form of external intuition in his inaugural dissertation of 1770. He now regarded Newton's infinite real space as a "mere fable" absurdly connecting the spiritual and the physical. In his new view, space is a preexisting faculty of the mind, without which the spatial order of things cannot be discerned in the first place. Space is an intuition (*Anschauung*), that is, a pre-conceptual condition of perception; it is pure for it does not contain any sensorial input. The truths of geometry, being obtained by constructing figures in this pure intuition, are necessary and a priori, whereas for a Leibnizian they have to be induced from our experience of the mutual relations of things. This notion of space and the similar notion of time as a pure form of internal intuition alimented the first part of Kant's *Critique of Pure Reason*, first published in 1881. Besides the passive, representative faculty of *intuition*, Kant posited the active, synthetizing and conceptualizing faculty of *understanding*. He believed the pure (pre-empirical) concepts of the understanding, alias *categories*, to be rigidly determined by the unity of our pre-perceptual thinking. He distinguished four chief categories: quantity, quality, relation, and modality, each containing three subcategories.[14]

The pure intuitions of space and time and the categories of quantity and quality are by themselves sufficient to build a concept of motion in a mathematical sense, but not yet in a physical sense because this would require a notion of the moving thing, that is, matter. The definition of matter and its consequences are the object of Kant's *Metaphysical Foundations of Natural Science*, published in 1786. This treatise has four parts, corresponding to the four categories of quantity, quality, relation, and modality.[15]

In the first part, called *phoronomy*, matter is considered from the exclusive point of movability, without implying any other property. Kant here introduces *empirical space* as a material space to which the motion of material bodies is referred. He also considers *absolute space* as an a priori condition of any spatial experience. He notes that this space coincides with the space obtained by abstracting the materiality of any of the empirical spaces. To put it in modern terms, he has in mind that space, as the necessary form of external intuition, defines the basic linear, homogeneous, Euclidean structure of which every empirical space must be a model. Phoronomy being defined under the flag of quantity,

---

[13] Kant 1746, 1758 (cited in Alexander 1956, pp. xlvi–xlvii), 1768. Cf. Alexander 1956, pp. xlv–xlix; Stan 2009.

[14] Kant 1770, 1781.

[15] Kant 1786. Cf. Friedman 2013.

Kant's chief concern is to define a concrete addition of velocities. This he does by composing the motion of a body in a given empirical space with the motion of this space with respect to another space.[16]

The second part of Kant's *Foundations* is the *dynamics*, based on the category of quality. Matter is now regarded as having the quality of force, through which a portion of matter attracts or repels another portion of matter. Repulsion is necessary to explain the impenetrability of material bodies; attraction to explain their cohesion as well as the empirically observed forces of gravitation. Matter is essentially continuous, and the elementary forces act on the line joining two material points. Empty space may exist between (extended) material bodies, action at a distance is a necessity, and contact action is a Cartesian abuse of geometry. Kant is here concerned with force per se, not with the motions it induces.[17]

Induced motion is the topic of the third part, the *mechanics*, which implements the category of relation according to its three subcategories of substance, causality, and community. Matter, as substance, should be conserved in some sense. This is how Kant justifies his first law of mechanics, the conservation of mass. Since, for Kant's continuous matter, mass cannot be defined by counting points or atoms, Kant defines it through the quantity of motion, which he assumes to be the measurable product of mass and velocity.

He cannot clearly articulate the definition, because that would require the two other laws he has not yet stated. He nonetheless seems to be anticipating Mach's later definition of the mass of a body through a balancing collision with another body of unit mass. In symbols, if $m\mathbf{v} + \mathbf{V} = \mathbf{0}$ holds for the velocities $\mathbf{v}$ and $\mathbf{V}$ of the two bodies, then the mass m of the first body is the velocity ratio V/v.[18]

Under the subcategory of causality, any change of the quantity of motion requires an external cause named force. In the absence of force, the velocity and direction of motion of a body cannot change. This is Kant's second law of mechanics, alias the law of inertia. Lastly, the subcategory of community requires the interaction between two bodies (material points) to be reciprocal. Namely, the two bodies should be globally at rest at any instant. In symbols, their masses and their velocities should permanently satisfy $m_1\mathbf{v}_1 + m_2\mathbf{v}_2 = \mathbf{0}$, which leads to complete rest for a binding, nonelastic collision. At every instant, a change of the quantity of motion in one body is compensated for by an opposite change in the other body. This is Kant's third law of mechanics, or the law of the equality of action and reaction.[19]

Strangely, Newton's second law does not figure among Kant's laws of mechanics, although he does associate force with change of motion.[20] A plausible explanation of this omission may be in Kant's tracing all evolution to the interaction of pairs of material points. In this context, his three laws of mechanics jointly give

$$\mathrm{d}(m_i\dot{\mathbf{r}}_i) = \sum_{j \neq i} (\mathbf{r}_i - \mathbf{r}_j)\varphi(|\mathbf{r}_i - \mathbf{r}_j|)\mathrm{d}t \qquad (3.1)$$

---

[16] Kant 1786, pp. 1–30.

[17] Kant 1786, pp. 31–105.

[18] Kant 1786, pp. 106–119.

[19] Kant 1786, pp. 119–121 (second law), 121–126 (third law).

[20] For instance in Kant 1786, p. 134.

for the variation of the quantity of motion of a given material point as a function of the positions of the other material points (the sum truly should be an integral). The relation $\mathbf{f}_i = m_i\ddot{\mathbf{r}}_i$ is a logically superfluous definition of force, not a law. This is the way Saint-Venant, Kirchhoff, Mach, and Poincaré later understood this relation, and this may be what Kant vaguely had in mind.

In the fourth and last part of his foundations, the *phenomenology*, Kant implements the category of modality under its three subcategories of possibility, reality, and necessity. Namely, he determines to what extent the motions defined in his phoronomy, dynamics, and mechanics have an experimental counterpart in material bodies and empirical spaces. In the phoronomy, Kant tells us, we can indifferently have a body move rectilinearly in a space, or the space move and the body be at rest in absolute space; relative motion is *possible*, and absolute motion is *impossible*. In the dynamics, the proper determination of forces excludes a rotation of the empirical space. Indeed, if the law of inertia holds in the empirical space, it cannot hold in the rotating space without introducing artificial forces. The rotation of a body is *real*; the rotation of an acceptable empirical space is an illusion. In the mechanics, the third law *necessitates* the lack of global motion of a closed system of bodies. Indeed, a motion of the center of mass of the system could only be an absolute motion, which is phoronomically impossible.[21]

In the preceding outline of Kant's treatise, I have tried to extract the essentials and to filter out confusing elements. This is not an easy task, for Kant's text is written in a convoluted style, labors to express simple mathematical notions in unsuitably informal language, and suffers from a loose and artificial imposition of the table of categories. Even for one willing to accept this table and the broader tenets of Kant's transcendental philosophy, close reading of the *foundations* reveals obscurities, contradictions, and erroneous arguments. Most confusingly, despite the intended four-step genealogy of his metaphysical foundations, Kant frequently mingles the four steps. In the phoronomy, he unfortunately anticipates the law of inertia (which belongs to the mechanics) and restricts the motions of bodies and spaces to rectilinear (implicitly uniform) motions. This is sufficient to construct the addition of velocities, but unnecessary from a purely kinematic point of view. The forces introduced in the dynamics vaguely presuppose a notion of equilibrium and thereby anticipate the relation between force and motion explored in the mechanics. In the phenomenology, Kant again pretends to be treating separately the modalities of phoronomy, dynamics, and kinematics, whereas in reality his three propositions about the phoronomic relativity of motion, the dynamic reality of rotation, and the mechanical necessity of global rest respectively appeal to his three laws of motion, all belonging to mechanics. Kant's desire to illustrate his three subcategories of modality induces him to mask this obvious contamination. Among his errors of reasoning, we may note his conflation of the balance of momentum variations with the balance of momenta of two interacting bodies, which misleads him to require zero momentum for a two-body system (unless he is here anticipating the third proposition of the phenomenology).[22]

[21] Kant 1786, pp. 138–145.

[22] Kant 1786, pp. 17 (inertia in phoronomy), 122–123 (error), 151–152.

Is there something to be fished out of these murky waters? One good catch is the intention to first treat motion independently of its causes, anticipating André Marie Ampère's *cinématique*. Another is the foreshadowing of Mach's idea of defining mass through the balance of momentum in collisions. Perhaps most important and most influential is the argument for reducing all phenomena to direct action at a distance in pairs of material points. Famously, Ruđer Josip Bošković anticipated a discrete version of this picture, and Laplace made it the basis of his grand-unification program of the early nineteenth century. Later in this century, Helmholtz derived a similar kind of mechanical foundation by arguments similar to those of Kant's dynamics.

Kant's first intention was to clarify the meaning of motion in Newtonian mechanics. In this register we may credit him with a few insights: the impossibility of absolute motion, the strict dynamical equivalence of all inertial frames, and the compatibility of "true," empirically detectable rotation with the impossibility of absolute rotation. The rotation of the earth, Kant tells us, is no absolute rotation because a rotating earth in a static sky is phoronomically equivalent to a static earth in a rotating sky. Yet it is true because it can be detected by experiments that involve local relative motions only. Absolute space and the space of the fixed stars are equally irrelevant to Kant's definition of the true rotation.[23]

If absolute space cannot serve to explain true rotation, and if absolute motion is impossible in general, what is Kant's absolute space good for? He needs it as a condition for the possibility of spatial experience. Kant's absolute space, unlike Newton's, is a strictly ideal means to express the relative character of all observable motions: since every bit of matter is intrinsically movable, no material space can be taken to be at rest, and this is best expressed by imagining an immaterial space through which all material bodies (including material spaces) can move. The choice of this absolute space is purely conventional, since it does not affect the relative motion of bodies that move in it.[24]

With the exception of a few praiseful references by late nineteenth-century German-speaking critics of mechanics,[25] Kant's *foundations* had little impact on later thinking on space, time, and inertia. The singularly difficult style of this text may not be the only reason for this neglect. Unlike Huygens, Kant did not pay sufficient attention to the effective selection of reference frames. By rejecting the real absolute space of Newton, Euler, and d'Alembert, he lost a way to justify inertial behavior. He did not offer anything in replacement, save for a merely nominal appeal to causality.

3.2 Criticism in the last third of the nineteenth century

Huygens, Leibniz, and Kant long were the only major thinkers who dared to challenge Newton's concept of space before the last third of the nineteenth century. Huygens and Leibniz agreed to reject any concept of absolute position and motion; Kant denied the reality of absolute space and made it a purely ideal, largely conventional construct. In the eighteenth and most of the nineteenth century, Newton's definition of absolute space as an

---

[23] Kant 1786, pp. 149–153.

[24] Kant 1786, pp. 146–147.

[25] For instance, by Heinrich Streintz and Ludwig Lange, discussed on pp. 70–72, this volume.

immovable, infinite, penetrable, real and yet imperceptible entity was generally accepted, even by the likes of Bélanger or Violle, who gave the principle of relativity a constructive, prominent role. When the authors of mechanical texts cared to define the space in which Newton's laws apply, they typically offered variants of this formal definition. As often happens, the tremendous success of Newton's mechanics prevented criticism of its fundamental concepts. The main purpose of the propagators of Newton's mechanics was to enable its easy, unambiguous application, not to analyze the deeper significance of its concepts.

From a mathematical and pragmatic point of view, it did not matter much whether the absolute space was real, as Newton and Euler wanted, or purely ideal, as Kant wanted. In his *Système du monde*, Laplace introduced this notion as follows:

> A body appears to us to be in motion when it changes its situation relatively to a system of bodies which we suppose to be in a state of repose. Thus in a vessel moving in a uniform manner bodies seem to us to move when they correspond successively to different parts of the vessel. This motion is only relative, for the vessel itself moves on the surface of the sea, which revolves round the axis of the Earth, the centre of which moves round the Sun, which is itself carried along the regions of space with the Earth and all the planets. To conceive a term to these motions, and to arrive at last at some fixed points from which we may reckon the absolute motion of bodies, we imagine a space without bounds, immovable and penetrable to matter. It is to parts of this space, real or imaginary, that we refer in imagination the position of bodies, and we conceive them in motion when they correspond successively to different places in this space.

Laplace thus recovers Kant's idea of absolute space as a convenient way to express the movability of all bodies, including the alleged reference bodies. He leaves it open whether this space is real or ideal. Still, for a geometer Laplace is unusually prolix on this topic of absolute space. His disciple Poisson does not offer any definition in the first edition of his *Cours de mécanique*. In the second edition, he simply writes:[26]

> A body is *in motion* when this body or its parts successively occupy different places in space. But *space* being infinite and everywhere identical, we cannot judge of the state of rest or motion of a body without comparing it to other bodies or to ourselves; and for this reason all the motions we observe are necessarily relative motions.

In France and elsewhere, a more critical attitude toward the concepts of space, time, and inertia began to occur in the late 1860s, in part because the recent promotion of the energy concept invited assessments of the foundations of mechanics but also, more broadly, because the rapid evolution of practices in all quantitative sciences challenged their systematic exposition. We will begin this section with a French isolated case, go on with the British reformers of mechanics, then dwell on the more abundant German and Austrian criticism, and finish with Poincaré.

---

[26]Laplace 1796, pp. 239–240; 1809, pp. 290–291; Poisson 1811, 1833, pp. 1–2.

Duhamel's rejection of absolute space and time

In 1870, the elderly French Academician Jean-Marie Duhamel, who had taught mechanics at the École Polytechnique some thirty years earlier, published the fourth installment of his *Des méthodes dans les sciences de raisonnement*, about the science of forces. Whereas at the Polytechnique he had maintained Newton's absolute space as a valuable induction from phenomena, he now declared:

> *Time* does not have any more real existence than space does; it may even be more elusive. These two alleged beings are fantastic creations of the imagination of human beings, who always want to go beyond what they can grasp or understand. . . . Can we attach a meaning to *absolute* rest or motion?—Those who talk about it assume a boundless space whose points have, as it were, a *personal* reality, and to which they convey an absolute immobility without seeing the vicious circle. . . . We can perceive only relative rest and motion and it is only by extension that we could dream absolute rest or motion. . . . Let us abandon this false notion. . . . What would we gain by starting from the relative in order to induce an absolute imaginary from which we would draw the principles applicable to the relative, which alone is real! Is it not better, after having established the principles of the relative, to apply them directly to the real, without invoking a fantastic absolute?

Duhamel goes on to explain that for astronomical purposes the system of the stars, for most terrestrial purposes the body of the earth, can serve as "immutable systems" of reference.[27]

After thus joining Huygens and Leibniz in their denial of absolute space and motion, Duhamel unfortunately fails to indicate what makes relative motion with respect to the systems of stars or to the earth so special. In his statement of the law of inertia, he does not specify the reference of motion. His statement of Galilean relativity, like Laplace's, concerns a common velocity added to a closed system of bodies (including the reference system) and does not refer to inertial systems of reference. It is only by going through the applications of these principles that Duhamel confirms his earlier announcement that the stellar system and the earth (in a certain approximation) are adequate references. No fundamental reason is given for these choices.[28]

Thomson and Tait on the essence of inertia

In 1867, two British heralds of the new energy-based physics, William Thomson and Peter Guthrie Tait, published the first volume of their monumental *Treatise on Natural Philosophy*. They accompanied their restatement of Newton's first law with the comment:

> The meaning of the term *Rest*, in physical science, cannot be absolutely defined, inasmuch as absolute rest nowhere exists in nature. If the universe of matter were finite, its centre of inertia might fairly be considered as absolutely at rest; or it might be imagined to be moving with any uniform velocity in any direction whatever through

---

[27] Duhamel 1870, pp. xvii–xix.

[28] Duhamel 1870, pp. 234–235 (Galilean relativity).

> infinite space. But it is remarkable that the first law of motion enables us to explain what may be called *directional* rest.

Thomson and Tait here agree with Newton and Poisson that only relative motion can be observed. They seem to be denying the existence of absolute rest, with two correctives: in a finite universe we may reasonably decide the center of mass to be at rest or in uniform rectilinear motion; and we may decide that the total angular momentum of the universe points in a fixed direction. This is because Newton's laws imply the conservation of the total momentum and total angular momentum in a space in which they are valid.[29]

Thomson and Tait recognize two components of the law of inertia: the uniformity of inertial motion, which they regard as a "convention" for the measurement of time as d'Alembert did in his dynamics; and a "great truth of nature," independent of any convention on the definitions of absolute space and time, that all inertial motions occur at the same rate:

> A curling-stone, projected along a horizontal surface of ice, travels equal distances, except in so far as it is retarded by friction and by the resistance of the air, in successive intervals of time during which the earth turns through equal angles. The sun moves through equal portions of interstellar space in times during which the earth turns through equal angles, except in so far as the resistance of interstellar matter, and the attraction of other bodies in the universe, alter his speed and that of the earth's rotation.

This empirical law by itself defines invariable directions. In order to show this, Thomson and Tait imagine two material points P and Q projected in different directions from a point A. The law implies that the positions (P, Q) and (P', Q') at two different times obey $AP/AP' = AQ/AQ'$, which implies that PQ is parallel to P'Q'. Then if O, P, Q, and R are four points projected from the same point, the directions OP, OQ, and OR define three fixed axes from a given origin. Like Huygens's older idea of four points in mutual rest, this construction amounts to an operational definition of an inertial frame in which Newton's laws are valid.[30]

James Thomson's inertial "reference frames"

Thomson and Tait's discussion of mechanical inertia had an echo in two notes communicated in 1884 by James Thomson, professor of engineering at Glasgow and elder brother of William. James insists on the difficulty of defining rectilinear, uniform motion with respect to "unmarked space" and purely ideal clocks. In his opinion, all we can measure is the mutual distance between two material points at a given instant. He defines a "reference frame," or just a "frame," as a system of material points whose mutual distances do not vary, and a "dial-traveller," that is, a pointer manually rotated on a dial. In order to illustrate the peculiar harmony of motion implied in the law of inertia, he attaches to a

---

[29]Thomson and Tait 1867, p. 179.

[30]Thomson and Tait 1867, p. 180.

given frame a dial-traveler and a system of pinions and racks that communicate to each member of a set of particles a translational motion in proportion to the rotation of the dial-traveler. Through an additional gearing mechanism, he connects the motion of another frame and the motions of another set of particles in this new frame to the original dial-traveler. Having thus prepared his audience for the kinematic possibility of mutually proportional rectilinear motions in a class of reference frames themselves in mutual, proportional rectilinear motion, Thomson states the principle of inertia as the existence of a frame and a clock in which the motion of any free particle with respect to this frame and this clock is rectilinear and uniform. There are infinitely many frames satisfying this condition, moving rectilinearly and uniformly with respect to each other. And there are infinitely many clocks satisfying this condition, giving mutually proportional times.[31]

Not content with a mere statement of existence of inertial frames, James Thomson expressed his desire to construct them from the measurable variation of the mutual distances of a small set of free particles. Tait, who had heard his communication, told him he knew how to do this with quaternions. In later communications, Thomson gave his own geometric construction, and Tait his quaternion-based construction. The engineer and the mathematician thus converged on a problem whose concreteness was purely imaginary.[32]

Maxwell's idea of absolute space

In his popular *Matter and Motion* of 1876, James Clerk Maxwell condemned Descartes's "error" of reducing matter to pure extension and defended Newton's absolute space and time: "We shall find it more conducive to scientific progress to recognise, with Newton, the ideas of time and space as distinct, at least in thought, from that of the material system whose relations these ideas serve to coordinate." Maxwell simply repeats Newton's definition of space and time, insisting even more than Newton that only relative space and time are accessible to observation:

> But as there is nothing to distinguish one portion of time from another except the different events which occur in them, so there is nothing to distinguish one part of space from another except its relation to the place of material bodies. We cannot describe the time of an event except by reference to some other event, or the place of a body except by reference to some other body. All our knowledge, both of time and place, is essentially relative.

In order to justify the law of inertia, Maxwell explores the consequences of the contrary assumption that the velocity vector of a free particle in a vacuum varies in time. The homogeneity of space and time requires this variation to take the form $d\mathbf{v}/dt = \mathbf{v}/\tau$. This law would imply a state of absolute rest (reached by the particle after a long time) and it would not hold in a moving frame. Maxwell concludes: "The denial of Newton's law [of inertia] is in contradiction to the only system of consistent doctrine about space and time which the human mind has been able to form." In essence, Maxwell is here showing that the

---

[31] J. Thomson 1884a. Cf. Martínez 2009, pp. 109–111.

[32] J. Thomson 1884b; Tait 1884.

principle of inertia results from a combined use of two symmetries: the homogeneity (and isotropy) of space and the Galilean equivalence of moving frames, as Euler and d'Alembert had nearly done under the principle of sufficient reason. Instead of this principle, Maxwell relies on the "general maxim of physical science" that "the same causes will always produce the same effects."[33]

Toward the end of his essay, Maxwell has a few articles in which he discusses the extent to which basic mechanical concepts are relative. In "The relativity of dynamical knowledge," he vividly reasserts the impossibility of detecting motion through absolute space:

> There are no landmarks in space; one portion of space is exactly like every other portion, so that we cannot tell where we are. We are, as it were, on an unruffled sea, without stars, compass, soundings, wind, or tide, and we cannot tell in what direction we are going. We have no log which we can cast out to take a dead reckoning by; we may compute our rate of motion with respect to the neighbouring bodies, but we do not know how these bodies may be moving in space.

In the following article, "The relativity of force," Maxwell asserts that "We cannot even tell what force may be acting on us; we can only tell the difference between the force acting on one thing and that acting on another." This is so because according to Newton's sixth corollary (not cited by Maxwell), the relative motion of a system of bodies remains the same when a common acceleration, as produced by a homogeneous gravitation field, is applied to the bodies. Maxwell's following articles concern the possibility of determining absolute direction and rotation in space, through Newton's bucket or through Foucault's pendulum.[34]

Carl Neumann's body Alpha

In 1869, the German mathematical physicist Carl Neumann chose the foundations of mechanics as the topic of his inaugural lecture of November 3 at the University of Leipzig. He had already proven his ability to bring mathematical clarity to the foundations of physical theories. He believed a well-formulated physical theory should proceed deductively and mathematically from a minimum of "ungraspable" and "arbitrary" principles. He rejected the rationalist attempts to prove the logical necessity of the principles, as well as the empiricist view of the principles as mere empirical generalizations. His goal was to identify the proper principles and to clarify their content.[35]

The bulk of Neumann's reflections concern mechanical inertia. Since inertial behavior cannot possibly hold in every reference frame, its meaningful expression requires a rigid body with respect to which the inertial motion is defined:

[33] Maxwell 1876, pp, 18, 20–21.

[34] Maxwell 1876, pp. 84–89, citation on p. 85. Maxwell's statement of the relativity of force is strikingly similar to Einstein's later view that the gravitational force varies with the acceleration of the frame to which it is referred. Maxwell (ibid., p. 88) extends this relativity to rotating frames, arguing that the centrifugal force in a rotating frame has the same static effects as an appropriate gravity field would in a nonrotating frame.

[35] Neumann 1870, pp. 12–13. On Carl Neumann, cf. Jungnickel and McCormmach 1986, Vol. 1, p. 181. On his inaugural lecture, cf. Martínez 2009, pp. 95–97.

As the *first principle* of the Galilei-Newton theory, we should therefore posit the existence of an unknown body in an unknown place of world space; this body should be *absolutely rigid*, its figure and dimensions should be invariable for all times.

Let us briefly call it the body *Alpha*. Neumann's second principle is the rectilinear character of free motion with respect to the body Alpha. The third principle, he remarks, cannot be stated as the uniformity of this motion without already knowing how to measure time. Like Thomson and Tait, he therefore requires the following empirical property:

> *Two* material points that are both left to themselves move in such a manner that equal path lengths for one correspond to equal path lengths for the other.

This principle can then be used to define the measure of time in d'Alembert's manner.[36]

The body Alpha, Neumann judges, is a *necessity* of Newtonian mechanics because the principle of inertia cannot be stated in terms of the relative motion of material points only. It is akin to Newton's absolute space since both serve to define the reference of inertial motion. In Neumann's eyes, it has the same right of existence as other nonmaterial entities such as the ether and electric fluids, on which physicists have not hesitated to base their theories. The choice of the body Alpha is no more univocal than the choice of electric fluids was, because any body moving rectilinearly and uniformly with respect to the body Alpha can serve as another body Alpha. Just as the electric fluid might perhaps be dispensed with (an allusion to the Faraday–Maxwell theory), perhaps a future mechanics could proceed without the body Alpha.[37]

The first Beneke prize

In the same year 1869, on April 2, the Göttingen Faculty of Philosophy advertised a prize for "a critical history of mechanics" from Galileo to energy-based physics. Five years earlier, the Royal Prussian consistorial counselor Carl Gustav Beneke had disappeared and his drowned body had been found in a Berlin pond. In his will he honored his brother Friedrich Eduard, a neo-Kantian philosopher whom Hegel had persecuted, by directing generous funds to the creation of a prize recompensing nonspeculative philosophical essays. The Göttingen Faculty of Philosophy hosted the prize and judged that historico-philosophical inquiries about the natural sciences would best fit Beneke's intentions. Mechanics was their first pick.[38]

An elite committee including the philosopher Hermann Lotze, the mathematician Alfred Clebsch, and the physicist Wilhelm Weber awarded the first prize to Eugen Dühring, a philosopher who had never done any work in physics. Dühring was probably unaware of Carl Neumann's lecture, and he could not have read Mach's *Erhaltung*, which appeared in

---

[36]Neumann 1870, pp. 15, 18. On d'Alembert's time, see pp. 75–76, this volume.

[37]Neumann 1870, pp. 20–22.

[38]*Königlich-Preussischer Stadts-Anzeiger*, 77 (1869), 1370–1371; *Nachrichten von der Königlichen Gesellschaft der Wissenschaften und der Georg-Augustus-Universität* (1870), pp. 131–140. Cf. Dühring 1873, pp. III–VIII; Zöllner 1894, pp. 580–581; Martínez 2009, pp. 90–92; Staley 2011, pp. 280–282.

print after the competition was closed. Despite the vast erudition demonstrated in his bulky submission, and despite the importance he gave to the introduction of systems of axes in mechanics (by MacLaurin, Euler, and others), Dühring did not say a word on absolute space or on the needed reference for inertial motion.[39]

The second prize went to Hermann Klein, a *Gymnasium* physics teacher in Dresden. Klein's discussion of the foundations of mechanics was based on Wilhelm Wundt's *physikalische Axiome* of 1866:

1) All causes in nature are causes of motion.

2) Every cause of motion lies outside the moved thing.

3) All causes of motion act in the direction of the line joining their origin and their target.

4) The action of every cause persists [*verharrt*].

5) Every action has an equal reaction.

6) Every action is equivalent to its cause.

Wundt believed he could rationally justify these six axioms. The fourth one concerns inertia: it stipulates first that the acquired motion of a material point persists after all causes of motion have ceased to act, and second that the motion also persists if a new cause of motion comes to act. Wundt regarded the first stipulation as a direct consequence of the principle of causality. For the second stipulation, he reasoned that before the new force comes to act and after all earlier causes of motion have disappeared, the material point is effectively at rest because there is nothing with respect to which it may move. This strange reasoning rests on a conflation between cause of motion and reference of motion, and it implicitly assumes some equivalence between rest and uniform motion. By the same conflation, Klein approved Carl Neumann's body Alpha as a needed cause of motion, but he neglected Neumann's tripartite statement of the law of inertia as a venture into "pure philosophy." No more than Wundt did he see the importance of a selected class of reference frames in Newtonian mechanics.[40]

Mach's relational inertia

The young Ernst Mach had also been reflecting on the foundations of mechanics, especially on the meaning of the principle of inertia, since 1868—early in his physics professorship at Prague University. When he read Neumann's inaugural lecture in 1870, he saw in it a confirmation of his own criticism of the received view based on absolute space. Yet he

---

[39] Benekesche Preisstiftung, *Nachrichten von der Königlichen Gesellschaft der Wissenschaften und der Georg-Augusts-Universität* (1872), pp. 145–154; Dühring 1873. Johann Bernoulli and Euler, rather than MacLaurin, pioneered the use of rectangular systems of axes in mechanics: cf. Truesdell 1960, p. 252.

[40] Klein 1872, pp. 54, 57; Wundt 1866.

disagreed with the solution proposed by Neumann based on the body Alpha. He first published his own suggestions in notes to his historic-critical essay of 1872 on the conservation of energy.[41]

Unlike Neumann and more like Huygens and Duhamel, Mach wanted to found mechanics on empirically testable hypotheses. This excludes any nonmaterial reference of motion in the statement of the basic principles. One might hope, Mach goes on, to find in the universe a system of bodies in mutual rest that can serve as reference for inertial motion. The fixed stars were long believed to offer a legitimate choice. But they are not so fixed after all (as Edmund Halley announced in 1718). If no specifically chosen masses in the universe can serve to determine inertial behavior, Mach reasons, then the totality of masses should do so. From experience, we know that distant masses contribute more than nearby masses. Mach sketches two reformulations of the law of inertia in agreement with this constraint:[42]

> We would fully bring out the known facts if for instance we just made the simple assumption that all bodies act in proportion with their mass and independently of their distance, or proportionally to their distance, and so forth. Another way of expression would be: as long as the bodies are so far from each other not to communicate to each other any significant acceleration, all distances vary in a mutually proportional manner.

In his influential *Mechanik* of 1883, Mach renewed his attack on absolute space and time, citing Newton extensively and charging him with "idle 'metaphysical' conceptions." Mach is now willing to replace Newton's absolute space with a material medium that would convey inertia to immersed particles somehow like water conveys additional mass to a rigid body moving through it. But he still prefers a mechanics formulated in terms of the relative distances of observable bodies only. He now gives a more precise idea of such a formulation, based on the equation

$$\frac{d^2}{dt^2}\left(\sum_i m_i r_i / \sum_i m_i\right) = 0 \qquad (3.2)$$

to be satisfied by the distances r_i of a given particle to the other particles of the universe in any given narrow cone centered on this particle (the index i labels these particles, and m_i is their mass). He does not explain in detail how this would work, but he indicates that special assumptions on the distribution of masses in the universe and fairly complex reasoning are needed to retrieve Newton's simple laws. He welcomes this complexity as a sign that the alleged inevitability of these laws depends on a misconceived metaphysics. The return to principally controllable quantities opens new vistas for future theories of motion:

---

[41] Mach 1872. The essay derived from a conference given on November 15, 1871 at the Böhmische Gesellschaft der Wissenschaften in Prag. On Mach's biography, cf. Blackmore 1972. On his science and philosophy, cf. Cohen and Seeger 1970; Pojman 2020. On his criticism of inertia, cf. Ray 1987, chap. 1; Barbour and Pfister 1995; Barbour 2001.

[42] Mach 1872, pp. 47–50, citation pp. 49–50; Halley 1718. Euler (1748, p. 328) had considered the possibility of making distant masses the cause of inertia, but judged such remote action absurd: see p. 57, this volume.

Here is the most important result of our considerations: *The mechanical laws that seem to be the simplest are in fact of a very complex nature; they rest on forever incomplete experience; to be true, they are sufficiently established in practice to serve as the basis of mathematical deductions in a sufficiently stable environment; but they are by no means to be themselves regarded as mathematically obtained truths; rather they are laws not only capable but also in need of sustained experimental control.* This insight is valuable as it favors scientific progress.

Mach thus means his criticism of the Newtonian concept of inertia to contribute to his broader debunking of any metaphysical hardening of the laws of mechanics.[43]

Streintz's reference systems

Mach's historico-critical analysis of the foundations of mechanics and the variety of positivism it exemplifies interested many of his contemporaries. Later discussions on the meaning of space, time, and inertia frequently referred to it. In 1883, Heinrich Streintz, former student of Carl Neumann and professor of mathematical physics at Graz University, published his remarkably deep and erudite *Physical Foundations of Mechanics*. The asserted object of this treatise is to address the long-ignored difficulty in defining the motion to which the principle of inertia applies. Streintz identifies three main answers to this question: Euler's, Neumann's, and Mach's, and he proposes his own. To Euler's argument for the reality of absolute space, he objects that a purely ideal construct cannot serve as the basis for a physical principle. To Neumann's body Alpha, he objects that a strictly invisible body cannot truly serve to define motion, thus echoing Newton's similar remark that a remote, invisible body at rest would not help assess absolute motion. As for Mach's way of tracing inertia to distant masses, he judges it most implausible:[44]

It is difficult to refrain from weighty objections against such a proposal. The hypothesis that the masses of remote celestial bodies affect the velocities of bodies around us while nearby masses have no such effect, is so incompatible with all the experience we have of natural laws and so far from being tested, that one easily understands why it did not catch on anywhere.

Streintz carefully distinguishes between mathematical coordinate systems and "systems of reference" (*Bezugssysteme*) made of a rigid combination of material bodies. In order to convey physical meaning to the law of inertia, he selects the nonrotating reference systems through the lack of Newton-bucket or Foucault-pendulum effect and calls them *Fundamental-Körper*. In a footnote, he also considers Thomson and Tait's definition through particles projected from a common origin, although he judges it "less natural." Streintz of course understands that any body of this class is in rectilinear uniform motion with any other, and he asserts the complete equivalence of all these inertial frames (as we would now call them). Most penetratingly, he compares this equivalence with the equivalence of all systems of coordinates in a given frame, noting that in both cases a change

---

[43] Mach 1883, pp. 209, 215, 218–219, 221–222 (citation).

[44] Streintz 1883, pp. 7–8, citation from p. 7; Newton 1687, pp. 7–8.

of reference means a change in the value of the integration constants for the equations of motion:

> We should get used to the idea that the substitution of a fundamental system [inertial frame] for another is completely analogous with a change of the origin of coordinates [in a given frame]. While in the latter case the initial values of the coordinates are changed, in the former case the initial values of both the coordinates and the velocities are affected in the determination of the integration constants. There is no other difference.

Streintz thus comes very close to the modern view of a Galilean space–time defined by the equivalence class of inertial systems. The only missing element is the idea that the coordinates in various inertial frames refer to the same point in an intrinsic space–time structure just as the coordinates measured from various origins refer to the same point in an intrinsic affine space.[45]

Lange's inertial systems

A reading of Streintz's book prompted the young Leipzig physicist Ludwig Lange to publish his own discussion of the principle of inertia in *Philosophische Studien*, Wundt's new journal for the philosophy of science. Lange agreed with Carl Neumann that according to the "axiom of the relativity of all motions" the principle of inertia needed a specification of the reference of motion, but he rejected Neumann's, Mach's, and Streinz's proposals for this reference. He agreed with Streintz that Neumann's body Alpha was too "transcendent" a notion to appear in the foundations of a physical theory; he disparaged Mach's proposal to trace inertia to all masses in the universe as contrary to Mach's own principle of the economy of thought; and he regarded Streinz's definition of inertial frames (*Fundamental-Körper*) through Newton's bucket or Foucault's pendulum as operationally and conceptually too complex. Indeed, it is only through the laws of mechanics, including the law of inertia, that we know that these devices detect absolute rotation.[46]

In his own proposal, Lange first *defines* an "inertial system" (of reference) as a coordinate system in which three particles projected from a common origin and left to themselves have noncoplanar, rectilinear trajectories. Second, he *hypothesizes* that in such a system the motion of any particle is rectilinear. Third, he *defines* equal times by equal inertial motions of a given free particle. Fourth, he *hypothesizes* that all free particles lead to the same equality of times.[47]

Lange observes that his definitions sample the basic objects (particles and frames) without presuming the empirical hypothesis or empirical law that they help to formulate. This is true for his first definition, because according to a kinematic theorem proven by Lange,

---

[45] Streintz 1883, pp. 27 (citation), 63n–64n. Streintz borrowed the word *Bezugssystem* from Narr 1875, an otherwise tedious treatise.

[46] Lange 1884, pp. 271–272.

[47] Lange 1885a, p. 544; 1885b, pp. 337–338. In an earlier version (1884, pp. 273, 277), Lange defined an inertial system through a free particle and the motion of a second free particle that does not pass through the origin. The mathematician Aurel Voss alerted him to the erroneousness of this definition.

for at most three particles moving arbitrarily from a common origin it is always possible to find a reference frame such that their three trajectories become rectilinear, and this frame is unique for three particles with a common origin. The *definition* enabled by this theorem therefore is pure "convention," and all empirical content is confined to the subsequent *hypothesis*. Lange recommends this strategy for any physical theory and calls it the "principle of particular determination" (*Princip der Particulardetermination*). He criticizes Thomson and Tait's earlier attempt to operationally determine an inertial frame through the motion of three particles because it does not satisfy this principle. Indeed, its defining operations presuppose an empirically testable harmony of inertial motions.[48]

Lange does not fail to note that his definition of inertial systems is "infinitely multi-valued" and he agrees with Streintz that this multivaluedness is essential. For these two physicists, Newton, Neumann, and Mach have erred in seeking an inconvenient univaluedness in different ways: Newton through absolute space, Neumann through the body Alpha, and Mach through the complete elimination of reference frames.[49]

In the second edition of his *Mechanik* (1889), Mach rejoiced over the rising interest in the meaning of the law of inertia and briefly discussed Streintz's and Lange's writings. Understandably, he was irked by Streintz's suggestion that Euler had anticipated Mach's proposal only to reject it, and he defended the originality and rationality of his position. He rejected Streintz's and Lange's attempts to refer the law of inertia to a privileged class of reference frames as disguising the true cause of inertial behavior (to be found in the stellar system in a first approximation). He nonetheless congratulated Lange for the clear and systematic character of his analysis, adding that he had himself pursued a similar line of thought before coming to his mature view.[50]

Budde's medium

Unlike Mach, the other critics of the principle of inertia did not see any difficulty in referring inertial behavior to a selected class of reference frames. They did not worry about the causal significance of this selection, or they believed it was sufficiently addressed by giving to the selected class an operational definition. One exception was the Berlin physicist-engineer Emil Budde, who in his mechanics textbook of 1890, tried to address this issue without falling back into absolute space or the body Alpha. Like James Thomson, Budde introduced the inertial frames as those in which the law of inertia is valid (he called them *Fundamentalsysteme*, probably echoing Streintz's *Fundamental-Körper*). Reflecting on the approximate experimental constructability of an inertial system ("the F system") through Foucault's pendulum, he wrote:

---

[48] Lange 1885b, p. 350; 1886, pp. 133–141.

[49] Lange 1884, p. 274.

[50] Mach 1889, pp. 481–485. In later editions, Mach also mentioned Budde 1890 for tracing inertia to a medium, MacGregor 1893 for a competent discussion (and partial anticipation) of German criticism of the foundations of mechanics, Friedländer 1896 for a proposal to experimentally decide between Mach's and other concepts of inertia, and Johannesson 1896 for a not-too-original review of various concepts of inertia.

The fundamental systems must be physically determined in the real universe. Indeed, if the F system could not be physically distinguished from the earth-bound system, there would be no reason why the determination of force would come out differently in the two systems.

Budde went on to assert that the physical character of the F system required it to be implemented by a physical body. He took this body to be a pervading penetrable medium responsible both for distance forces and for the inertia of matter. Mach, who had made a similar suggestion in the first edition of his *Mechanik*, approved Budde in the third edition, with the reservation that "the properties of this medium should be demonstrable physically in some other manner, and that they should not be assumed ad hoc." Of course, Mach still preferred his own way of tracing inertia to all matter.[51]

Poincaré's laws as conventions

The French mathematician Henri Poincaré rarely referred to Mach (and only for his "economy of thought"), and he never commented on Mach's criticism of mechanical inertia. As for the writings of others on the same topic, the only one Poincaré (obliquely) referred to was Carl Neumann's on the body Alpha. In fact, his own criticism of the law of inertia is strikingly at odds with his predecessors'. He first discussed the foundations of mechanics in 1897 in a review of Hertz's *Prinzipien* of 1894. He there approves of Hertz's objections to the traditional, anthropomorphic definitions of the concept of force, and more broadly deplores the "inextricable difficulty" of properly defining the concepts of mass, force, and acceleration. A proper definition, Poincaré tells us, should either include a method of measurement or build from other measurable quantities. Poincaré rejects the traditional way of measuring force through equilibrium, because it abusively separates the force from the body on which it is acting before or after the measurement. He prefers Kirchhoff's (and d'Alembert's) definition through the product of mass and acceleration.[52]

For measuring mass, he posits that the masses of two interacting bodies are inversely proportional to their accelerations (as Mach has earlier done), thus turning the equality of action and reaction into an implicit definition of mass. He is still unhappy with this definition, because it abusively isolates two bodies from the rest of the universe. In the end, the law of reaction and the law of acceleration become mere definitions and are therefore immune to experimental refutation. Poincaré nonetheless admits their experimental origin in the study of approximately isolated mechanical systems: "We now understand how experience could serve as a basis for the principles of mechanics and yet will never be able to contradict them." An apparent contradiction between these principles and observations, Poincaré explains, could always be solved by introducing interactions with unobserved masses.[53]

[51] Budde 1890–91, vol. 1, pp. 133–136, 332 (citation); Mach 1897, pp. 236–237.

[52] Poincaré 1897.

[53] Poincaré 1897, pp. 736–737.

Readers familiar with Poincaré's philosophy here recognize his so-called conventionalism, applied to the principle of mechanics. That a proper statement of mechanical principles involves "conventions" had become a commonplace after Thomson, Tait, Mach, Schell, Streintz, and Lange pushed this point. However, whereas these authors wanted only to separate the conventional and the empirically meaningful contents of the principles, Poincaré made them complete conventions. He amplified this view in his address to the International congress of philosophy in 1900. There he begins with denying absolute space, absolute time, and absolutely Euclidean geometry, referring the reader to his earlier writings on the foundations of geometry and on the measurement of time (to be discussed in a moment). Nonetheless, he "provisionally" assumes these notions in the following sections on the principles of inertia and acceleration.[54]

For the principles of acceleration and reaction, Poincaré summarizes the criticism already found in his Hertz review. For the principle of inertia, he rejects both a rationalist and an empiricist foundation. If the principle could be known by a priori means, then there is no reason why the Greeks would have overlooked it. Nor can it be a result of experience, because of the practical impossibility of producing a truly free body. Even the more modest requirement that "the acceleration of a body should only depend on its position and on the position of the neighboring bodies and on their velocities" could never be experimentally refuted because it could always be saved by invisible motions. We know the law to be true in a few simple experimental cases, and we assume it in more general cases, "because we know that in these cases experiment can neither confirm nor refute it."[55]

After thus reducing the three laws of Newtonian mechanics to experimentally motivated conventions, Poincaré at last comes to the postponed issue of absolute space and to the antagonistic "principle of relative motion." As discussed in Chapter 2, this name originated from a popular French derivation of the law of acceleration. Poincaré introduces the principle as follows:

> The motion of any system must obey the same laws, whether it be referred to fixed axes, or to movable axes carried along in a rectilinear and uniform motion. This is the principle of relative motion, which forces itself upon us for two reasons: first, the commonest experience confirms it, and second, the contrary hypothesis is singularly repugnant to the mind.

As he has already done for the other mechanical principles, Poincaré notes that the partial experimental confirmation of the relativity principle does not make it refutable. Regarding the rational necessity of the principle (or the repugnant character of the contrary hypothesis), Poincaré had earlier noted that the true "law of relativity" or the full exclusion of absolute motion would require the evolution of a system of material points to depend only on the initial values of the relative distances and their time derivatives, not on the absolute

---

[54]Poincaré 1901a, pp. 457–459. Poincaré 1897 does not have the word "convention" (only "definition"); Poincaré 1901a has many occurrences of it.

[55]Poincaré 1901a, p. 466.

position and orientation of the global system and their derivatives. He had then brought in the centrifugal flattening of the earth and Foucault's pendulum to observe:[56]

> Unfortunately, the law thus stated does not agree with the experiments, at least in their ordinary interpretation. . . . This is a shocking fact for a philosopher, and yet an incontrovertible one for the physicist.

In 1900 in front of philosophers, Poincaré enhances the difficulty by showing that all attempts to circumvent it, by assimilating the inertial forces with ordinary forces, by increasing the differential orders of the equation of motion, or by introducing an invisible body Alpha are equally artificial. So too is the difficulty itself:

> But, after all, the difficulty is artificial. Provided the future indications of our instruments can depend only on the indications they have given us or would have given us formerly, this is all that is necessary. Now as to this we may rest easy.

With this disarming conclusion, Poincaré means to tolerate absolute rotation and the concomitant selection of inertial frames as useful conventions for the convenient expression of physical laws, as long as the conventions do not affect the observable, instrumental correlations. On the one hand, he dismisses the reification of the conventions by Newton or Lange; on the other hand, he doubts the expediency of any attempt to minimize the conventions.[57]

3.3 The measurement of time

In Newton's mechanics, absolute time is merely theoretical, and the astronomical measures of time, for instance through the rotation of the earth, are justified by proving that Newton's laws of motion imply the approximate uniformity of the motion used in the measurement. There is no attempt to give a direct operational definition of the measure of time.

In his *Dynamique* of 1743, d'Alembert first proposed that the uniform rectilinear motion of a free body according to the law of inertia could serve as a basis for the measurement of time. In his opinion, this was "the simplest" and "the most natural" measure, for no motion is simpler than a uniform motion and no measurement is more basic than the measurement of space (traveled by the uniformly moving body). Concretely, d'Alembert goes on, we have two ways of judging the uniformity of a given motion: first, we may know that the impressed forces are negligible (which implies uniformity by the law of inertia); and second, we may verify that this motion is proportional to other potentially uniform motions occurring in parallel:[58]

---

[56] Poincaré 1901a, p. 477; 1899, p. 269.

[57] Poincaré 1901a, p. 488.

[58] D'Alembert 1743, pp. 9–12, citation from p. 12.

> If several bodies move in such a manner that the spaces they travel in the same time
> are always exactly or nearly in the same ratio, we judge that the motion of these bodies
> is exactly or at least nearly uniform.

In the second edition of his treatise, published in 1758, d'Alembert further requires the
ratio between the spaces traveled by the various bodies to be independent of the time at
which the motions are started. In the same spirit, he adds a third criterion of uniformity
(truly the first): that the spaces traveled by the moving body are equal in time intervals in
which perfectly similar effects occur; for instance, when the same quantity of water has
flown through a clepsydra. This is what Streintz later called "the principle of identical
processes."[59]

In his *Traité de mécanique* of 1811, Poisson relies on d'Alembert's second criterion: the
proportionality of various natural motions allows us to consistently define a measure of
time and then to test the uniformity of any other given motion with respect to this measure
without falling into circular logic. In the second edition of his treatise, published in 1833,
Poisson further indicates how proportional motions could be concretely produced, more
in lines with d'Alembert's third criterion: one just has to consider a system returning to
exactly the same state after some time, then have an identical copy of this system perform
the same motion etc., or, more simply, a single periodic system that returns exactly to the
same state after a period. Duhamel later relied on the same idea of defining the measure of
time through identical processes:[60]

> Two intervals of time will be said to be equal when two identical bodies at the begin-
> ning of each interval, when submitted to the same actions and influences of any kind,
> have run through equal spaces with respect to the immutable system.

Thomson and Tait combined two of d'Alembert's suggestions for the measurement
of time: through inertial motion and through proportional motions. Namely, they measured time through the spaces traveled in an inertial motion, but only after formulating the
law of inertia in a pre-metric manner as the mutual proportionality of free motions. Carl
Neumann did very nearly the same. Streintz devoted an entire chapter of his book to the
measurement of time and offered detailed discussions of d'Alembert's, Poisson's, Thomson
and Tait's, and Neumann's views. He sharply distinguished between the definition through
the law of inertia and the definition through the principle of identical processes, arguing
that the two definitions could conceivably lead to two different measures. Lange contradicted him by noting that the latter principle, when applied to the motion of a free body,
implies its uniformity. The inertial definition is just the simplest of all definitions through
identical processes.[61]

Lange emphasized the conventional character of all such measures of time: it was only
through a merely regulative principle of causality (à la Maxwell) that physicists agreed to

[59] D'Alembert 1758, pp. 14–15; Streintz 1883, p. 84 (discussed in a moment). Isaac Barrow already introduced
this principle in his *Lectiones* of 1674: cf. Jammer 2006, p. 71.

[60] Poisson 1811, pp. 264–265; 1833, pp. 205–206; Duhamel 1870, pp. xix–xx, 223–227.

[61] Thomson and Tait 1867, p. 380; Neumann 1870, p. 18; Streintz 1883, pp. 81–96; Lange 1884, pp. 287–297.

give equal duration to identical, successive processes. A different convention was always possible, though less convenient. As Streintz noted in his book, the conventional character of the measure of time had been earlier asserted by the Karlsruhe mathematician Wilhelm Schell in the second edition of a treatise in which he strove to produce a strictly mathematical and geometrical "theoretical mechanics" for the sake of clarity and easier applicability:[62]

> The choice of a uniformly flowing time as our foundation for the judgment of motion is only a matter of convention, it is not an absolute requirement, just as a uniform increase of the independent variable is not in mathematical analysis.

Lange's considerations appeared in his first essay of 1884 in *Philosophische Studien*. The following year, the Polytechnique-trained engineer Auguste Calinon published his own critical reflections on the foundations of mechanics, with special emphasis on the definition of time. Calinon there demolishes the principle of identical processes by noting that in order to judge the identity of two mechanical processes we need to compare the implied velocities, which cannot be done if the measure of time is not already given. The true rationale for time measurement in mechanics is the known proportionality of the rotations of celestial bodies and, most important, the fact that the time thus measured is the one for which the laws of mechanics have their simplest expression. As Calinon puts it in another essay in 1896:

> The measurable time is a variable, selected among all the variables that occur in the study of motions because it lends itself to the simple expression of the laws of motion.

Calinon also insists on the necessary specification of a reference frame (*repère*) in the definition of mechanical motion, and notes that the laws of motion, even the equality of action and reaction, are invalid in an arbitrary frame. Again, he associates the selection of frames with the simplicity of the laws.[63]

In 1898 Poincaré, to whom Calinon had communicated his reflections, published an essay on the measurement of time in the recently founded *Revue de métaphysique et de morale*. Poincaré agrees with Calinon that the principle of identical processes cannot serve to define the measure of time, for the reason evoked by Calinon but also because (as already noted by Lange) the causality implied in this principle is merely regulative: one could well imagine, without falling into contradiction, that two identical processes occurring in very distant regions of space and at widely different times would have different durations. As for the definition of time through the rotation of the earth, it can only be an approximation, owing to the slowing effect of tidal friction (a very well-known fact). As the principle of this correction rests on the laws of motion, it is as clear to Poincaré as it already was to Calinon that "time has to be defined so that the laws of motion have the simplest possible form."[64]

---

[62] Lange 1884, pp. 292–293; Streintz 1883, p. 92; Schell 1879, Vol. 1, pp. 6–7.

[63] Calinon 1884, p. 90; 1886, p. 26. Cf. Martínez 2009, pp. 196–200.

[64] Poincaré 1898, p. 6.

Poincaré notes he is not first to discuss the measurement of time, and moves on to the concept of simultaneity, which he believes to have escaped earlier critical attention. To be sure, in his reflections of 1884 on the law of inertia James Thomson had written:

> We can measure the distance between two material points; if, at least, we be content to waive the difficulty as to imperfection of our means of ascertaining or specifying, or clearly idealising, simultaneity at distant places. For this we do commonly use signals by sound, by light, by electricity, by connecting wires or bars, and by various other means. The time required in the transmission of the signal involves an imperfection in human powers of ascertaining simultaneity of occurrences in distant places. It seems, however, probably not to involve any difficulty of idealising or imagining the existence of simultaneity.

Thomson thus noted the difficulty of concretely asserting the simultaneity of distant events, but only to conclude that ideal simultaneity should not be too problematic. In 1885 Calinon made simultaneity and succession the basis of a pre-metric concept of time:

> We call *simultaneous* the positions [of two stars] when these positions are seen *à la fois*, without otherwise explaining the meaning of the locution *à la fois* about which everyone would agree. First consider two simultaneous positions of the two stars, then consider two different but still simultaneous positions: the latter positions are said to be *successive* with respect to the former.

Calinon also noted "a fundamental property of simultaneity"—that it is independent of the observer and of the reference frame. He did not see any difficulty there.[65]

In contrast, Poincaré spent much time criticizing the received concept of simultaneity. This God-given idealization, he argues, lacks direct empirical testability in the case of distant events. In order to date a distant event such as the birth of a new star, witnessed by Tycho Brahe in 1572, we implicitly assume the velocity of light to have a given constant value. There is no direct empirical verification of this constancy, since any measurement of the velocity of light presupposes its constancy. This constancy is only a convention. For asserting the simultaneity of distant events on earth (as is, for instance, required for longitude measurement), we have a variety of options: carry a clock from the location of the first event to the location of the second event, assert their simultaneity with a commonly visible third event (e.g., the phase of a moon of Jupiter), or send a telegraph signal from the first location to the second. The first procedure requires an ideal stability of the time keeper, and the two others assume instantaneous propagation for light or for electric signals. Any attempt to correct for the delay caused by this propagation would imply the convention of a constant velocity of light.[66]

---

[65] J. Thomson 1884, p. 569; Calinon 1885, p. 88. Cf. Galison 2003, pp. 86–87, 188–189; Martínez 2009, pp. 195–197.

[66] Poincaré 1898, pp. 6–9, 11–12. On pp. 9–10, Poincaré also argues that lack of causal order is not sufficient to establish simultaneity.

From this discussion, Poincaré concludes that our assessments of simultaneity depend on a multiplicity of implicit rules with no firm foundation, and that they are intimately related to the more quantitative problem of the measurement of time:

> We choose these rules not because they are true but because they are the most convenient ones, and we could summarize them by saying: "The simultaneity of two events, or the order of their succession, the equality of two durations, must be defined so that the enunciation of natural laws should be as simple as possible. In other words, all these rules, all these definitions are nothing but the fruits of an unconscious opportunism."

Poincaré's reflections were so original that one would like to know why they came to him and not to any earlier scrutinizer of the foundations of mechanics. Peter Galison has adduced a possible effect of the mathematician's involvement in the technology of the distribution of time at the head of the Bureau des longitudes. There is not much evidence for this connection besides the brief allusion to telegraphic synchronization. The astronomical context is much more pervasive: the dating of distant astronomic events; the shared observation of astronomic events. Calinon's remark about our uncritical acceptance of an *à la fois* even for distant stars may also have stimulated Poincaré's reflection.[67]

Conclusions

As we saw in Chapter 1, Newton had two reasons to introduce absolute space in his mechanical world: it was the naturally imagined container of matter in the atomist view, and it was the necessary reference of motion in a mechanics based on the principle of inertia. Similarly, absolute time was the linear container of all events, in harmony with the uniformity of inertial motion. Absolute space and time, in Newton's mind, were necessary physical entities, nearly substances, except that unlike ordinary matter they escaped direct detection. They also had metaphysical significance as conditions for God's omnipresence and immutability.

Nowadays, we would simply say that mechanics is first given as a symbolic, uninterpreted mathematical construct, and that the space and time variables in this construct acquire physical meaning indirectly, through the models of measuring devices permitted in this construct. This is roughly the view of mechanics defended by Ludwig Boltzmann in the late nineteenth century, in anticipative agreement with the semantic view of physical theory now popular among philosophers of science. In contrast, it would be very difficult to find anyone, before Boltzmann, who would accept the precedence of mathematical constructs over physical meaning. Some authors looked for a metaphysical or theological grounding of the basic concepts of mechanics; others looked for operational grounding; still others for empirically justified conventions.[68]

---

[67] Poincaré 1898, p. 13; Galison 2003, chap. 4. According to Max Jammer (2006, pp. 100–103), Poincaré was eager to find conventions everywhere after finding them in geometry.

[68] On Boltzmann's view, see p. 47, this volume. Neumann 1870 and Schell 1879 are the only anticipations of this view I am aware of.

There is one point on which all critics of mechanics from Newton to Poincaré agreed: the only observable motions are those of matter with respect to matter. The question was not whether absolute space could be a quasi-material observable entity, but whether unobservable entities were tolerable in the foundations of mechanics. Newton certainly thought so. He was happy with the indirect empirical evidence of absolute motion as given by the Newton bucket or by two balls connected with a rope. This did not completely determine the absolute space, since any space in rectilinear uniform translation with the privileged space would still pass the rotation test. But in Newton's eyes a real space had to be unique: it was meant to be a substance and not a modernist structure. He removed the ambiguity by requiring complete rest for the center of the universe.

Physicists or geometers till the end of the nineteenth century generally accepted Newton's absolute space and time, most of them in a passive, uncritical manner bordering on agnosticism. As we saw, the authors of most mechanical treatises were completely silent on the reference of motion in Newton's laws. Laplace did say a few words, but he left it to his reader to decide whether absolute space should be regarded as real or imaginary. Among those who strongly supported the reality of absolute space and time we find d'Alembert and Euler in the eighteenth century, and Maxwell in the nineteenth century. Euler forcefully argued that the laws of Newtonian mechanics could not be expressed without referring the described motions to real space and time. He, d'Alembert, and Maxwell asserted the homogeneity and isotropy of this space and time, and used it in purely rational derivations of the law of inertia. In modern terms, they regarded the law of inertia as a direct consequence of the underlying symmetries of space and time as physical entities. Among the symmetries exploited by d'Alembert and Euler and emphasized by Maxwell was Galilean relativity. The two geometers were aware of the conflict between this symmetry and the uniqueness of absolute space. Like Newton, they simply accepted that the basic theoretical representation might not reflect all phenomenal symmetries, just as Poincaré would later maintain an ether frame after introducing the complete relativity of electromagnetic phenomena.

Huygens was first among the few who dared reject Newton's absolute space. Through his correspondence published in the early eighteenth century, many later authors were aware of his insistence that mechanics should deal with relative motions only. However, his ingenious conciliation of this principle with the law of inertia remained unknown until the publication of the relevant manuscripts in the early twentieth century. His idea was to specify the reference frames in which the law of inertia applies as those in which a set of free particles originally at rest remain at rest. He simply posited the existence of this inertial class of frames, and then regarded all members of the class as strictly equivalent in harmony with his early constructive use of Galilean relativity.

In the last third of the nineteenth century, several authors unknowingly produced variants of Huygens's construction of the inertial class of reference frames. In their famous *Treatise*, Thomson and Tait constructed an inertial frame through the directions OP, OQ, and OR made by four material points O, P, Q, and R projected from a common origin. They justified this construction through a purely empirical statement of the law of inertia as the proportionality of the relative distance traveled in all free motions in nature. This "great truth of nature" implies the constancy of the directions OP, OQ, and OR in the former definition, as well as the possibility of consistently defining time through one of the free motions. Thomson and Tait thus admitted the existence of "directional rest" but denied translational rest as incompatible with Galilean relativity. They avoided "absolute" space and time.

Thomson's brother James introduced the word "reference frame" to mean concrete rigid frames to which motion is referred in practice. He simply posited the existence of inertial frames and inertial clocks for which the motion of free particles is rectilinear and uniform. Heinrich Streintz later defined the inertial frames in a more complex but more effective manner as those in which Newton's and Foucault's rotation tests (respectively through a bucket and through a pendulum) are negative. Streintz insisted on the full equivalence of all inertial frames and compared it with the equivalence of different systems of coordinates in a given space, thus foreshadowing the modern concept of an intrinsic space–time. Unlike his predecessors, Ludwig Lange wanted the definition of an inertial frame to be devoid of any empirical content. He gave it as a frame in which in which three particles projected from a common origin and left to themselves have noncoplanar, rectilinear trajectories. Such frames exist for merely kinematic reasons. What has genuine empirical content is the law that in such frames the motion of any additional free particle is rectilinear. All these suggestions by Huygens, Thomson and Tait, J. Thomson, Streintz, and Lange shared a common structure in the set of equivalent inertial frames and they all discarded Newton's absolute space and time in favor of an operational, multivalued definition of the reference of motion in conformity with Galilean relativity.

For one willing to break the Galilean symmetry, an easy substitute for Newton's absolute space is Carl Neumann's body Alpha: for Neumann, the law of inertia is valid when referred to an imaginary rigid body or to any other body in rectilinear, uniform motion with respect to this body. As Neumann's critics noted, this body plays a role very similar to Newton's absolute space and it suffers from the same defect of not being accessible to observation.

There were much more radical attempts to dispense with Newton's absolute space and time, typically coming from philosophers or mathematicians. Leibniz redefined space as the order of coexistence and time as the order of succession in a continuous plenum. In this view all motion is in essence relative, although (live) force may serve to determine the true intensity of motion. As Leibniz ignored Newton's laws and the implied concept of force, his philosophy could not serve to clarify the meaning of mechanical inertia.

In contrast, Kant wanted to justify the success of Newtonian mechanics in a new transcendental philosophy in which two preexisting faculties of the mind, understanding and sensibility, shape all our knowledge of natural phenomena. In this view, space and time are neither real physical objects (as Newton would have it) nor an intrinsic order of things (as Leibniz would have it). They are forms that our mind imposes on external and internal sense, respectively. Kant then redefines absolute space as a regulative idea serving to express the mobility of material bodies in a definite but conventional manner. From a kinematical ("phoronomic") viewpoint, motion is purely relative and we may indifferently have a body move with respect to a certain space or this space move with respect to the body. From a dynamical and mechanical viewpoint, the law of inertia selects a subclass of spaces that move rectilinearly and uniformly with respect to each other. From a phenomenological point of view, the selection of a concrete inertial frame is permitted by experiments performed in this frame. A concrete frame rotating with respect to an inertial space is *truly* rotating since experiments in this frame can detect the rotation, but it is not *absolutely* rotating since, kinematically, we could just as well have the inertial space rotate in the concrete frame.

Kant regarded Newtonian mechanics (properly reformulated) as a rational implementation of his table of categories to matter distributed in space and time. Nineteenth-century physicists rather saw it as the result of strong empirical inductions, not to be derived by any metaphysical or transcendental arguments. Pushing this empiricist attitude to its limits, Mach sought to formulate mechanics in terms of relative, verifiable motions only. He regarded Streintz's and Lange's contemporary formulations as pointless because their privileged frames dissimulated the true cause of inertial behavior, to be found in the distribution of matter in the universe. Mach's own proposal was only a sketch, it was inherently complex, it retrieved Newton's predictions as an approximation only, and it presupposed a distant causality that his contemporaries were unwilling to swallow.

Unlike Leibniz and Mach, Poincaré was not trying to replace Newton's mechanics with a new, better founded theory. Unlike Kant, he did not believe in its rational necessity. Instead he offered a radical criticism of the empirical content of Newton's laws. Whereas Newton and his followers regarded these laws as results of experience, and Kant as implementations of the subcategories of relation, Poincaré downgraded them to empirically motivated conventions. To be sure, Thomson, Tait, Streintz, and Lange had already isolated conventional components of the law of inertia in the definitions of inertial frames and inertial time, but these physicists still believed in a factual residuum of the law. In contrast, Poincaré denied that the laws of mechanics had any refutable content. They were nothing but implicit definitions of the involved terms; they were just conventions that had proved convenient in early applications of mechanics to quasi-isolated systems. Like any philosophical critic of Newtonian mechanics, Poincaré was puzzled by the incompatibility of its laws with the general relativity of motion. But he tolerated artificial conventions, such as Neumann's body Alpha, that could save this principle without hampering the instrumental efficiency of the theory.

As Newton presented his absolute space and his absolute time in parallel, those who rejected absolute space naturally rejected absolute time. D'Alembert's suggestion that time should be measured by inertial motion or by the repetition of identical processes gradually turned into the idea of *defining* time in this manner. From Mach's point of view, all we observe in nature is the correlation of diverse motions and we may take any motion as a time indicator.[69] As Schell and Streintz had earlier opined, the choice of the time indicator is a matter of convention. Simply, as Calinon and Poincaré later argued, we choose the convention for which the equations of mechanics take the simplest form. That the measure and even the definition of time intervals call for conventions easily comes to mind. Less evident is the case of the seemingly more qualitative concept of simultaneity. Common people and philosophers long took the simultaneity of two events to be an evident notion. For instance, Leibniz's definition of space as the order of coexistence presupposes a picture of all things existing *at the same time*. In 1885, Calinon insisted on this intimate, uncritically assumed correlation between spatiality and temporality. In 1884, James Thomson pointed to a possible difficulty in judging distant simultaneity through signals propagated

---

[69] Mach 1872, pp. 34–37, 56–57; 1883, pp. 208–209.

at a finite speed. In 1898, Poincaré multiply argued for the merely conventional character of simultaneity, thus denouncing a last, most entrenched absolutist component of Newtonian time.

All the criticisms of Newtonian mechanics described in this chapter concerned the basic question: how do we define motion in a mechanical context? The answer crucially depends on the law of inertia as a means to select the reference of motion. This reference evolved from Newton's absolute space and time to the inertial frames and clocks of Huygens, Thomson and Tait, Streintz, and Lange. What thus emerged, in modern terms, is a space–time structure defined by a class of inertial frames and clocks and the interrelating group of uniform translations. These authors did not reify the structure. Newton did so with his absolute space, Neumann with his body Alpha, and Budde (and Mach) with a fluid medium, at the price of breaking the symmetry of Galilean relativity. Mach rejected both the structure and its reifications, arguing that they missed the true cause of inertia, to be found in the distribution of masses of the universe. Poincaré drowned the structure in a conventionalist fog. A modern alternative would be to declare that the structure, despite its nonmateriality, is a causal power, itself causally conditioned by the distribution of matter. This way of thinking presupposes a kind of structuralism that emerged only after relativity theory was born.[70]

[70] On this last point, cf. Eddington's remarks discussed on pp. 349–350, this volume.

4

THE OPTICS OF MOVING BODIES

> The translational motion of the earth has no appreciable effect on the optical phenomena produced with a terrestrial source or with solar light. These phenomena do not give us the means to appreciate the *absolute* motion of a body and the *relative* motions are the only ones that we can reach.[1] (Éleuthère Mascart, 1874)

In any domain of physics, we may ask the question: do the observable phenomena depend on a common motion imparted to the involved material bodies? As we saw in the previous chapters, this question appeared early in the history of Newtonian mechanics, and a selectively positive answer conditioned later developments. As we will now see, this relativity question also occurred in the history of optics in the eighteenth and nineteenth century.

Unlike mechanics, optics is not primarily concerned with bodies in motion. On the contrary, most optical phenomena and experiments involve bodies and apparatus at rest (on earth). Moreover, there is no reason to expect any optical effect of bodily motion as long as its velocity is negligible compared with the velocity of light. This is why the optics of moving bodies was never more than a relatively small subsection of optics. This is also why its history started only in the late seventeenth century—with Ole Rømer's astronomical determination of the velocity of light—and why it was punctuated by long pauses. Yet this chapter is not the mere cherry-picking of aspects of the history of optics that informed the later history of relativity. We will observe, over long stretches of time, the continuity of problems and issues of this older optics of moving bodies, as well as its constant importance in astronomy. It affected astronomical observations through stellar aberration, and there were hopes that it could inform us on the velocity of celestial bodies, as happened in 1868 when William Huggins first measured a Doppler shift of starlight.

In a sense, considerations on the relativity of mechanical motion had a direct bearing on optics, since for most natural philosophers from Newton's times to the late nineteenth century, optics was to be subsumed under the laws of mechanics. If an optical system is in essence mechanical, from any permissible state of the system we may obtain another one by imparting a common motion to all mechanical parts, material and immaterial. The implications depend on the kind of mechanical reduction. Historically, there were essentially two kinds of dynamical theory of light. In the Newtonian kind, optical phenomena are reduced to the mechanical interaction between ordinary matter and light

---

[1] Mascart 1874, p. 420.

Relativity Principles and Theories from Galileo to Einstein. Olivier Darrigol, Oxford University Press.
© Olivier Darrigol (2022). DOI: 10.1093/oso/9780192849533.003.0004

corpuscles. In the medium-based kind, the reduction appeals to the motion of a mechanical medium, the ether. In the former kind, a uniform motion commonly imparted to all the material bodies of the optical system does not alter optical phenomena, because this imparted motion equally affects the motion of the light corpuscles: Galilean relativity smoothly extends to Newtonian optics. In the latter kind, the motion imparted to the material bodies does not necessarily concern the ether, and there is no reason to expect the invariance of optical phenomena unless the ether drag is specially adjusted.

A few natural philosophers of the eighteenth century (Clairaut, Euler, Wilson, and Robison) understood the Galilean relativity of optical phenomena in the Newtonian context. Early in the nineteenth century, François Arago performed a prism experiment that was later interpreted as a confirmation of the lack of effect of the earth's motion on optical refraction. The wave theorists of the next generation needed to adjust the ether drag to make it compatible both with this relativity and with stellar aberration. This turned out to be possible in two different ways, with tensions evolving into paradoxes at the end of the century.[2]

Section 4.1 of this chapter recounts the first determinations of the velocity of light: by Ole Rømer in 1676 through the apparent period of the eclipses of one of Jupiter's satellites, and more reliably by James Bradley in 1727 through the aberration of stars. The latter determination was originally understood as a confirmation of the corpuscular theory of light. As will be shown in Section 4.2, this conviction stimulated the development of a fuller Newtonian optics of moving bodies, including the aforementioned relativity. Section 4.3 is devoted to various discussions of stellar aberration in a wave-theoretical context: by Euler, Young, and Fresnel, based on a stationary ether; by Stokes, based on a fully dragged, irrotationally flowing ether; and by Doppler in a more interrogative mood. In his discussion of aberration, Doppler included the effects of the motion of the source and receptor of light (and sound) on frequency and on intensity. The Doppler effect thus became an important component of the optics of moving bodies. Section 4.4 recounts how Fresnel introduced a partial, index-dependent drag of the ether by moving transparent bodies in order to account for the lack of effect of the motion of the earth on optical refraction, and how Fizeau confirmed this partial drag in 1851. Section 4.5 describes the intensification of research in the optics of moving bodies in the 1860s with astronomical motivations, Mascart's and others' systematic investigation of the impossibility to detect effects of the motion of the earth under Fresnel's assumptions, and Michelson's attempts to discriminate between the competing views of ether motion.

4.1 The speed of light

The natural philosophers of the seventeenth century disagreed about the speed of light. The supporters of a corpuscular theory, including Beeckman, Gassendi, and Newton, naturally assumed a finite velocity. Descartes, on the contrary, believed the instantaneous propagation of light to be an essential consequence of his system: his reduction of matter to

---

[2] For historical overviews of the optics of moving bodies, cf. Mascart 1872; Lorentz 1887; Hirosige 1976, pp. 6–22; Janssen and Stachel 2004; Chappert 2004, chap. 10; Eisenstaedt 2005, chaps. 5–6, 9–13; Darrigol 2012 (of which a few extracts have been used later in the chapter); de Andrade Martins 2012.

extension implies the perfect rigidity of the contiguous globules that transmit the light impulses from the source to the eye. In addition, he was known to have argued, in a letter
to Beeckman of 1634, that a finite (and not too large) velocity of light would imply the
nonalignment of the sun and the observed position of the moon during a lunar eclipse. In
1638, Galileo, who favored a finite velocity, tried a first terrestrial measurement:

> Let each of two persons take a light contained in a lantern, or other receptacle, such
> that by the interposition of the hand, the one can shut off or admit the light to the vi-
> sion of the other. Next let them stand opposite each other at a distance of a few cubits
> and practice until they acquire such skill in uncovering and occulting their lights that
> the instant one sees the light of his companion he will uncover his own. . . . Having
> acquired skill at this short distance let the two experimenters, equipped as before, take
> up positions separated by a distance of two or three miles and let them perform the
> same experiment at night, noting carefully whether the exposures and occultations
> occur in the same manner as at short distances.

Galileo tried this at distances of less than a mile (although he trusted one could operate
at much larger distances with the help of telescopes), and could only conclude that the
velocity of light had to be much greater than that of sound.[3]

Rømer's pioneering determination

An astronomical determination of the speed of light turned out to be closer at hand. Soon
after his discovery of Jupiter's satellites, Galileo suggested that their periodic motion could
be used as a clock to determine longitude at sea. The practical urgency of this problem
prompted the Italian astronomer Giovanni Domenico Cassini to study and tabulate the
motion of these satellites. After being called on by Jean-Baptiste Colbert to found a new
observatory in Paris, he continued this activity with improved resources. This is how the
young Danish astronomer Ole Rømer came to assist him in the preparation of tables. In
1676, Rømer focused on a systematic anomaly in the measured period of Jupiter's first
satellite (Io): the successive emersions of this satellite from Jupiter's shadow occurred with
a period slightly larger than the successive immersions into the shadow. He explained this
shift by the varying distance the light had to travel between Jupiter and the earth (see
Figure 4.1). From the cumulated asymmetry observed during several revolutions of the
satellite, Rømer estimated that light would take about 22 minutes to travel the diameter of
the earth's orbit (which gives a velocity of about 220,000 km/s).[4]

   Although retrospective analysis of Rømers's data has vindicated his analysis, he failed
to convince most of his French colleagues. One reason may have been the contradiction

---

[3]Descartes to Beeckman, August 22, 1634, in Adam and Tannery 1897, pp. 307–312; Galilei 1638, p. 88 (transl.
by Crew and de Salvio).

[4]Rømer 1676. Cf. North 1983; Bobis and Lequeux 2008. Although Cassini authored the first report (August 22,
1676) of an explanation based on the finite velocity of light, he later made clear that Rømer was the true inventor
of the "ingenious hypothesis" (Cassini 1693, p. 52). There is an evident similarity between the Rømer effect and
the Doppler effect: in both cases, the motion of the observer affects the observed frequency of the periodic signal
emitted by the source.

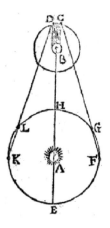

Fig. 4.1.   Rømer's determination of the velocity of light. From Rømer 1676, p. 234.
The larger circle represents the orbit of the earth around the sun (A). The
smaller circle represents the orbit of the first satellite of Jupiter (B); it enters
the shadow of Jupiter in C and emerges from it in D. L and K are the positions of the earth for two successive emersions of the satellite occurring near
one quadrature (time at which Jupiter and the sun are seen in perpendicular directions). F and G are the positions of the earth for two successive immersions
of the satellite occurring near the other quadrature. If the velocity of light has
a finite value c, the time between the two emersions exceeds the period of the
satellite by LK/c, whereas the time between the two immersions is inferior to
this period by FG/c.

with Descartes's system. Most important, Cassini rejected Rømer's "hypothesis" on the
ground that it failed for Jupiter's other satellites. In his opinion, orbital irregularities or
other unforeseen perturbations could explain the observed asymmetry. However, Rømer
received strong backing from Edmond Halley and Isaac Newton in England. His most
eloquent supporter was Christiaan Huygens, whose theory of light fundamentally relied
on a finite propagation of light in analogy with sound.[5]

Bradley's discovery

The last doubts about the finite propagation of light vanished in 1727, when the Oxford
astronomer James Bradley discovered stellar aberration. In the Copernican world, stars
are at a variable distance from the sun, and the closest stars could conceivably be seen in
a direction varying with the time of the year, owing to the earth's annual motion around
the sun. This is the so-called *parallax*. Astronomers had long searched for this effect, and
Robert Hooke pretended to have detected it on a bright star called γ *Draconis*. In late 1726,
Bradley and his friend Samuel Molyneux tested this claim with a (nearly) vertical telescope
whose pointing they could control to a fraction of a second. They found that the position
of γ *Draconis* was shifted by a few seconds in a few days, but in the direction opposite
to that expected for parallax. Bradley multiplied observations with several stars and an
improved telescope for several months. After eliminating instrumental errors or a possible
effect of the earth's nutation, he explained the observed regularities through the motion of
the observer during the journey of the light from the star.[6]

If the light from a star is observed through a narrow tube moving together with the
earth, he reasoned, the tube must be inclined away from the true direction of the star in
order that a particle of light traveling from the star through the moving tube does not hit
its walls (see Figure 4.2). From the 20.25″ he measured for the maximal aberration angle

[5]Huygens 1690, pp. 7–9. For a retrospective analysis, see Kristensen and Møller 2012. For the reception, cf.
Bobis and Lequeux 2008.
[6]Bradley 1728. Cf. Stewart 1964; Fischer 2010.

Fig. 4.2.   Stellar aberration according to Bradley. During the travel
of a light particle from one extremity of the observation tube AB
to the other, the earth carries this tube from the position AB to the
position A'B'. Consequently, the tube makes the (small) aberration
angle $\alpha = \angle AB'A' \approx (u/c) \sin\theta$ with the true direction of the star if
u denotes the velocity of the earth, c the velocity of light, and θ the
angle $\angle A'AB'$ between this motion and the direction of the star.

and from the accepted value of the velocity of the earth (around the sun), Bradley gave 16 minutes 20 seconds as the time needed by light to travel the diameter of the earth's orbit (which translates into 301,000 km/s). This first (1%) accurate measurement of the velocity of light convinced the remaining skeptics that light took time to propagate.[7]

4.2 The corpuscular approach

Stellar aberration

Bradley's derivation of the aberration angle employed Newton's corpuscular theory of light. For this reason, the observed stellar aberration was often regarded as evidence in favor of the corpuscular theory. For instance, in 1739 the French *géomètre* Alexis Clairaut cited Bradley's discovery against the neo-Cartesian theories still favored by his fellow countrymen. Two years earlier, he had compared the light falling on a moving telescope with rain falling into a moving tube, and he had elegantly derived all the geometrical circumstances of stellar aberration. Similar reasoning could be found in Thomas Simpson's *Essays* of 1740, in Pierre Fontaine des Crutes's *Traité complet sur l'aberration* of 1744, and in Ruđer Bošković's *Dissertationis de lumine* of 1748. In the "Aberration" entry of the *Encyclopédie* in 1751, d'Alembert presented Bradley's discovery as the most important astronomical discovery of the century. He derived the aberration angle by means of Bradley's tube, he cited Pierre Louis de Maupertuis's analogy with a hunter pointing his gun ahead of a flying bird, and he proposed a new demonstration based on the velocity of a light corpuscle with respect to the observer's eye. This velocity is the absolute velocity of the light corpuscle compounded with the (opposite of the) velocity of the earth, and its direction is the one in which the eye sees the star. In 1771, in the third volume of his authoritative *Astronomie*, Jérôme de Lalande reproduced Bradley's and d'Alembert's reasoning, and offered his own comparison with a stone thrown from a moving boat, after deploring that "ordinarily people had trouble figuring out [stellar aberration]."[8]

---

[7] Friedrich Bessel first succeeded in measuring a parallax in 1838 (0.3′ for 61 *Cygni*). Bradley privately introduced the name "aberration" in 1729 in order to distinguish the new effect from parallax: cf. Fisher 2010, pp. 42–46. For the motion of the earth around the sun, the ratio u/c is about 10^{-4}. The motion of a point on the surface of the earth owing to its diurnal rotation is two orders of magnitude smaller at moderate latitudes, so that the diurnal aberration is negligible.

[8] Clairaut 1737; 1739, p. 263; Simpson 1740; Fontaine des Crutes 1744; Bošković 1748a, p. 26; d'Alembert 1751; Lalande 1771.

Fig. 4.3. Diagram for Clairaut's original deduction of aberration (the aberration angle is grossly exaggerated).

Most of the early derivations of aberration relied on the trajectory of a light corpuscle and they assumed, without proof, that the axis of the optical instrument (the telescope or the eye) was aligned with this trajectory, as is trivially the case for the empty tube imagined by Bradley. An interesting exception is found in Clairaut's memoir of 1739. Let S and O represent the positions of the star and terrestrial observer at a given time. In the time taken by light to travel from the star to the earth, the earth moves from O to O' (see Figure 4.3). The direction SO' gives the true position of the star in the sky at the instant of observation. Now suppose that the star moves with the same velocity as the earth in absolute space, from S to S'. Evidently, the latter motion does not affect the absolute direction of the rays that reach the earth (since the star is extremely far from the earth). There being no detectable effect of a common motion of the earth–star system, the direction SO in which the light would be observed in the absence of this common motion is also the direction in which it is observed on the moving earth. Although Clairaut did not himself insist on the originality and power of this reasoning, we may remark that it is essentially based on the relativity principle and does not presuppose a specific theory of light. In a modern variant, we could apply the usual laws of optics (to the telescope) in a terrestrial frame in which the star moves at the opposite of the earth's velocity. Owing to the latter motion, the position of the star at the time of observation differs from the position it had when the observed light was emitted, by an amount corresponding to the aberration angle.[9]

Bošković's water-filled telescope

The superiority of this way of reasoning easily escaped Clairaut's contemporaries, who were more than willing to accommodate effects of the earth's motion on terrestrial optics. In 1766, Bošković wrote to the French astronomer Jérôme de Lalande to propose a way to measure the velocity of light by filling a telescope with water (between the objective and the cross-wire plate) and measuring the resulting shift in stellar aberration. In naive reasoning based on Bradley's tube, Bošković simply replaced the velocity of light in water with the velocity of light in a vacuum in the usual formula for the aberration angle. If this were justified, then the images given by a water-filled and an air-filled telescope would differ by an amount proportional to their common absolute velocity.

Lalande published an account of Bošković's letter in 1781, in the second edition of his *Astronomie*. The following year, the Scottish astronomer Patrick Wilson published an independent discussion of the water-filled telescope in which he reached the opposite

---

[9] Clairaut 1739, pp. 261–262.

conclusion: if Newton's theory of refraction holds in absolute space, then the aberration observed in a water-filled telescope is the same as in a normal telescope. Indeed, if **c** and **c′** denote the absolute velocities of a light particle before and after entering the water and **u** denotes the velocity of the earth, the relative velocity of this particle must be parallel to the axis of the telescope in order to be seen; Newton's theory of refraction implies the equality of the components of **c** and **c′** parallel to the interface between air and water; therefore, the relative velocity **c − u** is also parallel to the axis, in agreement with Bradley's value for the angle of aberration. Wilson concluded that a test of the water-filling independence of stellar aberration would provide "very strong additional evidence for [Newton's] principles."[10]

Bošković finally published his considerations in 1785, with a derogatory comment on Wilson and the additional suggestion of a diurnal aberration for terrestrial objects seen through the water-filled telescope. Again, his reasoning ignored refraction at the interface between air and water. The Scottish natural philosopher John Robison, who generally admired Bošković's writings, planned to test the diurnal aberration with a glass rod (a simple substitute for the water-filled telescope) but gave up after convincing himself that the effect should not exist. He intuitively grasped that there could be no aberration when there is no relative motion between the observer and the object. In conversations with his friend Wilson, he also understood that refraction at the interface between air and water annihilated Bošković's effect. In the ensuing memoir of 1790, he analyzed the motion of a light corpuscle when passing from a first medium at rest to a denser medium in motion under the normal attraction imagined by Newton, and derived the conservation of the tangential velocity both for the absolute and the relative motion of the corpuscle. Consequently, the relative motion of the light corpuscle must be along the axis of the telescope both for the empty and a water-filled telescope, and there is no Bošković effect. Robison concluded:

> But if the two aberrations shall be found to be the same, and if no terrestrial aberration shall be observed, we have a direct proof of the acceleration of light in the above mentioned proportion, and of its refraction being produced by forces acting perpendicularly to the refracting surface, and almost a demonstration that light consists of corpuscles emitted by the shining body.

In modern terms, Robison understood the Galilean invariance of the dynamics of Newton's light corpuscles. His trust in this invariance increased his trust in Newtonian optics so much that he ceased to look for the effects predicted by Bošković.[11]

Blair and Arago on the apparent velocity of light

In an unpublished memoir read to the Royal Society in 1786, the Scottish astronomer Robert Blair noted that light corpuscles emitted by a moving source should travel with

---

[10] Lalande 1781, p. 687; Wilson 1782, p. 58. Wilson may have been stimulated by Melvill's (unpublished) idea that stellar aberration depended on the velocity of light in the aqueous humor of the eye. Cf. Acloque 1991, pp. 43–74; Pedersen 1980; 2000, pp. 518–520 (Wilson), 530–533 (Bošković).

[11] Bošković 1785; Robison 1790, p. 95. Cf. Pedersen 1980; 2000, pp. 522–529; Cantor 1983, pp. 75–76. In modern terms, the invariance of the tangential velocity results from the equivalence of $(\mathbf{c} - \mathbf{u}) \times \mathbf{n} = (\mathbf{c'} - \mathbf{u}) \times \mathbf{n}$ and $\mathbf{c} \times \mathbf{n} = \mathbf{c'} \times \mathbf{n}$, if **n** denotes the normal to the interface.

a velocity obtained by compounding their normal velocity with the velocity of the source, and also that the motion of the earth implied a periodic variation of the apparent velocity. Before Robison, Blair understood that in the Newtonian theory of light, optical phenomena depended only on the relative motion of objects and observer. He even suggested a crucial test of this property by comparing the apparent velocities of light from candles placed in different directions: these velocities should be the same in the emission theory, whereas in the (stationary) ether theory they should vary in a manner depending on the velocity of the earth through the ether.[12]

For the velocity measurements, Blair suggested refraction through a series of achromatic prisms. For cases of altered velocity, he considered the light emitted from the opposite sides of a rotating planet (Jupiter), and the observation of two fixed stars directed along and against the motion of the earth. He thus hoped to confirm the Newtonian theory. His friend Robison publicized his suggestions, adding to them a test on Neptune's or Saturn's rings. In the early years of the nineteenth century, Robison's faith in Newtonian optics had somewhat declined. In 1804, shortly before his death, he pressed William Herschel to perform the test on Saturn's rings, which he believed to be "decisive of the question Is light an emission of matter, or is it an elastic undulation."[13]

In the same spirit, the budding French astronomer François Arago tried to determine whether the motion of the earth around its orbit affected the refraction of stellar light through an achromatic prism. On December 10, 1810, he reported to the French Academy that the dispersion of his measurements was inferior to the theoretical shift. At that time, he still believed in Newtonian optics and in the resulting correlation between velocity and refraction. He did his best to save this theory:

> There seems to be no other way of explaining [my result] than assuming that luminous bodies emit rays of every velocity, provided that these rays are only visible within certain limits. Under this assumption, indeed, the visibility of the rays depends on their relative velocity, and, as this velocity also determines the amount of refraction, the visible rays will always be equally refracted.

Arago mentioned the recent discovery of ultraviolet and infrared rays in support of this original hypothesis.[14]

---

[12] Blair [1786]. Cf. Cantor 1983, pp. 87–88; Eisenstaedt 2005a, chap. 9; 2005b.

[13] Robison to Herschel, April 14, 1804, quoted in Cantor 1983, p. 88. A velocity change implies a change of refraction, whether or not velocity is the parameter of color. Blair's recourse to achromatic prisms to measure velocity excludes its being the parameter of color, since in this case different velocities would yield the same deviation by the compound prism. According to modern wave theory, a regular prism yields a refraction modified by the Doppler shift of the stellar light, which is of the same order of relative magnitude as the Newtonian velocity effect; an achromatic prism is of course insensitive to this Doppler shift.

[14] Arago 1853 [read in 1810], p. 563. Laplace advertised Arago's results in the fourth edition of his *Système du monde* (1813, Vol. 2, p. 238). For this reason and also because he had lost his manuscript, Arago long delayed publication. According to Mascart (1872, pp. 158–159), Arago's apparatus did not have sufficient sensibility to detect an effect of the earth's motion. Cf. Acloque 1991, chap. 5; Eisenstaedt 2005a, chap. 10.

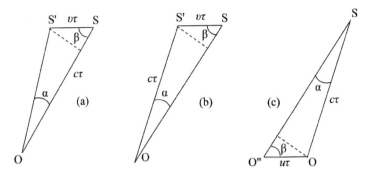

Fig. 4.4.   Diagrams for the aberration of a moving source in the wave theory (a), for the same in the
corpuscular theory (b), and for the aberration of a moving observer in both theories (c). In (a) and (b) the
star moves from S to S' during the time τ that the light takes to travel from S to the observer O. In (c) the
observer moves from O'' to O during the time τ that the light takes to travel from S to O.

4.3 Stellar aberrations in the wave theory

Euler's optical kinematics

In 1739, the young Leonhard Euler discussed the aberration of stars and planets in both
a Newtonian and a wave-theoretical context. Although his memoir supporting the wave
theory only appeared seven years later, he was already inclined in its favor. He refused to
believe that the aberration phenomenon required the light corpuscles, and he showed how
to reason with waves.[15]

Euler first considers the case in which the observer is at rest and the celestial object
moves rectilinearly and uniformly at the velocity **v**. During the time τ, the source travels the
distance $v\tau$ from S to S', and the light travels the distance $c\tau$ from the source S to the observer
O, because in the wave theory the velocity of light is a characteristic of the medium and
does not depend on the velocity of the source (see Figure 4.4a). Call α the angle between the
direction of observation SO and the true direction of the source at the time of observation,
and call β the angle between the apparent direction SO and the direction of motion SS' of
the source. Simple trigonometry yields

$$\tan \alpha = \frac{v \sin \beta}{c - v \cos \beta}. \tag{4.1}$$

Euler next addresses the case of a fixed star as seen by an observer moving at the constant velocity **u**. He first assumes that a common motion imparted to the source and the
observer does not alter the appearances. Then the common motion **v** = −**u** transforms
this second problem into the first problem and the aberration angle should still be given
by Equation (4.1), if β now denotes the angle between the velocity of the observer and
the apparent direction of the source. Euler next treats the same case by a direct method,
simply compounding the velocity **c** of the light with the opposite −**u** of the velocity of the

[15]Euler 1739, 1746a, 1746b. Cf. Maltese 2000; Pedersen 2008.

observer in order to get the apparent direction of the light rays with respect to the observer (see Figure 4.4c). This gives the result

$$\tan \alpha = \frac{u \sin \beta}{\sqrt{c^2 - u^2 \sin^2 \beta}},$$

(4.2)

which differs from Equation (4.1) (with $v = u$) at second order only.

Euler explains this discrepancy by noting that in the first method of calculation, the common motion imparted to the source and observer alters their motion with respect to the ether. In modern terms, there is no reason to expect optical phenomena to depend only on the relative motion of the implied material bodies, since the ether defines a privileged reference frame in which light travels at the velocity c. In contrast, Euler notes, in the corpuscular theory of light the velocity of the light corpuscles has the constant value c with respect to their source, and the desired symmetry is obtained. Whereas in the wave theory light travels from the moving source S to the fixed observer at the velocity c as indicated in Figure 4.4a, in the corpuscular theory it travels with the velocity $\mathbf{c} + \mathbf{v}$, so that in Figure 4.4b it is S′O, not SO, that has the length $c\tau$. Consequently, the diagrams of Figures 4.4b and 4.4c are perfectly similar, and the aberration angle has the expression of Equation (4.2) in both cases of aberration. For this reason, Euler judges the corpuscular theory to be "better adapted" to the calculation of aberrations, even though he globally favors the wave theory.[16]

To summarize, Euler claimed that the aberration of a moving observer was the same in the wave theory and in the corpuscular theory, that the aberration of a moving source differed in the two theories at second order, and that in the corpuscular theory only there was a perfect symmetry between these two kinds of aberration. He did not truly prove the first claim. The concept of a ray velocity with respect to the observer, on which it is based, did not have a definite meaning in the wave theory of light as Euler knew it. He may have unconsciously imported a corpuscular intuition into the wave-theoretical context.

Young and Fresnel

In the early nineteenth century, the chief promoter of the wave theory of light, Thomas Young, agreed with Euler that this theory easily accommodated stellar aberration. In his Bakerian lecture of 1804, Young made clear that the explanation, since it rested on the rectilinear propagation of light, required the ether to be unperturbed by the earth's travel though it:

> Upon considering the phenomena of the aberration of the stars, I am disposed to believe, that the luminiferous ether pervades the substance of all material bodies with little or no resistance, as freely perhaps as the wind passes through a grove of trees.

In his famous lectures at the Royal Institution, Young recycled Bradley's tube, just replacing the trajectory of the light corpuscles in the tube with light rays. No more than Euler did

---

[16] Euler 1739, p. 165.

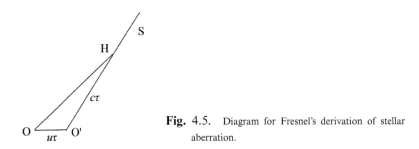

Fig. 4.5.   Diagram for Fresnel's derivation of stellar aberration.

he worry about the propagation of light in the instruments truly used for astronomical observation.[17]

In his first private speculations on the nature of light in the summer of 1814, young Augustin Fresnel was already leaning toward the system of "vibrations," even though he was unaware of Young's writings. Fresnel then deplored that stellar aberration seemed to confirm the competing system: "The strongest evidence in favor of Newton's opinion is, I think, the aberration of stars. . . . I do not see how one could explain aberration under the hypothesis of vibrations." This opinion then was the dominant one, as witnessed by Laplace's statement in the fourth edition of his *Système du monde*: "The phenomena of double refraction and of the aberration of stars seem to give to the system of the emission of light, if not a complete certainty, at least an extreme probability. These phenomena cannot be explained under the hypothesis of the undulations of an ethereal fluid." Fresnel had learned about aberration from René Juste Haüy, who, in his *Traité élémentaire*, derived this phenomenon in d'Alembert's manner through the relative velocity of the light rays from the star when impacting the moving eye. This impact being based on a corpuscular intuition, Fresnel believed aberration to elude the wave theory, until, in a letter written the next day, he rediscovered an explanation similar to Bradley's, based on the motion of the telescope during the travel of light from the objective to the eye.[18]

In substance, Fresnel replaces the telescope with a hole H on a screen and a comoving eye O (see Figure 4.5). When the light from a star reaches the hole, a pencil of light emerges in the true direction of the star. During the time τ that the light takes to travel from the hole to the eye, the eye moves from O to O'. Therefore, the true direction HS of the star differs from the apparent direction OH. Rectilinear propagation at finite speed is all that is needed in this derivation. The only defect of this reasoning, Fresnel comments, is the lack of proof of rectilinear propagation in the wave theory. Fresnel wrote up these "rêveries" in a memoir he sent to André Marie Ampère for comments, to no avail.[19]

---

[17] Young 1804, pp. 12–13; 1807, p. 436.

[18] A. Fresnel to L. Fresnel, July 5 and 6, 1814, in Fresnel 1866–1870, Vol. 2, pp. 822, 825–826; Laplace 1813, p. 327; Haüy 1806, pp. 502–503. Haüy also remarked (p. 139) that aberration was "very difficult to conceive" in the wave theory.

[19] A. Fresnel to L. Fresnel (November 3, 1814) and Mérimée to Fresnel (December 20, 1814), in Fresnel 1866–70, Vol. 2, pp. 829–831; Laplace 1813, p. 327. Variants of Fresnel's reasoning based on light passing through two holes on a box are found in Robison 1790 and in Lorentz 1887 (pp. 104–106).

In the two following years, Fresnel conquered fame through his outstanding work on diffraction. He returned to the optics of moving bodies in 1818, when his friendly mentor Arago asked him whether the wave theory of light could account for the negative result of his prism experiment of 1810 regarding the effect of the motion of the earth on the refraction of stellar light. As we will see in a moment, Fresnel then introduced the partial ether drag through which the laws of refraction are very nearly unaffected by a common motion of transparent bodies. No one paid much attention at the time. During the next twenty years, research in optics focused on the phenomena for which the superiority of the new wave theory was most glaring, namely diffraction, polarization, and refraction.[20]

When, in 1839, the outstanding ether theorist Augustin Cauchy came to address Arago's prism experiment, he ignored Fresnel's analysis and instead proposed that the ether accompanied the earth in its motion as the oceans and the atmosphere obviously do:

> Under this assumption, all the phenomena of reflection and refraction observed at the surface of the earth will be the same as if the earth had lost its motion of diurnal rotation and its motion of annual translation around the sun. These movements can affect only the direction of the wave planes, therefore the direction of the luminous ray, thus producing the aberration phenomenon as we know.

A fellow Academician told Cauchy that Fresnel had already considered this possibility in a letter to Arago of 1818, only to reject it for being unable to reproduce the known aberration angle. Cauchy took note but judged this difficulty might not be great enough to abandon a most probable hypothesis.[21]

Doppler's aberrations

In 1842, a virtually unknown professor of mathematics in Prague, the Austrian Christian Doppler, published a treatise in which he claimed to offer a general theoretical frame for three kinds of effects of the motion of sources and observers in the wave theories of light and sound: directional aberration, frequency shift, and modified intensity. The second effect, a frequency shift, had entirely escaped the acumen of earlier writers on sound and light. First consider the case of an observer moving toward the source S. Doppler reasoned that after a given wave crest reaches the observer in O, the next wave crest A reaches him after he has moved from O to O' (see Figure 4.6a). Calling c the velocity of the waves, v the velocity of the observer, T the true period of the oscillation, and T' the apparent period, we have

$$cT' + vT' = cT, \quad \text{or} \quad T' = (1 + v/c)^{-1}T. \tag{4.3}$$

In the case of a source moving toward the observer, a wave crest travels at the velocity c from S to A while the source moves from S to the location S' at which the next wave crest

[20] Fresnel 1818. See pp. 102–104, this volume.

[21] Cauchy 1839, p. 327.

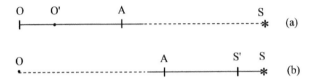

Fig. 4.6.   Diagrams for Doppler's derivation of his effect, for a moving observer (a), and for a moving source (b).

is emitted (see Figure 4.6b). This gives

$$cT' + vT = cT, \quad \text{or} \quad T' = (1 - v/c)T. \tag{4.4}$$

As Doppler remarks, the two cases are not symmetric to each other.[22]

In the case of sound, Doppler did not bother to verify his prediction. The Dutch physicist Christoph Buijs Ballot did so in 1845 by having a band of trumpeters play the same note from a running train. In the case of light, Doppler believed that the observed color difference in binary stars demonstrated his principle: since one of the stars recedes and the other approaches us, the average frequency of their lights must differ. In retrospect, this is doubly incorrect: the velocity difference is too small to produce a (then) measurable frequency shift, and the global shift of a spectrum extending from the infrared to the ultraviolet does not produce any alteration of the light in the visible part of the spectrum. The first observation of a Doppler shift in starlight was made by the British astronomer William Huggins in 1868. Doppler had died of tuberculosis in Venice fifteen years earlier, after losing a fight about the reality of the frequency shift in his home country.[23]

Doppler regarded his theory as an extension of "Bradley's theorem" from direction to frequency and intensity. Strangely, he believed that these effects of motion existed only for longitudinal waves. He (wrongly) reasoned that for transverse waves there could be no effect of the motion of the observer for a star at the zenith because this motion, then being in the direction of the optical vibrations, could not affect the perceived direction of propagation. In a prolix review of stellar aberration in 1845, he found irreparable defects in all available derivations of this phenomenon. In particular, he agreed with Cauchy about the implausibility of Fresnel's assumption of a free flow of the ether through the earth. He added that even in this case there would be no aberration because the motion of the earth with respect to the incoming waves does not affect their orientation and because the direction of propagation is always normal to the waves (unlike Fresnel, he did not understand that the *apparent* direction of propagation need not be normal to the waves). As a possible way out of these difficulties, he briefly considered Cauchy's speculation that a fully dragged ether might be able to curve the light rays to the extent required by aberration (he had himself explored the consequences of a diurnal ether drag in 1844). No more than

---

[22] Doppler 1842, pp. 9–10. Cf. Eden 1992; Schuster 2005; Nolte 2020.

[23] Buijs Ballot 1845; Doppler 1842; Huggins 1868. Cf. Wolfschmidt 2005; Hearnshaw 2014. On Huggins, cf. Becker 2011.

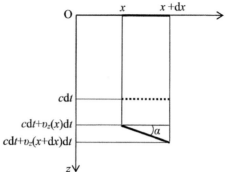

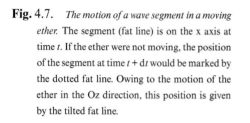

Fig. 4.7.   *The motion of a wave segment in a moving ether.* The segment (fat line) is on the x axis at time *t*. If the ether were not moving, the position of the segment at time *t* + d*t* would be marked by the dotted fat line. Owing to the motion of the ether in the Oz direction, this position is given by the tilted fat line.

Fresnel and Cauchy, however, could he see how to make it work. He concluded that no satisfactory explanation of stellar aberration was yet available in the wave theory of light.[24]

Stokes's theory, and a polemic with Challis

Young and Fresnel both allowed the ether to freely penetrate the earth, so that the motion of starlight in the ether would not be affected by the motion of the earth. In the spring of 1845, George Gabriel Stokes, a brilliant fellow of Pembroke College, Cambridge, judged this hypothesis "very startling" and declared:

> I shall suppose that the earth and planets carry a portion of the æther along with them so that the æther close to their surfaces is at rest relatively to those surfaces, while its velocity alters as we recede from the surface, till, at no great distance, it is at rest in space.

Under this assumption, Stokes computed how the orientation of a small (approximately plane) portion of wave changed during its travel from the star to the earth. Consider this segment at a given instant *t*, draw the z axis in the direction of its normal motion when the ether does not move, and the x axis in the direction of the projection of the ether motion on the plane of the wave portion (see Figure 4.7). In a stationary ether, the orientation of this plane would be the same at time *t* + d*t*. In a moving ether, the *z* coordinate of a point of the ether increases by $v_z(x)dt$, depending on the initial value of the *x* coordinate of this point and on the velocity component $v_z(x)$ of the ether at this point. Consequently, a wave segment extending between *x* and *x* + d*x* rotates by the angle $d\alpha = (\partial v_z / \partial x)dt$ (the component of the ether motion in the Ox direction of course does not contribute to the rotation).[25]

---

[24] Doppler 1845, pp. 6–7 (against Fresnel), 9 (orientation of the waves), 14 (longitudinal waves), 15–16 (curved rays); Doppler 1844. Cf. Hirosige 1976, pp. 9–10.

[25] Stokes 1845b, 1845c. On Stokes, Wallis, and aberration, cf. Wilson 1972; 1987, chap. 6; Darrigol 2019a, pp. 65–73.

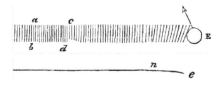

Fig. 4.8.   Stokes's drawing of the rotation of a wave segment (*ab*, *cd*, . . .) when approaching the earth E, and the orthogonal line *ne*. From Stokes 1845c, p. 13 (simplified).

The velocity of the ether being small compared with the velocity of light, we may take $dt \approx dz/c$, and the total rotation is given by

$$\alpha = \int \, (\partial v_z / \partial x) dz/c. \tag{4.5}$$

In general, the result of this integration does not agree with the observed aberration. Stokes removes this difficulty by assuming the ether motion to be such that $\mathbf{v} \cdot d\mathbf{r}$ is an exact differential. Then we have $\partial v_z / \partial x = \partial v_x / \partial z$, and the integration yields $\alpha = \bar{v}_x / c$ if $\bar{v}_x$ denotes the velocity of the ether at the surface of the earth in the direction Ox. This velocity is reckoned with respect to the absolute space of the fixed stars. It is equal to the velocity of the observer in absolute space since, by assumption, the ether close to the earth is at rest with respect to the earth. By the same assumption, the normal to the wave front that reaches the observer is the direction in which the star is seen. The star is therefore seen in a direction inclined by $(u/c) \sin \theta$ toward the motion of the earth, wherein u denotes the velocity of the earth at the point of observation and θ the angle between this velocity and the direction of the star. This is the received law of stellar aberration.[26]

Stokes represented the successive positions *ab*, *cd*, etc. of a wave front, and the orthogonal trajectory *ne* on Figure 4.8. Far from the earth, the successive wave fronts are parallel and their common normal gives the true direction of the emitting star. Close to the earth, the orthogonal trajectory of these fronts is curved and its angular deviation gives the aberration. As Stokes later understood, any temptation to confuse this orthogonal trajectory with a ray of light should be avoided. In essence, Stokes's reasoning relies on wave fronts only, and it depends on two assumptions: (1) the ether motion is irrotational, and (2) the ether is at rest with respect to the earth near the earth. The first assumption yields the desired angular deviation, and the second is necessary for the direction of observation to be perpendicular to the wave front. In 1845, Stokes did not offer any justification for the irrotational character of the ether motion. For the second assumption, he invoked the ether "being entangled with the earth's atmosphere."[27]

At first glance, Stokes's new explanation of aberration seems completely disconnected from Bradley's older explanation. It appeared in the July 1845 issue of the *Philosophical Magazine*. James Challis, director of the Cambridge observatory and Plumian Professor of astronomy and experimental philosophy, published his own theory in the November issue of the same journal, together with a reflection on its compatibility with the wave theory

---

[26]Stokes 1845c, pp. 9–10. The reasoning is here given in the simplified version of Stokes 1880–1905, Vol. 1, pp. 139–140.

[27]Stokes 1845b, p. 9.

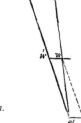

Fig. 4.9.   *Challis's diagram for his derivation of stellar aberration.*
From Challis 1845b, p. 321 (simplified).

of light. Challis traced stellar aberration to the comparison between the direction of the
observed star and the direction of an object moving together with the observer, namely,
the cross-wire of the astronomer's telescope. After the light from the star has reached the
cross-wire w, it takes a certain time to reach the eye e of the observer. During this time,
the cross-wire has moved together with the earth in absolute space. Therefore, the true direction of the cross-wire at the time of observation differs from the direction it had when
the light from the star reached it (see Figure 4.9). The angle between the line *swe'* and the
line *e'w '* defines the aberration. In essence, this reasoning is only a variant of Bradley's:
Challis's cross-wire and the eye of his observer play the role of the upper and lower extremities of Bradley's tube. In both ways of reasoning, the crucial ingredient is the rectilinear
propagation of light. The ether and its possible motion play no role whatsoever. Challis
commented:[28]

> According to Mr. Stokes's views, the phenomenon of aberration is entirely owing to
> the motion which the earth impresses on the æther, and which at the earth's surface
> he supposes to be equal to the earth's motion. On the contrary, I have to show that
> the amount of aberration will be the same whatever be the motion of the æther, and
> if this cannot be shown, the undulatory theory, and not the foregoing explanation, is
> at fault.

Challis proceeded to show that by a geometric version of Stokes's "very ingenious and
original mathematical reasoning," the propagation of light in a moving ether did remain
rectilinear (with respect to absolute space) whatever the motion of the ether. Challis's
demonstration has two steps. First, he argues that owing to the ether motion, the absolute direction of propagation of light at a given point of space is altered by the motion of
the ether by the amount $\alpha' = -v_x/c$, v_x being the projection of the velocity of the ether on the
plane perpendicular to the original direction of propagation. This is a trivial consequence
of the vector composition of velocities. Second, by erroneous geometry Challis finds that a
small portion of a wave front rotates by the amount $d\alpha = (\partial v_x / \partial z)dt$ during propagation
in the moving ether in the time dt. Taking the integral of this expression for $dt \approx dz/c$, he
finds $\alpha = v_x/c$ for the total rotation of the wave front from a star to a given point of space
at which the ether velocity is v_x. This rotation of the wave front exactly compensates for

[28] Challis 1845a, 1845b, p. 323. On the ensuing debate, cf. Powell 1846.

the direct effect of the ether drag (namely, $\alpha + \alpha' = 0$). Therefore, the propagation remains rectilinear whatever the ether motion and Challis's derivation of aberration remains fully justified.[29]

Stokes soon reacted to Challis's claims. First, he criticized the idea that aberration should be traced to a misevaluation of the direction of the cross-wire. In his own theory, for instance, the cross-wire is evidently seen in the true direction. What truly defines the direction of observation of a star is just the direction of the line joining the cross-wire (in the focal plane of the objective) and the center of the objective of the telescope, as implicitly assumed in Bradley's reasoning or later variants. Second, Stokes corrected the error in Challis's derivation of the rotation of a wave portion and reasserted the necessity of the condition that the motion of the ether should be irrotational.[30]

Although Challis soon admitted the latter error, he insisted that his derivation of aberration was independent of any theory of light and therefore superior to one based on hypothetical ether motions. Stokes denied this and the polemic turned sour. Yet he learned something from this exchange. In July 1846, he combined his own derivation of the rotation of a wave segment with Challis's remark that the absolute motion of a wave segment was altered by the motion of the medium. Now identifying the direction of the net motion of the wave segment with the direction of a ray of light, he found that a ray of light kept a constant direction as long as the motion of the ether was irrotational:

> Hence the direction of the light coming from a star is the same as that of a right line drawn from the star, not merely at such a distance from the earth that the motion of the æther is there insensible, and again close to the surface of the earth, where the æther may be supposed to move with the earth, but *throughout the whole course* of the light; so that a ray of light will proceed in a straight line even when the æther is in motion, provided the motion be such as to render $[\mathbf{v} \cdot \mathbf{dr}]$ an exact differential.

As Stokes admitted, he had originally been unaware of "this curious consequence of [his] theory." But he failed to credit his unpleasant interlocutor for the insight. He did so many years later, in the first volume of his collected papers:

> Some remarks made by Professor Challis in the course of discussion suggested to me the examination of the path of a ray, which in the case in which $[\mathbf{v} \cdot \mathbf{dr}]$ is an exact differential proved to be a straight line, a result which I had not foreseen when I wrote the above paper [of July 1845]. . . . The rectilinearity of the path of a ray in this case, though not expressly mentioned by Professor Challis, is virtually contained in what he wrote.

The rectilinear propagation of light in the moving ether being thus established, the aberration of stars can be derived in Bradley's or similar manners. As Stokes noted in July

---

[29] Challis 1845b, p. 324.

[30] Stokes 1846a.

1846, the assumption that the ether does not move with respect to the earth near its surface now becomes unnecessary. The only condition on the ether motion is its irrotational character.[31]

For the sake of the modern reader, Pierre Fermat's principle of least time can be used to simplify Stokes's proof of rectilinear motion in an irrotationally moving ether. Call d**l** an element of the path of light in absolute space, ds its length, and **v**(**r**) the absolute velocity of the ether. The time needed for light to travel from one end of this element to the other is

$$dt = \frac{ds}{c + \mathbf{v} \cdot d\mathbf{l}/ds} \approx \frac{ds}{c} - \frac{\mathbf{v} \cdot d\mathbf{l}}{c^2}. \qquad (4.6)$$

For an irrotational ether motion, the second term of this expression is an exact differential, so that the motion of the ether has no effect on the ray propagation of light.

The main purpose of Stokes's memoir of July 1846 was to unravel the physical circumstances that might justify the irrotational character of the ether drag. Ignoring the effect of the atmosphere in a first step of the reasoning, he considers the ether motion induced by the translation of the earth through the ether. In conformity with speculations aired in his famous memoir on fluid friction, he assumes a jelly-like ether that behaves like an incompressible fluid at the macroscopic scale and like a rigid elastic body at the time scale of luminous vibrations. Ignoring any shear stress and imagining the motion of the earth to start from rest in ether at rest, a theorem by Lagrange implies that the induced ether motion should be irrotational. In reality, Stokes tells us, there may be temporary shear stresses in the medium but any resulting shear will be quickly propagated away as light.[32]

Based on this remark, Stokes expects the induced ether motion to be irrotational at least when the effect of the atmosphere is ignored. From his pendulum studies, he must know that the motion of a perfect liquid induced by a moving sphere implies a shift of the fluid along the walls of the sphere. As for the effect of the atmosphere, in the absence of any established model for the interplay of ether and matter molecules, he offers the conjecture that the molecules of the air act as a dense cloud of solid corpuscles forcing the ether to be nearly at relative rest in the denser part of the atmosphere.[33] The motion being still irrotational between the air molecules by the former reasoning, Stokes thus reconciles irrotational motion through space with lack of relative motion near the surface of the earth. At the same time, he notes that the latter assumption is unnecessary for the derivation of stellar aberration.[34]

---

[31] Challis 1846a, 1846b, 1846c, 1846d, 1848; Stokes 1846c; 1846d, p. 9; Stokes 1880–1905, Vol. 1, p. 138.

[32] Stokes 1845a, 1846d. By analogy with ordinary fluids of small viscosity, Stokes expects the irrotational motion around the sphere to be unstable. But again his ether differs from an ordinary fluid by quickly dissipating any shear stress that would appear when the velocity of the sphere is gradually increased from zero to its actual value. See also Stokes 1848.

[33] This conjecture was not too farfetched in the 1840s, because the modern kinetic theory of gases, according to which the molecules of a gas occupy a negligible fraction of its volume, was still marginal in Britain. Lorentz later criticized the associated reasoning for ignoring diffraction by the molecules: Lorentz 1887, pp. 115–116.

[34] Stokes 1846d, pp. 7–9.

This assumption becomes important as an explanation of the lack of effects of the motion of the earth on optical experiments. It is now time to return, as Stokes did, to Arago's old prism experiment and to Fresnel's explanation of its negative result.

4.4 The Fresnel drag

Fresnel's theory

Remember that in 1810, Arago, then reasoning in a Newtonian framework, expected the refraction of starlight to depend on the motion of the earth, because the original velocity of the light corpuscles had to be compounded with the velocity of the earth and because in Newton's theory of refraction the angle of refraction obviously depends on the relative velocity of the impacting corpuscles. For a terrestrial source, the motion of the earth cannot affect the law of refraction as a direct consequence of the relativity principle applied to the mechanical interactions between matter and light corpuscles. As Fresnel noted, the situation is different in the wave theory. The velocity of light is defined with respect to the ether, which Young and Fresnel assume to be stationary in order to account for the aberration of stars. Therefore, there should be no difference between the laws of refraction for a stellar source and for a terrestrial source. In both cases, the prism is moving with the earth through the ether and one should expect an effect of this motion on the angle of refraction, unless the velocity of light within the prism is properly adjusted. In a letter to Arago published in 1818, Fresnel reasoned as follows.[35]

The prism of Figure 4.10 moves through the ether in a direction perpendicular to one side of the prism, and two rays enter this side perpendicularly. When the light of the inferior ray reaches the oblique side of the prism at B, the light of the superior ray is still inside the prism, at point A. This light exits the prism at a later time for which the facing point D of the oblique side of the prism has traveled to the position D'. According to a simple rule for the construction of refracted rays in the wave theory, at this later time the light of the inferior refracted ray reaches the point C' such that C'D' is perpendicular to BC'. Call t the elapsed time, V the velocity of light in the moving glass, c the velocity of light in the vacuum (to which Fresnel assimilates the air around the prism), u the velocity of the prism, i the angle of incidence, r' the angle of refraction for an observer at rest in the ether, and I the foot of the perpendicular from D to BC'. Since $i = \angle ABD$ and $r' = \angle BDI$, we have

$$IC' = DD'\cos(r' - i), \quad \sin i = AD/BD, \quad \sin r' = BI/BD. \tag{4.7}$$

Together with BC' = ct, DD' = ut, and AD' = Vt, this implies the relation

$$\frac{\sin r'}{\sin i} = \frac{BC' - IC'}{AD' - DD'} = \frac{c - u\cos(r' - i)}{V - u}, \tag{4.8}$$

which implicitly determines r' as a function of i. The refraction angle r measured by a terrestrial observer (moving with the prism) is the angle that the line BJ resulting from

---

[35] Fresnel 1818. Cf. Whittaker 1951, pp. 109–113.

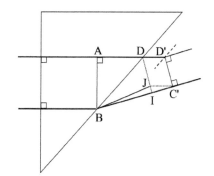

Fig. 4.10.   Variant of Fresnel's figure for refraction in a
moving prism.

the vector composition of BC' (**c**t) and D'D (−**u**t) makes with the normal to the refracting
surface. It is therefore given by the relation

$$\sin(r - r') = (u/c) \sin(r - i). \tag{4.9}$$

To first order in u/c, the two previous equations lead to

$$\sin r = \left(\frac{c}{V-u} - \frac{u}{c} \right) \sin i, \tag{4.10}$$

which differs from the law

$$n \sin i = \sin r \tag{4.11}$$

for a prism at rest in the ether, unless the velocity of light inside the prism is taken to be[36]

$$V = \frac{c}{n} + u \left(1 - \frac{1}{n^2} \right). \tag{4.12}$$

Readers who do not like geometrical reasoning may want to take a look at a later reasoning by Alfred Potier, based on Fermat's principle. In the most general problem of terrestrial
refraction, light travels through transparent matter that has the variable optical index $n(\mathbf{r})$
and the uniform velocity **u** in absolute space (with $u \ll c$). The velocity of light with respect
to the ether is c/n. The ether being dragged at the velocity $\alpha\mathbf{u}$ in absolute space, it moves at
the velocity $(\alpha - 1)\mathbf{u}$ with respect to the transparent matter. The velocity of light along the
element d**l** of this matter therefore is

---

[36] Fresnel 1818a. Cf. Whittaker 1951, pp. 109–113. There is an error in Fresnel's figure: HH' should be parallel
to DD'.

$$\frac{\mathrm{d}s}{\mathrm{d}t} = \frac{c}{n} + (\alpha - 1)\mathbf{u} \cdot \frac{\mathrm{d}\mathbf{l}}{\mathrm{d}s}. \tag{4.13}$$

To first order in u/c, the time needed for light to travel this element is

$$\mathrm{d}t \approx n\frac{\mathrm{d}s}{c} + (1 - \alpha)n^2\frac{\mathbf{u} \cdot \mathrm{d}\mathbf{l}}{c^2}. \tag{4.14}$$

By Fermat's principle, a light ray takes the path of least time between two fixed points of the medium. For the choice $\alpha = 1 - 1/n^2$ of the dragging coefficient, the contribution of the second term to the integral time is a constant, and the light rays are the same as if the matter were not moving.[37]

As Fresnel remarks, the stationary ether needed for his theory of stellar aberration is compatible with the partial drag, because for air the optical index differs little from unity and the dragging coefficient nearly vanishes. Fresnel also argues that in his theory, just as in the corpuscular theory, Boškovič's water-filled telescope should give the same aberration angle as an empty telescope, because the law of refraction holds for a terrestrial observer. Again, in the wave theory it does not matter whether the object is a star or a terrestrial point source, because in both cases the velocity of light is defined with respect to the ether. Later in the century, several astronomers ignored Wilson's, Robison's, and Fresnel's denials of the Boškovič effect and tried to test it experimentally. Most famously, in 1867 the Göttingen astronomer Wilhelm Klinkerfues announced a positive result confirming his own theoretical prediction. In 1869, the Dutch astronomer Martin Hoek criticized Klinkerfues's theory and found no terrestrial aberration for a water-filled telescope. In Cambridge, the senior Astronomer Royal George Biddell Airy judged the matter to be of "great physical importance" and he refuted Klinkerfues's claim through his own meticulous observations in 1871.[38]

Stokes on Fresnel

An obvious advantage of Stokes's assumption of a fully dragged ether is that it immediately implies the absence of effects of the motion of the earth on optical experiments performed on earth. In contrast, the Young–Fresnel theory of aberration implies an ether wind that could in principle disturb such optical experiments. Fresnel's partial ether drag is meant to compensate for the ether wind's effect on refraction. In 1846, after completing his own theory of aberration, Stokes carefully studied Fresnel's assumption of a partial drag of the ether by transparent moving matter.[39]

As Stokes knew, Fresnel justified the partial drag of the ether by assuming that a transparent body moving through the ether carries with it the excess of ether needed to explain

---

[37] Potier 1874, building on Veltmann's proof (1870b) of the invariance of phase differences (see pp. 108–109, this volume). Cf. Lorentz 1887, pp. 132–135; Mascart 1889–93, Vol. 3, pp. 95–96.

[38] Fresnel 1818, pp. 64–66. Klinkerfues 1867; Hoek 1869; Airy 1871. Cf. Acloque 1991, pp. 67–73, 129–140; Antonello 2014.

[39] Stokes 1846b. The phrase "ether wind" is a bit anachronistic as it does not seem to have been used before the late nineteenth century (by Oliver Lodge, for instance in Lodge 1893).

the slower velocity of light, and that waves travel in a transparent body as in a single medium moving at the average velocity of the normal ether and the excess ether. In Stokes's simpler reasoning, the equality of the mass fluxes of the ether on both sides of the air/glass interface reads $-\rho u = \rho' w$, where ρ and ρ' denote the densities of the ether in air and in glass, respectively, and w denotes the velocity of the ether in glass with respect to the glass. Since, according to Fresnel, the elasticity of the ether is the same in any medium, we have $\rho = \rho'/n^2$. Therefore, the ether in the glass has the absolute velocity $u + w = u - u/n^2$, in agreement with the desired dragging coefficient.[40]

Stokes also offers an alternative proof of the first-order invariance of the laws of refraction for a prism subjected to the ether wind at the surface of the earth. And he shows that the Fresnel drag explains the lack of fringe shift in an experiment in which Jacques Babinet brought the two parts of a split light beam to interfere after traveling through two plates of equal thickness oriented in different directions with respect to the earth's motion. Knowing that his own assumption of a vanishing ether wind directly explains Arago's and Babinet's negative results, Stokes concludes:[41]

> This affords a curious instance of two totally different theories running parallel to each other in the explanation of phænomena. I do not suppose that many would be disposed to maintain Fresnel's theory, when it is shown that it may be dispensed with, inasmuch as we would not be disposed to believe, without good evidence, that the æther moves quite freely through the solid mass of the earth. Still it would have been satisfactory, if it had been possible, to have put the two theories to the test of some decisive experiment.

Fizeau's experiments

A relevant, though not decisive experiment, was completed in 1851 by the rising star of French optics, Hippolyte Fizeau. Twelve years earlier, Arago had published a plan to measure the velocity of light using Charles Wheatstone's method of the rotating mirror, with sufficient precision to decide whether light moved faster in air than in water (as the wave theory predicted) or vice-versa (as the emission theory predicted). Arago, with Louis Bréguet's help, had made significant progress in setting up the apparatus in the late 1840s. However, his vision declined and he left the project in the hands of Bréguet, who hired Fizeau and his former school friend Léon Foucault. In 1849–50, Fizeau and Foucault both measured the velocity of light in air, the first using his own method of the rotating toothed wheel, and the second with a rotating mirror. They also determined that the propagation was slower in water, as required by the wave theory.[42]

While Fizeau was preparing the first terrestrial measurement of the speed of light, he was also thinking about the measurable consequences of a finite speed. One of them is

---

[40] Fresnel 1818, pp. 631–632; Stokes 1846b, pp. 76–77.

[41] Fresnel 1818; Stokes 1846b, p. 81; Babinet 1839.

[42] Arago 1838, 1839; Fizeau 1849: Fizeau and Bréquet 1850a, 1850b; Foucault 1850, 1853, 1854. Cf. Frercks 2000; Tobin 2003.

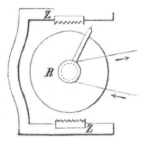

Fig. 4.11.   Fizeau's device for verifying the Doppler effect. The pin on the rotating wheel R produces a sound when running over the teeth in Z, at a frequency proportional to the angular speed and to the periodicity of the teeth. For a listener on the right side of the device, the pitch of the sound from the upper rack is lower than that the pitch of the sound from the lower rack. From Ketteler 1873, p. 21.

stellar aberration, another is the Doppler effect. In 1847, a major French importer of foreign physics novelties, Abbot François Moigno, began the second volume of his *Répertoire de l'optique moderne* with the question of the effects of the earth's motion through the ether, including Fresnel's and Cauchy's views on this matter. He announced he would soon discuss Doppler's "singular opinion" on the origin of the variable color of stars. In a communication of December 1848 to the Société Philomatique, Fizeau rederived the Doppler effect, described its acoustic verification through a clever mechanical contraption (see Figure 4.11), and indicated how the frequency shift of the Fraunhofer lines may someday be used to measure the radial velocity of stars.[43]

Another question raised by Moigno in his *Répertoire* was about the lack of effects of the earth's motion on the laws of refraction in Arago's and Babinet's relevant experiments. As Moigno put it in 1847: "Why and how is it that the velocity of the earth does not at all modify the phenomena of refraction and interference caused by the velocity of light?" The learned abbot, who admired Cauchy's science no less than his faith, approved of his idea of a totally dragged ether as a natural and simple explanation for the lack of effects of the earth's motion (he was not yet aware of Stokes's theory). In contrast, Fizeau originally favored a stationary ether, because the ether could easily be separated from the mercury in a Torricelli tube, and because of stellar aberration. In November 1847, he imagined an ether-drag test in which the light from the same source would be split into two beams traveling along and against running water, and then recombined to form interference fringes (see Figure 4.12). At that time, he judged he could not produce a flow of water fast enough to produce a detectable fringe shift. In 1848–49, his work on the old question of terrestrial aberration brought him to favor a fully dragged ether. In this case, water may be advantageously replaced with air in the former experiment. After two years of work on this plan, Fizeau concluded that the air flow did not produce the expected fringe shift.[44]

By that time, Fizeau had probably seen the two last volumes of Moigno's *Répertoire*, which contained, besides lavish praise of Fizeaus's measurement of the speed of light, a summary of Doppler's "slightly tenebrous" review of aberration theories, as well as short accounts of Fresnel's, Stokes's, and Challis's theories. In this light, Fizeau understood that the negative result of Arago's prism experiment excluded a completely stationary ether.

---

[43] Moigno 1847, Vol. 2, pp. 401–404; 1850, pp. 1165–2003 (details contributions by Doppler, Bolzano, Buijs Ballot, Scott Russell, and Fizeau); Fizeau 1848.

[44] Moigno 1847, Vol. 2, p. 401. Cf. Frercks 2001, 2005 for a lucid, fully documented account of Fizeau's efforts.

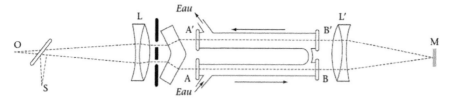

Fig. 4.12.   Fizeau's ether–water experiment. One portion of the light from the source S travels along running
water in the tube AB, and along the running water of the tube A'B' after reflection on the mirror M. The
other portion travels twice against the running water. The two portions interfere in O, with a phase
difference depending on the velocity of the water.

He therefore opted for Fresnel's partially dragged ether, for which the drag is negligible in
an air current but significant in a water current. Returning to the running-water scheme of
1847, after a few months of work he could announce a phase shift compatible with Fresnel's
theoretical value for the dragging coefficient. He prudently concluded:[45]

> The success of this experiment could induce us to admit Fresnel's hypothesis, or at
> least the law he has found to express the change in the velocity of light in a moving
> body. Even though the truth of this law strongly confirms the hypothesis of which it
> is only a consequence, maybe Fresnel's conception will appear so extraordinary, and,
> in some respects, so hard to believe, that one will demand still other proofs and the
> *géomètres'* close scrutiny before adopting it as the expression of reality.

In France, Fizeau's experiment was commonly regarded as a confirmation of Fresnel's
nearly forgotten views on the motion of the ether. In Britain, a single, not so accurate (10%
error) experiment could not weigh much against Stokes's authority. In 1864–67, James
Clerk Maxwell performed a variant of Arago's old prism experiment, with a negative result. He wrongly expected Fresnel's partial drag to imply a positive result and therefore
believed that he had contradicted both Fresnel and Fizeau. After Stokes corrected him in
private, he still did not regard Fizeau's experiment as decisive:

> This experiment seems rather to verify Fresnel's theory of the ether; but the whole
> question of the state of the luminiferous medium near the earth, and of its connexion
> with gross matter, is very far as yet from being settled by experiment.

In his "Ether" article of 1878 for the *Encyclopædia Britannica*, Maxwell still judged Stokes's
hypothesis for the motion of the ether "very probable." Indeed, as we will see in Chapter
5, Maxwell's electromagnetic theory implied a fully dragged ether. In 1885, his disciple Richard Glazebrook still favored Stokes's theory in a British Association report that
nonetheless gave proper attention to Fresnel's theory.[46]

[45] Moigno 1850, pp. 1159–1203 (Fizeau), 1756–1760 (aberration); Fizeau 1851, p. 355; 1859a.
[46] Maxwell [1867], p. 535; Maxwell 1878, p. 571.

4.5 Toward an optical relativity

Optical detection of the earth's motion in the 1860s

Stokes's theory immediately and generally implies the lack of effect of the earth's motion on terrestrial optics. In Fresnel's theory, this motion implies an ether wind and such effects seem plausible, despite the compensatory effects imagined by Fresnel himself. Fizeau kept looking for them. In 1859, he had polarized light pass obliquely through a pile of glass plates and found the rotation of the plane of polarization to depend on the angle between the light beam and the velocity of the earth. In 1862, Babinet predicted that the motion of the earth would affect the position of the fringes produced by a diffraction grating. Two years later, the Swedish physicist Anders Jonas Ångström predicted the same effect and claimed experimental confirmation. These results mattered to astronomers, for they provided a novel access to the velocity of the earth and of the solar system. In 1867, Klinkerfues offered his water-filled telescope for this purpose. In 1868, Huggins detected a Doppler shift in starlight, which gave him a means to measure the radial velocities of stars. In the same year, Hoek criticized Klinkerfues and verified the absence of effect of the earth's motion in a variant of Babinet's interference experiment of 1839. He thus believed that he had confirmed Fresnel's dragging coefficient with a precision of 1%. Interest in the optics of moving bodies was rapidly rising.[47]

German theorems

In the early 1870s, the Bonn mathematics professor Wilhelm Veltmann gave a general proof, based on the Huygens–Fresnel principle, that the Fresnel drag implied the lack of first-order modification of the laws of refraction. In the case of interference, he elegantly proved that the motion of the earth did not affect the phase difference on a closed light path after any number of reflections, refractions, or diffractions. This path is made up of a series of segments s_i, with a fixed value n_i of the optical index in each segment. Owing to the refraction theorem, the shape of this broken path is unaffected by a global motion of the interference setup at the velocity $\mathbf{u}$ for the earth. Along the segment s_i, Veltmann obtains the relative light velocity w_i by vector-composing the specific velocity c/n_i with the velocity $(1 - \alpha_i)u$ of the relative ether drag. In modernized notation, this yields

$$w_i = c/n_i + (1 - \alpha_i)\mathbf{u} \cdot \mathbf{s}_i/s_i. \qquad (4.15)$$

The traveling time along this segment is

$$s_i/w_i \approx n_i s_i/c - (n_i/c)^2(1 - \alpha_i)\mathbf{u} \cdot \mathbf{s}_i. \qquad (4.16)$$

For Fresnel's value of the dragging coefficient α_i, the product $n_i^2(1 - \alpha_i)$ is a constant, so that the second term does not contribute to the total traveling time around the closed

[47] Fizeau 1859b, 1860; Babinet 1862; Ångström 1864; Klinkerfues 1867; Huggins 1868; Hoek 1868. Cf. Hirosige 1976, pp. 12–17. On Huggins, cf. Becker 2011.

broken path: this time and the resulting phase difference are unaffected by the common motion.[48]

Veltmann's Bonn colleague Eduard Ketteler soon treated a broader range of optical devices involving the Doppler effect, reflection on mirror and gratings, anisotropic propagation, dispersion, and intensity rules. These German studies were merely theoretical, despite their empirical motivation. In contrast, the French Academy of sciences, under Fizeau's probable influence, encouraged experimental investigations of the optics of moving bodies by making them the subject of the most prestigious *Grand prix des sciences mathématiques* for the year 1870. The winner was Éleuthère Mascart, then an assistant professor at the Collège de France.[49]

Mascart's optical relativity

In order to decide questions that were still open or undecided, Mascart pushed the limits of precision of contemporary optics through innovative designs and technical virtuosity. Beginning with Babinet's diffraction experiment, he determined that the predicted fringe shift did not occur for a terrestrial source, be it a lamp or reflected solar light (as in Ångström's realization). He still hoped to find other ways to optically detect the motion of the earth. He did not repeat Fizeau's polarization experiment of 1859, which he judged too problematic. Instead he tried effects of the motion of the earth on double refraction and on optical rotation, with negative results. Next, he considered the refraction of the light from a sodium laboratory source by a simple prism, and again found that the orientation of the global setup did not matter. Lastly, he found the same absence of effect of the earth's motion on Newton's rings and on Young's mixed-plate interference.[50]

On the theoretical side, Mascart used the Huygens–Fresnel principle to prove that the motion of the earth did not affect the law of reflection by a mirror or the position of the fringes produced by a diffraction grating both for a terrestrial source and for reflected sunlight. For the prism experiment, he generalized Fresnel's reasoning based on the Fresnel drag as Veltmann had already done. He expected a positive effect for stellar light as a consequence of the Doppler shift (he initially did not realize that Arago's old experiment, being done with an achromatic prism, was unable to detect this shift). For the interference experiments, he simply combined the Fresnel drag with his earlier considerations on reflection. For double refraction and optical rotation, he had the dragging coefficient in the effective velocity formula (Equation (4.12)) depend on the polarization. He noted that this dependence excluded Fresnel's explanation of the drag of waves by a drag of the ether, and

---

[48] Veltmann 1870a; 1870b, pp. 140–144 (interference theorem); 1873. Fizeau (1851, p. 354) had combined reflection and the Fresnel drag to explain the negative result of Babinet's interference experiment (which involved a mirror): cf. Acloque 1991, p. 351.

[49] Ketteler 1873. Cf. Hirosige 1976, pp. 17–22. The French prize question read: "Sur les modifications qu'éprouve la lumière par suite du mouvement de la source lumineuse et du mouvement de l'observateur" (*CR*, 71 (1870), p. 1219; *CR*, 79 (1874), 1531–1534).

[50] Mascart 1872, 1874. Cf. Pietrocola 1992.

he left it "to the mathematicians" to better determine the effect of moving matter on the propagation of light. Here is his last word on the matter:[51]

> The general conclusion of this memoir would then be (if we ignore Fizeau's experiment on the rotation of the plane of polarization through a pile of glass plates) that the translational motion of the earth has no appreciable difference on the optical phenomena produced with a terrestrial source or with solar light, that these phenomena do not give us the means to appreciate the *absolute* motion of a body and that the *relative* motions are the only ones that we can reach.

Boussinesq's derivation of the Fresnel drag

One "mathematician," Joseph Boussinesq, had supplied a better theory of the Fresnel drag a few years before Mascart's call, in a powerful memoir read in 1867. At that time Maxwell's electromagnetic theory of light was still in limbo and the optical medium was still understood as some kind of elastic body. In the first dynamical theories of the 1830s by Cauchy, Stokes, George Green, James MacCullagh, and Franz Neumann, there was only one medium with variable density and elastic constants, and optical dispersion was traced to the discreteness of this medium. Cauchy and Neumann soon considered double-lattice theories in which ether and matter were treated as separate, interacting dynamical entities. In the more refined picture developed by Cauchy and his disciples in the 1850s and 1860s, the molecules of matter modified the properties of the continuous ether between them, in a manner designed to explain dispersion, double refraction, and optical rotation. In 1867, Boussinesq drastically simplified these theories by assuming that the molecular action on the ether did not extend beyond their immediate vicinity.[52]

In analogy with the problem of a suspended rigid particle in a moving fluid, Boussinesq assumes that the molecules of matter are displaced by a fraction α of the instantaneous displacement $\mathbf{d}$ of the surrounding ether (in the simplest case in which the finite dimension of the molecules can be neglected). By the principle of action and reaction, the molecules impress a force $-\sigma\ddot{\mathbf{d}}$ per unit volume on the ether, wherein the effective inertia σ is proportional to the density of matter. The effective equation of motion of the ether of density ρ and elastic constant K reads:

$$\rho\ddot{\mathbf{d}} = K\Delta\mathbf{d} - \sigma\ddot{\mathbf{d}} \tag{4.17}$$

Now suppose the molecules of matter travel through the ether at the common velocity $\mathbf{u}$. To the corresponding motion $\mathbf{r}_0 + \mathbf{u}t$ of a molecule, Boussinesq adds the modulation $\alpha\mathbf{d}(\mathbf{r}_0 + \mathbf{u}t)$ caused by the ether's motion. The molecule's acceleration now being $(\partial/\partial t + \mathbf{u} \cdot \nabla)^2\mathbf{d}$, the equation of motion of the ether becomes

$$\rho\ddot{\mathbf{d}} + \sigma(\partial/\partial t + \mathbf{u} \cdot \nabla)^2\mathbf{d} = K\Delta\mathbf{d}. \tag{4.18}$$

---

[51] Mascart 1874, p. 420.

[52] Boussinesq 1868, 1873. Cf. Darrigol 2012, pp. 256–258.

For a plane monochromatic wave $e^{i(\omega t - \mathbf{k} \cdot \mathbf{r})}$, this leads to the dispersion formula

$$\rho\omega^2 + \sigma(\omega - \mathbf{k} \cdot \mathbf{u})^2 = Kk^2, \tag{4.19}$$

whose first-order approximation in u gives

$$V = \frac{c}{n} + \left(1 - \frac{1}{n^2}\right) u \cos\theta \tag{4.20}$$

for the wave velocity $V = \omega/k$ as a function of the velocity c in vacuum, the index $n = \sqrt{1 + \sigma/\rho}$, and the angle θ between the direction of the wave normal and the velocity of the body. Boussinesq thus pioneered the derivation of the Fresnel drag through the interaction between a stationary ether and point-like molecules moving through it. A few years later, after Mascart's call for a mathematical derivation of the Fresnel drag, Potier published a variant of Boussinesq's theory in which condensed ether replaced the immersed molecules. The basic idea was still the same as Boussinesq's: divide the optical medium into a stationary ether component and a convected, discrete, material component.[53]

The Michelson–Morley experiments

In his "Ether" article of 1878 for the *Encyclopaedia Britannica*, Maxwell raised the question of the motion of the ether and considered measurements of the velocity of light as the first, obvious means to determine this motion:

> If it were possible to determine the velocity of light by observing the time it takes to travel between one station and another on the earth's surface, we might, by comparing the observed velocities in opposite directions, determine the velocity of the aether with respect to these terrestrial stations. All methods, however, by which it is practicable to determine the velocity of light from terrestrial experiments depend on the measurement of the time required for the double journey from one station to the other and back, again, and the increase of this time on account of a relative velocity of the aether equal to that of the earth in its orbit would be only about one hundred millionth part of the whole time of transmission, and would therefore be quite insensible.

Maxwell then recalled the null result of all past ether-drift experiments, except for Fizeau's dubious experiments of 1869 with the polarizing pile of glass plates. As a more promising venue, he suggested measuring the time light takes to travel from Jupiter to the earth for different positions of Jupiter in the sky. On March 19, 1879 he wrote to the director of the Nautical almanac, David Peck Todd, to ask him whether his Jupiter satellite tables could be used to determine the velocity of the solar system through the ether. He thereby repeated his order-of-magnitude argument against using terrestrial measurements of the velocity of light in different directions.[54]

---

[53] Boussinesq 1868; Potier 1876.

[54] Maxwell 1878, p. 570. On Blair's similar unpublished suggestion [1786], see p. 91, this volume.

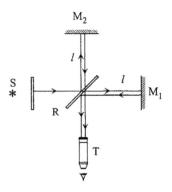

Fig. 4.13.   Michelson's interferometer.

The young American Albert Michelson, who had already collaborated with the astronomer Simon Newcomb on an improvement of Alfred Cornu's method for measuring
the velocity of light, saw Maxwell's letter (posthumously published in *Nature* in 1880) and
judged that what Maxwell had deemed impossible could easily be done through interferom-
etry. During a visit to Helmholtz's laboratory in Berlin in 1881, he constructed the device
shown in Figure 4.13 in which the light from the source S travels through two different
paths and then recombines to form interference fringes seen through the telescope T. The
light on the first path crosses the silvered plate R, bounces on the mirror M_1, and then on
the rear of plate R. The light on the second path bounces on the front of plate R, bounces
again on mirror M_2, and then crosses R. Michelson thus combined his expert knowledge
of French interferometric methods with his signature reliance on a semi-reflecting plate as
a light-splitting device.[55]

If the interferometer moves through a stationary ether, for instance in the direction SM_1
of the first arm, then the mirror M_1 moves forward at the velocity u during the light's travel
from R to M_1 in the ether, and the time needed is $l/(c - u)$ (Figure 4.14a). Similarly, the
time needed for the return trip is $l/(c + u)$, and the round trip takes $(2l/c)(1 - u^2/c^2)^{-1}$. After
moving his experiment to quieter Postdam, Michelson reached the needed stability and
found that a rotation of the interferometer did not cause the expected fringe shift (of about
2% of the fringe spacing). He concluded that the earth dragged the ether along, against
Fresnel's assumption of a stationary ether. At the end of his report, Michelson judged it
"not out of place" to quote Stokes's call of 1846 for a decisive experiment between his and
Fresnel's theory of aberration.[56]

As Potier soon told him in a conversation in Paris, Michelson had overlooked the increase of the light path in the perpendicular arm of the interferometer (see Figure 4.14b),
by a factor $(1 - u^2/c^2)^{-1/2}$. This correction halves the expected fringe shift, making it even
harder to detect. Michelson maintained his conclusion, but failed to convince the French
and British masters of optics (Cornu, Mascart, Thomson, and Rayleigh). Under Rayleigh's
suggestion, he decided to work on a variant of Fizeau's running-water experiment of 1851.

[55] Michelson 1881. Cf. Swenson 1972; Staley 2008, chap. 2.

[56] Michelson 1881, p. 129. For Stokes's call, see p. 105, this volume.

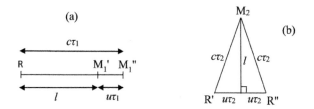

Fig. 4.14.   Light paths in the two arms of Michelson's interferometer, as seen from the ether frame; (a) for the
arm parallel to the motion, (b) for the arm perpendicular to the motion of the interferometer in the ether.
R denotes the position of the semi-reflecting plate; M_1' and M_1'' the positions of the first mirror at the
beginning and end of the light's travel from R to this mirror. R' and R'' denote the initial and final positions
of the plate for a round trip of the light in the perpendicular arm, and M_2 the position of the second mirror
at the time of reflection. τ_1 is the time of the light's travel from R to M_1, τ_2 the time from R to M_2.

The phase difference being of first order in u/c in this case, the measurement promised to
give an easier answer to the question of the drag of ether by matter. With improved interfer-
ometry and the help of the chemist Edward Morley, in 1886 Michelson confirmed Fresnel's
partial drag with a 5% margin of error and concluded that "*the luminiferous ether* [*was*] *entirely unaffected by the motion of the matter which it permeates*," because he assumed, like
Boussinesq and Potier, "that the ether within a moving body remains stationary with the
exception of the portions which are condensed around the particles."[57]

A few months later, in what Michelson rightly called a "very searching analysis" of
the optical effects of the earth's motion, the Dutch theorist Hendrik Lorentz judged the
Fresnel–Fizeau drag in moving transparent bodies to be compatible with two otherwise acceptable assumptions for the motion of the ether near the earth: Fresnel's stationary ether,
and an ether flowing irrotationally around the earth. The second option of course derives
from Stokes's theory of aberration. In a detailed criticism of this theory, Lorentz notes
that Stokes's assumption of a vanishing relative velocity of the ether near the surface of
the earth is kinematically impossible, since an irrotational flow around a rigid body cannot have zero velocity everywhere on the surface of the body. He further excludes Stokes's
appeal to the atmosphere as a means to check the motion of the ether around the earth (the
distance between the molecules of air being too large for that). To Stokes's remark that his
theory after all does not require a nonvanishing ether wind on earth, he objects that the
ether wind would perturb the trajectory of the light rays in the telescope used for observing
the aberration of stars. However, this perturbation can be avoided by assuming Fresnel's
partial ether drag within the lenses of the telescope (or else by making the opaque walls
of the telescope impermeable to the ether). Only future experiments, Lorentz concludes,
could decide between this modified Stokes theory and Fresnel's theory.[58]

[57]Michelson 1882 (with Potier's remark); Michelson and Morley 1886, pp. 386, 378. On pp. 378n–379n,
Michelson offered a kinematic variant of Potier's reasoning, based on the smaller speed of light in the condensed
ether.

[58]Lorentz 1886, 1887.

In England, Michelson and Morley's confirmation of Fizeau's result at last tilted the balance in Favor of Fresnel's theory. In a paper written in 1887, Lord Rayleigh wrote: "The Fresnel hypothesis of a stationary ether appears to be at the present time the more probable ; but the question must be considered to be an open one." Rayleigh called for a repeat of Michelson's experiment of 1881, and also for a test of ether motion induced by the rapid motion of heavy masses.[59]

In the same year, 1887, Michelson and Morley repeated Michelson's experiment of 1881 with a larger interferometer mounted on a gravestone floating in a mercury bath in order to ease rotation and minimize vibrations. They determined that the expected fringe shift was at least twenty times smaller than implied by a stationary ether, with an embarrassing conclusion:

> It appears, from all that precedes, reasonably certain that if there be any relative motion between the earth and the luminiferous ether, it must he small; quite small enough entirely to refute Fresnel's explanation of aberration. . . . Lorentz proposes a modification [of Stokes's theory] which combines some ideas of Stokes and Fresnel, and assumes the existence of a [velocity] potential, together with Fresnel's coefficient. If now it were legitimate to conclude from the present work that the ether is at rest with regard to the earth's surface, according to Lorentz there could not be a velocity potential, and his own theory also fails.

In England, Oliver Lodge picked on Rayleigh's suggestion of testing ether motion near fast-moving masses. He built an "ether whirling machine" in which a Michelson interferometer was sandwiched between two heavy spinning disks of steel. The negative result again favored a stationary ether, while the Michelson–Morley experiment favored an ether dragged by heavy bodies. As Lodge foresaw, physicists would need a new electromagnetic theory of light and matter in order to escape from the accumulating paradoxes of ether motion.[60]

Conclusions

Bradley's discovery of stellar aberration naturally led to relativity considerations in the Newtonian theory of light. Clairaut assumed, as evident, the invariance of the aberration angle when a common velocity is added to the source and the receptor. This enabled him to reduce the aberration caused by the moving earth to the more easily understood aberration caused by a moving star (by choosing the opposite of the velocity of the earth for the commonly added velocity). Another way in which relativity entered corpuscular optics was the consideration of Bošković's and Wilson's water-filled telescope: as Wilson and Robison understood, in the Newtonian theory there cannot be any effect of the water filling because the laws of refraction within the telescope are unaffected by the motion of the earth. Despite Bošković's impermeability to Wilson's reasoning, this invariance was an easily demonstrable consequence of the Galilean invariance to the mechanical laws ruling

---

[59] Rayleigh 1892 [1887], p. 502.
[60] Michelson and Morley 1887, p. 341; Lodge 1892, 1893. Cf. Hunt 1986.

the interaction between matter and light corpuscles. As we will see in Chapter 7, in the early twentieth century Einstein and Ritz similarly capitalized on the Galilean relativity of "emission theories" in which the velocity of the source adds to the velocity of the emitted light.

In the corpuscular theory of light, stellar aberration is so easily understood that it was long regarded as a proof of this theory. In the wave theory, aberration is more difficult to account for, and its proper derivation was still debated in the 1840s by Doppler, Stokes, and Challis. Early derivations all assumed a stationary ether; they figured a simple tube or a goniometer for the star pointing; and they often relied on an implicit analogy between the trajectory of light rays and the trajectory of light corpuscles. Euler, who gave the first wave-theoretical derivation of stellar aberration in 1739, understood that the case of the moving earth and the case of a star moving with the opposite velocity were not exactly symmetric: the aberration angles agree only to first order in the ratio u/c of the earth–star velocity to the velocity of light. The origin of this asymmetry is obvious: in one case the star moves through the ether, in the other the earth moves through the ether. In modern terms, one should not expect the relativity principle to hold in a stationary-ether theory, because the ether then provides a privileged reference of motion. What ought to be more surprising is the approximate, first-order relativity of aberration. Euler was not overly surprised because he knew that this relativity exactly held in the corpuscular theory.

A similar situation occurs in the case of the Doppler effect. As Doppler showed in 1842, this effect depends only on the relative motion of the source and the receptor to first order in the relative velocity. The symmetry is exact for the analogous Blair–Arago effect in the corpuscular theory: the velocity of the source and the velocity of the receptor affect the apparent velocity (and the refraction) of the light corpuscles in exactly opposite manners. The negative result of Arago's experiment of 1810 for detecting this corpuscular effect meant, once reinterpreted in wave-theoretical context, that the laws of refraction on earth were not affected by the velocity of the earth. Again, this seems incompatible with the intuition of a stationary ether through which the earth is moving. In 1818, Fresnel nonetheless showed that the motion of the earth had no first-order effect on the laws of refraction if the ether was partially dragged by moving transparent matter, in the proportion $1 - 1/n^2$ depending on the optical index n.

In the cases of stellar aberration and Doppler shift, first-order relativity was known to derive from the wave theory, but the means to test it were lacking. In the case of refraction, the lack of effect of the earth motion was a surprising experimental result before Fresnel and Stokes found theoretical explanations. If we believe Mascart's retrospective judgment, Arago's experiment did not have the precision required to exclude an effect of the order u/c.[61] Arago's null result nevertheless motivated Fresnel's invention of the partial drag. After a long period of neglect, this concept informed debates on the effect of the motion of the earth on terrestrial optics. It became clear that in Fresnel's theory the motion of the earth should have no first-order effect on refraction, interference, and diffraction at least as long as polarization and dispersion are ignored. Fragmentary proofs of this theoretical fact could be found in Fresnel's and Stokes' writings; more general proofs were given by

---

[61] Mascart 1872, p. 159.

Veltmann, Ketteler, and Mascart around 1870. These authors explained the negative result of a variety of ether-drift experiments: Babinet 1839 regarding glass-shifted interference, Hoek 1869 and Airy 1871 regarding a water-filled telescope, and Mascart 1874 regarding Newton's rings and Young's mixed plates. The few positive results did not withstand criticism: Fizeau doubted his own difficult experiment of 1859 with a polarizing pile of glass plates; Ångström's result of 1864 with a diffraction grating and Klinkerfues's result of 1867 with a water-filled telescope failed to be replicated by Mascart and Airy, respectively. In the few cases left undecided by the theory, concerning double refraction and optical rotation, Mascart obtained negative results. Having confirmed all previous negative experiments and discredited all previous positive experiments, he concluded that the relative motions of bodies were the only ones that could be detected optically.

By the 1870s, Fresnel's theory had succeeded in accounting for all observations and experiments of the optics of moving bodies, including stellar aberration and the absence of effects of the motion of the earth on terrestrial optics. At what price? As Cauchy, Doppler, and Stokes deplored, Fresnel's assumption of the stationary ether required the enormous mass of the earth to be fully permeable to the ether. Young's wind-in-the-grove analogy did not help much until, toward the end of the century, evidence accumulated in favor of a highly lacunar structure of matter. As for the ether's partial drag in transparent bodies, it could easily pass for an ad hoc hypothesis, even after Stokes derived the amount of drag through the conservation of the flux of ether at the interface between two media. The situation began to change when, in 1851, Fizeau's running-water experiment confirmed the Fresnel drag experimentally. In addition, in 1867 Boussinesq showed that this drag resulted from a dynamical theory in which the ether was regarded as stationary everywhere except in the immediate vicinity of the molecules of the transparent body. Variants of this reasoning by Potier in 1876 and by Michelson in 1886 strengthened the idea that the hypothesis of the stationary ether, when properly extended at the molecular scale, accounted for all optics of boding bodies.

In the stationary ether theory, aberration is relatively easy to understand (despite the aforementioned difficulties) because the optical vibrations of the ether are not disturbed by any convective effect, whereas the absence of (first-order) effects of the motion of the earth on terrestrial optics comes at the end of a fairly complex reasoning. The reverse is true in Stokes's theory of 1845. Out of distaste for an ether-permeable earth, Stokes proposed that the ether moved together with the earth in its vicinity. In this case, the ether is at rest in our laboratories and strictly no effect of the earth's motion should be expected in terrestrial optics. Aberration can still be explained if the flow of the ether is everywhere irrotational in the space between the star and the earth. As Stokes emphasized, this is a striking case of equally successful explanations of a wide class of phenomena by two utterly different theories.

As Lorentz noted in 1886 and as Stokes must have known, the irrotational flow of an incompressible fluid around a sphere necessarily involves a shift of the fluid on the surface of the sphere. This did not bother Stokes, for he assumed the molecules of air around the earth to be crowded enough to prevent the shift. British physicists favored his theory over Fresnel's until Michelson and Morley confirmed the Fresnel–Fizeau drag in 1886. As Lorentz then noted, there was an additional reason to favor Fresnel's theory: in the now well-established theory of gases, the molecules of the air around the earth are too far from each other to block the ether shift in the manner imagined by Stokes. There still was a possible escape from contradiction with experience: one could assume the Fresnel drag to occur in optical apparatus subjected to the local ether wind on earth. In this modified Stokes theory, the ether cannot penetrate the earth and it necessarily circulates around it, but it penetrates transparent bodies to the extent admitted by Fresnel. Although this option was logically possible, it was more complex than Fresnel's and therefore less appealing.

In a portion of space subjected to a constant ether wind, the propagation of light becomes anisotropic: it is faster alongside the wind, slower against the wind. When, in 1881, Michelson failed to detect the resulting second-order effect for the round trips of light in an interferometer, he concluded in favor of Stokes's fully dragged ether (he had not yet encountered Lorentz's argument against shiftless drag). After becoming aware of flaws in this experiment and after confirming the Fresnel–Fizeau drag with Morley in 1886, he converted to Fresnel's stationary ether and its molecular-level generalization à la Boussinesq–Potier. The following year, with Morley, he perfected his first experiment and confirmed the inexistence of the second-order anisotropy of the velocity of light on earth. Now he was totally perplexed: the Fresnel–Fizeau drag experiment seemed to confirm the stationary ether, whereas the anisotropic-propagation experiment did not detect the expected ether wind! The modified Stokes theory was not entirely excluded, because the ether shift could have been negligible at the time and place of observation, but this was not a probable option.

To sum up, the altogether most satisfactory theory, Fresnel's, accounted for the first-order invariance of optical phenomena with respect to the motion of the earth, but predicted a second-order effect on the propagation of light. Mascart and many others confirmed the former prediction; Michelson and Morley refuted the latter. The year of this paradoxical refutation, 1887, was also the year of Heinrich Hertz's confirmation of Maxwell's electromagnetic theory. In the following years there was a growing suspicion that the key to the paradoxes of ether motion should be sought in the electromagnetic theory of light. In other words, the optics of moving bodies had to be part of a broader electrodynamics of moving bodies, which is the subject of the next chapter.

5

THE ELECTRODYNAMICS OF MOVING BODIES

> These considerations clearly show against what difficulties the treatment of the
> electrodynamics of a system of moving bodies has to fight today. Seemingly well-
> established ideas, which we used to count as facts, appear to be unreliable as
> soon as we take a critical look at the material, and we begin to suspect that dur-
> ing the future development of science perhaps much of what we now regard as
> the unshakeable foundations of our vision of Nature might lose the aureole of
> unconditional validity. And yet it is precisely on these difficulties that we should
> base our confident hope for new insights.[1] (August Föppl, 1894)

Electrodynamics, in André Marie Ampère's original definition, is the science of the forces
exerted by electricity in motion. These forces induce a mechanical motion of conductors
and magnets. Reciprocally, the motion of a conductor near a magnet induces an electric
current. Electrodynamics is, therefore, in conformity with its name, a science of motion.
Questions about the reference of motion, in particular whether a common motion imparted
to all the bodies of a system affect the phenomena, emerged spontaneously in this domain.

Just as in the case of optics, physicists until the late nineteenth century sought to reduce
electrodynamics to the better-known science of mechanics. They did so in two different
ways: through direct action at a distance in France and in Germany, and through the internal motions of a mechanical ether in Britain. In the first kind of reduction, the theory
automatically satisfies Galilean relativity and even a stronger relativity according to which
all observable (electromotive and electrodynamic) forces depend only on the relative motion of the bodies of the system. In the second kind of reduction, the situation is similar to
that of wave optics: the phenomena a priori depend on the bodies' motion with respect to
the invisible ether, unless compensation mechanisms operate. Historically, the compensation was sometimes anticipated in the theory, and sometimes artificially imposed in order
to comply with an observed invariance. Toward the end of the century, optics became
part of electrodynamics, and the latter theory therefore had to accommodate the relativity
of optical phenomena. As we will see in the two following chapters, this accommodation
largely contributed to the emergence of relativity theory.

Section 5.1 of this chapter recalls the origins of electrodynamics in the 1820s: Ørsted's
current–magnet action, Ampère current–current interaction, and Faraday's electromagnetic rotation and induction. Section 5.2 covers German attempts to unify electrodynamics
physically and mathematically in the 1840s: through a fundamental law for the force between moving electric particles in Weber's case, and through the concept of electrodynamic

---

[1]Föppl 1894, p. 311.

Relativity Principles and Theories from Galileo to Einstein. Olivier Darrigol, Oxford University Press.
© Olivier Darrigol (2022). DOI: 10.1093/oso/9780192849533.003.0005

potential in Neumann's case. In both cases relativity appeared directly in the foundations, with electromagnetic induction in mind. Section 5.3 is about British field theories from the 1850s to the 1880s: Maxwell's theory based on Faraday's lines of force and on Thomson's mechanical analogies, and British Maxwellian theories, with an emphasis on Heaviside's full electrodynamics of moving bodies. Section 5.4 recounts the early German reception of Maxwell's theory: Helmholtz's reformulation in a distance–action context in 1870, and Hertz's confirmation, purification, and extension in the late 1880s, leading to equations similar to Heaviside's for the electrodynamics of moving bodies. Maxwell, Heaviside, and Hertz broadly commented on the relativity of motion in their theories. Maxwell and Hertz assumed ether and matter to form a single compound, continuous medium, with a common velocity varying from place to place. Although this conception implied the relativity of a large class of electrodynamic phenomena, it still allowed for effects of absolute motion. Section 5.5 details specific proposals for detecting such effects, either in thought experiments or in real experiments. Section 5.6 shows how, in the last third of the century, Maxwell's assumption of a single ether–matter medium proved insufficient when theorists confronted electrolysis, discharge in rarefied gases, and a variety of optical phenomena implying dispersion, moving bodies, and magneto-optics. The atomistic structure of matter and electricity, already assumed by Weber and his disciples, now entered field theories. In the 1890s, Lorentz, Wiechert, and Larmor assumed a strictly stationary ether in which atoms, ions, and electrons freely circulated, with field-mediated interactions regulated by the Maxwell–Lorentz equations. This image permitted a new take on the electrodynamics and moving bodies, with original ideas and techniques for complying with the observed relativity of phenomena.

5.1 Early electrodynamics

In 1820, the Copenhagen physics professor Hans Christian Ørsted observed the deflection of a magnetic needle by an electric current, a phenomenon soon called electromagnetism. Since a current and a magnet both act on a magnet, Ampère then reasoned, a current might also act on a current, and a magnet might be nothing but an assemblage of microscopic currents. Ampère verified the force between currents, and, by a clever mix of experience and rational guess, arrived at the law we would now write as

$$d^2\mathbf{f} = -ii'\frac{\mathbf{r}}{r}\left[\frac{d\mathbf{l}\cdot d\mathbf{l}'}{r^2} - \frac{3}{2}\frac{(\mathbf{r}\cdot d\mathbf{l})(\mathbf{r}\cdot d\mathbf{l}')}{r^4}\right],\tag{5.1}$$

for the force $d^2\mathbf{f}$ between the current element $i\,d\mathbf{l}$ and the current element $i'\,d\mathbf{l}'$, $\mathbf{r}$ being the vector joining the second element to the first. Ampère compared or measured his forces by static means, and he initially ignored the resulting motions.[2]

[2] Ørsted 1820; Ampère 1820a, 1820b, 1820c, 1820d. Cf. Blondel 1982; Steinle 2005.

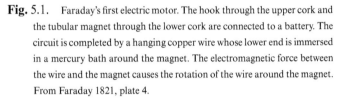

Fig. 5.1.   Faraday's first electric motor. The hook through the upper cork and the tubular magnet through the lower cork are connected to a battery. The circuit is completed by a hanging copper wire whose lower end is immersed in a mercury bath around the magnet. The electromagnetic force between the wire and the magnet causes the rotation of the wire around the magnet. From Faraday 1821, plate 4.

Ampère thus started an electrodynamics of bodies at rest: in this domain the conductors and magnets do not move, although electricity moves within them. In the fall of 1821, Michael Faraday carefully examined the action of a horizontal magnetic needle on a vertical current-carrying wire, and determined that the wire tended to rotate around each pole of the needle. He soon modified the setup to allow the wire to freely and continuously circulate around the pole (see Fig. 5.1). This was the first electric motor ever built, as well as the birth of the electrodynamics of moving bodies.[3]

Ten years later, in 1831, Faraday wrapped two coils around the same iron ring, connected the first coil to a switch and a battery, and the second coil to a primitive galvanometer made of a straight wire above a compass needle. He observed a brief current in the secondary when opening or closing the primary. Through various mutations of this device, he determined that a variable current, a moving magnet, and a moving current carrier could just as well induce a current in a secondary circuit. He also found that an electromotive force was induced in a wire moving near a magnet at rest. These were natural expectations for Faraday because, since the beginning of his electric researches, he had been representing the action of magnets and currents through "magnetic curves" everywhere tangent to a test compass needle. In every case of electromagnetic induction, he traced the induced current in a wire to the crossing of the magnetic curves by this wire. He later determined that the electromotive force in the wire was proportional to the number of magnetic lines of force crossed by the wire per unit time, provided that the lines of force are spaced in inverse proportion to the intensity of the magnetic force. The lines of force accompany a magnet in its motion, so that the induced current in a nearby wire depends only on the relative motion of the magnet and the wire.[4]

[3] Faraday 1821. Cf. Gooding 1985.

[4] Faraday 1831, 1832. Cf. Darrigol 2000a, pp. 31–37, and further references therein.

5.2 German action at a distance

Neumann's theory

Until the mid-1840s, there were four classes of electric or magnetic phenomena: electro-static forces, magnetic forces, electrodynamic forces, and electromagnetic induction. Save for Faraday's field-based rules—which no one took seriously—there was no unified theory. In 1845, the Königsberg professor Franz Neumann offered a mathematical unification of the two last classes of phenomena. He starts with Emil Lenz's moderation law, according to which the electromotive force in a moving wire is directed so that the magnetic force on the induced current counteracts the motion of the wire. As a refined expression of this law, Neumann proposes the relation

$$\mathbf{E}_d \cdot d\mathbf{l} = -\mathbf{v} \cdot d\mathbf{f}_1 \qquad (5.2)$$

between the electromotive force $\mathbf{E}_d$ in a wire element $d\mathbf{l}$ moving at the velocity $\mathbf{v}$ and the electromagnetic force $d\mathbf{f}_1$ that would act on the wire element if it were fed by a unit current. For a linear inducing current i', Neumann obtains the force $d\mathbf{f}_1$ by integrating Ampère's force law over this circuit. He then computes the electromotive force $\oint \mathbf{E}_d \cdot d\mathbf{l}$ in a closed circuit, with the result

$$e = \frac{dP(1, i')}{dt}, \quad \text{with} \quad P(i, i') = -\oint \oint \frac{ii'd\mathbf{l} \cdot d\mathbf{l}'}{|\mathbf{r} - \mathbf{r}'|}. \qquad (5.3)$$

Neuman calls $P(i, i')$ the electrodynamic potential of the two linear currents i and i', because its variation during a deformation of the first circuit yields the electrodynamic force on the elements of this circuit.[5]

The argument is so far confined to induction in a moving circuit. For induction in a circuit at rest by a moving circuit or magnet, Neumann simply assumes that induction depends on relative motion only:

> Induction can only depend on the relative motion of the elements [of the conductors]. Indeed, to the motion of two elements one may superimpose a shared motion such that the one or the other element remains at rest, and this shared motion cannot have any inductive effect; otherwise the motion of the earth would by itself induce a current in a conductor at rest near another current [also at rest].

For induction by a current varying from i' to $i' + di'$, he imagines the circuit to be sent to infinity with the intensity i', and then brought back from infinity with the intensity $i' + di'$, thus reducing this case to the former case. Altogether, the time derivative of the potential gives the desired value of the electromotive force, whatever the cause of the variation of the potential.[6]

---

[5]Neumann 1846. Cf. Jungnickel and McCormmach 1986, Vol. 1, pp. 148–152; Olesko 1991; Darrigol 2000a, pp. 43–49. For a modern proof of Equation (5.3), see p. 123, this volume.

[6]Neumann 1846, p. 22.

Weber's theory

While Neumann was unifying electrodynamics through a noninterpreted potential between noninterpreted currents, his Leipzig colleague Wilhelm Weber was working on a fuller unification through a picture of the electric current as a symmetric double flow of positive and negative electricity. Following a suggestion by Gustav Fechner, Weber reduced all electrostatic and electrodynamic interactions to distance forces between two electric fluid particles. In this picture, electrostatic and electrodynamic forces are indirect consequences of the forces between the electric fluids contained in matter. Weber assumes the elementary force between two particles of electricity e and e' to be a function of their distance r and of the time derivatives $\dot{r}$ and $\ddot{r}$. As he shows, the Ampère force law between two arbitrary current elements holds if only the elementary force is a central force of intensity

$$f = \frac{ee'}{r^2}\left(1 - \frac{\dot{r}^2}{2c^2} + \frac{r\ddot{r}}{c^2}\right). \tag{5.4}$$

The first term in the parenthesis yields the Coulomb force, and the two other terms yield the electrodynamic force. The constant c has the dimension of velocity. Weber interprets it as the ratio between the electromagnetic and electrostatic units of electric charge, for which there is no dimensioned coefficient in Ampère's and Coulomb's laws, respectively. This constant can be measured by discharging a capacitor of known electrostatic charge into a ballistic galvanometer. With Rudolph Kohlrausch's help in the delicate electrostatic measurement, in 1857 Weber found 3.10×10^8 m/s to be the value of c. He noted the proximity to the velocity of light, only to emphasize the disparity of meanings.[7]

When the current carriers move or when the currents vary, the Weber forces acting on the positive and the negative electricity at a given point of the conductor no longer balance each other. The difference between these two forces represents the electromotive force of induction. As Weber proves, this way of reasoning exactly reproduces Neumann's predictions for closed linear currents, the only accessible ones in those times. In addition, Weber's theory explains electric resistance as some kind of friction between the electric fluids and the matter through which they circulate. Magnetism is addressed through Amperean microcurrents. Weber could thus claim a unifying picture of all phenomena of electricity and magnetism, based on a single fundamental law. This law was not as simple as Coulomb's law, since it involved the mutual velocity and acceleration of the electric fluid particles. Weber and Gauss then regarded this dependence as a symptom of a mediated interaction, with a medium that might serve to propagate light. Weber further noted a welcome property of his law:[8]

> The fundamental electric law . . . makes the action of an electric mass on another depend only on their *relative* distance, velocity, and acceleration. . . .As a consequence, the same induction must result from the same relative motion of the two elements [of

---

[7] Weber 1846; Weber and Kohlrausch 1857. Cf. Whittaker 1951, pp. 201–205; Darrigol 2000a, pp. 54–66. Weber's c differed by a factor $\sqrt{2}$ from the modern c used in this paragraph.

[8] Weber 1846, p. 143.

current], no matter which of the two elements is in [a state of] *absolute* rest. As is well known, this result is also in agreement with experiments.

Bridging Neumann's and Weber's theories

In 1857, Neumann's disciple Gustav Kirchhoff completed Weber's proof of a connection between the two German electrodynamics, thereby introducing the useful vector potential.[9] In modern notation, Neumann's theory naturally leads to the vector potential

$$\mathbf{A}(\mathbf{r}) = \int \frac{\mathbf{j}(\mathbf{r}')}{|\mathbf{r} - \mathbf{r}'|}\,d\tau' \tag{5.5}$$

for the current density $\mathbf{j}(\mathbf{r})$. The variation of the total potential $-i \oint \mathbf{A} \cdot d\mathbf{l}$ of a linear current during a deformation $\delta\mathbf{l}$ of the circuit gives

$$\delta P = -i \oint \delta\mathbf{l} \cdot [d\mathbf{l} \times (\nabla \times \mathbf{A})], \tag{5.6}$$

in agreement with the Amperean expression

$$d\mathbf{f} = i\,d\mathbf{l} \times (\nabla \times \mathbf{A}) \tag{5.7}$$

of the electromagnetic force on the current element $i\,d\mathbf{l}$.

Neumann's relation (Equation (5.2)) between induction in a moving conductor and electrodynamic force then gives

$$\mathbf{E_d} \cdot d\mathbf{l} = -\mathbf{v} \cdot (d\mathbf{l} \times \mathbf{B}) = \mathbf{B} \cdot (d\mathbf{l} \times \mathbf{v}), \quad \text{with } \mathbf{B} = \nabla \times \mathbf{A}. \tag{5.8}$$

The latter expression represents the flux of the vector $\mathbf{B}$ across the surface element swept by the element $d\mathbf{l}$ during a unit time. Integrating over a closed circuit, and using the divergenceless character of $\mathbf{B}$, we get

$$e = \oint \mathbf{E_d} \cdot d\mathbf{l} = -\frac{d}{dt} \iint \mathbf{B} \cdot d\mathbf{S}, \tag{5.9}$$

in conformity with Faraday's rule of the cut lines of force. For any kind of electromagnetic induction, Neumann's theory requires

$$e = -\frac{d}{dt} \oint \mathbf{A} \cdot d\mathbf{l} = \oint \left(-\frac{\partial \mathbf{A}}{\partial t} + \mathbf{v} \times (\nabla \times \mathbf{A})\right) \cdot d\mathbf{l}, \tag{5.10}$$

in agreement with

$$\mathbf{E_d} = -\frac{\partial \mathbf{A}}{\partial t} + \mathbf{v} \times (\nabla \times \mathbf{A}) \tag{5.11}$$

for the general expression of the electromotive force of induction.

[9] Kirchhoff 1857.

Weber and Kirchhoff found Weber's law to imply the electromotive force

$$\mathbf{E} = -\nabla\phi - \frac{\partial \mathbf{A_K}}{\partial t} + \mathbf{v} \times (\nabla \times \mathbf{A_K}), \qquad (5.12)$$

with

$$\phi(\mathbf{r}) = c^2 \int \frac{\rho(\mathbf{r'})}{|\mathbf{r}-\mathbf{r'}|}d\tau' \text{ and } \mathbf{A_K}(\mathbf{r}) = \int \frac{\mathbf{r'}}{r'^3}[\mathbf{r'} \cdot \mathbf{j}(\mathbf{r}-\mathbf{r'})]d\tau'. \qquad (5.13)$$

Besides the electrostatic contribution $-\nabla\phi$, Weber's theory departs from Neumann through a different expression of the vector potential. As is easily seen, the two expressions differ by a gradient only, so that the difference affects open currents only. Despite the disparity of approach, the two theories are both based on direct action at a distance, they are both compatible with Ampère's force law, and they both have electromagnetic induction depend on relative motion only, at least for closed circuits. In both cases, a relativity principle contributed to the theoretical construction, at the macro-level of current carriers and magnets for Neumann, and at the micro-level of electric fluid particles for Weber.

5.3 British field theories

While in Germany Neumann and Weber were designing post-Newtonian theories of electrodynamics, in Britain William Thomson and James Clerk Maxwell took Faraday's field conception seriously and turned it into a mathematical theory of electricity and magnetism. For Faraday, electric and magnetic phenomena essentially depend on the lines of force between the source charges, currents, and magnets. These lines of force represent a "polarization" (or "induction") in the "field" or intermediate space, to an extent depending on the presence of matter in the field and on the electric or magnetic state of the sources. In the 1840s, William Thomson compared these lines with the streamlines of a fluid (heat or other) circulating in a heterogeneous medium between sources (in a more literal sense), and used this analogy to reformulate the continental distance–action theories in harmony with Faraday's views.[10]

On Faraday's lines of force

In 1855–56, Maxwell perfected this analogical method to introduce four basic electromagnetic field quantities: the electric force $\mathbf{E}$, the magnetic force $\mathbf{H}$, the electric polarization $\mathbf{D}$, and the magnetic polarization $\mathbf{B}$. In a rough operational definition, $\mathbf{E}(\mathbf{r})$ is the force that would act on a unit-point charge at $\mathbf{r}$ in the absence of matter, and $\mathbf{H}(\mathbf{r})$ the force that would act on a unit magnetic pole (extremity of a uniformly magnetized needle) at $\mathbf{r}$ in the absence of matter. The electric polarization is directly related to the charge density ρ through $\nabla \cdot \mathbf{D} = \rho$, and the magnetic polarization satisfies $\nabla \cdot \mathbf{B} = 0$ as there are no

---

[10]Cf. Wise 1981.

magnetic charges. The relations $\mathbf{D} = \varepsilon\mathbf{E}$ and $\mathbf{B} = \mu\mathbf{H}$ hold in a linear medium of "dielectric permittivity" ε and "magnetic permeability" μ. In a medium of conductivity σ, Maxwell also assumes the linear relation $\mathbf{j} = \sigma\mathbf{E}$ expressing Ohm's law for the conduction current $\mathbf{j}$.

From Ampère's law and from Ampère's representation of a magnetic dipole by a small current loop, Maxwell derives

$$\oint \mathbf{H} \cdot \mathrm{d}\mathbf{l} = \iint \mathbf{j} \cdot \mathrm{d}\mathbf{S} \tag{5.14}$$

for the magnetic force associated with the electric current density $\mathbf{j}$, the line integral being taken along the boundary of the surface across which the current is flowing. This gives

$$\nabla \times \mathbf{H} = \mathbf{j} \tag{5.15}$$

for an infinitesimal loop. Maxwell next expresses Faraday's rule of the cut lines of force as

$$\oint \mathbf{E} \cdot \mathrm{d}\mathbf{l} = -\frac{\mathrm{d}}{\mathrm{d}t} \iint \mathbf{B} \cdot \mathrm{d}\mathbf{S} = -\frac{\mathrm{d}}{\mathrm{d}t} \oint \mathbf{A} \cdot \mathrm{d}\mathbf{l}, \text{ with } \mathbf{B} = \nabla \times \mathbf{A}. \tag{5.16}$$

Since for Maxwell there is only one ether–matter medium, the material points of the loop over which the electromotive force is integrated travel at the velocity $\mathbf{v}$ of the medium (which may vary from point to point). Confusing a convective derivative (more on this in a moment) with a total derivative, Maxwell takes this to imply the expression $-\mathrm{d}\mathbf{A}/\mathrm{d}t$ for the electromotive force of induction. To recapitulate, by 1856 Maxwell had the system[11]

$$\mathbf{D} = \varepsilon\mathbf{E}, \quad \mathbf{B} = \mu\mathbf{H}, \quad \mathbf{j} = \sigma\mathbf{E}, \quad \nabla \cdot \mathbf{D} = \rho, \quad \nabla \cdot \mathbf{B} = 0 \tag{5.17}$$

$$\nabla \times \mathbf{H} = \mathbf{j}, \quad \mathbf{B} = \nabla \times \mathbf{A}, \quad \mathbf{E} = -\mathrm{d}\mathbf{A}/\mathrm{d}t - \nabla\psi. \tag{5.18}$$

On physical lines of force

So far Maxwell's theory was evidently limited to closed currents, since the relation $\nabla \times \mathbf{H} = \mathbf{j}$ implies $\nabla \cdot \mathbf{j} = 0$. Moreover, Maxwell deplored the lack of a mechanical model for Equation (5.18). Unlike Faraday, who regarded his lines of force as independent, primitive, physical entities, Thomson and Maxwell had them represent the state of a mechanical medium of yet unknown constitution. Maxwell designed the wanted model of the electromagnetic medium in a memoir of 1861–1862 entitled "On physical lines of force."[12]

As Maxwell first recalls in this memoir, a few years earlier Faraday had discovered the rotation of the plane of polarization of light when passing through a magnetized glass, and Thomson had interpreted this rotation as a consequence of tiny eddies in the magnetic

---

[11] Maxwell 1856. Cf. Wise 1979; Harman 1998; Darrigol 2000a, pp. 139–147. Maxwell introduced $\mathbf{A}$ as the mathematical expression of the "electrotonic state" that Faraday believed to precede induced currents.

[12] Maxwell 1861–63. Cf. Siegel 1991.

medium. In a fluid densely populated with locally aligned eddies, Maxwell reasons, the pressure has to be anisotropic owing to the centrifugal force in a direction perpendicular to the axis of rotation. Calling **H** the local rotation velocity, the pressure system takes the form

$$p_{ij} = -p\delta_{ij} + \mu H_i H_j, \tag{5.19}$$

where p is the hydrostatic pressure, and μ is proportional to the density of the fluid. The resulting force density, $f_i = \partial_j p_{ij}$, is given by

$$\mathbf{f} = (\nabla \cdot \mu\mathbf{H})\mathbf{H} + (\nabla \times \mathbf{H}) \times \mu\mathbf{H} + \mu\nabla(\tfrac{1}{2}H^2) - \nabla p. \tag{5.20}$$

Identifying **H** with the magnetic force and μ with the magnetic permeability, the first term vanishes, and the second represents the electromagnetic force on the current $\nabla \times \mathbf{H}$.[13]

In order to connect the rotations of consecutive eddies, Maxwell somewhat rigidifies them and introduces a layer of particles that play the role of idle wheels between them, in the manner indicated on Fig. 5.2. The hexagonal shape of the rotating cells is just for space filling. When the magnetic force varies in space, the rotations of two consecutive cells differ, the particles roll between them with a linear velocity proportional to $\nabla \times \mathbf{H}$. Identifying this drift with the electric current, Maxwell gets $\nabla \times \mathbf{H} = \mathbf{j}$. On average, the cells exert a force **E** on the particles, to be interpreted as the electromotive force. By the equality of action and reaction, the particles exert a tangential force proportional to $-\mathbf{E}$ on each side of a cell. When **E** varies from place to place, these forces produce a torque increasing or diminishing the rotation of the cell. Maxwell thus gets

$$\nabla \times \mathbf{E} = -\frac{\partial(\mu\mathbf{H})}{\partial t}, \tag{5.21}$$

in conformity with his earlier expression (5.16) of Faraday's induction law.

Again the theory is restricted to closed currents, since $\nabla \times \mathbf{H} = \mathbf{j}$. In order to allow for divergent currents, Maxwell relaxes this kinematic relation by means of a slight elastic deformation of the cells. This move has two other advantages: it warrants the rolling without sliding of the particles between the cells, and it allows for elastic waves that may perhaps represent light. The tangential force $-\mathbf{E}$ of the particles causes a proportional (orthoradial) elastic deformation $-\varepsilon\mathbf{E}$ of the cells. Owing to this deformation the current of particles becomes

$$\mathbf{j} = \nabla \times \mathbf{H} - \frac{\partial \varepsilon\mathbf{E}}{\partial t}. \tag{5.22}$$

Taking the divergence of this equation, we get

$$\nabla \cdot \mathbf{j} + \frac{\partial\rho}{\partial t} = 0, \quad \text{with } \rho = \nabla \cdot (\varepsilon\mathbf{E}). \tag{5.23}$$

---

[13] The pressure system p_{ij} is defined so that, in modern tensor notation, $p_{ij}\mathrm{d}S_j$ represents the i-component of the oblique pressure on the surface element $\mathrm{d}\mathbf{S}$. The indices i and j vary from 1 to 3, and summation over the repeated index j is understood.

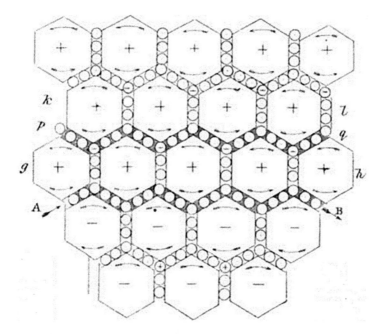

Fig. 5.2. Maxwell's cellular model. From Maxwell 1861, plate.

The model now permits an accumulation of particles to be interpreted as electric charge, and this charge agrees with the electrostatic definition if only the elasticity constant ε^{-1} of the cells is interpreted as the dielectric permittivity. In addition, the elastic medium of the cells allows for elastic waves propagating at the velocity $(\varepsilon\mu)^{-1/2}$. In a vacuum, Maxwell's expressions for the electromagnetic and electrostatic forces give another interpretation of the same quantity as the ratio c between the electromagnetic and the electrostatic unit of electric charge, which Weber and Kohlrausch have found to be close to the velocity of light. Maxwell concludes: "We can scarcely avoid the inference that *light consists in the transverse undulations of the same medium which is the cause of electric and magnetic phenomena.*"[14]

Lastly, Maxwell makes his medium move together with the matter it contains. By construction, the current **j** of particles, the rotation **H** of the cells, and their deformation $-\varepsilon$**E** should accompany the medium in its motion. Therefore, in the integral equations

$$\oint \mathbf{E} \cdot d\mathbf{l} = -\frac{d}{dt} \iint \mathbf{B} \cdot d\mathbf{S} \quad \text{and} \quad \oint \mathbf{H} \cdot d\mathbf{l} = \frac{d}{dt} \iint \mathbf{D} \cdot d\mathbf{S} + \iint \mathbf{j} \cdot d\mathbf{S} \quad \text{(with } \mathbf{D} = \varepsilon\mathbf{E}\text{)},$$

(5.24)

the surfaces over which the double integrals are taken should follow the motion of the medium at the local velocity **v**. This gives

---

[14] Maxwell 1861–1862, p. 22.

$$\nabla \times \mathbf{E} = -\frac{D\mathbf{B}}{Dt}, \quad \nabla \times \mathbf{H} = \mathbf{j} + \frac{D\mathbf{D}}{Dt}, \quad \text{with} \quad \frac{D\mathbf{X}}{Dt} = \frac{\partial \mathbf{X}}{\partial t} - \nabla \times (\mathbf{v} \times \mathbf{X}) + \mathbf{v}(\nabla \cdot \mathbf{X})$$

$$(5.25)$$

for the convective derivative D/Dt of a flux $\mathbf{X}$. Having no interest in moving dielectrics, Maxwell derives only the first of these equations. Had he also given the second, he would have obtained the full electrodynamics of moving bodies later formulated by Hertz and Heaviside.[15]

Maxwell's final theory

Although Maxwell insisted on the possibility of a consistent mechanical representation of electromagnetic interactions, he did not take every feature of his cellular model seriously. The hexagonal cells and their gearing mechanism seemed an implausible candidate for the true constitution of the electromagnetic medium. Yet there are two features of the model in which Maxwell never ceased to believe: the cellular rotation associated with the magnetic field, and the mechanical connection between electric current and cellular rotation. In a third memoir of 1865, Maxwell proved that these two features were sufficient to derive his field equations when combined with Faraday's concept of electric charge and current.[16]

The idea was to apply Lagrangian dynamics to the connected system defined by the currents and the cellular rotation. According to Lagrange, for a connected system whose energy is purely kinetic, it is enough to know the kinetic energy as a function of the generalized coordinates and velocities of the system in order to derive the equations of motion. In the simplest case of a system of kinetic energy $T(q, \dot{q})$ subjected to the generalized force f_q, Lagrange's equation reads

$$f_q = \dot{p} - \frac{\partial T}{\partial q}, \quad \text{with} \quad p = \frac{\partial L}{\partial \dot{q}}. \tag{5.26}$$

That is to say, any change of the generalized momentum p calls for the inertial force $-\dot{p}$. In Maxwell's hidden field mechanism, the generalized velocities are electric currents of density $\mathbf{J}$, and they generate a hidden rotation of velocity $\mathbf{H}$ through the relation $\nabla \times \mathbf{H} = \mathbf{J}$. The associated kinetic energy is

$$T = \frac{1}{2} \int \mu H^2 d\tau = \frac{1}{2} \int \mathbf{H} \cdot (\nabla \times \mathbf{A}) d\tau = \frac{1}{2} \int \mathbf{A} \cdot (\nabla \times \mathbf{H}) d\tau = \frac{1}{2} \int \mathbf{J} \cdot \mathbf{A} d\tau, \tag{5.27}$$

in which the vector potential $\mathbf{A}$ is defined through $\nabla \times \mathbf{A} = \mu \mathbf{H}$. The current $\mathbf{J}$ being divergenceless, it can be decomposed into closed current filaments. Maxwell takes the

---

[15] On convective derivatives, cf. Darrigol 2000a, pp. 406–409.

[16] Maxwell 1865. Cf. Buchwald 1985, pp. 60–64; Siegel 1991, chap. 6.

intensity of these filament as independent velocities in the Lagrangian. For a given filament of intensity i, the associated momentum is:

$$p = \frac{\partial T}{\partial i} = \oint \mathbf{A} \cdot d\mathbf{l}. \tag{5.28}$$

This is why Maxwell calls $\mathbf{A}$ the electromagnetic momentum. The inertial force $-\dot{p}$ gives the electromotive force e in the elementary circuit as

$$e = -\frac{d}{dt} \oint \mathbf{A} \cdot d\mathbf{l}. \tag{5.29}$$

Again, the current being part of a mechanism that follows the bulk motion of the medium, the elements of the loop of the former integral travel with the velocity $\mathbf{v}$ of the medium. For the electromotive force $\mathbf{E}$ of which e is the integral, this gives

$$\mathbf{E} = -\frac{\partial \mathbf{A}}{\partial t} + \mathbf{v} \times (\nabla \times \mathbf{A}) - \nabla \psi. \tag{5.30}$$

The kinetic energy of the system is also a function of the spatial coordinates of the current carriers. Calling q such a coordinate, the associated force is $\partial T / \partial q$ according to Lagrange. For a filamentary current of intensity i, the force $d\mathbf{f}$ acting on the element $d\mathbf{l}$ of the circuit is obtained by varying the shape of the filament. The part $i \oint \mathbf{A} \cdot d\mathbf{l}$ of the Lagrangian that contributes to this variation being the sign-reversed Neumann potential, this gives

$$d\mathbf{f} = i d\mathbf{l} \times (\nabla \times \mathbf{A})$$

in agreement with Equation (5.7).

So far the theory concerns closed currents only. In order to generalize to open currents, Maxwell cannot relax the mechanical connections without leaving the province of the Lagrangian method. Instead he appeals to Faraday's concept of electricity, according to which every current is closed even when the chain of conductors is interrupted. As was already mentioned, Faraday made a polarization in the space between conductors the essence of electricity. He regarded this polarization as a primitive concept, and not, as we would do now, as a microscopic displacement of elastically bound electricity. He *defined* electric charge as the termination of the lines of force representing the polarization. Typically, a charge appears at the surface of a conductor because the electric lines of force cannot penetrate the conductor. In addition, Faraday assumed that every variation of the polarization constituted an electric current. In a conductor, an electromotive source can temporarily induce a polarization, but this polarization promptly decays with the production of heat. A conduction current corresponds to a rapid alternation of polarization and depolarization. When an electric condenser discharges through a condenser, the current in the conductor is continued by a dielectric current in the gap of the condenser because the polarization in this gap decreases. Every current is closed.

In conformity with this profoundly original view of electricity, Maxwell now defines electric charge as a consequence of polarization, through the equation $\rho = \nabla \cdot \mathbf{D}$ (a discontinuity of the polarization $\mathbf{D}$ across a surface thereby implies a distribution of charge on this surface).[17] He defines the total current $\mathbf{J}$ as the sum of the conduction current $\mathbf{j}$ and of the time derivative of the polarization:

$$\mathbf{J} = \mathbf{j} + \frac{\partial \mathbf{D}}{\partial t}, \quad \text{with} \quad \mathbf{j} = \sigma \mathbf{E} \quad \text{and} \quad \mathbf{D} = \varepsilon \mathbf{E}. \tag{5.31}$$

The current thus defined is divergenceless as a consequence of the definition $\rho = \nabla \cdot \mathbf{D}$ of the electric charge and of the conservation of electricity $\nabla \cdot \mathbf{j} + \partial \mathbf{D}/\partial t = 0$. There now is, in addition to the kinetic energy of the hidden magnetic rotation, a potential energy $V = \frac{1}{2} \int \varepsilon E^2 d\tau$ of the polarization $\mathbf{D}$, and there is a force $\rho \mathbf{E}$ acting on electrically charged matter. In this more abstract version of this theory, Maxwell no longer needs the elastic cells to propagate light. Instead he shows that proper combinations of his field equations in a homogeneous medium lead to wave equations for the vectors $\mathbf{E}$ and $\mathbf{H}$, with the propagation velocity $(\varepsilon\mu)^{-1/2}$.

We may now recapitulate the equations of Maxwell's system, as Maxwell himself did in his *Treatise* of 1873 (in quaternion notation):

$$\nabla \times \mathbf{H} = \mathbf{J}, \quad \mathbf{J} = \mathbf{j} + \frac{\partial \mathbf{D}}{\partial t}, \quad \mathbf{B} = \nabla \times \mathbf{A}, \quad \mathbf{E} = -\frac{\partial \mathbf{A}}{\partial t} + \mathbf{v} \times (\nabla \times \mathbf{A}) - \nabla \psi$$

$$\rho = \nabla \cdot \mathbf{D}, \quad \nabla \cdot \mathbf{B} = 0$$

$$\mathbf{f} = \mathbf{j} \times \mathbf{B} + \rho \mathbf{E}$$

$$\mathbf{D} = \varepsilon \mathbf{E}, \quad \mathbf{B} = \mu \mathbf{H} + \mathbf{M}, \quad \mathbf{j} = \sigma \mathbf{E}.$$

$$\tag{5.32}$$

Here $\mathbf{f}$ is the force density, and $\mathbf{M}$ the rigid magnetic polarization needed to represent magnets. The potential ψ is indirectly determined through $\nabla \cdot \mathbf{D} = \rho$. Maxwell recognizes the gauge freedom in the choice of $\mathbf{A}$, although he usually sets $\nabla \cdot \mathbf{A} = 0$. In addition, he shows that the force density derives from the stress system

$$p_{ij} = D_i E_j - \tfrac{1}{2}\delta_{ij} \mathbf{E} \cdot \mathbf{D} + B_i H_j - \tfrac{1}{2}\delta_{ij} \mathbf{B} \cdot \mathbf{H}, \tag{5.33}$$

in conformity with Faraday's intuition of a tension along the lines of force and a mutual pressure across them.[18]

---

[17] In 1865, Maxwell wrote $\rho = -\nabla \cdot \mathbf{D}$ using a misleading analogy with the Poisson–Thomson–Mossotti concept of polarization as a microscopic displacement of electric charge. In his *Treatise* of 1873, he corrected the sign but still introduced an imaginary incompressible, space-pervading fluid of which $\mathbf{J}$ is the current (cf. Buchwald 1985, chap. 1). In a dielectric (including vacuum), the shift of this fluid is elastically resisted (Maxwell then calls it "displacement"), whereas in a conductor the elastic resistance continually breaks down, thus permitting a permanent circulation of the fluid. Maxwell confused many of his readers by calling this fluid "electricity," whereas it does not carry any electric charge. This is why I avoid this notion in the following.

[18] Maxwell 1873, secs. 237–238. Maxwell inadvertently used $-\nabla\psi$ instead of $\mathbf{E}$ in the force formula.

As an electrodynamics of moving bodies, Maxwell's system of equations is incomplete, for he does not take into account the convection of the polarization $\mathbf{D}$ (which generates the convection current $\rho\mathbf{v}$). He nonetheless investigates how his equation for electromagnetic induction transforms when passing to a moving system of axes. In a frame whose points move at the velocity $\mathbf{u}$ with respect to the original frame, he finds

$$\mathbf{E}' = -\frac{\partial \mathbf{A}'}{\partial t} + \mathbf{v}' \times \mathbf{B}' - \nabla(\psi - \mathbf{u} \cdot \mathbf{A}'), \tag{5.34}$$

as should be expected from the interpretation of the electromotive force as a convective derivative of the vector potential. Maxwell concludes: "In all phenomena relating to closed circuits and the currents in them, it is indifferent whether the axes to which we refer the system are at rest or in motion." To draw this conclusion, Maxwell implicitly assumes that the relation between the vector potential and the conduction current remains the same in the moving frame. This is true to the extent that the displacement current can be neglected.[19]

British Maxwellians

After the publication of Maxwell's *Treatise* in 1873 and even more after his untimely death in 1879, Maxwell's theory spread widely in Britain thanks to the efforts of a few diligent disciples. Two of them, the Liverpool experimentalist Oliver Lodge and the Irish theorist George Francis FitzGerald, offered mechanical illustrations of Maxwell's difficult concepts of electric charge and current. FitzGerald developed models of the electromagnetic ether compatible with Maxwell's equations, and used them to complete the electromagnetic theory of light. Two other disciples of Maxwell, the Cambridge-educated physicist John Henry Poynting and the self-educated telegraphic engineer Oliver Heaviside, shunned mechanical modeling and focused on energetic relations.[20]

In 1884, Poynting introduced the energy flux $\mathbf{E} \times \mathbf{H}$ that bears his name. Around the same time, Heaviside simplified Maxwell's equations by "murdering" the potentials, by eliminating all the 4π's (as I have already done above), and by using a vector notation of his own. By balancing the "activity" (work developed in a unit time) $\mathbf{J} \cdot \mathbf{E}$ of the force $\mathbf{E}$ with the activity $\mathbf{v} \cdot \mathbf{f}$ of the electromagnetic force density $\mathbf{f}$ and by analogy between the electric and magnetic fields, he obtained the "duplex equations"

$$\nabla \times (\mathbf{H} - \mathbf{D} \times \mathbf{v} - \mathbf{h}) = \mathbf{j} + \frac{\partial \mathbf{D}}{\partial t} + \rho\mathbf{v}, \quad \nabla \times (\mathbf{E} - \mathbf{v} \times \mathbf{B} - \mathbf{e}) = -\frac{\partial \mathbf{B}}{\partial t}, \tag{5.35}$$

in which $\mathbf{v}$ is the velocity of matter when there is any, $\mathbf{e}$ is the electrochemically impressed electric force, and $\mathbf{h}$ the magnet-impressed magnetic force. These equations provide a fully consistent electrodynamics of moving bodies. Through ingenious mathematical techniques of his own, even anticipating distribution theory, Heaviside solved many problems of telegraphic and electrotechnic interest.[21]

---

[19] Maxwell 1873, secs. 600–601. See pp. 139–140, this volume, for Des Coudres's different conclusion.

[20] Cf. Hunt 1991.

[21] Poynting 1884; Heaviside 1885–87, 1886–87, 1888–89. Cf. Yavetz 1995.

In the electrodynamics of moving bodies, one historically important problem was the electrically charged particle in uniform rectilinear motion. The Cambridge physicist J. J. Thomson first attacked this problem in 1881, having in mind the charged molecules that William Crookes took to be projected from the cathode in a vacuum tube. In Thomson's view, the motion of an electrified particle implies a proportional displacement current. The magnetic energy of this current being proportional to the square of the velocity of the particle, it contributes to the mass of the particle. When the particle is immersed in an external magnetic field, the cross-term of the magnetic field energy implies an electromagnetic force through the Lagrangian method. Unfortunately, Thomson used a wrong expression of the displacement current and the resulting formulas for the electromagnetic mass and force had incorrect numerical coefficients. FitzGerald and Heaviside corrected him, finding $\mu q^2/6\pi a$ for the electromagnetic mass of a uniformly electrified spherical shell of radius a and charge q, and $q\mathbf{u} \times \mathbf{B}$ for the electromagnetic force on a particle moving at the velocity $\mathbf{u}$.[22]

Unlike the other Maxwellians, Heaviside doubted that the electromagnetic ether should accompany matter in its motion, and he did not use this assumption in his derivation of the duplex equations. In reaction to J. J. Thomson's reliance on this assumption, he wrote:

> This is somewhat beyond me. I do not yet know certainly that the ether can move, or its law of motion if it can. Fresnel thought the earth could move through the ether without disturbing it; Stokes, that it carried the ether along it, by giving irrotational motion to it. Perhaps the truth is between the two. Then there is the possibility of holes into the ether, as suggested by a German philosopher. When we get into one of these holes, we go out of existence. It is a splendid idea, but experimental evidence is much wanting.

In 1889, Heaviside simplified the problem of the moving charge by assuming that the ether remained completely at rest while the electrification moved at the velocity $\mathbf{u}$ of the particle. The corresponding field equations read

$$\nabla \times \mathbf{H} = \rho\mathbf{u}, \quad \nabla \times \mathbf{E} = -\frac{\partial \mathbf{B}}{\partial t}. \tag{5.36}$$

Whereas earlier computations of the field of the moving charge all assumed a small velocity, Heaviside solved this system exactly. For this purpose, he introduced the comoving frame in which the equations take the form

$$\nabla \times (\mathbf{H} - \mathbf{D} \times \mathbf{u}) = \mathbf{0}, \quad \nabla \times (\mathbf{E} - \mathbf{u} \times \mathbf{B}) = \mathbf{0}. \tag{5.37}$$

Axial symmetry and $\nabla \cdot \mathbf{B} = 0$ imply $\nabla \cdot (\mathbf{H} - \mathbf{D} \times \mathbf{u}) = -\mathbf{u} \cdot (\nabla \times \mathbf{D}) = 0$, which, combined with the first duplex equation, gives $\mathbf{H} = \mathbf{D} \times \mathbf{u}$. Injecting this relation into the second duplex equation, we get

[22] J. J. Thomson 1881; FitzGerald 1881; Heaviside 1885–87, p. 466; 1889, pp. 505–506. Cf. Darrigol 1993b, pp. 303–309.

$$\nabla \times (\gamma^2 E_x, E_y, E_z) = \mathbf{0}, \quad \text{with } \gamma = (1 - u^2/c^2)^{-1/2}. \tag{5.38}$$

Introducing the primed coordinates and the primed electric field such that $x' = \gamma x$ and $E'_{x'} = \gamma E_x$ (the other components being unchanged), the former equation and the equation $\nabla \cdot \mathbf{E} = (q/\varepsilon)\delta(\mathbf{r})$ for the electric field of a point charge transform into

$$\nabla' \times \mathbf{E}' = 0, \quad \nabla' \cdot \mathbf{E}' = (q/\varepsilon\gamma)\delta(\mathbf{r}'), \quad \text{hence } \mathbf{E}' = q\mathbf{r}'/4\pi\varepsilon\gamma r'^3 \text{ and } \mathbf{E} = q\mathbf{r}/4\pi\varepsilon\gamma r'^3. \tag{5.39}$$

The field of the moving point charge therefore differs from the field of a charge at rest by a contraction in the direction of motion, in the proportion $(1 - u^2/c^2)^{1/2}$. Heaviside thus pioneered the use of simplifying coordinate transformations in the electrodynamics of moving bodies.[23]

5.4 Maxwell in Germany

Helmholtz's potential theory

Despite the importance of the British reception of Maxwell's theory, the first important physicist who publicly addressed this theory was Hermann Helmholtz, in the early 1870s in Berlin. In electro-physiological experiments performed in the late 1860s, Helmholtz had confirmed that electric shocks did not reach the deeper-lying nerves of the body. With a Du Bois induction coil, a bath of salted water, and the naked nerve of a frog leg (as a current detector), he showed that the electric penetration diminished for faster impulses. These observations called for a theory of open currents. Helmholtz was aware of three candidates: Neumann's, Weber's, and Maxwell's. In a large memoir of 1870, he compared these three theories by focusing on the vector potential $\mathbf{A}$ from which the induced currents derives. From Neumann and Kirchhoff, he knew that the motion of electricity in a conductor generally obeys

$$\mathbf{j} = \sigma\left(-\nabla\phi - \frac{\partial \mathbf{A}}{\partial t}\right), \tag{5.40}$$

with

$$\phi(\mathbf{r}) = c^2 \int \frac{\rho(\mathbf{r}')}{|\mathbf{r} - \mathbf{r}'|} d\tau', \quad \mathbf{A}(\mathbf{r}) = \int \frac{\mathbf{j}(\mathbf{r}')}{|\mathbf{r} - \mathbf{r}'|} d\tau' + \frac{1-k}{2}\nabla\xi, \quad \xi = -\int \nabla \cdot \mathbf{j}(\mathbf{r}') |\mathbf{r} - \mathbf{r}'| d\tau'. \tag{5.41}$$

The $\nabla\xi$ contribution to the vector potential vanishes for closed currents, and it has no effect on a closed circuit. As Helmholtz proves, the cases $k = 1, -1, 0$ correspond to the predictions of Neumann's, Weber's, and Maxwell's theories, respectively (for the motion of electricity in an infinitely extended conductor). Helmholtz excludes the Weber option as it leads to an unstable equilibrium of electricity in conductors.[24]

---

[23] Heaviside 1888–89, p. 497(citation), Part 4 (moving charge).

[24] Helmholtz 1870a. Cf. Darrigol 2000a, pp. 214–233.

In the rest of his memoir, Helmholtz considers the effect of dielectrics (and magnetized matter) around the conductors. As William Thomson had done long ago, he imagines that the bound electricity of the dielectric is displaced from its equilibrium position by electric force, and then acts as a source for the electric field. In addition, he assumes that a variation of the polarization contributes to the electric current on which the vector potential depends. Lastly, he considers the case of a polarizable vacuum. Taking the limit of infinite polarizability and renormalizing electric charge to compensate for its screening by the polarized medium, he finds that the polarization vector obeys the same equations as Maxwell's **D**. He concludes:[25]

> The remarkable analogy between the motions of electricity in a dielectric and those of the luminiferous ether do not depend from the special form of Maxwell's hypotheses; it can be obtained in an essentially similar manner if we maintain the older view of electric actions at a distance.

Helmholtz's exclusion of Weber's theory prompted responses from Weber and his allies. In France, Joseph Bertrand defended Ampère's law against Helmholtz's extension of the bare potential theory (without the polarized medium) to open currents. In this extension, the variation of the potential $P = -i\int_1^2 \mathbf{A} \cdot d\mathbf{l}$ of an open linear current yields not only the force density $i d\mathbf{l} \times (\nabla \times \mathbf{A})$ along the current but also the forces $-i\mathbf{A}(\mathbf{r}_1)$ and $i\mathbf{A}(\mathbf{r}_2)$ at its extremities. These forces are unknown to Ampère's theory. Also, the time derivative of $-\int_1^2 \mathbf{A} \cdot d\mathbf{l}$ should yield the electromotive force of induction $\int_1^2 \mathbf{E}_d \cdot d\mathbf{l}$ whatever be the cause of variation of this integral. This implies

$$\mathbf{E}_d = -\frac{\partial \mathbf{A}}{\partial t} + \mathbf{v} \times (\nabla \times \mathbf{A}) - \nabla(\mathbf{v} \cdot \mathbf{A}). \tag{5.42}$$

The gradient term was unknown to Neumann's and Weber's theory. As a consequence, the net electromotive force in a blade rotating in a uniform magnetic field parallel to the axis of rotation should vanish, whereas in any other theory it had the value given by Faraday's rule of the cut lines of force. Helmholtz performed this crucial experiment in 1875 (see Fig. 5.3), and he managed to detect a small electric charge at the end of the blade, in conformity with Faraday's rule. He thus refuted his own preferred theory. Two options were left to him: either the potential theory supplemented with a large vacuum polarization (in order to retrieve Maxwell's predictions) or a more complex theory based on Ampère's law. Helmholtz favored the first option and privately developed a full electrodynamics of moving bodies based on it. In the limit of infinite vacuum polarizability, the equations were the same as the yet unknown Hertz–Heaviside equations.[26]

---

[25] Helmholtz 1870a, p. 558.

[26] Helmholtz 1872, 1873a, 1873b, 1874a, 1874b, 1875a. For a full reconstruction of Helmholtz's theory, cf. Darrigol 2000a, pp. 412–419.

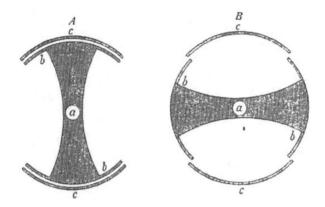

Fig. 5.3. Helmholtz's crucial experiment. The part *bb* rotates around the axis *a*. Thanks to a rotating commutator, the fixed curved plates *c* and *c* are grounded whenever the rotating blade is in the position *A*, so that the apparatus acts as a condenser *bc* charging under a given electromotive force. In the position *B*, they are instead connected to a high capacity condenser, which accumulates the charges developed in the position *A*. After a large number of cycles, the charge of the latter condenser is tested with an electrometer. From Helmholtz 1875, p. 583.

Hertz's discovery and theory

In Helmholtz's reinterpretation of Maxwell's theory, the two crucial assumptions are the magnetic action of polarization currents and the possibility of polarizing a dielectric (even a vacuum) through an electromotive force of induction. In 1879, the Berlin Academy proposed a prize for an experimental test of these two effects. Helmholtz's star pupil, Heinrich Hertz, considered the problem but quickly determined that the frequencies needed to produce appreciable polarization currents were much higher than could be realized at that time. A few years elapsed before he discovered, in 1886–1887, how to build an oscillator at frequencies we would now call Hertzian. He was thus able to verify the first effect of the prize question, and then to verify a joint consequence of the first and second effects: that electromagnetic induction takes time to propagate from the source to the detector. In February 1888, he was able to produce stationary electromagnetic waves and to measure their length. As the results confirmed the limiting case in which Helmholtz's theory retrieves the predictions of Maxwell's theory, Hertz now favored the latter theory for its greater simplicity.[27]

In two large theoretical memoirs of 1890, Hertz reformulates Maxwell's theory as a system of equations for the two vectors **E** and **H** in a medium characterized by the parameters ε, μ, σ, and the velocity **v**. He discards any picture of electric charge and current, and defines them formally as $\nabla \cdot (\varepsilon \mathbf{E})$ and $\sigma \mathbf{E}$, respectively, after defining the vectors **E** and **H** operationally as the forces on a unit test-charge and a unit pole, respectively. For the electrodynamics of moving bodies, he has the field (its lines of force) follow the motion of the medium, leading to the equations

---

[27] Hertz 1887a, 1887b, 1888a, 1888b, 1888c, 1889. Cf. Buchwald 1994.

$$\frac{1}{c}\frac{D\mu\mathbf{H}}{Dt} = -\nabla \times \mathbf{E}, \quad \frac{1}{c}\frac{D\varepsilon\mathbf{E}}{Dt} = \nabla \times \mathbf{H} - \frac{\sigma}{c}\mathbf{E}, \quad \text{with} \quad \frac{D\mathbf{X}}{Dt} = \frac{\partial\mathbf{X}}{\partial t} - \nabla \times (\mathbf{v} \times \mathbf{X}) + \mathbf{v}(\nabla \cdot \mathbf{X})$$

$$\text{(5.43)}$$

for the convective derivative D/Dt of a flux. As Hertz notes, the convective derivative of $\varepsilon\mathbf{E}$ generates the convection current $\rho\mathbf{v}$ as well as the polarization current $\nabla \times (\mathbf{D} \times \mathbf{v})$, whose magnetic action had been verified by Henry Rowland in 1875 and Wilhelm Röntgen in 1888, respectively.[28]

In order to derive the force density of the medium, Hertz examines the variation in time of the electromagnetic energy in a volume element of the medium. The postulated energy density being $\frac{1}{2}\varepsilon E^2 + \frac{1}{2}\mu H^2$, one term of the variation corresponds to the flux of the Poynting vector $\mathbf{E} \times \mathbf{H}$ across the surface of the element, and the other to the work $p_{ij}\partial_i v_j$ of the Maxwell stresses (compare with Equation (5.33))

$$p_{ij} = \varepsilon(E_i E_j - \tfrac{1}{2}\delta_{ij}\mathbf{E}^2) + \mu(H_i H_j - \tfrac{1}{2}\delta_{ij}\mathbf{H}^2). \tag{5.44}$$

The force density is the divergence of these stresses. It contains a term $c^{-2}\,\partial(\mathbf{E} \times \mathbf{H})/\partial t$ that may act on the medium even where there is no electric charge or current. As Hertz notes, when the medium contains even a small amount of matter (rarefied air), its inertia is large enough to make the resulting motion undetectable. In a strict vacuum, the ether would need to have a finite inertia in order to respond to this force, a property Hertz judges to have "little intrinsic probability." In a subsequent note to his memoir, he recognizes that in his theory a radiating body could lose momentum even in a vacuum, thus violating the conservation of momentum unless the ether carries the missing momentum. He comments: "Such consequences seem little probable, but in this domain we have no right to judge from probabilities only, so complete is our ignorance of eventual motions of the ether."[29]

For the time being, Hertz followed Maxwell in assuming electromagnetic processes to reside in a single ether–matter medium of variable velocity. This was the crux of his derivation of the field equations, and this implies that observable phenomena can depend only on the relative motion of the various parts of the medium, not on their absolute motion. Hertz explained:

> Our derivation of the [field] equations did not require that the system of coordinates should be absolutely at rest in space. We may therefore, without change of form, transform our equations from the original coordinate system to any other system moving arbitrarily in space, if only by [**v**] we understand the velocity [of the medium] with respect to the new system of coordinates, and if at any instant we refer the direction-dependent constants ε, μ, σ to this system. It follows that the absolute motion of a rigid system of bodies has no influence on any of the internal electrodynamic processes, as long as all the bodies under consideration, including the ether, take part in the motion.

---

[28] Hertz 1890a, 1890b.

[29] Hertz 1890b, p. 284; 1892, p. 295. Hertz takes the example of a magnetized steel sphere rotating around an axis different from the axis of magnetization. The emitted radiation causes the sphere to lose angular momentum. Helmholtz later attempted and failed to compensate for the Hertz force with a pressure gradient of the ether (Helmholtz 1893b).

Hertz knew that his and Maxwell's hypothesis of a single ether–matter medium, although it had never been contradicted by electrodynamic phenomena, was incompatible with the partial ether drag observed in Fizeau's experiment. For this reason and also because of the undesired ether momentum, Hertz judged his theory to be merely provisional:[30]

> This theory shows how the electromagnetic phenomena in moving bodies can be treated under certain arbitrary limitations. There is little chance that nature would comply with these limitations. The true theory is much more likely to be one that distinguishes the states of the ether from the states of the imbedded matter at every point. However, it seems to me that a theory building on this intuition would require more numerous and more arbitrary assumptions than the theory I have developed.

After reading Hertz's memoir, Heaviside changed his mind about the relation between ether and matter. He admitted that Hertz's simple, unitary assumption permitted a direct derivation of the duplex equations and removed any ambiguity in the determination of electromagnetic forces. He agreed that in a vacuum the momentum $\partial\,(\mathbf{D} \times \mathbf{B})/\,\partial\,t$ was communicated to the ether in a unit time, but in his mind this did not necessarily imply a translational motion of the ether, because other forces could simultaneously result from the structure of the ether (elastic forces, etc.). More problematic in his eyes was the conflict of this assumption with the optics of moving bodies:[31]

> Now, a really difficult and highly speculative question, at present, is the connection between matter (in the ordinary sense) and ether. When the medium transmitting the electrical disturbances consists of ether and matter, do they move together or does the matter only partially carry forward the ether which immediately surrounds it? Optical reasons may lead us to conclude, though only tentatively, that the latter may be the case; but at present, for the purpose of fixing the data, and in the pursuit of investigations not having especially optical bearing, it is convenient to assume that the matter and ether in contact move together. This is the hypothesis made by Hertz in his recent treatment of the electrodynamics of moving bodies.

5.5 Effects of absolute motion

As we have seen, all electrodynamic theories from Neumann to Hertz had electromagnetic induction depend on relative motion only. Neumann assumed this property to construct his theory, Weber derived it from the form of his fundamental force law, Helmholtz from the existence of the electrodynamic potential, and Maxwell from the transformation properties of his field equations. As was sometimes remarked on, its negation would imply the possibility of detecting absolute motion through a circuit and magnet in mutual rest. Did this relativity of induction phenomena extend to other electrodynamic phenomena? Yes in Weber's theory, since Weber's force law depends only on the relative motion of the implied

---

[30] Hertz 1890b, pp. 261–262, 285.

[31] Heaviside 1891–1892, pp. 524 (citation), 557–558 (on the Hertz force). Around the same time, J. J. Thomson ascribed the momentum $\mathbf{D} \times \mathbf{B}$ to moving tubes of electric force; cf. Darrigol 2000a, pp. 295–298.

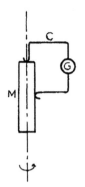

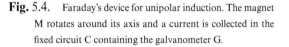

Fig. 5.4. Faraday's device for unipolar induction. The magnet M rotates around its axis and a current is collected in the fixed circuit C containing the galvanometer G.

particles of electricity. Yes too in Helmholtz's bare potential theory (without a polarizable medium), because the electrodynamic potential depends on the relative position of the implied currents only and because in this theory electric charge does not participate in electrodynamic interactions (convection currents do not have any magnetic effects). In contrast, in Maxwell's theory there is a medium that is reduced to the ether outside material bodies. One should therefore expect effects of a global motion of the material system with respect to the unmoved parts of the ether.

Unipolar induction

A special kind of electromagnetic induction is created by having a cylindrical magnet rotate around its axis. The magnetic field is unaffected by the rotation, and there is an electromotive force $(\boldsymbol{\omega} \times \mathbf{r}) \times \mathbf{B}$ in the substance of the magnet if $\boldsymbol{\omega}$ denotes its angular velocity and $\mathbf{r}$ the radius vector from the axis. The resulting current produces minute electric charges on the north and south ends of the magnet. In 1831, Faraday believed that he had proven this effect of absolute rotation by collecting the induced charges through sliding contacts in the manner of Fig. 5.4. This is what Weber later called unipolar induction, because only one pole of the magnet seems to be involved. In 1885, the telegraphic engineer Samuel Tolver Preston noted that the same induced current would flow in the sliding wire if the magnetic lines of force were rotating together with the magnet. Under this assumption, it becomes obvious that no current occurs when the collecting circuit C and the magnet M rotate together (this is also true in Maxwell's theory, since the magnetic flux in the closed circuit C+M is invariable): "I only suppose," Preston wrote, "that in reference to inductive effects with currents we have to do with *relative* and not at all with *absolute* motions." Also, no charge should appear at the ends of the magnet when it rotates by itself, with no collecting circuit: the effect imagined by Faraday does not exist. Although electric engineers often espoused Preston's naïve idea of comoving lines of force, Maxwellian field theorists rejected it and therefore admitted the reality of this effect of absolute rotation.[32]

[32] Faraday 1832, pp. 64–65; Weber 1839; Preston 1885, p. 133. Cf. Miller 1981b; Darrigol 1993b, pp. 298– 301. The absolute translation of a magnet also induces surface charges: see ibid. p. 333, and p. 139, this volume, for FitzGerald's consideration of this case.

The twin charges

Another effect of absolute motion occurred to Maxwell while he was comparing the two main systems of electric units for his *Treatise* of 1873. Having derived the equality of the velocity of light with the ratio between the electromagnetic and electrostatic charge units, he surmised that something peculiar should happen when an electric charge moves at this velocity. For the sake of computability, he considered two parallel uniformly electrified sheets and had them move at the same constant velocity in their own plane. Although he had not introduced the convection current in his field equations, he assumed that the two current sheets interacted magnetically just as conduction currents do. Consequently, the Coulomb repulsion of the two planes is counteracted by an Ampère attraction that increases with their common velocity. When this velocity reaches the velocity of light, the net force vanishes.[33]

Maxwell of course did not hope this experiment would be realized. Instead he suggested measuring the magnetic action of a fast rotating, electrified disk. This is what the American physicist Henry Rowland did two years later in Helmholtz's laboratory in Berlin. The result was important because only one theory, Weber's, clearly implied the magnetic activity of convection currents. Helmholtz understood (before anyone) that in a polarizable medium the magnetically active current includes the convective derivative of polarization, which contains the convection currents, but his own potential theory (without the polarizable vacuum) excluded them.[34]

In 1882, FitzGerald cited Maxwell's thought experiment with the twin charges as further proof of the existence of the ether:

> Two quantities of electricity moving in the same direction with the velocity of light would have no action on one another, their electrostatic action being balanced by an equal and opposite electrokinetic action. As it is unlikely that anything depends on absolute motion, the motion here spoken of must be with respect to something, and this something can hardly be any other thing than what is known as the ether in space.

FitzGerald further noted that the motion of the earth through the ether, under Fresnel's assumption of a stationary ether, should affect electrostatic experiments. Yet he did not expect this motion to produce a force between an electric charge and a magnet in relative rest: as he showed, the electromagnetic interaction between the magnet and the moving charge is exactly compensated for by the electrostatic interaction between this charge and the charges electromagnetically induced by the motion of the magnet in its own field.[35]

Des Coudres's ether-drift experiment

In 1889, the Leipzig experimentalist Theodor Des Coudres had a new idea to detect the astronomical motion of the earth electrodynamically, in analogy with Fizeau's old idea to

---

[33] Maxwell 1873, secs. 769–770.

[34] Rowland 1878; Helmholtz 1874a, 1876 (report on Rowland). Cf. Buchwald 1985, pp. 74–77.

[35] FitzGerald 1882, p. 319. Cf. Darrigol 1993b, pp. 334–335.

compare the intensities of the light from a terrestrial source at equal distance in opposite directions. In his experiment, Des Coudres interrupted the current flowing in two similar parallel coils with opposite windings, and looked for a current induced in a third coil in mid-space. Owing to the retarded propagation of the electrodynamic potential, he expected an effect of the order u/c, u being the velocity of the system of coils through the ether. Through refined measurements, he found that the velocity u could not exceed 1/200th of the velocity of the earth on its orbit. He thus believed that he had confirmed Michelson and Morley's recent optical result. Contrary to Fresnel's view, the ether seemed to be fully dragged by the earth.[36]

Heaviside riding a light wave

A most evident though merely imagined effect of the ether wind on electromagnetic phenomena concerns electromagnetic waves. In 1891, Heaviside examined how the electromagnetic field of a plane wave would appear when referred to a space moving with the velocity $\mathbf{c}$ of the wave. The field becomes static, and it obeys the duplex equations in the ether wind $-\mathbf{c}$:

$$\nabla \times (\mathbf{H} + \mathbf{D} \times \mathbf{c}) = \mathbf{0}, \quad \nabla \times (\mathbf{E} + \mathbf{c} \times \mathbf{B}) = \mathbf{0}, \tag{5.45}$$

in conformity with the well-known relations of the electric and magnetic vectors in an electromagnetic plane wave. The only difference is that these vectors, instead of being functions of $\mathbf{r} - \mathbf{c}t$, are functions of $\mathbf{r}$. This is what we would expect in an ether theory in which field vectors are intrinsic properties of the ether. Like Hertz, Heaviside insisted that the "fixed space" to which we refer electromagnetic phenomena could only be arbitrary, as a consequence of the "relativity of motion (the absolute motion of the universe being quite unknown, if not inconceivable)." In the present problem, a fixed space bound to the source of the electromagnetic wave is more customary, but one moving together with the wave is equally permissible.[37]

Föppl's perplexities

Heaviside had a German admirer, August Föppl, professor at the Technische Universität in Munich. The first edition of his influential book on Maxwell's theory appeared in 1894. Like Heaviside, Föppl constructs the theory by means of energetic principles, and he refrains as much as possible from speculating on ether motion. Yet he begins his chapter on the electrodynamics of moving bodies with an extensive discussion of the problems of space and motion in electrodynamic context. In his view, optics and electrodynamics alter the received concept of space, because space can never be emptied of its ether content. Whether ether follows the motion of matter or remains stationary then is "perhaps the

---

[36] Des Coudres 1889; Fizeau 1852. In 1895, Lorentz proved the theoretical prediction to be erroneous (see p. 145, this volume).

[37] Heaviside 1893, pp. 42–43 [March 1891], 53 [May 1891].

most pressing question of contemporary science." At any rate, absolute motion acquires a precise physical meaning as motion with respect to the undisturbed parts of the ether (far away from any matter). The ether can thus fulfill the role of Carl Neumann's body Alpha in accounting for mechanical inertia. Most important, electrodynamic phenomena do not have to depend on relative motion only, unless experience teaches us the contrary. Föppl admits the relativity of electromagnetic induction as an experimental truth to be used in the construction of the theory. At the same time, the force between two electrically charged bodies should theoretically depend on their absolute motion. The disparity of these two cases baffled Föppl:[38]

> These considerations clearly show against what difficulties the treatment of the electrodynamics of a system of moving bodies has to fight today. Seemingly well-established ideas, which we used to count as facts, appear to be unreliable as soon as we take a critical look at the material, and we begin to suspect that during the future development of science perhaps much of what we now regard as the unshakeable foundations of our vision of Nature might lose the aureole of unconditional validity.

5.6 The separation of ether and matter

In their electrodynamics of moving bodies, Hertz and Heaviside went as far as one could without facing the conflict with the optics of moving bodies. As we saw, they were aware of the revival of discussions of the latter topic in the wake of Michelson and Morley's experiments. In particular, they knew that the Maxwellian assumption of a fully dragged ether did not agree with the Fresnel–Frizeau drag. In February 1889, Heaviside wrote to Hertz:

> A point to which I should like to direct your attention is
> Aberration!
> or, the influence of motion through ether of a transparent dielectric carrying a wave on the motion of the wave itself.

Six months later, he insisted:

> Then there is the vexed question of the motion of the ether. Does it move when "bodies" move through it, or does it remain at rest? We know that there is an ether; the question is therefore a legitimate physical question which must be answered. . . . I believe that in the next few years some very great advance will take place in our knowledge of the connection of "matter" and "ether".

To this prophecy Hertz replied: "The motion of the ether relatively to matter – this is indeed a great mystery."[39]

---

[38] Föppl 1894, pp. 309–311. On Neumann's body Alpha, see pp. 66–67, this volume.

[39] Heaviside to Hertz, February 14 and August 14, 1889; Heaviside to Hertz, September 3, 1889, in O'Hara and Pricha 1987.

Electromagnetic theories of dispersion

We saw that in elastic-solid theories of the optical ether, there was a growing tendency to treat ether and matter as two separate, interacting systems. In the 1860s, the discovery of anomalous dispersion revived Young's idea that dispersion resulted from the coupling of the ether's vibrations with material oscillators. Maxwell in a Tripos question of 1869, Wolfgang Sellmeier in a laborious essay of 1871, and Helmholtz in a clearer article of 1875 showed that such coupling resulted in a dispersion formula of the form

$$n^2 - 1 = \alpha \frac{\omega_0^2}{\omega_0^2 - \omega^2}, \tag{5.46}$$

where α denotes the coupling constant and ω_0 the proper frequency of the material oscillators. Normal/anomalous dispersion (refraction increasing/decreasing with the frequency) occurs when the frequency ω of the waves is smaller/larger than the absorption frequency ω_0.[40]

By the end of the decade, Hertz's confirmation of Maxwell's theory induced many physicists to adopt the electromagnetic theory of light in place of the older elastic-solid theories of the ether. At the same time, there was a growing consensus that at least in electrolytes and in rarefied gases electric conduction implied the circulation of ions, namely, of atoms or groups of atoms carrying the electrolytic quantum of charge. Helmholtz, the most eloquent supporter of this view, also imagined that dielectric polarization corresponded to the displacement of elastically bound ions from their position of equilibrium. In 1892, he gave an electromagnetic theory of dispersion in which the electric force of the incoming electromagnetic waves produced a forced vibration of the ions, and this vibration in turn acted as a source in the Maxwell–Hertz field equations. The following year, Richard Reiff extended this picture to moving transparent bodies, in which the ions have a global translational motion through the ether in addition to their field-induced oscillation. He thus retrieved the Fresnel dragging coefficient, as Boussinesq had done in 1873 by similarly permitting the free circulation of the molecules of matter through a mechanical-elastic ether.[41]

Lorentz's grand program

Unknown to Helmholtz and Reiff, the Dutch theorist Hendrik Lorentz had already given an electromagnetic theory of dispersion in 1878 and of the Fresnel drag in 1892. Early in his career, Lorentz conceived a grand conciliation of Maxwell's theory with Weber's corpuscular concepts of electricity and magnetism. In his Leiden dissertation of 1875, he dwelt on a footnote of Helmholtz's large memoir of 1870 stating without proof that Maxwell's theory of light naturally gave the correct boundary conditions at the interface between two different transparent media. Helmholtz and Lorentz regarded this result as a major advantage of

---

[40] Maxwell 1869; Sellmeier 1871; Helmholtz 1875b. Equation (5.46) does not take the damping of the material oscillators into account (Helmholtz did). Cf. Darrigol 2012a, pp. 250–252.

[41] Helmholtz 1893a; Reiff 1893. Cf. Buchwald 1985, pp. 234–239; Darrigol 2000a, pp. 319–322. On Boussinesq's theory, see pp. 110–111, this volume.

Maxwell's theory, because the earlier elastic-solid theories tended to yield boundary conditions incompatible with the transverse character of the electromagnetic waves. Lorentz familiarized himself with Helmholtz's version of Maxwell's theory, and fully solved the reflection–refraction problem. In addition, he began to integrate the molecular structure of matter into this theory.[42]

In particular, Lorentz assumed that in a gas the ether had "exactly the same properties as in a vacuum." This implies, in the Maxwell–Helmholtz theory, that the polarization of the gas is the sum of an ether and a matter component, the latter being proportional to the number of molecules per unit volume. The optical index thereby depends on the density of the gas, in conformity with an empirical law by Arago and Biot. Three years later, Lorentz proposed a theory of dispersion based on elastically bound, electrically charged particles (not yet called ions), thus anticipating the results of Helmholtz's later theory of dispersion. Although these Dutch writings remained unknown abroad, they won Lorentz the Leiden chair of theoretical physics at age twenty-five.[43]

Lorentz returned to his atomistic reworking of Maxwell's theory in the 1890s, after Hertz's confirming experiments. From continental electrodynamics, especially Weber's, he adapted the basic corpuscular picture of electric and magnetic phenomena: the electric current as a migration of electric particles, electric charge as their accumulation, dielectric polarization as their elastic displacement, and magnetism as their circulation in tiny loops. From Maxwell's electrodynamics, he borrowed the concept of field-mediated interaction, the Maxwell–Hertz equations in the pure ether, and the concept of electric charge as a discontinuity of displacement (at the molecular level). Yet he betrayed both Weber and Maxwell in some ways: he discarded Weber's electric fluid particles in favor of charged material particles (ions or electrons), and he contradicted Maxwell's concept of electrification as a macro-phenomenon emerging from purely mechanical micro-processes. Most important, Lorentz completely separated matter from the ether. He had the charged particles mediate every interaction between ether and matter, while the ether between the particles remained strictly at rest. The latter hypothesis had two immediate advantages: it agreed with Fresnel's stationary ether, and it simplified the Maxwell–Hertz equations by getting rid of the motional fields $\mathbf{v} \times \mathbf{B}$ and $\mathbf{D} \times \mathbf{v}$.[44]

In Hertz's units, Lorentz's field equations read:

$$\nabla \times \mathbf{e} = -\frac{1}{c}\frac{\partial \mathbf{b}}{\partial t}, \quad \nabla \times \mathbf{b} = \frac{1}{c}\left(\rho\mathbf{v} + \frac{\partial \mathbf{e}}{\partial t}\right), \quad \nabla \cdot \mathbf{e} = \rho, \quad \nabla \cdot \mathbf{b} = 0, \tag{5.47}$$

for the microscopic fields $\mathbf{e}$ and $\mathbf{b}$ in the presence of ions carrying the charge density ρ and moving with the velocity $\mathbf{v}$. The electromagnetic force density within the charged particles is

$$\mathbf{f} = \rho(\mathbf{e} + c^{-1}\mathbf{v} \times \mathbf{b}). \tag{5.48}$$

[42] Lorentz 1875; Helmholtz 1870a, pp. 558n–559n. Cf. Hirosige 1969, pp. 171–173.

[43] Lorentz 1875, pp. 279–280; 1878.

[44] Lorentz 1892a, chap. 4. Cf. Hirosige 1969.

Lorentz first derived standard electrostatic and electrodynamic laws from these microscopic equations by averaging their solutions over volume elements containing many particles. At the micro-level, he investigated the motion of an elastically bound particle under the Lorentz force from an incoming ether wave, and then added the secondary waves produced by the vibrating particles in a volume element to get the net macroscopic field in a dielectric. In the case of a moving dielectric, the result agreed with the Fresnel–Fizeau drag.[45]

In this difficult calculation, Lorentz relied on a remarkable trick. The Lorentz field equations, when combined and applied to a plane wave in the direction of the x axis, involve the wave operator $(\partial/\partial x)^2 - c^{-2}(\partial/\partial t)^2$ in a reference frame bound to the ether. In a frame moving together with the transparent body at the velocity u, the time derivative $\partial/\partial t$ is replaced with $\partial/\partial t - u\,\partial/\partial x$. The resulting wave operator being too complicated for further calculation, Lorentz tried to simplify it by a change of coordinates. He thus found that the transformation

$$x' = \gamma x, \quad t' = \gamma^{-1}t - \gamma ux/c^2 \quad \text{with} \quad \gamma = (1 - u^2/c^2)^{-1/2} \tag{5.49}$$

restored the initial form of the wave operator. The product of this transformation with the former Galilean transformation $x \rightarrow x - ut$ is what we would now call a Lorentz transformation.[46]

Lorentz's mature theory

These results appeared in a large memoir of 1892 in French. Lorentz soon simplified his deduction of the Fresnel drag by forming and solving the average field equations in a frame bound to the transparent body. He also remarked that for an electrostatic system in a terrestrial laboratory, the field equations in the earth frame could be brought back to the form of the field equations of a body at rest in the ether through a transformation involving the dilation $x' = \gamma x$ of the coordinate x along the velocity $\mathbf{u}$ of the earth in the ether. He extended this property to the molecular equilibrium of a rigid body S, although it could imply molecular forces of nonelectric origin. Then the former transformation, when applied to S, generates a fictitious x-dilated body S_0 that has the dimensions of a body of the same constitution when at rest in the ether. Therefore, the motion of the body S through the ether implies its contraction in the direction of motion. Lorentz used this effect to compensate for the longer light trip in the $\mathbf{u}$-parallel arm of Michelson's interferometer. Unknown to him, in 1889 FitzGerald had already given a similar explanation of the null result of the

[45] Lorentz 1892a, chaps. 5–7, p. 319 (drag).

[46] Lorentz 1892a, p. 297. In 1887, Woldemar Voigt had already given the Lorentz transformations as a means to derive the Doppler effect for transverse waves in an elastic medium from a moving source (Voigt 1887). Voigt sent his paper to Lorentz in July 1908, and Lorentz promised to refer to it in the future (Lorentz to Voigt, July 30, 1908, in Kox 2008, p. 254). Minkowski mentioned this forgotten fact in Voigt's presence during a discussion of Bucherer's experiment at the *Naturforscherversammlung* of September 1908 in Cologne (*PZ*, 9: 762). Cf. Heras 2017. On earlier occurrences of the Lorentz transformations in mathematics (hyperbolic geometry, theory of quadratic forms, etc.), cf. the rich Wikipedia entry "History of Lorentz's transformations."

Michelson–Morley experiment, on the basis of Heaviside's finding that the electric field of a moving charge is contracted in the direction of motion.[47]

In the famous *Versuch* of 1895, Lorentz systematically developed the consequence of his theory for standard electrodynamic phenomena, optical dispersion, crystal optics, magnetism, magneto-optics, and the optics and electrodynamics of moving bodies. In this last register, he developed his earlier transformation for electrostatic phenomena, recovered FitzGerald's compensation mechanism for the interaction between an electric charge and a current at rest on earth, and argued against Des Coudres that any effect of the motion of the earth on electromagnetic induction could only be of second order. For the optics of moving bodies, he obtained a general derivation of the lack of first-order effects of the motion of earth through an insight into the transformation properties of his equations. Namely, he showed that in a first-order approximation the macroscopic field equations in the moving earth frame regained their original form in the ether frame when referred to the "local time" (*Ortszeit*) $t' = t - \mathbf{u} \cdot \mathbf{r}/c^2$, with a concomitant field transformation. In his words and with his emphasis:[48]

> *If, for a given system of bodies at rest, a state of motion is known for which [the original fields] are certain functions of x, y, z, and t, then in the same system drifting with the velocity u, there exists a state of motion for which [the transformed fields] are the same functions of x, y, z, and t'.*

Instead of using the average, macroscopic field equations, we may (as Lorentz himself did in 1899) directly exploit the transformation properties of the microscopic field equations (Equation (5.47)) and the Lorentz force (Equation (5.48)). Passing from the ether frame to the earth frame involves the substitutions

$$\mathbf{v} \to \mathbf{v} + \mathbf{u} \text{ and } \frac{\partial}{\partial t} \to \frac{\partial}{\partial t} - \mathbf{u} \cdot \nabla. \tag{5.50}$$

The resulting equations are too complicated to be solved directly. This is why Lorentz introduces the new coordinates and fields

$$t' = t - \mathbf{u} \cdot \mathbf{r}/c^2, \quad \mathbf{e}' = \mathbf{e} + c^{-1}\mathbf{u} \times \mathbf{b}, \quad \mathbf{b}' = \mathbf{b} - c^{-1}\mathbf{u} \times \mathbf{e}. \tag{5.51}$$

Together with $\partial / \partial t' = \partial / \partial t$ and $\nabla' = \nabla + \mathbf{u}c^{-2} \partial / \partial t$, this transformation restores the original form of the field and force equations to first order in u/c. In Lorentz's parlance, the *corresponding states* defined by the primed fields are the same as the states of a system at rest in the ether. Since the prime fields and the unprimed fields vanish together, and since all optical experiments are based on stationary patterns of brightness and darkness (light beams, interference fringes) for which the local time shift has no effect, the motion of the earth has no first-order effect on these experiments.

[47] Lorentz 1892a, 1892b, 1892c; FitzGerald 1889. For Heaviside's finding, see pp. 132–133, this volume. Cf. Buchwald 1988; Nersessian 1988; Hunt 1991, pp. 189–197; Janssen 1995, pp. 190–198.

[48] Lorentz 1895, p. 84. On the local time and corresponding states, cf. Ryniasiewicz 1988; Janssen 2019. On FitzGerald and Des Coudres, see pp. 139–140, this volume.

Lorentz also uses his theorem of corresponding states for a new, simpler derivation of the Fresnel drag. For a transparent body moving along the x axis with the velocity u through the ether, the theorem implies that a plane wave with the phase $k(x - ct'/n)$ should be a solution of the field and force equations. Inserting the value $t' = t - ux/c^2$ of the local time in this phase, the phase velocity of light in the transparent body is

$$V = \frac{c/n}{1 + u/nc} \approx \frac{c}{n} - \frac{u}{n^2}, \tag{5.52}$$

in conformity with the Fresnel drag.[49]

In the electrostatic case, Lorentz finds that the field and force equations in the earth frame can be transformed back to the equations in the ether frame through the transformation

$$\mathbf{r}' = (\gamma, 1)\mathbf{r}, \quad \mathbf{e}' = (1, \gamma)(\mathbf{e} + c^{-1}\mathbf{u} \times \mathbf{b}), \quad \mathbf{b}' = (1, \gamma)(\mathbf{b} - c^{-1}\mathbf{u} \times \mathbf{e}), \quad \rho' = \gamma^{-1}\rho, \quad \mathbf{f}' = (1, \gamma)\mathbf{f}, \tag{5.53}$$

where (α, β) in front of a vector means the multiplication by α of its component parallel to $\mathbf{u}$ and the multiplication by β of its perpendicular component. Consequently, the effects of the motion of the earth on electrostatic phenomena can be only of second order. In particular, by the last transformation formula the force between two electric charges at rest on earth is reduced by the factor $\gamma^{-1} = \sqrt{1 - u^2/c^2}$ when the line joining the charges is perpendicular to the velocity $\mathbf{u}$. Lorentz judges this second-order effect to be too small to be detectable.[50]

In 1899, Lorentz combined the rescaling of the moving fields and the local time shift to get the transformations

$$\mathbf{r}' = \varepsilon(\gamma, 1)\mathbf{r}, \quad t' = \varepsilon(\gamma^{-1}t - \gamma\mathbf{u} \cdot \mathbf{r}/c^2), \tag{5.54}$$

$$\mathbf{e}' = \varepsilon^{-2}(1, \gamma)(\mathbf{e} + c^{-1}\mathbf{u} \times \mathbf{b}), \quad \mathbf{b}' = \varepsilon^{-2}(1, \gamma)(\mathbf{b} - c^{-1}\mathbf{u} \times \mathbf{e}). \tag{5.55}$$

Combined with the Galilean transformation $\mathbf{r} \rightarrow \mathbf{r} - \mathbf{u}t$, this gives what is now called a Lorentz transformation. In the case of a dielectric for which the sources are electrons vibrating with a very small amplitude, Lorentz derives

$$\rho' = \varepsilon^{-3}\gamma^{-1}\rho \quad \text{and} \quad \mathbf{j}' = \varepsilon^{-3}(\gamma, 1)\mathbf{j} \tag{5.56}$$

for the transformed charge density and convection current. These relations, combined with the coordinate transformations, bring back the field equations exactly to the form they would have for a system at rest in the ether. Lorentz's intention is to prove that a variant of the Michelson–Morley experiment in which the interferometer is filled with a dielectric would still give a null result. For the transformed system to be exactly similar to a system

[49] Lorentz 1895, pp. 95–97.
[50] Lorentz 1895, pp. 35–39.

at rest, the equation of motion of an electron should also be invariant. If $\mathbf{f}$ denotes the binding force of the electron, m its mass, q is charge, and $\mathbf{r}$ is its position, and if its relative velocity $\mathbf{v}$ remains small compared with the velocity of the earth, this equation reads

$$\mathbf{F} \equiv q(\mathbf{e} + c^{-1}\mathbf{u} \times \mathbf{b}) + \mathbf{f} = m\ddot{\mathbf{r}}. \tag{5.57}$$

Lorentz assumes the binding force $\mathbf{f}$ to transform like the electromagnetic force. In order to extend the lack of effect of the earth's motion on optical experiments to any order in u/c, he further speculates that the electromagnetic inertia of the electron might modify the product $m\ddot{\mathbf{r}}$ in accordance with this common transformation of the forces. For small relative velocities, the coordinate and field transformations lead to

$$\mathbf{F}' = \varepsilon^{-2}(1, \gamma)\mathbf{F} \quad \text{and} \quad \ddot{\mathbf{r}}' = \varepsilon^{-1}\gamma^2(\gamma, 1)\ddot{\mathbf{r}}. \tag{5.58}$$

In order that the usual $\mathbf{F}' = m_0\ddot{\mathbf{r}}'$ holds for the corresponding electron at rest, one must take

$$\mathbf{F} = m_0(\varepsilon\gamma^3, \varepsilon\gamma)\ddot{\mathbf{r}}. \tag{5.59}$$

Lorentz is willing to admit this velocity dependence of the mass as well as the difference between longitudinal and transverse mass, "since the effective mass of an electron may depend on the state of the ether [around it]."[51]

As we have just seen, in 1899 Lorentz was calling the charged particles of his theory electrons. In 1895, he called them ions, in reference to the then widely accepted concept of the electrolytic current as a migration of ions. In 1896–1897, he explained Pieter Zeeman's magnetic splitting of the D lines of sodium through the precession induced by the Lorentz force $q\mathbf{v} \times \mathbf{B}$ on a vibrating "ion" of charge q. Around the same time, cathode ray experiments by Emil Wiechert and J. J. Thomson demonstrated the existence of negatively charged particles with a charge to mass ratio almost 2,000 times larger than that of the hydrogen ion. This estimate agreed with the value of the mass of the D-line emitter as inferred from equating the magnetic precession frequency qB/mc with the Zeeman splitting. Lorentz soon made the newly discovered particle (and its charge conjugate) the fundamental mediator between ether and matter. As we will see in a moment, he borrowed the name electron from Joseph Larmor.[52]

Wiechert's and Larmor's theories

Lorentz was not alone in proposing a new electrodynamics of atomistic entities moving through a stationary ether. In 1894, unaware of Lorentz's works, the Königsberg physicist

---

[51] Lorentz 1899a, p. 154. Alfred Liénard had suggested the variant of the Michelson–Morley experiment (Liénard 1898b). As we will see in Chapter 7 (p. 210, this volume), Lorentz's mass formulas agree with those given by special relativity, save for the ε factor.

[52] Zeeman 1896, 1897a, 1897b; Lorentz 1897a, 1897b. Cf. Kox 1997; Darrigol 1998; Arabatzis 2006, chap. 4. Most histories of the discovery of the electron have underplayed Wiechert's and Zeeman's contributions.

Emil Wiechert proposed to reduce matter to a myriad of "atoms of electricity" carrying the electrolytic quantum of charge (with both signs) and endowed with a mass of purely electromagnetic origin. These atoms, like Lorentz's ions, moved freely through the ether and served as mediators of all electromagnetic interactions. In 1896, Wiechert speculated that the cathode rays, whose nature was then being fiercely debated, could be projections of the electric atoms. His measurements of the velocity and magnetic curving of these rays soon provided the first evidence of an electric particle much lighter than the lightest ion. Wiechert believed he had discovered the electric atom. He went on developing the consequences of his picture of matter, though only for bodies at rest. He admired Lorentz's fuller theory, and remained active on the theory of electrons well after his appointment, in 1898, to a new chair of geophysics in Göttingen.[53]

In Cambridge, the Irish theorist Joseph Larmor also tried to integrate the atomistic structure of matter in the electromagnetic ether. Unlike Lorentz and Wiechert, he did not content himself with the abstract picture of an ether as a medium ruled by the Maxwell–Hertz equations. He wanted a quasi-mechanical substratum, which he found in the rotational ether of a fellow countryman. In 1839, James MacCullagh had invented a special kind of elastic medium, resisting only the rotation of its elements and better able to represent the optical ether than earlier elastic-solid theories. In 1879, FitzGerald showed that this medium exactly satisfied Maxwell's equations if only the electric polarization $\mathbf{D}$ were identified with the elementary rotation $\nabla \times \xi$, and the magnetic force $\mathbf{H}$ with the velocity $\partial \xi / \partial t$ (ξ being the linear displacement of a particle of the medium). In 1893–94, Larmor tried and failed to graft William Thomson's vortex atoms on MacCullagh's ether. One basic difficulty was the incompatibility of the relation $\mathbf{D} = \nabla \times \xi$ with electric charge (the divergence of a rotational vanishes). In the summer of 1894, Larmor gave up Thomson's vortices and decided to build all matter from point-like singularities, namely, centers of radial twist carrying the electrolytic quantum of charge. He first called these singularities monads, and then electrons, the name given to the electrolytic quantum of charge by his fellow countryman George Stoney.[54]

In principle, Larmor's picture of ether and matter had very nearly the same implications as Lorentz's more abstract scheme, because it shared the idea of reducing all charges, current, polarization, and magnetism, to the motion of electric particles in a stationary ether. However, it is only after reading Lorentz's *Versuch* of 1895 that Larmor was able to fully develop his theory, thanks to Lorentz's averaging technique and to his strategy of corresponding states. More originally, Larmor relied on what he called "the completeness of the aethereal scheme": for him, matter is made of singularities in the ether, the electrons (positive and negative) whose mass and interactions in principle derive from the structure of the rotational ether: "The electron *taken by itself* must be in any conceivable theory a simple singularity of the aether whose movements . . . are traceable through the differential equations of the surrounding free aether alone." In this picture, the Lorentz

---

[53] Wiechert 1894, 1896a, 1896b, 1897,1898a, 1898b, 1899. Cf. Darrigol 2000a, pp. 344–347.

[54] MacCullagh 1848 [1839]; FitzGerald 1880; Larmor 1894, 1895; Stoney 1891. Cf. Buchwald 1985, Part 3; Warwick 2003, chap. 7; Darrigol 1994.

contraction and the invariance of optical phenomena result from the transformation properties of the free-field equations alone, without any further assumption on the binding
forces of electrons or on their internal structure. In 1897, Larmor already gave the exact
Lorentz transformations, without realizing they were exact. He stated and exploited the
full invariance of his field equations under these transformations in his *Aether and Matter*
of 1900. No more than Lorentz did he believe in the complete lack of effect of the ether
wind. He predicted a first-order force between an electric charge and a nonconducting
magnet (nonconducting to block the FitzGerald–Lorentz compensation mechanism), and
a second-order alteration of electric conductivity.[55]

The difficulties of the corresponding-states strategy

Larmor and Lorentz used the transformations properties of the field equations in essentially the same way, through the concept of corresponding states. One may distinguish
three formal steps in their strategy: (1) write the field equations in the ether; (2) rewrite
these equations in the earth frame by means of a Galilean transformation of coordinates
(the field values, being intrinsic properties of the ether, are unchanged during this transformation); and (3) perform an additional, purely formal transformation of coordinates and
fields, the Lorentz "correspondence," in order to bring the field equations back to the form
they had in the ether frame. Thanks to this second transformation, to the true field state of
a system on earth, Lorentz and Larmor associate a state that the same system could have
if it were at rest in the ether. Conversely, from a possible state of a system at rest, they use
the inverse of the second transformation to generate a possible state of the system when it
moves at the velocity of the earth through the ether.[56]

There are two difficulties in this strategy. First, under given boundary conditions the
field state is not entirely determined by the field equations (and boundary conditions): it
also depends on nonelectromagnetic forces of cohesion and on the equation of motion of
the electrons (to a lesser extent for Larmor). Second, the field state is not directly observable (e.g., the force on a test-charge on earth does not measure the electric field, since this
field is defined with respect to the ether and since the test-charge moves together with the
earth). As a first example, consider the length of a rod and the rate of a clock (e.g., a spectral source) on earth. The second difficulty is here irrelevant, and Lorentz avoids the first
by assuming that the nonelectromagnetic forces in the rod and clock transform like electromagnetic forces. Provided that the forces at play uniquely determine the equilibrium length
of a rod and the rate of a clock, rods should contract and the vibrations of spectral sources
should slow down in the ether wind. Lorentz and Larmor both predicted these two effects (with little emphasis on the latter).[57] As a second example, take the force between two
charged particles at rest on earth. The second difficulty is here very mild, since the relation
between the force on a charged particle and the field is given by Lorentz's formula. In 1895

[55] Larmor 1896, 1897a; 1897b, pp. 54–55 (Lorentz transformations); 1900, pp. 165, 171–172.

[56] Cf. Nersessian 1986, pp. 218–221; Janssen 1995, pp. 157–179.

[57] Larmor 1897b, p. 41; Larmor 1900, pp. 79, 176–177; Lorentz 1899a, p. 155.

Lorentz circumvents the first difficulty by implicitly assuming the existence of dynamometers unaffected by the ether wind. This is why he then predicts a second-order correction to the Coulomb force.

As a third example, take an optical system. To the extent that optical experiments depend only on patterns of darkness (and not on subtler gradations of intensity), the second difficulty is solved by noting that darkness (zero field energy) is conserved by the Lorentz correspondence. The first difficulty occurs whenever the system includes dielectrics, since the vibrations of the electrons of the dielectric involve the equation of motion of an electron under both electromagnetic and nonelectromagnetic forces. Lorentz and Larmor escape this difficulty by tentatively assuming that the Lorentz correspondence preserves the form of the equation of motion. For Lorentz, there is an additional difficulty: the source terms of the field equations do not necessarily transform like the other terms. As we saw, in 1899 he worked in the dipolar approximation, in which he found the transformed source terms to have the desired form. As we will later see, this is no longer true for the general expression he gave in 1904 for the transformed source terms.[58]

Lorentz's growing success

For all these reasons, Lorentz's correspondence strategy was much trickier than the relativistic reasoning to which we are now accustomed. In the last years of the century, Lorentz's theory was nonetheless the only one that successfully covered the entire range of known optical, electric, and magnetic phenomena. With Larmor's and Wiechert's theories, this theory also had the great merit of easily integrating and partially anticipating the new experimental microphysics that thrived after Wilhelm Röntgen's discovery of X-rays in 1895. In 1897 the electron captured the attention of many experimentalists and theorists as the first atomistic entity ever confirmed by experience and as a plausible building block of matter. The theories of Lorentz and Larmor, soon called electron theories, then renewed the ether-drag question, to which the two experiments of Michelson and Morley and Lodge's ether-whirling experiment had given contradictory answers. In 1898, the committee for the physics section of the 70th *Naturforscherversammlung* in Düsseldorf chose the problem of ether motion as the topic of the traditional "special discussion." A German authority in radiation theory, Willy Wien, was in charge of a general review, and Lorentz was invited to respond to this review.[59]

Wien described all relevant experiments and pondered the relative advantages of the two main options: an ether moving like an incompressible fluid in Hertz's theory, and a stationary ether in Lorentz's theory. In the first option, one may either assume the ether to be without inertia (Hertz and Helmholtz), or to withstand impressed forces with a finite inertia (Heaviside). Wien questioned the ether motions deduced by Helmholtz in the first case (by balancing the Maxwell stresses with a fluid pressure), but did not exclude finite inertia, with gravity possibly explaining why a heavy body would drag the ether and a light one would not (as Des Coudres had recently suggested). That

---

[58] On the theory of 1904, pp. 173–175, this volume.

[59] Wien 1898; Lorentz 1898. Cf. Hirosige 1976, pp. 33–36.

said, Wien preferred the stationary ether, at least because Lodge had failed to stir the ether in his whirling experiment. Under this assumption his only qualm was the contradiction with the principle of reaction: there was nothing, in Lorentz's theory, to compensate for the Hertz force on the ether. In his response Lorentz belittled this difficulty: the principle of reaction did not have to apply at every scale, or it could be saved by assuming small, unobservable ether motions. He dwelt on his many successes in explaining negative ether-wind experiments, and he insisted that a fully dragged ether was incompatible with stellar aberration.

It would be excessive to say that the Düsseldorf meeting marked the victory of Lorentz's theory over all alternatives. A few ether-wind experiments, mainly Fizeau's old positive result with the polarizing pile of glass plates, and Mascart's null result with optical rotation, still contradicted this theory. The following year Lorentz accommodated Mascart's result, and Rayleigh confirmed it experimentally in 1902. Three more years elapsed before the American experimentalist DeWitt Bristol Brace disconfirmed Fizeau's polarizing-plates result. In the meantime, Fizeau's old idea of a possible asymmetry in the intensity of emitted light resurfaced. In 1902, Lorentz and Alfred Bucherer agreed that this asymmetry did not exist in Lorentz's theory. Bucherer's student Paul Nordmeyer confirmed their prediction in the following year. In the end, whenever Lorentz's theory implied a null effect of the ether wind, an experiment confirmed the prediction. In contrast, two expected (second-order) effects of the ether wind were refuted: the induced anisotropy of transparent media by Rayleigh in 1902 and Brace in 1904, and the induced torque on a charged condenser by Trouton and Noble in 1904. We will later return to these effects. For the moment, it is sufficient to observe that between 1895 and 1904 there never was a complete agreement of Lorentz's theory with known experiments. Some contradictions went away, but others took place.[60]

At the Düsseldorf meeting of 1898, another obstacle to the acceptance of Lorentz's theory was its microphysical character. With Boltzmann's exception, the attending elite, Woldemar Voigt, Max Planck, Paul Drude, and Gustav Mie, favored a more phenomeno-logical approach to physics. They soon grew more tolerant of microphysics, and two of them, Planck and Drude, adopted the new electron theory. Around 1900, Drude shone with a new theory of metals in which an electron gas was held responsible for both electric and thermal conductivity. Planck began to relate his radiation theory with Boltz-mann's and Lorentz's theories, and his former student Max Abraham soon started work on electron theory. The winds were turning in favor of atomistic theories combining waves and electrons in a powerful synthesis of Maxwell's theory, Weber's electrodynamics, and Boltzmann's kinetic theory of gases.[61]

[60] Fizeau 1852; Mascart 1874; Rayleigh 1902a; Brace 1905; Fizeau 1849; Lorentz 1902; Bucherer 1903; Nord-meyer 1903; Rayleigh 1902b; Brace 1904; Trouton and Noble 1904. Cf. Laub 1910; Hirosige 1976, pp. 36–39; Janssen 1995. On the Rayleigh–Brace experiments, see p. 173, this volume. On the Trouton–Noble experiments, see p. 173, this volume.

[61] Drude 1900a, 1900b, 1900c. Cf. Kaiser 1987; Eckert, Schubert, and Torkar 1992; Taltavull 2013. Planck briefly speculated that the motion of an ether compressed by gravitation could be irrotational (as required by Stokes's theory of aberration), but soon agreed with Lorentz's objections to this idea (see Lorentz 1899b). Mie later studied ether motion (in reaction to Wien's Düsseldorf review), with little profit.

Conclusions

In addition to the already known laws electrostatics, magnetostatics, and Voltaism, in the 1820s and 1830s Ørsted, Ampère, and Faraday discovered a flurry of new phenomena implying electric conduction, electrodynamic forces, electromagnetic induction, dielectric permittivity, and magnetic susceptibility. Moving bodies played an essential role in the resulting electrodynamics of Ampère and Faraday, as well as in the derived technology of electric motors and generators. In the rest of the nineteenth century, there were five major attempts to cover all these phenomena in a single, unified theory.

The first attempt was the potential theory of Neumann, Kirchhoff, and Helmholtz, based on the concept of electrodynamic potential for currents and magnets. The second was Weber's theory, based on a fundamental law for the forces between any two particles of electricity, and on atomistic pictures for charge, current, and polarization. The predictions of these two theories agreed for the closed or quasi-closed currents to which experiments were confined until the 1870s. They assumed instantaneous action at a distance, and they did not survive Hertz's proof of retarded action in the late 1880s. The third attempt was the Maxwell–Hertz electrodynamics of moving bodies, based on Faraday's and Maxwell's field concept and on the simple assumption that matter, whenever it moves, carries along the ether within it. In this view, motional induction and electric convection are direct consequences of the ether's convection. In the fourth attempt, defended by Heaviside and Föppl, the field equations are the same as in the Maxwell–Hertz theory, but the motion of the ether within matter is left undetermined; the terms representing motional induction and electric convection are justified by energy and relativity arguments. The fifth attempt is the electron theory of Lorentz, Wiechert, and Larmor, developed in the 1890s. It is based on Weber's atomistic picture of charge, current, and polarization; on a stationary ether through which atoms, ions, and electrons freely move; on Maxwell's equations for this free ether; and on the Lorentz force. As Hertz and Heaviside feared, the fully dragged ether did not survive the integration of optics into electrodynamics. In contrast, Lorentz's simple atomistic picture successfully embraced the totality of optical, electrodynamic, and magneto-optical phenomena known at the close of the century, with a few glitches only. In addition, this theory integrated the newly flourishing microphysics of X-rays, ions, electrons, and radioactivity.

Relativity considerations entered these five theories in different ways. The potential theory directly implies the relativity of induction phenomena and any other motion-dependent phenomenon by tracing them to the mutual potentials of the conductors and magnets. Weber's theory indirectly implies the same perfect relativity by summing elementary forces that depend only on the relative distance of two electric particles and its two first time derivatives. While Neumann based his general potential law on the observed relativity of induction, Weber was a priori convinced that the elementary force should depend on relative motion only. In the Heaviside–Föppl theory, the observed relativity of induction is used to justify motional induction, just as in Neumann's theory. The velocity entering the field equations is the velocity of matter, and there is no concept of ether velocity. The velocity of matter is defined with respect to absolute space, to be perhaps identified with the ether.

In the Maxwell–Hertz theory, Maxwell shows that in the absence of retardation (when the displacement current is negligible) and for closed currents, his field equations are invariant under any change of frame. Hertz extends this invariance to the exact field equations, granted that the ether velocity transforms like the velocity of matter, even when there is no matter. In more concrete words, electrodynamic and electromotive forces remain the same when a common rigid motion (translation, rotation, etc.) is impressed on the system, including the ether. As Maxwell and Hertz understood, in reality the ether at some distance of the material bodies more plausibly remains at rest when a common motion is impressed on these bodies. Strictly speaking, there is nothing in the Maxwell–Hertz theory to determine the motion of the ether outside matter. This deficiency, which Helmholtz tried to correct in 1893, is related to another difficulty: as Hertz saw, the force $c^{-2}\, \partial\, (\mathbf{E} \times \mathbf{H})/\, \partial\, t$ acts on the ether in vacuum, with nothing to counterbalance it unless inertia is attributed to the ether itself. For electromagnetic radiators, this force implies an uncompensated change of momentum.[62]

In Lorentz's theory, the strictly stationary ether dissolves the ether-motion problem and aggravates the uncompensated-momentum problem. To the extent that Lorentz's equations, after averaging over macroscopic volume elements, agree with Maxwell's equations, they imply as much relativity as the latter equations do for electromagnetic phenomena. In the optical regime, they imply the Fresnel wave drag and the consequent first-order relativity. In addition, they imply a second-order longitudinal contraction of static fields on the moving earth. Assuming that all static forces, including those responsible for the internal equilibrium of rigid bodies, undergo the same contraction, Lorentz derived the eponymous contraction, in the amount needed to explain the negative result of the Michelson–Morley experiment of 1887.

For all missing effects of the earth's motion through the ether, Lorentz thus imagined a compensating mechanism. In addition, he invented the technique of *corresponding states* through which the absence of effect could be deduced in a more direct and general manner. Namely, he discovered a simple coordinate and field transformation that turns the states of a system at rest in the earth frame into states of the same system at rest in the ether frame. He did not interpret the transformed quantities as the ones measured by moving observers, as we would now do. Instead he argued that the transformation did not affect the effectively observed patterns of light, and therefore implied the invariance of optical phenomena. Although Lorentz originally limited his reasoning to a limited class of optical systems (to first order) and to electrostatic systems (to second order), in 1899 he discovered the exact Lorentz transformations (except for the source terms) and he generalized his theorem of corresponding states to any optical system at any order. In the case of dielectrics, he thereby assumed that the inertial forces and the elastic forces on a dielectric's electrons transformed like the Lorentz force. This implied a specific velocity dependence of the inertial mass, whatever might be the internal structure of the electron. That said, Lorentz's reasoning implied approximations (the dipolar approximation for the sources in the Maxwell–Lorentz equations), and it did not cover all electrodynamic phenomena. In general, Lorentz expected effects of the earth motion, although these were too small to be detected by contemporary means. Larmor believed the same.

---

[62] More on this point in Chapter 6, pp. 162–167, this volume.

It is instructive to observe how the relativity or nonrelativity of a few simple phenomena appeared in all these theories. Electromagnetic induction was the prime empirical incitation to assume that only the motion of matter with respect to matter has observable effects in electrodynamics. As Neumann put it, the motion of the earth would otherwise suffice to produce an induced current, without any relative motion of the circuit and magnet. In Neumann's theory, the induced current hinges on the time variation of a potential that depends on the relative position of circuit and magnet. In Weber's theory, it hinges on the time variation of the relative distance of the electric particles of the circuit and magnet. In both cases, the theoretical representation contains the observed symmetry of induction phenomena.

The situation is more complex in the ether theories. Heaviside and Föppl directly inject the relativity of induction into their theory, and they leave the motion of the ether undefined. Maxwell and Hertz assume the ether and matter to be parts of a single, continuous medium. The ether therefore follows the earth and the atmosphere in their motion, and the motion of the earth does not affect the ethereal circumstances of a magnet and circuit in relative rest. The case of a magnet and circuit in relative motion is less obvious. It should in principle matter whether the magnet or the circuit moves through the ether. In practice it does not because for closed circuits in the quasi-stationary approximation, Maxwell's field equations imply the same electromotive force as in Neumann's theory. As long as the ether accompanies the matter of the circuit and the matter of the magnet, and as long as field retardation is ignored, the predicted phenomena are the same whatever the ether motion may be in the intervening space. This is no longer true for open circuits. For instance, for a magnet moving by itself through the ether, an electrostatic charge should be generated at the surface of the magnet. This is easily seen in the case of a rotating magnet because, according to Faraday and Maxwell, the conducting mass of the magnet cuts the internal lines of force of the magnet, thus generating an internal transitory current. Faraday regarded this internal induction in a rotating magnet as responsible for the unipolar induction current observed in the contraption of Fig. 5.4 (above, p. 138). Others denied internal induction, either because there was no ether in their theory (Weber and Neumann) or because they assumed the magnetic lines of force to rotate together with the magnet (Preston). That said, everyone agreed that unipolar induction depended on the relative motion of the magnet and the sliding wire only.[63]

Lorentz's theory mostly agrees with Maxwell's theory in its predictions for induction phenomena, because, as was already said, it generates Maxwell's macroscopic field equations through an averaging process in a sufficient approximation (in the absence of dielectric polarization). In particular, the electromotive force of induction in a moving conductor has the same expression $\mathbf{v} \times \mathbf{B} - \partial \mathbf{A}/\partial t$ as in Maxwell's theory, the first term being generated by the motion of the conductor, and the second by the motion of the magnet. There is an important difference, however. In Maxwell's theory, the two terms are essentially of the same nature: they belong to the convective variation $-D\mathbf{A}/Dt$ of the "electromagnetic momentum" $\mathbf{A}$ and they both are genuine electric fields contributing to

---

[63] More on this in Darrigol 1993b, pp. 298–301, 332–333.

the electric field energy. In Lorentz's theory, only the second term is a genuine electric field, while the second derives from the Lorentz force acting on the electrons of the conductor. Thus, there is a fundamental difference between the case of a moving magnet and the case of a moving conductor, despite the relativity of the predicted induction current. This is the asymmetry that Einstein later deplored in the received electrodynamics. One may wonder why Lorentz and other electron theorists were completely silent on this issue. Most likely, they reasoned that in a stationary ether theory there was a real, principally detectable asymmetry between the ether frame and other reference frames. For induction as for other phenomena, they saw the lack of observed effects of the motion of the earth through the ether as a temporary consequence of our incapacity to detect the most delicate predictions of the theory.

Another interesting, though purely imaginary effect of common motion is the alteration of the force between two charged particles when they move together uniformly in the direction perpendicular to the line joining them. In the Neumann–Helmholtz potential theory, there is no such effect because a moving charge does not constitute an electric current. In Weber's theory the effect is also lacking, since the Weber force between two charges depends on their relative motion only. In Maxwell's theory and in any nineteenth-century field theory after him, the force depends on the common motion and it vanishes when the common velocity reaches the speed of light, because the electrodynamic attraction between the attending convection currents compensates for the electrostatic repulsion. This was an often-discussed effect of "absolute" motion, here meaning motion with respect to the ether in which the two charges are immersed. As we will see in Chapter 6, the alleged effect is akin to the lack of invariance that Alfred Liénard established in 1898 for the Lorentz force, and its later elimination employed the velocity dependence of inertial mass.[64]

Lastly, let us consider three major facts regarding the optics of moving bodies, stellar aberration, the Fresnel–Fizeau drag, and the repeated failure to detect effects of the earth's motion on terrestrial optics (terrestrial relativity). Evidently, Neumann and Weber had nothing to say on this topic since their theory did not cover optics. Maxwell, Hertz, and also Heaviside (after reading Hertz) could easily accommodate terrestrial relativity, since they assumed ether and matter to be comoving components of a single medium. By further requiring the ether's motion to be irrotational between the stars and the earth's atmosphere, they could explain aberration in Stokes's manner. Their stone in the shoe was the Fresnel–Fizeau drag. After Michelson and Morley confirmed this effect in 1886, they could no longer doubt its existence. Moreover, in the 1890s the accumulated evidence for a lacunar structure of matter made the full drag of the ether by matter highly implausible. Hertz and Heaviside adopted this hypothesis for the sake of simplicity, with a warning that it would probably not survive the future integration of the optics of moving bodies. This is indeed what happened in the subsequent theories of Lorentz, Wiechert, and Larmor.

At the close of the century, Lorentz's theory surpassed all competing theories by encompassing the greatest variety of phenomena. With impressive virtuosity, Lorentz responded to numerous experimental challenges, and many of his predictions were confirmed. The atomistic character of his thinking, originally a handicap in the eyes of German phenomenologists, soon turned in his favor as a new experimental microphysics rose. His theory had roots in Fresnel's optics of moving bodies, in Maxwell's theory, in Helmholtz's, Hertz's, and

---

[64]See p. 167n, this volume.

Heaviside's reformulations of this theory, and in Weber's atomistic theory of electricity and magnetism. The multiplicity and depth of these roots, the formal simplicity of the resulting picture, the power of the deductive machinery, and Lorentz's just-do-it temperament gave him the certitude that his theory was there to last, with perhaps a few gradual improvements. Yet, for one who was reading Lorentz without knowing the origins of his thinking, much of his theory could seem strange and artificial. How could the ether, as physical substratum of the fields, enjoy some mechanical attributes (energy, stress) and not others (velocity, momentum)? How could equations written in the privileged ether frame lead, after some mathematical wizardry, to the observational equivalence of all Galilean frames? Such questioning and its consequences are the subject of the next two chapters.

6

POINCARÉ'S RELATIVITY THEORY

> Experience has revealed a large number of facts that may be summarized as
> follows: it is impossible to detect the absolute motion of matter, or, better, the
> relative motion of ponderable matter with respect to the ether. All we can de-
> tect is the motion of ponderable matter with respect to ponderable matter.[1]
> (Henri Poincaré, 1895)

Let us begin with a platitude: it often takes an outsider to see the defects of a reigning
intellectual construct and to imagine improved, alternative constructs. All major contributors to late-nineteenth-century optics and electrodynamics were physicists by training
and profession, with the exception of Henri Poincaré, a profound mathematician with an
unusual interest and competence in physics, engineering, and philosophy.

Poincaré studied at the École Polytechnique and at the École des Mines, briefly acted as
a mining engineer, and soon published major work in mathematics. He favored geometrical, visual methods, with a special flair for shared structures, for instance when he employed
non-Euclidean geometry to study the transformation properties of the class of functions
he called Fuchsian. He usually accompanied his research with philosophical remarks on
foundation and method. These often concerned the relationship between mathematics and
physics, for he believed the physical world to be the *occasion* (motivating circumstance)—
not the foundation—of our chief mathematical constructs. Measurement being essential
in this relationship, he tracked the implicit conventions of physical measurement. His insistence on the free choice of conventions led to his so-called conventionalism, first in
geometry, then in mechanics and in physics. At the same time, he recognized invariant, stable, convention-independent structures (the *rapports vrais*) in physical theories. As we will
see, this triple concern with measurement, conventions, and structures shaped his approach
to fin-de-siècle electrodynamics.[2]

In physics, Poincaré continued the strong French tradition of a *physique mathématique* done by mathematicians. Perhaps owing to his engineering background, he had a
much greater interest in experiments and technology than his forerunners. His handling of
physical theories was not only mathematically superior, it was also sensitive to the conditions of experimental observation, and it promoted physical understanding through images
and models. His published courses could compete with or even surpass the best treatises

---

[1] Poincaré 1895b, p.14.

[2] On Poincaré's biography, cf. Gray 2013; Galison 2003, chap. 2. On his philosophy, cf. Stump 1989; Friedman
1995; Paty 1993, pp. 250–263; Ben-Menahem 2001; Darrigol 2007, 2018; Príncipe 2012; Heinzmann and Stump
2017; Doran 2018. On mechanics, see pp. 73–75, this volume. On geometry, see p. 253, this volume.

on their topic, in France and abroad. Being born in the country of Laplace and Fresnel, he entered physics through astronomy and optics. He got his first notions of optics at the *lycée* in Nancy, and much more at the École Polytechnique, where the two physics professors Jules Jamin and Alfred Cornu and the physics *répétiteur* Alfred Potier were leading experts in optics. In an obituary for Cornu, Poincaré later observed:

> He has written much on light. Even though he left his mark on every part of physics, optics was his favorite topic. I surmise that what attracted him in the study of light was the relative perfection of this branch of science, which, since Fresnel, seems to share both the impeccable correction and the austere elegance of geometry. In optics better than in any other domain, he could fully satisfy the natural aspiration of his mind for order and clarity.

It seems reasonable to assume that Poincaré was here projecting the reasons for his own predilection for optics.[3]

At the Nancy *lycée*, Poincaré already learned about stellar aberration and how it could be explained by composing the velocity of light with the velocity of the earth. This fact is not so minor, since he later traced the dilemmas of the electrodynamics of moving bodies to the discovery of stellar aberration: "Astronomy raised the question by revealing the aberration of light." At the École Polytechnique, he heard more about stellar aberration both in Cornu's physics course and in Hervé Faye's astronomy course. Cornu dwelt on Fizeau's experiment as a proof that matter could not be the sole medium for the propagation of light. As we saw in Chapter 4, the *répétiteur* Potier was the man who would later correct Michelson and Morley for miscalculating the path difference in their moving interferometer. In these student years, Poincaré performed an ether-drift experiment he remembered late in his life with a touch of irony:[4]

> I was long ago a student at the École Polytechnique. I must concede that I am extraordinarily clumsy and that since then I have felt I should better stay away from experimental physics. At that time, however, I was helped by a fellow student, Mr. Favé, who is manually very adroit and who, in addition, has a very resourceful mind. We jointly tried whether the translational motion of the earth affects the laws of double refraction. If our investigation had led to a positive result, that is, if our light fringes had been shifted, this would only have shown that we lacked experimental skills and that the buildup of our apparatus was defective. In reality the outcome was negative, which proved two things at the same time: that the laws of optics are not affected by the translational motion, and that we were quite lucky on this matter.

Section 6.1 of this chapter recounts how Poincaré's early interest in optics materialized in a series of courses he gave at the Sorbonne on optical and electrodynamic theories from 1888 to 1899. In this context he introduced the relativity principle and he diagnosed a major crisis in contemporary physics. The alleged symptoms of this crisis were the conflict

---

[3] Poincaré 1905b, p. 146. On the *lycée* physics course, cf. Walter 1996b. On optics and astronomy in nineteenth-century France, cf. Davis 1986.

[4] Poincaré 1904a, p. 320; 1910b, p. 104; Cornu 1875, p. 101; Faye 1874, pp. 170–174.

of the best theory of the late century, Lorentz's, with this principle and with the principle of reaction. Section 6.2 is about the deeper analysis Poincaré gave of these symptoms in his contribution to the Lorentz jubilee of 1900. Section 6.3 retraces the emergence of the electromagnetic worldview, Abraham's and Lorentz's ensuing models of the electron, and Poincaré's updated criticism in his Saint-Louis conference of 1904. Section 6.4 is devoted to Poincaré's invention of a Lorentz-invariant dynamics of the electron and to his subsequent reflections on the measurement of time and on the ether problem.

6.1 Critical teaching

Theories of light

In 1886, the Sorbonne faculty elected Poincaré to the chair of mathematical physics and probability calculus. For the topic of his first course, he chose "the mathematical theories of light," knowing that many such theories had been proposed since Fresnel's first intimation of a molecular ether. Instead of professing a single favored theory, as a normal person would have done, Poincaré taught no less than five theories:

> The theories propounded to explain optical phenomena by the vibrations of an elastic medium are very numerous and equally plausible. It would be dangerous to confine oneself to one of them. That would bring the risk of blind and therefore deceptive confidence in this one. We must therefore study all of them. Most important, comparison tends to be highly instructive.

The ether could be discrete (molecular) in one theory, continuous in another; the optical vibration could be either in the plane of polarization or in the perpendicular direction; and the connection between ether and matter could vary. Poincaré explained the multiplicity of surviving theories by their sharing the same mathematical structure: the differential equations of motion of one theory could be turned into those of another theory through a simple transformation, with concomitant adjustment in the definition of the optical vibration.[5]

Through this comparative approach, Poincaré saw much arbitrariness in any mechanical model of the vibrating ether. This is why, even though he defended such modeling for the sake of clarity, he predicted the ultimate downfall of the ether (p. I):

> It matters little whether the ether really exists; that is the affair of the metaphysicians. The essential thing for us is that everything happens as if it existed, and that this hypothesis is convenient for the explanation of phenomena. After all, have we any other reason to believe in the existence of material objects? That, too, is only a convenient hypothesis; only this will never cease to be so, whereas probably the ether will someday be thrown aside as useless. On this very day, however, the laws of optics and the equations that express them analytically will remain true, at least in a first approximation. It will therefore be always useful to study a doctrine that interconnects all these equations.

[5] Poincaré 1889, p. II. Cf. Darrigol 2012b.

This was a daring pronouncement at a time in which almost every physicist firmly believed in the reality of the ether. Yet there is some ambiguity in Poincaré's position: on the one hand the ether may become "useless" as an explanatory hypothesis; on the other it may still be useful as a constructive, unifying model of the differential equations of motion. We will see that Poincaré never fulfilled his own prediction of an etherless theory.

Poincaré devoted the last chapter of his optical lectures to stellar aberration and other optics of moving bodies. He introduced the Fresnel drag directly in the discussion of stellar aberration, as the drag value compatible with Airy's finding that water-filling does not affect the stellar aberration observed in a reflecting telescope. Poincaré then described Fizeau's running-water experiment, and gave the Veltmann–Potier proof that the earth's motion does not affect optical experiments on earth if the ether is dragged by transparent bodies according to Fresnel's hypothesis. He concluded: "In one word, optical phenomena can provide evidence only for the relative motion of the luminous source and of the ponderable matter with respect to the observer."[6]

The physics of principles

Although Poincaré originally confined his lectures to the elastic-solid theories of light, he also meant to teach Maxwell's electromagnetic theory. This he did in the summer of 1888, just after Hertz's groundbreaking experiments. One aspect of Maxwell's theory most retained his attention: the mature Maxwell avoided any specific ether mechanism and contented himself with requiring the Lagrangian form of the field equations. Poincaré described "Maxwell's fundamental idea" as follows:

> In order to prove the possibility of a mechanical explanation of electricity, we need not worry about finding this explanation itself, we only need to know the expression of the two functions T and U which are the two components of the energy, to form the Lagrange equations for these two functions, and then to compare these equations with the experimental laws.

Poincaré supported this assertion with the mathematical demonstration that any Lagrangian system admits an infinite number of mechanical realizations. He generally admired the lofty abstraction he saw in Maxwell's treatise:[7]

> The same spirit pervades the entire work. The essential, namely, what must remain in common in all the theories, is brought to light. Anything that would concern only a particular theory is almost always kept silent. The reader thus faces a form nearly void of matter, a form which he at first tends to take for a fleeting and elusive shadow. However, the efforts to which he is thus condemned prompt him to think, and he at last becomes aware of the somewhat artificial character of the theoretical constructs that he formerly admired.

---

[6]Poincaré 1889, p. 391. Poincaré also gave Boussinesq's theory of the Fresnel drag (see p. 110–111, this volume). On Airy, Veltmann, and Potier, see pp. 104, 108–109, 103, this volume.

[7]Poincaré 1890, pp. XV (Poincaré's emphasis), XVI.

In Poincaré's eyes, Maxwell's theory thus exemplified a physics based on general principles, without a sustaining mechanical picture. So too did Clausius's and Thomson's thermodynamics, to which Poincaré devoted his next course. He later insisted on the general evolution of physics from a constructive mechanical approach to a "physics of principles":

> A day arrived when the conception of central forces no longer appeared sufficient
> .... What was done then? The attempt to penetrate into the detail of the structure of
> the universe, to isolate pieces of this vast mechanism, to analyze one by one the forces
> which put them in motion, was abandoned, and we were content to take as guides
> certain general principles the express object of which is to spare us this minute study.

The principles could be mechanical principles such as the energy principle or the principle of least action, or they could be more direct expressions of experience as the second principle of thermodynamics was. They all had an inductive origin:

> The principles are results of experiments boldly generalized; but they seem to derive
> from their very generality a high degree of certainty. In fact, the more general they
> are, the more frequent are the opportunities to check them, and the verifications mul-
> tiplying, taking the most varied, the most unexpected forms, end by no longer leaving
> place for doubt.

Poincaré occasionally argued that the principles, despite their empirical origin, could be elevated to the rank of definition and convention for the construction of theories and would thus become immune to refutation. At the same time, he made clear that it would be unwise to save a principle by multiplying invisible entities.[8]

Violated principles

After teaching and praising Maxwell's theory, Poincaré studied and taught the related theories of electrodynamics by Helmholtz (1870), Hertz (1890), Helmholtz-Reiff (1893), Larmor (1894), and Lorentz (1895). As these theories, unlike the older optical theories, widely differed in range and predictions, Poincaré needed a way to assess their relative merits. He did that mainly by testing their compatibility with a few general principles of mechanical origin: the energy principle, the principle of least action, the principle of reaction, and the relativity principle; and also two electric principles: charge conservation, and the unity of the electric force. Charge conservation, the energy principle, and the principle of least action were not truly selective, because the competing theories had been built so as to satisfy these principles. The truly decisive principles were the principle of reaction and the principle of relativity, also the unity of the electric force. This last principle, introduced by Hertz in 1884, requires all sources of electric force to have the same effects (e.g., an

---

[8]Poincaré 1904a, pp. 299–301. See also Poincaré 1900b, pp. 1168–1170, and pp. 49–51, 171, this volume.

electric charge and a variable current should both act on a charged body). Poincaré agreed with Hertz that Helmholtz's theory violated this principle.[9]

Now consider the principle of reaction. Through its stress system Maxwell's theory implies a yet unobserved radiation pressure, and all derived theories imply the Hertz force density $c^{-2} \partial (\mathbf{E} \times \mathbf{H})/ \partial t$ acting on the ether in vacuum. Consequently, in order to comply with the reaction principle these theories must have the ether contribute its share of momentum in the global momentum balance. While Hertz, coming from an action-at-a-distance tradition, was very reluctant to admit as much, the British ether theorists easily accommodated this sort of materiality. As Lorentz himself noted in 1895, his theory aggravated Hertz's difficulty, since in principle his ether could not move. He had an easy escape: "As far as I can see, nothing forces us to elevate the principle [of reaction] to the rank of a fundamental law of unlimited validity."[10]

Poincaré judged differently. In a critical essay published the same year in *L'éclairage électrique*, he denounces Lorentz's violation of the equality of action and reaction as a "grave difficulty." In this theory, Poincaré remarks, electromagnetic waves from a remote source could move a charged particle without compensatory recoil of the source or of the medium. Lorentz's ether, being essentially immobile and divorced from matter, is too immaterial to carry any momentum. In contrast, Poincaré judges that Hertz's theory, being based on a single ether–matter medium, necessarily endows the ether with the needed inertia. In his last exposition of Hertz's theory in 1899, Poincaré will simply erase the ether and entirely reduce Hertz's medium to matter.[11]

At this point, Poincaré names three conditions that any decent electrodynamic theory should meet: (1) it should account for Fizeau's partial ether drag, (2) it should satisfy the conservation of charge (and magnetism), and (3) it should satisfy the principle of reaction. In his opinion, Hertz's theory satisfies (2) and (3), but of course not (1) since it implies a complete ether drag; the Helmholtz–Reiff theory of 1893 satisfies (1), perhaps (3), but not (2); Lorentz's theory satisfies (1) and (2), but not (3). In sum, none of the available theories satisfies the three conditions. Poincaré then asks himself whether the three conditions are mutually compatible. In reply he proves that in the approximation of small, slowly varying fields, Hertz's equations are the only ones compatible with (2) and (3). Consequently, any theory satisfying (2) and (3) necessarily violates (1), and the three conditions are mutually incompatible. Poincaré concludes:

> We could not hope to escape this difficulty without deeply modifying the generally admitted ideas. It is not clear, however, in what sense this modification should be made. We must therefore renounce to develop a perfectly satisfactory theory and provisionally content ourselves with the least defective theory, which seems to be Lorentz's.

[9]Poincaré 1901b. Cf. Darrigol 1995. Hertz did not care to derive his theory from the principle of least action, but Helmholtz did it for him. Poincaré found spurious contradictions between Maxwell's stress system and the energy principle, and also between the Helmholtz–Reiff theory of dispersion and charge conservation.

[10]Lorentz 1895, p. 28. See also Wien–Lorentz in 1898, p. 151, this volume.

[11]Poincaré 1895a, p. 294; 1901b, pp. 345–420.

After summarizing the contents of this theory, Poincaré insists:[12]

> Needless to say that this theory, even if it may help us somehow in clarifying our ideas, cannot fully satisfy us, or be regarded as definitive.
>
> I find it difficult to admit that the principle of reaction would be violated, even seemingly, and that it would no longer be true when we consider only the actions on ponderable matter and discard the reaction of this matter on the ether.
>
> Therefore, we will someday have to modify our ideas in some important way and to break the frame into which we try to jointly fit optical and electrical phenomena.

Right after this call for a conceptual revolution, Poincaré turns to another difficulty of Lorentz's theory, its conflict with the following law:

> Experience has revealed a large number of facts that may be summarized as follows: it is impossible to detect the absolute motion of matter, or, better, the relative motion of ponderable matter with respect to the ether. All we can detect is the motion of ponderable matter with respect to ponderable matter.

In Poincaré's opinion, Lorentz's theory accounts for this law, but only to first order in u/c (and with additional approximations). He goes on:

> This is not enough: the law seems to be true even without these restrictions, as was proved in a recent experiment by Mr. Michelson.
>
> We thus find here an additional lacuna that may not be unrelated with the one that the present article is meant to bring out.
>
> Indeed, the impossibility to detect the motion of matter with respect to the ether, and the equality which probably holds between action and reaction without taking into account the action of matter on the ether, are two facts whose proximity seems evident.
>
> Perhaps these two lacunas will be filled at the same time.

Poincaré means that Lorentz's ether, if it had enough materiality to contribute to the momentum balance, would plausibly affect the behavior of systems moving through it. In his eyes, the ether just cannot have this crude materiality.[13]

Poincaré returned to these difficulties in 1899, while lecturing at the Sorbonne on the theories of Hertz, Lorentz, and Larmor. There he insisted that Lorentz's theory violated the principle of reaction. From a formal point of view, Lorentz's field and force equations imply the relation

$$\int \mathbf{f} d\tau + \frac{1}{c}\frac{d}{dt}\int (\mathbf{e} \times \mathbf{b}) d\tau = 0 \qquad (6.1)$$

---

[12] Poincaré 1895b, pp. 385–392, 12 (citation), 14 (citation).

[13] Poincaré 1895b, p. 14.

between the total force acting on matter and the fields **e** and **b**. Concretely, this force does not vanish when a material oscillator emits radiation in one direction only. There is a recoil of the emitter, which Poincaré compares with the recoil of a piece of artillery. The comparison is not meant to justify the recoil of the radiator. On the contrary, Poincaré judges this recoil "difficult to admit."[14]

Later in his course, after giving a detailed proof that in Lorentz's theory the motion of the earth has no first-order effect on optical experiments, Poincaré mentions the explanation of the null result of the Michelson–Morley experiment through the "supplementary hypothesis" of the Lorentz contraction, with the comment:

> This strange property would seem a true *coup de pouce* [a nudge] given by nature to avoid that the absolute motion of the earth be revealed by optical phenomena. This could not satisfy me and I will not hide my feeling: I regard it very probable that optical phenomena depend only on the relative motions of the implied bodies, light sources, and optical apparatus, and *this not only to first or second order* [in *u/c*] *but rigorously*. When the experiments will become more exact, this principle will be verified with more precision.
>
> Will we need a new *coup de pouce*, a new hypothesis, at each approximation? Of course not: A well-wrought theory should allow us to demonstrate the principle in one stroke in all rigor. Lorentz's theory is not yet able to do that. Of all theories that have been proposed, it is the one that comes closest to this aim. We may therefore hope to make it perfectly satisfactory in this respect without altering it too deeply.

In this citation, we observe that Poincaré now calls the dependence of optical phenomena on relative motion a "principle," whereas in 1895 he introduced it as an empirical "law." This agrees with his contemporary definition of principles as elevated laws whose validity will no longer be questioned. The principle should guide the final theoretical construction, although Poincaré does not seem to regard it as logically primitive: it will have to be "demonstrated" (*démontré*) by the construction.[15]

In his lectures of 1899, Poincaré does not relate this shortcoming of Lorentz's theory to the violation of the reaction principle. However, he remarks that the validity of the reaction principle in Hertz's theory results from energy conservation combined with the invariance of Hertz's equations when changing the systems of axes to which they are referred. Call J the total energy of the electromagnetic field, Q the energy density communicated to the medium by external sources minus the Joule energy lost to the medium in a unit time, **f** the density of mechanical force, and **v** the velocity of the medium. The conservation of energy gives:

$$\frac{\mathrm{d}J}{\mathrm{d}t} = \int (Q + \mathbf{f} \cdot \mathbf{v})\mathrm{d}\tau. \tag{6.2}$$

The energy balance in a system of axes moving at the velocity **u** is obtained by replacing the velocity **v** with **v** + **u** for the same value of the intrinsic quantities J, Q, **f**. Subtracting

[14] Poincaré 1901b, pp. 453, 600.

[15] Poincaré 1901b, p. 536.

the two versions of the energy balance gives $\int \mathbf{f} \cdot \mathbf{u} d\tau = 0$ for any $\mathbf{u}$, hence $\int \mathbf{f} \cdot d\tau = \mathbf{0}$: the forces acting on the medium sum to zero, in conformity with the reaction principle.[16]

Poincaré returned to the optics of moving bodies a year later, in the summer of 1900, in his address to the International Congress of Physics in Paris. After raising the subversive question, "And our ether, does it really exist?", he cites two known facts in favor of this entity: the propagation of light waves at a finite speed, and Fizeau's partial-drag experiment through which "we seem to be touching the ether with our finger." He then imagines two ways one might someday "touch the ether even closer": observe a violation of the equality of action and reaction when applied to matter alone, or detect effects of the motion of the earth on optical experiments. The ether would be needed in the first case to provide for the missing momentum; in the second case as a concrete reference of motion. Many physicists, Poincaré goes on, have tried to detect the ether wind experimentally: "Their hope was illusory." Lorentz's theory, though it is the best available, cannot be the last word because it contradicts the principle of reaction and because it has accumulated hypotheses to explain the lack of effects of the ether wind: "Of hypotheses there is never lack." If the ether exists, it can only be as a ghost devoid of any materiality.

6.2 For the Lorentz jubilee

In the same summer, Poincaré also spoke at the International Congress of Philosophy (both congresses were connected to the Paris Exposition of 1900), this time on the foundations of mechanics. We earlier saw that on this occasion he discussed the "principle of relative motion," the French attempts to derive Newton's second law of motion from this principle, and the difficulty of justifying the limitation of this principle to inertial reference frames. This takes us to the next international event in which Poincaré participated: the publication of a collective volume celebrating Lorentz's doctoral jubilee in 1900. With proper apologies, Poincaré contributed a criticism of Lorentz's theory with regard to the principle of reaction.[17]

The fictitious fluid

In this memoir, Poincaré first derives a few formal consequences of Lorentz's field and force equations, concerning energy, momentum, and center of mass. Calling m the mass density of matter and j the energy density $(\mathbf{e}^2 + \mathbf{b}^2)/2$ of the field, the Maxwell–Lorentz equations imply the local energy relation

$$\frac{\partial j}{\partial t} + \nabla \cdot (c\mathbf{e} \times \mathbf{b}) = -\rho \mathbf{v} \cdot \mathbf{e} \tag{6.3}$$

[16] Poincaré 1901b, pp. 420–421.

[17] Poincaré 1901a; 1900a. On the former, see pp. 40–41, 74–75, this volume.

and the global momentum relation[18]

$$\int m\mathbf{v}d\tau + \int c^{-1}\mathbf{e} \times \mathbf{b}d\tau = \text{constant}. \tag{6.4}$$

In turn, these two relations imply

$$\frac{d}{dt}\int c^{-2}j\mathbf{r}d\tau + \int m\mathbf{v}d\tau + \int (c^{-2}\rho\mathbf{v} \cdot \mathbf{e})\mathbf{r}d\tau = \text{constant}. \tag{6.5}$$

When there is no energy transfer between matter and field, the third term vanishes and the theorem of the uniform motion of the center of mass of the system is saved by associating with the electromagnetic field the flow of a fictitious fluid of mass density j/c^2. In the general case, Poincaré adds the fiction of a latent, ether-bound fluid locally converted into free fluid at the rate $\rho\mathbf{v} \cdot \mathbf{e}/c^2$, so that the center of mass of matter, free fluid, and latent fluid moves uniformly. It must be emphasized that he introduces these fictitious entities only to show more precisely how Lorentz's theory violates the theorem of the center of mass.[19]

Poincaré then illustrates this violation by the recoil of a directional radiator (a Hertz oscillator at the focus of a parabolic mirror). He recalls that all theories (Maxwell's, Hertz's, and Lorentz's) lead to the same recoil and that Maxwell, in his treatise, has indeed computed the momentum imparted to a mirror by reflecting radiation. Yet there is no violation of the reaction principle in Hertz's theory, because its material medium (typically the air) furnishes the missing momentum. One could hope for a similar compensation to occur in Lorentz's theory when the radiator is immersed in air. Poincaré shows this is not the case: the air of optical index n takes only the fraction $(n^2-1)/(n^2+1)$ of the missing momentum.[20]

Why should we care so much about a violation of the reaction principle when applied to matter alone? To answer this question, Poincaré first recalls that in ordinary mechanics the inequality of action and reaction between two bodies would imply the possibility of perpetual motion: the system obtained by joining the two bodies with a rigid bar would forever accelerate under the net force, granted that the forces only depend on the relative position of the two bodies. More generally, Poincaré goes on, for a conservative mechanical system of potential energy U, velocities $\mathbf{v}$ and masses m, we have

$$\sum \frac{1}{2}m\mathbf{v}^2 + U = \text{constant} \tag{6.6}$$

in a given inertial frame. According to the "principle of relative motion," a similar relation also applies in a frame moving at the constant velocity $\mathbf{u}$ with respect to the former frame:

$$\sum \frac{1}{2}m(\mathbf{v} + \mathbf{u})^2 + U = \text{constant}. \tag{6.7}$$

---

[18] In 1900, Poincaré had no reason to doubt the expression $m\mathbf{v}$ of the momentum density of matter. He assumed that eventual nonelectromagnetic forces balanced each other out.

[19] Poincaré 1900a, pp. 253–260.

[20] Poincaré 1900a, pp. 260–267; Maxwell 1873, secs. 792–793.

The difference between these two relations holds for any $\mathbf{u}$ if and only if the total momentum $\sum m\mathbf{v}$ is also a constant. In brief, the energy principle and the principle of relative motion together imply the principle of reaction.[21]

The recoiling radiator

In Lorentz's theory, we do not only have material bodies, we also have the ether. This is why, Poincaré remarks, physicists originally did not expect the principle of relative motion to apply to matter alone. Yet, the principle seems to hold a posteriori: ether-wind experiments have consistently given a null result. By the former reasoning, this should imply that the principle of reaction also holds. Yet it does not in Lorentz's theory. Poincaré solves this paradox by arguing that Lorentz's theory truly involves a first-order violation of the principle of relative motion. For this purpose, he again considers a Hertzian oscillator placed at the focus of a parabolic mirror and emitting radiation at a constant rate. This system moves with the absolute velocity $\mathbf{u}$ in the direction of emission, and is so heavy that the change of its velocity can be neglected during recoil. For an observer at rest in the ether, the conservation of energy reads

$$S = J + (-J/c)u, \tag{6.8}$$

where S is the energy spent by the oscillator in a unit time, J the energy of the emitted wave train, and $-J/c$ the recoil momentum according to Lorentz's theory. For an observer moving at the velocity $\mathbf{u}$ of the emitter, the recoil force does not work, and the spent energy S is obviously the same. According to the Lorentz transformations for time and fields (to first order), this observer should ascribe the energy $J(1 - u/c)$ to the emitted radiation and the value $(-J/c)(1 - u/c)$ to the recoil momentum. Hence the energy principle is satisfied for the moving observer, but the (time integral of the) electromagnetic force acting on the emitter is modified by Ju/c^2. Poincaré regards this difference as a first-order violation of the relativity principle, the desired counterpart of the first-order violation of the principle of reaction.[22]

The apparent time

In this calculation, Poincaré uses Lorentz's first-order field transformations, including the local time $t' = t - ux/c^2$. Unlike Lorentz, he interprets the transformed fields and coordinates as those measured by the moving observers under natural conventions. In particular, he defines the local time in the following manner:[23]

---

[21] Poincaré 1900a, pp. 270–271. The first known reasoning of this kind is in Schütz 1897; see p. 50, this volume.

[22] Poincaré 1900a, pp. 271–278. As Poincaré knew, in 1898 Alfred Liénard had obtained the complementary force $-\rho(\mathbf{v} \cdot \mathbf{e})\mathbf{u}/c^2$ by applying Lorentz's first-order field transformation to the Lorentz force. This expression agrees with Poincaré's Ju/c^2 since $\rho(\mathbf{v} \cdot \mathbf{e})$ represents the work done by the electric force density $\rho\mathbf{e}$ on the moving charge, which is the opposite of the radiated energy in the case of a radiating electron. Cf. Darrigol 2000b.

[23] Poincaré 1900a, p. 272.

> I suppose that observers placed in different points [of the moving frame] set their
> watches by means of optical signals; that they try to correct these signals by the trans-
> mission time, but that, ignoring their translatory motion and thus believing that the
> signals travel at the same speed in both directions, they content themselves with cross-
> ing the observations, by sending one signal from A to B, then another from B to A.
> The local time is the time given by watches adjusted in this manner.

Poincaré states this *en passant*, gives no proof, and appears to assume that Lorentz already
knew it. The proof goes as follows. When B receives the signal from A, he sets his watch to
zero (for example), and immediately sends back a signal to A. When A receives the latter
signal, he notes the time τ that has elapsed since he sent his own signal, and sets his watch
to the time $\tau/2$. By doing so he commits an error $\tau/2 - t_-$, where t_- is the time that light really
takes to travel from B to A. This time and that of the reciprocal travel are given by

$$t_- = AB/(c + u) \quad \text{and} \quad t_+ = AB/(c - u), \tag{6.9}$$

since the velocity of light is c with respect to the ether (see Figure 6.1). The time τ is the sum
of these two traveling times. Therefore, to first order in u/c the error committed in setting
the watch A is

$$\tau/2 - t_- = (t_+ - t_-)/2 = uAB/c^2. \tag{6.10}$$

At a given instant of the true time, the times indicated by the two clocks differ by uAB/c^2,
in conformity with Lorentz's expression of the local time.

One may wonder how and why Poincaré arrived at this interpretation of the local time.
That it eluded Lorentz and any other student of his theory is not a surprise: most physicists
accepted Newton's absolute time uncritically, and Lorentz was satisfied with a merely formal interpretation of the corresponding states. In contrast, Poincaré was among the few
critics of mechanical time, and two years earlier he had produced a most searching analysis of the measurement of time. As we saw in Chapter 3, this analysis included the optical
definition of simultaneity based on the convention that the velocity of light is a constant.
When, in 1900, Poincaré's thought experiments on radiation brought him to consider the
electromagnetic fields measured by moving observers, he could not fail to notice that the
measured fields agreed with the Lorentz-transformed fields, save for the local time-shift.
For instance, observers moving at the velocity $\mathbf{u}$ see the field $\mathbf{e} + c^{-1}\mathbf{u} \times \mathbf{b}$ since they rely on
the motion of a comoving test charge. It then becomes natural to inquire about the time
measured by the moving observers. According to the principle of relativity, these observers

Fig. 6.1. Cross-signaling between two observers moving at the
velocity u through the ether. The points A, A', A'', B, B', and
B'' represent the successive positions of the observers in the
ether when the first observer sends a light signal, when the
second observer receives this signal and sends back another
signal, and when the first observer receives the latter signal.

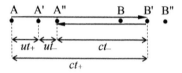

have no way to detect their common motion through the ether. Therefore, in optically synchronizing their clocks they must do as if the velocity of light had the constant value c. The result is the local time.

Critical conclusions

No matter how important we may retrospectively judge this insight, Poincaré promptly returns to his discussion of the principle of reaction. He concludes by asserting that Lorentz's theory implies "intimately connected" violations of the reaction principle and the principle of relative motion. He also notes that in Fizeau's experiment, the velocity of light with respect to the moving water has the value in water at rest only if this velocity is reckoned with respect to Lorentz's local time. In other words, the Galilean transformation must be "corrected" to include the time transformation. This correction, by analogy with the radiation case, should imply a violation of the reaction principle. Poincaré concludes: "All theories respecting the principle would seem to be condemned in bulk, *unless we consent to deeply modify our ideas on electrodynamics.*"[24]

In retrospect, this memoir marks a major turning point in the prehistory of relativity theory, for at least five reasons. First, Poincaré reinterprets Lorentz's corresponding states, which were merely fictitious in Lorentz's approach, as the states measured by moving observers. This move considerably simplifies the application of Lorentz's transformations in deducing the invariance of electrodynamic phenomena, since the transformed states now directly concern the moving system and attached observers. Second, Poincaré understands that Lorentz's local time is the time measured by observers who, in conformity with the relativity principle, act as if the propagation of light on earth were unaffected by the motion of the earth. Third, in his discussion of the violation of the principle of reaction, Poincaré formally (not really) ascribes the momentum density $c^{-1}\mathbf{e} \times \mathbf{b}$ to the electromagnetic field. Fourth, he combines the energy principle and the relativity principle in an original manner, thus detecting a paradox that Einstein would later solve by assuming the inertia of energy. Fifth, Poincaré gives a name to the impossibility of detecting the motion of the earth through the ether: "the principle of relative motion," which would soon become "the relativity principle." The name being borrowed from mechanics, he thereby fuses two kinds of relativity: the relativity of optical phenomena and the relativity of mechanical motion. This implies an extension of the mechanical principle of relativity to electromagnetism, and a restriction of optical-electrodynamical relativity to inertial frames.

Instead of highlighting these multiple innovations, Poincaré apologizes to his readers "for having spent so much time on ideas so little new." This modesty, and perhaps too the singularity of his worries regarding the reaction principle, reduced the impact of his memoir. No one before 1904–1905, not even Lorentz, seems to have appreciated the game-changing reinterpretation of the corresponding states. As we will see in a moment, the only way Poincaré's memoir impacted contemporary research was by inspiring the concept of electromagnetic momentum—which he then meant to reject![25]

---

[24] Poincaré 1900a, p. 278. For the local time in Fizeau's experiment, see p. 146, this volume.

[25] Poincaré 1900a, p. 252.

6.3 Inside the electron

The electromagnetic worldview

In the short term, a more influential contribution to the Lorentz jubilee volume was a memoir in which Willy Wien launched a program often called "the electromagnetic world view." In 1881 J. J. Thomson had remarked that the interaction of an electrically charged particle with its own magnetic field implied an additional inertia. In the 1890s, Wiechert and Larmor had speculated that all matter could be made of electrons that were mere singularities in the electromagnetic ether. They believed the field equations implicitly contained the electron's laws of motion, including the Lorentz force, inertia, and the law of acceleration. In order to justify the Lorentz contraction, Lorentz assumed molecular forces to behave like electromagnetic forces under the ether wind. In a memoir of 1899, he proposed that the law of motion of an electron might be Lorentz invariant, in which case the mass of the electron would be a velocity-dependent quantity. A few months later, he revived an old idea by Ottaviano Mossotti and Wilhelm Weber; namely, the Newtonian attraction between two bodies would be the resultant of two slightly different electrostatic forces, an attraction between the positive electric fluid particles of one body and the negative particles of the other, and a slightly inferior repulsion between electric particles of the same sign. Lorentz replaced the electric particles with positive and negative electrons, and the electrostatic force with an electromagnetic action propagated at the velocity of light through the stationary ether. He then tried and failed to explain the anomaly in Mercury's perihelion as a second-order effect of the delayed propagation.[26]

In his contribution to the Lorentz jubilee, Wien combined these suggestions to propose a grand program of reduction of all physics to electromagnetism. He now adopted Lorentz's stationary ether and the electrons migrating through it. In addition, he reduced all the inertia of an electron to self-induction. On the basis of a calculation done in 1897 by George Searle for an ellipsoidal electron model (dilated by $(1 - v^2/c^2)^{-1/2}$ in the normal plane), he predicted that the electromagnetic mass should increase when the velocity of the electron approaches the velocity of light, and found partial confirmation in Philipp Lenard's experiments on fast cathode rays. He also remarked that Lorentz's electromagnetic theory of gravitation automatically explained the mysterious proportionality of the inertial and gravitational masses.[27]

The following year in Göttingen, the cathod-ray expert Walter Kaufmann studied the velocity dependence of the electron's mass by cleverly combining the electric and magnetic deflection of beta rays from radium in an evacuated vessel. Comparing his results with the electromagnetic mass derived from Searle's calculation for a spherical electron, he concluded that this mass was of the same order as the "true" nonelectromagnetic mass. The Göttingen theorist Max Abraham soon developed this spherical model as a basis for a purely electromagnetic electron. He meant the spherical charge distribution to be

---

[26]Wien 1900; Lorentz 1900. On J. J. Thomson, Larmor, and Wiechert, see pp. 132, 147–149, this volume. On the electromagnetic worldview, cf. McCormmach 1970. Friedrich Zöllner had earlier tried to explain the Mercury-perihelion theory in the modified Weber theory: cf. Kragh 2012, p. 15.

[27]Wien 1900; Searle 1897.

strictly rigid, by a purely kinematic constraint. Then the energy and momentum of the electron entirely belong to the accompanying electromagnetic field. Abraham interpreted Poincaré's imaginary-fluid reasoning of 1900 as implying the existence of the momentum density $c^{-1}\mathbf{e} \times \mathbf{b}$ in the electromagnetic field. He therefore computed the momentum of his electron by integrating the momentum of the associated field in the quasi-stationary approximation in which retardation is neglected. The result reads

$$\mathbf{p} = \frac{q^2}{6\pi ac^2} \left[\frac{1+\beta^2}{2\beta} \ln\left(\frac{1+\beta}{1-\beta}\right) - 1 \right] \mathbf{v}, \tag{6.11}$$

where q denotes the charge of the electron, a its radius, and β the velocity ratio v/c. Equating force and momentum variation, Abraham derived the longitudinal and transverse mass of the electron. Kaufmann resumed his experiments and obtained results in reasonable agreement with Abraham's theory.[28]

The revolutionary flavor of Abraham's theory, its mathematical sophistication, and its apparent confirmation attracted much attention. A few other Göttingen theorists including Karl Schwarzschild, Gustav Herglotz, Paul Hertz, and Arnold Sommerfeld became involved. In the summer of 1905, the mathematicians David Hilbert and Hermann Minkowski ran a seminar on electron theory with Wiechert's help.[29] In 1904, Lorentz proposed his own model of an electron with purely electromagnetic momentum, on which more in a moment. These developments impressed Poincaré. At the International Congress on Arts and Science in Saint-Louis later in the same year, he integrated them into his diagnosis of a major crisis in contemporary physics.[30]

The Saint-Louis lecture of 1904

After presenting the general evolution of physics from detailed mechanical modeling to a physics of principles, Poincaré shows that all the major principles are being threatened by new theories and discoveries. Carnot's principle seems incompatible with Boltzmann's theory and Brownian motion. The "principle of relativity" seems unlikely to be true in Lorentz's theory of a stationary ether. To Poincaré's relief, no effect of the ether wind has so far been observed. Lorentz has explained these null results at the cost of accumulated hypotheses: the local time, the Lorentz contraction, and the analogy of all forces with electromagnetic forces with respect to the Lorentz transformations.[31]

The principle of reaction is even more threatened: Lorentz's theory not only implies the recoil of radiation emitter and radiation pressure, but this pressure has recently been confirmed in delicate experiments. One could still save the principle of reaction by having the ether carry the missing momentum:

---

[28] Kaufmann 1901, 1902, 1903; Abraham 1902a, 1902b, 1903. Cf. Miller 1981a, pp. 45–61; Cushing 1981; Hon 1995.

[29] Schwarzschild 1903c; Herglotz 1903; Hertz 1904; Sommerfeld 1904a, 1904b, 1905. Cf. Pyenson 1979.

[30] Lorentz 1904; Poincaré 1904a.

[31] Poincaré 1904a, pp. 309–312.

We can also assume that the motions of genuine matter are exactly compensated by those of the ether. But . . . the principle thus understood will explain everything: whatever be the visible motions, we will always be able to imagine hypothetical motions that compensate them. If it can explain everything, it can predict nothing. . . . Thus it becomes useless.

This is why I have long thought that the consequences of the theory contrary to the principle of reaction would someday be abandoned. Yet, the recent experiments on the electrons from radium rather seem to confirm them.

Kaufmann's experiments indeed confirmed a consequence of the ether's momentum: the velocity dependence of the electron's mass. This meant a violation of yet another principle: the conservation of mass. Not only would the mass of an electron be a function of its velocity, but the mass of any other particle would obey the same law according to Lorentz's assumption that all forces transform like electromagnetic forces during a change of inertial frame:[32]

Of all these results, if they were confirmed, would emerge an entirely new mechanics that would principally be characterized by this fact: no velocity can exceed the velocity of light (since bodies would oppose an increasing inertia to the causes of their acceleration, and this inertia would become infinite when approaching the velocity of light), just as no temperature can fall under the absolute zero. Moreover, for an observer undergoing a translation of which he is unaware, no apparent velocity could exceed the velocity of light. This would be a contradiction if one did not remember that this observer would not use the same clocks as an observer at rest, but would instead use clocks giving the "local time."

Poincaré is now willing to accept a violation of the principles of reaction and mass conservation, as long as the relativity principle is preserved for apparent phenomena. He next moves to the energy principle and its possible violation in radioactivity: Marie Curie's radium seems to be an inexhaustible source of energy (Poincaré nonetheless mentions Ramsay and Soddy's fresh discovery of accompanying chemical transformations). The only principle not yet challenged is the principle of least action, perhaps because of its higher abstraction and generality. Altogether, this is a serious crisis. Poincaré suggests several exit strategies: more experiments on electrons and radiations, alternative models of the electron, or modifications of Lorentz's theory, for instance assuming a contraction of the ether instead of the Lorentz contraction: "Let us take, then, the theory of Lorentz, turn it in all senses, modify it little by little, and perhaps everything will arrange itself." This is indeed what Poincaré would himself do after studying an "extremely important" memoir by Lorentz.[33]

[32] Poincaré 1904a, pp. 314, 317–318. For the confirmation of radiation pressure, see Lebedev 1901; Nichols and Hull 1903.

[33] Poincaré 1904a, p. 319; Poincaré to Lorentz, c. May 1905, in Walter 2007a, p. 255.

Lorentz's contractible electron

In the letter he wrote in 1901 to thank Poincaré for his jubilee contribution, Lorentz downplayed the difficulty with the reaction principle. The principle did not have to apply at the electronic scale, or, better, it could be saved by regarding the $c^{-1}\mathbf{e} \times \mathbf{b}$ in Poincaré's Equation (6.4) as a quantity "equivalent" to a momentum. The equivalence, Lorentz added, would become an identity if matter ended up being reduced to a "modification of the ether." This is exactly what Abraham soon did in his purely electromagnetic model of the electron, and this is what Lorentz himself tried to do in 1904 with a contractible electron.[34]

Remember that in 1899 Lorentz had used the correspondence

$$\mathbf{r}' = \varepsilon(\gamma, 1)\mathbf{r}, \quad t' = \varepsilon(\gamma^{-1}t - \gamma\mathbf{u} \cdot \mathbf{r}/c^2), \tag{6.12}$$

$$\mathbf{e}' = \varepsilon^{-2}(1, \gamma)(\mathbf{e} + c^{-1}\mathbf{u} \times \mathbf{b}), \quad \mathbf{b}' = \varepsilon^{-2}(1, \gamma)(\mathbf{b} - c^{-1}\mathbf{u} \times \mathbf{e}) \tag{6.13}$$

to turn the field equations in the earth frame (moving at the velocity $\mathbf{u}$ with respect to the ether frame) into the field equations for a system at rest in the ether (in the dipolar approximation for the sources). Lorentz further contemplated the possibility that all forces on an electron, including the inertial force, would transform like the Lorentz force through this correspondence. In this case, all optical phenomena would be unaffected by the earth's motion, at any order in u/c. Lorentz's motivation then was to extend the null result of the Michelson–Morley experiment to a variant in which a dense dielectric replaces the air in the interferometer. By 1904 Lorentz had two additional reasons to extend his correspondence theorem: the negative result of Rayleigh's and Brace's attempts to detect the birefringence induced by the Lorentz contraction, and Trouton and Noble's failure to detect a torque on a charged condenser. After finding flaws in Rayleigh's trial of 1902, the American experimentalist DeWitt Bristol Brace had recently confirmed the absence of induced birefringence. His conclusion: either the ether was dragged along by the earth, or the ether wind did not induce the Lorentz contraction. In 1902, the Irish physicist Frederick Trouton had argued that the plates of a charged condenser should tend to orient themselves perpendicularly to the direction of motion of the earth, because the magnetic energy of the charges of the moving plates is minimal for this orientation. In 1904, in collaboration with Henry Noble, he found the orienting torque to be lacking. Again, Lorentz could explain these null results at the atomic level by requiring all forces to transform like electromagnetic forces under the Lorentz transformations. As he knew, for Abraham's electron the inertial force $-d\mathbf{p}/dt$ does not transform like the Lorentz force $q[\mathbf{e} + c^{-1}(\mathbf{u} + \mathbf{v}) \times \mathbf{b}]$. It was time for him to imagine a model of the electron with the desired transformation property.[35]

Lorentz assumes the electron to be the uniformly electrified ellipsoid of semi-axes $(a\varepsilon^{-1}\gamma^{-1}, a\varepsilon^{-1}, a\varepsilon^{-1})$, which the correspondence $\mathbf{r}' = \varepsilon(\gamma, 1)\mathbf{r}$ turns into a sphere of

[34] Lorentz to Poincaré, January 20, 1901, in Walter 2007a, pp. 252–254; Lorentz 1904.

[35] Lorentz 1899a; Rayleigh 1902b; Brace 1904; Trouton 1902, pp. 383–384; Trouton and Noble 1904; Lorentz 1904, pp. 172–173.

radius *a*. He then computes the electromagnetic momentum of the attached field using Abraham's recipe

$$\mathbf{p} = c^{-1} \int (\mathbf{e} \times \mathbf{b}) d\tau. \tag{6.14}$$

For an electron moving at the constant velocity **v**, the electromagnetic field of the corresponding spherical electron at rest reduces to the electrostatic field $\mathbf{e}' = q\mathbf{r}/4\pi r^3$ ($r > a$). Using the field transformation (Equation (6.13)) and calling $\mathbf{e}'_\perp$ the component of this field perpendicular to **v**, this gives

$$\mathbf{p} = c^{-2} \varepsilon \gamma \int \mathbf{e}'_\perp d\tau' = m_0 \gamma \varepsilon \mathbf{v} \quad \text{with } m_0 = q^2/6\pi a c^2. \tag{6.15}$$

In Abraham's quasi-stationary approximation, this expression still holds when the velocity varies in the equation of motion $\mathbf{f} = d\mathbf{p}/dt$. Lorentz thus obtains

$$\mathbf{f} = (m_\parallel, m_\perp)\ddot{\mathbf{r}} \quad \text{with } m_\parallel(v) = m_0 d(\varepsilon\gamma v)/dv \quad \text{and} \quad m_\perp(v) = m_0 \varepsilon\gamma. \tag{6.16}$$

In a detailed analysis of Kaufmann's data, Lorentz find them compatible with his formulas for the parallel and transverse masses (for $\varepsilon = 1$).

For an electron moving on earth at a small relative velocity **v**, Lorentz uses an approximation in which the longitudinal and transverse masses have the constant values $m_\parallel(u)$ and $m_\perp(u)$. In the corresponding states in the same approximation, the Lorentz force and the acceleration are given by

$$\mathbf{f}' = \varepsilon^{-2}(1, \gamma)\mathbf{f} \quad \text{and} \quad \ddot{\mathbf{r}}' = \varepsilon^{-1}\gamma^2(\gamma, 1)\ddot{\mathbf{r}}, \quad \text{with } \gamma = (1 - u^2/c^2)^{-1/2}. \tag{6.17}$$

The corresponding equation of motion will reduce to the equation of motion $\mathbf{f} = m_0\ddot{\mathbf{r}}$ for a slowly moving electron if and only if

$$\varepsilon^2 m_0 = \varepsilon\gamma^{-3} m_\parallel(u) \quad \text{and} \quad \varepsilon^2\gamma^{-1} m_0 = \varepsilon\gamma^{-2} m_\perp(u). \tag{6.18}$$

Taking into account the identity $d(\gamma u)/du = \gamma^3$, this condition is met whenever the global scale factor ε is a constant. The constant must be *one*, since the Lorentz correspondence must reduce to the identity for $\mathbf{u} = \mathbf{0}$. The momentum of Lorentz's electron thus reduces to $m_0\gamma\mathbf{v}$.

Putting together this result, the validity of the Maxwell–Lorentz equations for the corresponding states, and the transformation rule assumed for all forces, Lorentz obtains the lack of effects of motion of the earth on all electrostatic and optical phenomena, including the Trouton–Noble and Rayleigh–Brace experiments. However, he still relies on various approximations for computing the radiation field and for deriving the desired correspondence of the moving system with a system at rest. In particular, the simple expressions he uses for the transformed charge and velocity,

$$\rho' = \varepsilon^{-3}\gamma^{-1}\rho \quad \text{and} \quad \mathbf{v}' = (\gamma^2, \gamma)\mathbf{v} \tag{6.19}$$

work only in the approximation $v \ll u$, for which all $\mathbf{v} \cdot \mathbf{u}$ terms can be ignored in the transformed quantities. Beyond these approximations, Lorentz does not exclude effects of the ether wind. At the end of his memoir, he admits that a sudden charge or discharge of a condenser moving through the ether should imply an impulsive force, since the electromagnetic momentum of the drifting condenser is proportional to its electrostatic energy. Trouton had performed this experiment in 1901 under FitzGerald's suggestion, with a negative result. Lorentz judged that the method of measurement had been too imprecise to detect the effect.[36]

Lorentz introduced the deformation of his electron as a necessary condition for extending his correspondence theorem, and he said nothing on the deeper nature of the deformation. All we know is that he regarded the energy and momentum of his electron as purely electromagnetic and thus ignored any role of nonelectromagnetic forces in the buildup of the electron. Abraham soon criticized Lorentz on this point. For the rigid electron, Abraham noted, the energy variation of the electron and its momentum variation are related through the equation

$$\frac{dE}{dt} = \mathbf{v} \cdot \frac{d\mathbf{p}}{dt}, \tag{6.20}$$

which is an expression of energy conservation: the work of the force acting on the electron serves to increase its energy. For Lorentz's electron, the electromagnetic energy (obtained by integrating Maxwell's energy density) and the electromagnetic momentum do not satisfy this equation. As Abraham explains, the reason is that the product $\mathbf{v} \cdot d\mathbf{p}/dt$ no longer represents the work done on the electron by external forces. There is an additional work owing to the deformation of the electron. Abraham assumes the deformation to work against nonelectromagnetic forces that keep the electron in internal equilibrium. He therefore equates $\mathbf{v} \cdot d\mathbf{p}/dt$ to the variation of the total energy of the electron, which is the sum of the electromagnetic energy and of the potential energy of the additional forces. He of course regards this implicit appeal to nonelectromagnetic forces and energy as an infraction of the electromagnetic worldview.[37]

6.4 The postulate of relativity

The dynamics of the electron

Poincaré was already aware of Lorentz's new memoir when he prepared his Saint-Louis conference. Back home he studied it and found it to be the key to a theory in complete agreement with the principle of relativity. In a series of letters to Lorentz in the spring of 1905, he gave the exact transformations of charge and current, the Lorentz group of

[36]Trouton 1902; Lorentz 1904, pp. 196–197. This effect disappears when the mass-energy equivalence is taken into account; cf. Janssen 2003.

[37]Abraham 1904, p. 578; Abraham to Lorentz, January 26, 1905, cited in Miller 1981a, p. 77; Abraham 1905, pp. 203–208. See also Lorentz 1905, p. 100n; 1909, p. 213. More on this objection on pp. 227–229, this volume.

transformations, and the necessity of complementary, nonelectromagnetic forces to keep the deformable electron together. Poincaré announced his main results in the June 5 issue of the *Comptes rendus*, and published them in full the following year in the *Rendiconti del Circolo Matematico di Palermo* under the title *Sur la dynamique de l'électron*.[38]

Poincaré's first step was to correct and improve Lorentz's correspondence theorem. Remember that Lorentz first performed a Galilean transformation from the ether frame to the earth frame and then transformed back the resulting field equations into the equations for a system at rest in the ether through an ad hoc "correspondence." Instead, Poincaré directly considered the product of these two transformations, which associates a system at rest in the ether to a system moving through the ether (at the earth's velocity). This makes more mathematical sense, because the new transformations, which Poincaré generously calls the "Lorentz transformations," leave the Lorentz–Maxwell equations invariant and therefore belong to a group. They comprehend the boosts

$$\mathbf{r}' = \varepsilon(\gamma, 1)(\mathbf{r} + \mathbf{u}t), \quad t' = \varepsilon\gamma(t + \mathbf{u} \cdot \mathbf{r}/c^2), \tag{6.21}$$

$$\mathbf{e}' = \varepsilon^{-2}(1, \gamma)(\mathbf{e} - c^{-1}\mathbf{u} \times \mathbf{b}), \quad \mathbf{b}' = \varepsilon^{-2}(1, \gamma)(\mathbf{b} + c^{-1}\mathbf{u} \times \mathbf{e}) \tag{6.22}$$

as well as all spatial rotations. Assuming the (positive) scale factor ε to depend on the velocity of the boost only and combining a boost with a rotation of 180° around an axis perpendicular to $\mathbf{u}$, Poincaré gets $\varepsilon(\mathbf{u}) = \varepsilon(-\mathbf{u})$. The inversion of a boost gives him $\varepsilon(-\mathbf{u}) = \varepsilon^{-1}(\mathbf{u})$. These two relations together imply $\varepsilon = 1$. Poincaré observes that the coordinate transformations then leave the form $c^2t^2 - \mathbf{r}^2$ invariant, and he determines their Lie algebra. He shows that the product of two parallel boosts of velocities u and v is a boost of velocity[39]

$$w = \frac{u + v}{1 + uv/c^2}. \tag{6.23}$$

Poincaré obtains the exact invariance of the Lorentz–Maxwell equations thanks to the expressions

$$\rho' = \gamma\rho(1 + \mathbf{u} \cdot \mathbf{v}/c^2), \quad \rho'\mathbf{v}' = (\gamma, 1)\rho(\mathbf{v} + \mathbf{u}) \tag{6.24}$$

for the transformed charge and current required by charge conservation. For the transformed Lorentz force, he gets

$$\mathbf{f}' = (\gamma, 1)[\mathbf{f} + c^{-2}\mathbf{u}(\mathbf{f} \cdot \mathbf{v})]. \tag{6.25}$$

Given that all forces, including the inertial force of electrons, transform in the same way, the Lorentz invariance of the theory will be perfect. For Poincaré, this invariance is the formal expression of the "postulate of relativity":

---

[38] Poincaré to Lorentz, c. May 1905, in Walter 2007a, pp. 255–259; Poincaré 1905a, 1906a. For analyses of the Palermo memoir, cf. Cuvaj 1968, 1970; Miller 1980; Provost and Bracco 2006; Bracco and Provost 2009; Le Bellac 2010; Damour 2017.

[39] Note that for Poincaré $\mathbf{u}$ denotes the velocity of a boosted system (active transformations), so that the signs in front of $\mathbf{u}$ are reversed in comparison with the now usual Lorentz transformations.

> Lorentz's idea may be summarized as follows: the reason why we may, without any change of the apparent phenomena, impress on a system a common translational motion, is that the equations of an electromagnetic medium are not altered by certain transformations, which we shall call *transformations de Lorentz*; two systems, one in translation, the other at rest, thus become the exact image of one another.

As usual for him, Poincaré credits Lorentz with an insight that is truly his own. For Lorentz, the formal agreement of the corresponding states with the states of a system at rest was only approximate and it did not directly imply the invariance of apparent phenomena. In contrast, Poincaré wants the Lorentz invariance to be exact and to directly express the relativity postulate. Unfortunately, his memoir does not dwell on this (retrospectively) essential point. We will see in a moment how he later addressed it.[40]

As the title suggests, Poincaré's memoir is mostly about the construction of a Lorentz-invariant electron dynamics. The goal is to prove that a properly modified version of Lorentz's contractible electron complies with the relativity postulate. The means are the principle of least action—whose solidity Poincaré has earlier emphasized—and nonelectromagnetic forces that keep the electron together during its deformation. Poincaré's developments are difficult to follow for at least two reasons: he relies on an ill-suited expression of the electromagnetic field action, and he covers a wide range of conceivable models of the electron in order to prove that the modified Lorentz model is the only one compatible with the relativity postulate. Indeed, at the time of Poincaré's writing there were three competing models: Abraham's, Lorentz's, and the constant-volume model of Paul Langevin and Alfred Bucherer. In the following, I will use the now preferred version of the electromagnetic action, and I will consider the Lorentz–Poincaré electron only.[41]

In this case, the total action of electron and field reads

$$S = \frac{1}{2}\int (\mathbf{E}^2 - \mathbf{B}^2)\mathrm{d}^4 x - \int (\rho\varphi - \rho c^{-1}\dot{\mathbf{r}}\cdot\mathbf{A})\mathrm{d}^4 x - \alpha\int \chi\mathrm{d}^4 x$$

with $\mathbf{E} = -\nabla\varphi - c^{-1}\,\partial\mathbf{A}/\partial t$ and $\mathbf{B} = \nabla\times\mathbf{A}$. (6.26)

The vectors $\mathbf{E}$ and $\mathbf{B}$ here denote the microscopic fields, φ and $\mathbf{A}$ the potentials. The electron is assumed to be a conducting shell of yet undetermined shape. The density $\rho(\mathbf{x} - \mathbf{X})$ denotes the charge distribution on the shell, $\mathbf{X}$ the instantaneous position of the center of the shell, $\mathbf{r}$ the position of an element of the shell, and $\chi(\mathbf{x} - \mathbf{X})$ the function that takes the value *one* inside the shell and zero outside. The last term of the action thus represents the time integral of the potential energy of the negative pressure $-\alpha$ within the electron. As Poincaré shows (for his version of the action), the field action is invariant through the Lorentz transformations, and so is too the pressure term, because it is equal to the time integral of the electron's volume and because the dilation of time exactly compensates for the contraction of this volume.[42]

[40] Poincaré 1906a, p. 130.

[41] Bucherer 1904, 1905; Langevin 1905. On the various electron models, cf. Janssen and Mecklenburg 2006.

[42] In modern covariant notation the electromagnetic action is $-(1/4)\int F_{\mu\nu}F^{\mu\nu}\mathrm{d}^4 x - \int A_\mu j^\mu \mathrm{d}^4 x$ with $j^\mu = \rho\dot{x}^\mu$. On the history of least action in electrodynamics, cf. Darrigol 2000a, pp. 422–428. On Poincaré and least action, cf. Bracco and Provost 2009.

The quantities to be varied in this action are φ, $\mathbf{A}$, $\mathbf{X}$, and $\mathbf{r}$. The variation with respect to the potentials yields the Maxwell–Lorentz equations, the variation with respect to the deformation $\delta\mathbf{r}$ yields the condition of internal equilibrium of the electron, and the variation with respect to the position $\mathbf{X}$ yields its equation of motion. After performing the first variation, we may use the resulting field equations to arrive at Poincaré's effective action for the internal equilibrium and the motion of the electron:[43]

$$S_{\text{eff}} = -\tfrac{1}{2}\int (E^2 - B^2)\mathrm{d}^4x - \alpha\int \chi \mathrm{d}^4x. \tag{6.27}$$

For an electron moving at the constant velocity $\mathbf{v}$, this action may be rewritten in terms of the variables of the corresponding electron at rest. This gives

$$S_{\text{eff}} = -\tfrac{1}{2}\int E^2\mathrm{d}^3\xi \mathrm{d}\tau + \alpha\int\chi_0\mathrm{d}^3\xi \mathrm{d}\tau = -(U_s + U_p)\int\sqrt{1-v^2/c^2}\,\mathrm{d}t \tag{6.28}$$

if U_s and U_p denote the electrostatic and cohesive energy of the electron. For internal equilibrium, the variation of this action during a deformation of the electrified shell must vanish. This implies the balance between the centrifugal pressure $E^2/2$ and the centripetal pressure $-\alpha$ on the surface of the electron. The latter pressure being uniform, the shell must be a sphere of radius a. Hence we have

$$q = 4\pi a^2 E, \quad U_s = q^2/8\pi a, \quad U_p = \alpha(4/3)\pi a^3, \tag{6.29}$$

and the balance of pressures gives[44]

$$\alpha = q^2/32\pi^2 a^4 \quad \text{and} \quad U_p = q^2/24\pi a = U_s/3. \tag{6.30}$$

Now suppose that other electrons are present (with additional terms in the action) or that additional nonelectromagnetic forces act on the given electron. In the quasi-stationary approximation, we may assume the proper field to "follow" the electron during a moderately accelerated motion and we may still use the effective action (Equation (6.28)) to describe the behavior of this electron. This leads to the Lagrangian

$$L = mc^2\sqrt{1-v^2/c^2}, \quad \text{with} \quad m = (4/3)(U_s/c^2) = q^2/6\pi ac^2. \tag{6.31}$$

The corresponding momentum is

$$\mathbf{p} = \frac{\mathrm{d}L}{\mathrm{d}\mathbf{v}} = \frac{m\mathbf{v}}{\sqrt{1-v^2/c^2}}. \tag{6.32}$$

The inertial force $-\mathrm{d}\mathbf{p}/\mathrm{d}t$ transforms like the Lorentz force in Equation (6.25).

---

[43] Searle 1897 and Abraham 1903 already had the electromagnetic part of S_{eff}.

[44] As Lorentz noted in 1906 in his Columbia lectures, the equilibrium is unstable (Lorentz 1909, pp. 214–215, 324–325). This did not worry Lorentz too much, because he believed additional forces could restore stability. Cf. Damour 2017.

The former expression of the electron's momentum agrees with the value Lorentz obtained by integrating the electromagnetic momentum. Without the cohesive pressure, the Lagrangian would be reduced to $L = \gamma^{-1}U_s$, which is incompatible with the equation $\mathbf{p} = dL/d\mathbf{v}$ and the expression $\mathbf{p} = (4/3)U_s\gamma\mathbf{v}$ of the electromagnetic momentum. Poincaré notes this contradiction early in his text, and he announces it will be solved thanks to the cohesive pressure. Yet, after introducing the pressure term in the action, he forgets to modify the effective action as I have done in Equation (6.28). This slip does not affect the rest of his discussion, since he absorbs the mass factor of the Lagrangian in a redefinition of units.[45]

Beyond the quasi-stationary approximation, Poincaré returns to the full action of the field and the electrons. This action being Lorentz invariant, the equations of motion, no matter how complicated they may be, are necessarily covariant and the relativity postulate is satisfied. Having already admitted the Lorentz covariance of the cohesive forces of the electron, Poincaré goes on to extend this requirement to gravitational forces, so as to make gravitation compatible with the relativity postulate. This part of this memoir will be discussed in Chapter 9. Here it is sufficient to note that in order to build the necessary invariants, Poincaré uses the imaginary time coordinate ict through which the form $\mathbf{r}^2 - c^2t^2$ becomes Euclidean, and he builds Euclidean four-vectors and their scalar products.

Even though Poincaré thus demonstrates the constructive power of Lorentz invariance, he is not satisfied with a merely formal interpretation of this requirement. In his introduction, he writes:

> If the propagation of [gravitational] attraction occurs at the velocity of light, this cannot be by a fortuitous agreement. This must be the consequence of a certain function of the ether. We will have to penetrate the nature of this function and connect it to the other functions of this fluid. We cannot content ourselves with simply juxtaposed formulas that would agree only by some happy coincidence. The formulas should, so to say, penetrate each other. Our mind will be satisfied only when we believe that we perceive the reason of this agreement, so that we may fancy we have predicted it.

> In the event that all interactions would truly propagate at the same velocity, there would be two ways of explaining this coincidence:

> - Either there is nothing in the world that is not of electromagnetic origin.

> - Or this common part of all physical phenomena would only be an appearance, something that would pertain to our methods of measurement. How do we perform our measurements? By superposing objects that are regarded as rigid bodies, would be one first answer; but this is no longer true in the present theory, if one assumes the Lorentz contraction. In this theory, two equal lengths are, by definition, two lengths which light takes equal time to travel through. Perhaps it would be sufficient to renounce this definition so that Lorentz's theory would be as completely overturned [*bouleversée*] as Ptolemy's system was through Copernicus' intervention.

---

[45] Poincaré 1906, pp. 53–54 (contradiction), 158 (effective action), 160 (mass factor absorbed). The contradiction is related to Abraham's earlier criticism of the Lorentz electron; see p. 175, this volume.

Poincaré does not pursue any of these two vague suggestions in his memoir. He regards his theory as fragmentary, provisional, and hypothetical. With his usual eagerness to detect crises, he ends his introduction with the remark that the entire theory might be jeopardized by the recent discovery of magneto-cathodic rays.[46]

The light ellipsoid

To contemporaries and still to us, Poincaré's memoir on the dynamics of the electron was a highly mathematical, unusually long and difficult work published in a mathematical journal. At that time Poincaré did not address the problem of measurement in relativity theory, save for the speculation just cited and for a brief reference to stellar aberration and the Michelson–Morley experiment. He did not even mention his earlier argument that Lorentz's local time represented the time measured by moving observers. One may imagine several reasons for this silence: Poincaré had recently repeated the argument in his Saint-Louis lecture; he perhaps did not yet have the generalization of this argument to higher orders in u/c; he was focusing on the dynamics of the electron, not on the effects of the earth's motion; and he did not discuss any change of reference frame, for he interpreted the Lorentz-transformed field states as the states of a boosted physical system in the ether frame.[47] In this active view of the Lorentz transformation, one may imagine the space- and time-measuring agencies to belong to the boosted system, in which case the invariance of the equations describing the global system implies that the transformed field and coordinates are the measured ones. But there is no trace of such reasoning in the Palermo memoir.

Poincaré first addressed the metric aspects of relativity theory in his Sorbonne lectures of winter 1906–1907 on "The limits of Newton's law." There he argued that the Lorentz contraction was needed for the transitivity of the optical synchronization of distant clocks. This reasoning did not reoccur in his later discussions of time measurement, because he must have realized that the desired transitivity holds for any amount of contraction (or even for no contraction at all). One ingredient of his reasoning, a peculiar ellipsoidal light shell, survived as a means to prove that the Lorentz contraction generally implies the isotropy of the apparent propagation of light on earth.[48]

Poincaré imagines an observer that moves with the constant velocity **u** through the ether and emits a flash of light at time zero. At the value t of the true time, this light is located on a sphere of radius ct centered at the emission point. When measured with Lorentz-contracted rulers, this sphere appears as an ellipsoid of revolution, the half-axes of which have the values $a = \gamma ct$ and $b = ct$ (see Figure 6.2). The eccentricity being $e = \sqrt{1 - b^2/a^2} = u/c$, the focal distance $f = ea = \gamma ut$ is equal to the apparent distance traveled by the observer during

[46] Poincaré 1906, pp. 131–132. The "magneto-cathodic rays" refer to Paul Villard's spurious discovery of magnetically guided rays in cathode ray tubes: cf. Carazza and Kragh 1990.

[47] Cf. Provost and Bracco 2006, who first pointed to this important characteristic.

[48] Poincaré 1906–1907, pp. 217–219; 1908, p. 392; 1909, p. 172; 1913a, pp. 45–46. Cf. Damour 2005, chap. 1; Darrigol 2012b.

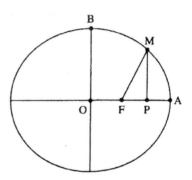

Fig. 6.2. Poincaré's light ellipsoid (a = OA, b = OB, f = OF).

the time t. Therefore, the Lorentz contraction is the contraction for which the position of the observer at time t coincides with the focus F of the light ellipsoid he has emitted.

A second observer traveling with the same velocity **u** receives the flash at time t if and only if his position M belongs to the ellipsoid. The distance FM represents the apparent distance between the two observers, which is invariable. According to a well-known property of ellipses, we have

$$FM + eFP = b^2/a, \tag{6.33}$$

where P denotes the projection of M on the larger axis. The length FP being equal to the difference x' of the apparent abscissae of the two observers, this implies

$$t = \gamma FM/c + \gamma ux'/c^2. \tag{6.34}$$

The apparent time t' being the time for which the velocity of light appears to be c to the moving observers, we have FM = ct' and

$$\gamma t' = t - \gamma ux'/c^2. \tag{6.35}$$

We may add (Poincaré does not) that for the coordinate x measured in the ether frame from the origin F, we have

$$x' = \gamma(x - ut), \tag{6.36}$$

which, together with the former equation, implies

$$t' = \gamma(t - ux/c^2). \tag{6.37}$$

Poincaré's ellipsoid can thus be used to derive the Lorentz transformations. Poincaré rather sees the light ellipsoid as a simple geometrical means to prove that the Lorentz contraction implies the apparent isotropy of light propagation for observers using optically synchronized clocks and contracted rulers. In turn, this isotropy implies the invariance of optical phenomena.[49]

---

[49] Poincaré 1908, p. 393.

This optical measurement strategy allows Poincaré to interpret the Lorentz transformations in a more concrete manner. As he explains in his Sorbonne lectures of 1906–1907, they connect the field equations in the ether frame to the apparent field equations that relate the apparent field quantities and the apparent space and time coordinates in a moving frame. Apparent quantities are those measured by moving observers ignoring their motion through the ether and therefore misestimating the true quantities defined in the ether frame. Poincaré thus assumes the complete invariance of *observable phenomena* when passing from one inertial frame to another, but he does not require the invariance of the *theoretical description* of the phenomena. In his view, the ether frame is a privileged frame in which true space and time are defined. The Lorentz-transformed quantities in another frame are only "apparent." Of course, the choice of the ether frame can only be conventional, since the relativity principle excludes any empirically detectable difference between the various inertial frames.[50]

This attitude explains some features of Poincaré's light ellipsoid that may seem very odd to a modern, Einsteinian reader. For Poincaré, as for Lorentz, the Lorentz contraction is meant to be a physical effect of the motion of a material body through the ether. Measuring a light pulse at a given value of the true time with contracted rods is a natural operation because the pulse is meant to be a disturbance of the ether and because the contraction is caused by the motion of the rulers through the ether. Although this way of mixing quantities pertaining to two different frames of reference may baffle a modern relativist, it is self-consistent and it leads to the correct expression of the Lorentz transformations.

The ether

One may still wonder why Poincaré retained a strictly undetectable ether. Had he not written, some twenty years earlier, that "probably the ether will someday be thrown aside as useless"? As a supporter of Felix Klein's definition of geometry through a group of transformation, was he not prepared to define a new geometry based on the Lorentz group? Poincaré answered this question in a talk given in London toward the end of his life. By that time, he was undoubtedly aware of the possibility, exploited by Hermann Minkowski, of defining the geometry of space–time through the Lorentz group; at the same time, he still believed that the choice of the group defining a geometry was largely conventional and that ancestral habits were important in judging the convenience of a convention:[51]

> What should be our position with regard to these new conceptions? Shall we be compelled to modify our conclusions? No, assuredly: We had adopted a convention because it seemed convenient to us and we were saying that nothing could force us to abandon it. Today some physicists want to adopt a new convention. Not that they are forced to do so. They simply judge this convention to be more convenient. Those who do not share this opinion can legitimately keep the old convention not to disturb their old habits. *Entre nous,* I believe they will do so for a long time.

[50] Poincaré 1906–1907, p. 221.

[51] Poincaré 1912, pp. 159–170, on 170. On Klein and Minkowski, see pp. 252–253, 215–220, this volume.

In Poincaré's eyes, the ether offered a means to preserve the old convention: it defined a reference frame for true space and time. He had other reasons to preserve the ether. Until 1900, his ether served only to illustrate propagation phenomena, and it could not have any detectable motion or momentum. In the following years, Kaufmann's, Lebedev's, and Nichols' experiments forced him to change his mind and accept electromagnetic inertia. Also, he was eager to correct exaggerations of the conventionalism he had earlier promoted:

> "Did you not write," you might say if you were seeking a quarrel with me, "did you not write that the principles, though they are of experimental origin, are now beyond the possibility of experimental attack, because they have become conventions? And now you come to tell us that the triumphs of the most recent experiments put these principles in danger." Very well, I was right formerly, and I am not wrong today. I was right formerly, and what is taking place at present is another proof of it.

In a hardened conventionalism, the necessity of certain conventions and the superfluity of others are exaggerated. In 1902, in a refutation of Edouard Le Roy's nominalism, Poincaré defended the ether against the latter sort of exaggeration:

> It can be said, for instance, that the ether does not have less reality than any external body. To say that this body exists is to say that there is an intimate, robust, and persistent relation between its color, its flavor, and its odor. To say that the ether exists, is to say that there is a natural relationship between all kinds of optical phenomena. Evidently, one proposition does not weigh more than the other.

Poincaré had offered the same comparison in 1888, with the nearly opposite conclusion that the ether, unlike ordinary bodies, would probably someday be rejected.[52]

Poincaré now admitted the possibility that the ether could carry momentum and even that all matter could be represented as a set of singularities of the ether. As we saw, in 1904 he contemplated the possibility that all inertia would be of electromagnetic origin, in which case the ether would be the universal momentum carrier. In 1906, he cautiously announced "the end of matter" in a popular conference:

> One of the most astonishing discoveries announced by physicists during the past few years is that matter does not exist. I hurry to say that this discovery is not yet definitive . . . [If Kaufmann's results and the Lorentz–Abraham theory are correct], every atom of matter would be made of positive electrons, small and heavy, and of negative electrons, big and light. . . . Both kinds have no mass and they only have a borrowed inertia. In that system, there is no true matter; there are only holes in the ether.

This seemingly naive description of the ether reflected a move toward a more substantial concept. Late in his life, in a conference on materialism, Poincaré insisted: "Thus the active role is removed from matter to be transferred to the ether, which is the true seat of

---

[52] Poincaré 1904a, p. 322; 1902b, p. 293.

the phenomena that we attribute to mass. Matter no longer is; there are only holes in the ether."[53]

To summarize, after Kaufmann's experimental confirmation of electromagnetic inertia and after his own discovery that moving observers measure Lorentz-transformed fields and coordinates, Poincaré employed the ether in two manners: as a momentum carrier in which momentum was globally conserved, and as a privileged reference frame in which the conventions of the usual chrono-geometry could be maintained. The ether thus offered a way to conciliate relativity theory with older intuitions and conventions. This is not to say that Poincaré was unaware of the possibility of changing the conventions. He just did not believe in the necessity of such a radical move.

Conclusions

From his Sorbonne chair, Poincaré took a God's-eye view of the still competing theories of optics and electrodynamics. Having no pet theory, he compared the existing theories to extract common structures and to relativize the attached mechanical picture. In particular, the ether, having been the subject of so many contradictory mechanical constructs, was not to be compared to the usual bodies of mechanics. Poincaré then decided that it should not carry momentum, and that motion through it should not be detectable. Perhaps, he prophesied in 1888, the ether would someday completely disappear from physics. These were unique, daring pronouncements at a time when every physicist believed the ether to exist just as much as the sun and the moon.

Another characteristic of Poincaré's criticism was his reliance on general principles. Like Maxwell and Helmholtz, he renounced detailed mechanical pictures of everything in favor of general principles of mechanical origin: the energy principle, the principle of reaction, the relativity principle, and the principle of least action. These principles were the result of broad empirical inductions, and they could be applied directly to the controllable aspects of phenomena that were not necessarily mechanical. Denying the energy principle or the principle of reaction would have implied the possibility of perpetual motion. Denying the relativity principle would have hurt common sense and contradicted experience. Poincaré perceived an intimate connection between these three principles; namely, the energy principle and the Galilean relativity principle together imply the reaction principle. Being the result of broad empirical inductions and being then elevated to the status of necessary conventions, the principles could be used to judge existing theories and to guide the construction of new theories, independently of any mechanical picture. This is the physics of principles Poincaré promoted at the turn of the century.

Principle-based criticism is ambiguous when applied to theories involving the ether. In applying the reaction principle and the relativity principle, should the ether be regarded as any other body, or should these principles be restricted to ordinary matter? Poincaré chose the second alternative, in part because of his general skepticism toward the idea of a mechanical, substantial ether, but also for empirical and methodological reasons. No one, in the nineteenth century, had been able to verify the radiation pressure predicted by

---

[53] Poincaré 1906b, pp. 201–202; 1913c, p. 65.

Maxwell. In Lorentz's theory, this pressure implied a violation of the reaction principle unless the ether was allowed to carry momentum. Poincaré regarded this way of saving the reaction principle as artificial and improbable. As for the relativity principle, Poincaré invoked the repeated failures to detect effects of the ether wind from Arago to Michelson and Morley. He found it strange that in a theory like Lorentz's these failures would be accounted for only in successive approximations, and he betted on a strict validity of the relativity principle when applied to matter alone in optics and electrodynamics. He first did so in 1895, a few years before anyone else challenged Lorentz's contrary opinion.

The aforementioned connection between the reaction principle and the relativity principle does not smoothly extend to ether theories because it presupposes direct, unretarded action at a distance. This is why Poincaré, in 1900, appealed to a thought experiment to extend this connection to the radiation case. He considered the recoil of a unidirectional radiator in two different frames, and found that Lorentz's theory implied different values for the recoil force. This force violates the reaction principle (provided the radiation cannot have momentum), and its frame dependence implies a first-order violation of the relativity principle. In this reasoning, Poincaré made two important side discoveries. First, he showed that the reaction principle could be *formally* saved by associating with the radiation an imaginary fluid of momentum density $c^{-1}\mathbf{e} \times \mathbf{b}$. Second, he showed that Lorentz's corresponding states truly were the states measured by moving observers. He thereby interpreted Lorentz's local time as the time given by optically synchronized clocks. The (first-order) Lorentz transformations thus became a relation between the phenomena observed in two different inertial frames, and the (first-order) Lorentz symmetry became the direct expression of the (first-order) invariance of optical phenomena.

Save for this partial symmetry, Poincaré still believed Lorentz's theory to be violating both the relativity principle (at first order already) and the reaction principle. He nonetheless judged Lorentz's theory to be the best available, and he expected the future theory to be a modified version of Lorentz's. Although he hesitated on the magnitude of the needed modifications,[54] he did not expect an easy way out. The major stumbling block was Fizeau's running-water experiment. With it, "we seem to be touching the ether with our finger" because if there were no ether, the velocity of light with respect to the running water could not possibly depend on the velocity of the water flow. As Poincaré noted in 1900, the observed partial drag could be conciliated with the relativity principle by having the comoving observers measure the velocity of light in water with optically synchronized clocks. Fizeau's result still implied a violation of the reaction principle, as Poincaré had argued since 1895.

In retrospect, Poincaré was right to apply the relativity principle to electrodynamic phenomena, and wrong to apply the reaction principle to matter alone, even though there seemed to be an intimate connection between the two principles. New events forced him to change his mind on the reaction principle. Early in the twentieth century, the supporters of a new electromagnetic worldview not only accepted the concept of electromagnetic momentum, but they also traced the inertia of matter to the momentum of the constitutive electromagnetic field. Kaufmann's experiments confirmed the velocity dependence that

---

[54] As we saw, he used the adverb "deeply" in 1895 and 1900, "not too deeply" in 1899, and "little by little" in 1904.

Abraham predicted for the mass of the electron in this picture. In addition, Lebedev and Nichols confirmed the existence of radiation pressure. By 1905 Poincaré understood that the concomitant violations of the reaction principle (when applied to matter alone) were after all compatible with the relativity principle. At the electronic level, the reason is that the lack of invariance of the Lorentz force is compensated for by the requirement that all forces, including cohesive and inertial forces, should transform like the Lorentz force under the Lorentz transformations. Had Poincaré reconsidered his radiative-recoil experiment of 1900 in the same light, as Einstein later did for him, he would have seen that in this case the noninvariance of the force causing the recoil is compensated for by a diminution in the mass of the radiator in proportion with the emitted energy.[55]

In 1899, Lorentz had already remarked that the noninvariance of the Lorentz force could be compensated for by requiring the same transformation for all forces, including inertial forces. He also had the exact Lorentz transformations, except for the source terms in the field equations. In 1904, in reaction to Abraham's fully electromagnetic electron and also in response to Poincaré's criticism regarding the *coups de pouce* of his earlier theory, Lorentz designed the contractible electron for which the inertial force transforms like the Lorentz force, and he thus generalized his correspondence theorem to any optical experiment involving dielectrics and their oscillating electrons. The transformation he thereby used for the field equations still differed from the now admitted transformations in the source terms. He still expected violations of the relativity principle for nonoptical phenomena.

In early 1905, Poincaré obtained the exact "Lorentz transformations" for the source terms too. He defined these transformations as those turning a system globally at rest in the ether into the same system in uniform translation in the ether (whereas Lorentz's own transformations corresponded to the inverse of a transformation of the former kind, combined with a Galilean transformation). With this new definition, the transformations and the spatial rotations of the system together form a group. Poincaré regarded the symmetry of the fundamental equations of a theory under this group as the expression of the "relativity postulate." He then relied on an invariant action to construct a relativistic electron model involving both electromagnetic forces and a cohesive pressure. He also proposed a covariant generalization of Newton's law of gravitation.

In the resulting memoir we may recognize essential aspects of special relativity as we know it: the postulated validity of the relativity principle, the Lorentz group, the requirement of the invariance of all the laws of physics under this group, and even the idea of using invariant scalars and covariant four-vectors in the construction of invariant laws. In addition, Poincaré soon explained how to generalize his earlier considerations about quantities measured by moving observers. He now understood that the Lorentz-transformed coordinates and fields were exactly those measured under conventions compatible with the relativity principle. He also understood that the Lorentz transformations directly resulted from these metric considerations, and that the relativistic dynamics of the electron was the only covariant generalization of the Newtonian dynamics—a result anticipated by Lorentz in 1899 and obtained by Einstein in 1905.

The similarities end there. Until the end of his life Poincaré retained the ether as the momentum carrier for matter and fields, and as the privilege frame in which true space

---

[55] On this last point, see pp. 210–211, this volume.

and time are defined. In this view, the space and time measured in moving frames are only apparent, even though the relativity principle forbids any empirical selection of the ether frame among the inertial set of frames. Poincaré did not address simple perspectival implications of relativity theory such as time dilation or the relativistic Doppler effect, he used a now undesirable mix of true and apparent space–time coordinates in some of his reasoning, and he shared Lorentz's view of the Lorentz contraction as a dynamical effect of the motion of a body through the ether. Lastly, he did not regard the relativity postulate as a self-sufficient tool in the parallel construction of diverse phenomena (electromagnetism, cohesion, gravitation): he hoped for a homogeneous ether-based construction of all phenomena that would account for the shared Lorentz-group structure.

In brief, Poincaré created one form of relativity theory, but not the one to which we are now accustomed. He has rarely been given proper credit for his contribution to relativity theory. He is either regarded as a marginal contributor to the prehistory of this theory or as its sole inventor. These two misconceptions partly result from the early reception of his and Einstein's works, which will be discussed in Chapter 7. It is also a consequence of a few peculiarities of Poincaré's style. He liked walking on the tightrope of paradoxes, he presented his thoughts in flux, with frequent counterfactual detours, and he did not care to give his theories a completely polished, systematic form. His way of expression was lively, thought-provoking, but also perplexing for those who did not care to penetrate his motivations. He often attributed his own innovations to others, or he presented them as minor "details" added to the great works of forerunners. This is strikingly the case for his most essential contributions to relativity theory in the Lorentz jubilee volume and in the Palermo memoir. He introduced the physical interpretation of the corresponding states and the local time as if it could already be found in Lorentz, he called transformations he truly invented "Lorentz transformations,"and he presented the relation between the relativity postulate and the Lorentz invariance as the precise expression of an idea by Lorentz. By way of compensation, Abraham credited him for the concept and formula of the electromagnetic momentum, which he meant to reject.

Poincaré saw himself as a lucid outsider, able to clarify, criticize, correct, and recast the theories of master physicists, but not in charge of building entirely new foundations. This explains his humble attitude and his incapacity to project the magnitude of his achievements in physics. In addition, his readers may have felt uncomfortable with the odd mix of revolutionary and conservative spirit that permeates his writings in mathematical physics. On the one hand, he was eager to criticize received theories and to anticipate radical change. On the other hand, he resisted Einstein's and Minkowski's etherless space–time. His most important moves had both a conservative and a revolutionary side: the optical measurement of time was meant to preserve the relativity principle, and the exact Lorentz invariance of the Maxwell–Lorentz equation was meant to bring the finishing touch to Lorentz's strategy. While, as a mathematician, he understood better than anyone that the relativity postulate brought with it an essentially new group structure, he tried to preserve as much as possible of the old conventions for the definition of space and time.[56]

---

[56]More on this on pp. 182, 230–231, this volume.

7

THE RELATIVITY THEORY OF EINSTEIN, MINKOWSKI, AND LAUE

> It was enough to realize that a certain auxiliary quantity introduced by Lorentz and called "local time" could be simply defined as "time."[1] (Albert Einstein, 1907)
>
> From now on, space for itself and time for itself should be completely reduced to shadows, and only a kind of union of them should preserve its autonomy.[2] (Hermann Minkowski, 1908)
>
> Already in Newtonian mechanics, it has often been said that it is more logical to place the dynamics of continua before the dynamics of point-masses. It seems to me that the two mentioned problems [consistent electron model and Trouton–Noble null result] suggest that in relativity theory the advantages of this order over the reverse one are even greater than in the old theory.[3] (Max Laue, 1911)

For most of us, the advent of relativity theory was a great intellectual revolution for it affected the most basic concepts through which we order the events of the world: time and space. As Poincaré resisted change of this kind, he cannot be regarded as the sole inventor of relativity theory. No one can. The theory is best seen as the end result of the constructive efforts of a few penetrating, innovative thinkers. Arguably, the most important figures were Lorentz for building the electromagnetic scaffolding of the theory, Poincaré for introducing the relativity principle and the Lorentz group, Einstein for redefining time and accepting the complete equivalence of inertial frames, Minkowski for revealing space–time as the mathematical structure behind this equivalence, and Laue for providing the relativistic continuum dynamics without which relativity theory would largely remain an empty shell.

The magnitude of Einstein's innovations in the relativity papers of 1905 is obvious to anyone. His contemporaries resisted them or applauded them accordingly. Yet the theory was not quite ripe in this early version. In 1908, Minkowski altered it in an essential manner. Whereas Einstein had insisted on the relativity of our definitions of space and time relations, Minkowski unveiled the intrinsic objects of the theory in a nonfactorable space–time and he built a tensor calculus to manipulate them. His formulation of relativity theory played a decisive role in its early reception, because in the eyes of Göttingen-style

---

[1] Einstein 1907c, p. 413.

[2] Minkowski 1909, p. 104.

[3] Laue 1911c, pp. 525–526.

Relativity Principles and Theories from Galileo to Einstein. Olivier Darrigol, Oxford University Press.
© Olivier Darrigol (2022). DOI: 10.1093/oso/9780192849533.003.0007

electron theorists it revealed a formal simplicity and beauty that was lacking in the competing theories. The new converts still found something to be missing in relativity theory. As Laue observed, the principles of this theory did not lead to any new empirical prediction in the case of pure electromagnetism, and for the concrete dynamics of objects that were not purely electromagnetic it could not predict anything without presupposing the asymptotic validity of Newtonian mechanics. Laue filled this gap in 1911 by postulating an energy–momentum tensor for nonelectromagnetic forces. The resulting continuum dynamics convinced the last reticent electron theorists, and it soon informed Einstein's quest for a relativistic gravitation theory—as we will see in Chapter 9.[4]

These are the reasons why in this chapter prominence is given to the contributions of Einstein, Minkowski, and Laue. Yet we should not isolate their works from other contemporary efforts in the electrodynamics of moving bodies and in electron theory. A selective hero-worshiping account would deform our understanding of the most important breakthroughs, and it would falsely suggest that relativity theory was the only reasonable answer to the questions it purported to answer in the years 1905–1910. In earlier chapters, we saw how Lorentz, Larmor, and Wiechert prepared the ground for relativity theory with a deep reform of the received electrodynamics, and we saw how Poincaré introduced his own version of relativity theory. In this chapter, we will see that Einstein's theory was competing with other innovative theories by Cohn, Abraham, and Bucherer, and that besides Minkowski and Laue, there were a few other notable contributors to early relativity theory: particularly Planck, who was Einstein's earliest, strongest supporter and the pioneer of relativistic thermodynamics, but also Mosengeil, Born, Ehrenfest, Herglotz, Nordström, Sommerfeld, and a few others.

In particular, we should not underestimate the importance of electron theory in the formation and evolution of relativity theory. Historians have often insisted that Einstein's success depended on his emancipation from contemporary electron modeling. Although there is some truth to this, we should not forget that Lorentz's and Poincaré's essential contributions belonged to electron theory, that Einstein himself was addressing problems of contemporary electron theory, and that Laue's important contribution largely resulted from his concern with electron modeling. In the history of relativity, the electron often acted as a toy model for developing new relativistic concepts and methods. That the equation of motion of the electron could be derived just from Lorentz covariance and Newtonian correspondence did not mean that nothing could be learned from relativistic models of the electron.

In investigating the origins of Einstein's spectacular breakthrough, historians have to face the fragmentary character of documentary evidence. Almost no pre-1905 manuscripts of Einstein's have survived, and we have only a few letters to friends and to his future wife Mileva Marić. Commentators have often relied on his recollections, or they have proposed personal reconstructions of his path. I will avoid these two ways of filling the gaps. One reason for my reticence is that Einstein's recollections are often inconsistent and unreliable. The alleged trigger of his breakthrough varies from place to place: it could be the Michelson—Morley experiment, Lorentz's local time, the Fizeau experiment, or the

---

[4]Laue's observation is in Laue 1913, pp. 174–175.

relativity of electromagnetic induction. In one case, Einstein wants us to believe that at age sixteen he imagined himself riding a light wave, thus seeing a static field, and inferring the necessary relativity of optical phenomena.[5] The conclusion is contradicted by ample epistolary evidence that Einstein did not adopt the relativity principle before 1901. In another case, Einstein asserts that the Michelson–Morley experiment played no role in his thinking, whereas we know he early read authors who gave it a prominent role.[6] As for reconstructions, they betray the historians' excessive intolerance of indeterminacy. There is no need to know in every detail how Einstein thought about the electrodynamics of moving bodies between 1895 and 1905. It is enough to show that by 1905 he had all the keys he needed in his hands. This can be inferred from the precious little information we have (mostly from letters to Marić) and from contemporary attempts of a similar or related kind.

There are two common myths about the young Einstein: that his revolutionary endeavors were absolutely unparalleled, and that he worked in isolation, with little or no awareness of contemporary developments. In reality, Poincaré, Cohn, Bucherer, and Abraham all wanted a radical break from received theories, and we will find instructive similarities between the breaks they conceived and Einstein's own revolution. Moreover, there is plenty of evidence that the young Einstein read much of the best physics literature of his time and learned from it, although he often overlooked work that was directly relevant to his projects. He heard lectures on Kant and he read philosophy by Mach, Poincaré, Hume, Pearson, Spinoza, and Stuart Mill with his friends of the Akademie Olympia. Most important, the physicists Einstein read before he read much philosophy, principally Helmholtz, Hertz, and Drude, had a philosophico-critical approach to physical theory, they promoted the very style of theory construction that Einstein implemented in relativity theory, and they identified the electrodynamics of moving bodies as a domain in need of deep reform.[7]

Section 7.1 of this chapter describes how the young Einstein first groped toward a theory similar to Lorentz's before reading Lorentz. Section 7.2 is about a few (roughly) contemporary attempts, by Cohn, Bucherer, and Ritz to replace Lorentz's theory with radically new constructs. Cohn thereby gave up the ether; Bucherer and Ritz assumed exact validity of the relativity principle. Sometime after 1901, Einstein adopted the relativity principle and tried to satisfy it in an emission theory partially similar to Ritz's, in which the velocity of light depends on the velocity of its source. Section 7.3 recounts this episode (based on Einstein's recollections, for once), and shows how, in a sudden inspiration in the spring of 1905, he realized he could reconcile Lorentz's theory and the relativity principle by properly redefining time. Section 7.4 is about the early reception of the resulting theory of relativity, first by Planck and his disciples, then by Minkowski in the last year of his life. In the eyes of electron theorists, Minkowski's reformulation of relativity theory provided new concepts and

---

[5]Einstein 1949, pp. 52–53. Cf. Norton 2013.

[6]Einstein to Davenport, February 9, 1954, Einstein Archive, Jerusalem: "In my own development Michelson's result has not had a considerable influence. I even do not remember if I knew of it at all when I wrote my first paper on the subject." Cf. Holton 1969.

[7]On Einstein's readings, see pp. 191–192, 204, this volume.

tools to address the internal structure of the electron. Section 7.5 describes two resulting developments: Born's relativistic definition of rigidity, and Laue's relativistic continuum dynamics, which will both play a role in the genesis of general relativity. Section 7.6 is a brief attempt to explain why relativity theory, in those years, remained a mostly German affair.

7.1 The young Einstein's ventures in electrodynamics

Early musing and reading

In 1895, the sixteen-year old Albert Einstein sent his maternal uncle Caesar Koch an essay about determining the structure of the mechanical ether through the effects of a magnetic field on the velocity of light. In his view, the magnetic field implied an elastic deformation of the ethereal medium as well as a modification of its elastic constant, on which the velocity of elastic waves depends. Being born into a family who ran an electrotechnical company at an age of massive electrification of the world, the young Einstein was naturally interested in electromagnetism. From casual reading or from conversations with his paternal uncle, the engineer, he knew that Hertz's experiments had confirmed the existence of a medium responsible both for electromagnetic forces and for the propagation of light. He relied on the still popular idea of the ether as a kind of elastic solid, without knowing that the magnetic field was usually interpreted as an internal motion of the ether.[8]

In the fall of 1896, Einstein entered the prestigious ETH (*Eidgenössische Technische Hochschule*) in Zürich. As the physics program did not include the recent theories he judged most exciting, he read much by himself. In 1898, he studied Paul Drude's *Physik des Aethers*, then the prominent German text on Maxwell's theory and Hertz's experiments. Being a heir of Neumann's and Kirchhoff's phenomenological tradition as well as an admirer of Mach's philosophy, Drude adopted a most "economical" presentation in which he entirely avoided mechanical pictures of the ether, electric charge, and current. He introduced the basic magnetic and electric concept inductively, through Faraday's lines of force and the associated rules. Like Hertz, he regarded the propagated character (*Nahewirkung*) and the unity of electric and magnetic forces as the fundamental principles of the theory, and he defined the electric charge and current through the corresponding electric and magnetic fields of force.[9]

As suggested by the title of his treatise, Drude regarded the ether as the carrier of the electromagnetic field. But he did not assume the mechanical nature of this medium, not even in the abstract Lagrangian guise promoted by Maxwell, Helmholtz, and Poincaré, and he left it open whether mechanics or electromagnetism would be the better foundation of the physics of the future. He ignored Hertz's concept of a moving ether, and the only velocities appearing in his equations were those of the material carriers of charge, current,

---

[8]Einstein 1895. On Einstein's biography, cf., e.g., Pais 1982; Fölsing 1993, 1997. On the young Einstein, cf. Holton 1973; Pyenson 1980, 1982, 1985; Cahan 2000; Bracco 2017.

[9]Einstein to Marić, April 1898, in *ECP* 1, p. 213; Drude 1894 (treatise), 1895 (praising Mach). Cf. Darrigol 1993a, pp. 264–267. On Einstein's early readings and reflections on electrodynamics, cf. Stachel et al. 1987, pp. 223–225.

and magnetization. His ether being strictly stationary, he proposed to regard it as space endowed with physical properties:

> Just as well as we ascribe the mediation of forces to a particular, space-filling medium, we could also dispense with the medium, and attribute to space itself the physical properties which are currently attributed to the ether. Physicists have not yet considered the latter view, because by "space" they mean an abstract representation without physical properties.

In conformity with his phenomenological bent, Drude avoided atomistic theorizing, although unlike Mach he recognized the need for molecules and atoms to explain some phenomena, especially anomalous dispersion of which he gave a theory similar to Helmholtz's.[10]

Einstein next read from the third volume of Helmholtz's collected works, which contains his famous Faraday lecture on the necessity of ions in electrolysis as well as his late introduction of vibrating ions in optics. This reading must have counteracted Einstein's earlier exposition to the Maxwellian tendency, continued by Drude, to avoid atomistic concepts of electric charge and current. Another difference between Drude and Helmholtz was the latter's founding of electrodynamics on the principle of least action. As Einstein had trouble understanding the relevant memoirs, he decided to study Hertz's more straightforward approach. Remember that Hertz simply postulated the field equations after operationally defining the fields $\mathbf{E}$ and $\mathbf{H}$, dispensed with any mechanical foundation of the ether, reduced electric charge and current to "names" given to certain mathematical expressions ($\nabla \cdot \varepsilon \mathbf{E}$ and $\sigma \mathbf{E}$), and had ether and matter move together at a variable velocity in the electrodynamics of moving body, although he recognized that the optics of moving bodies might require a separate motion of the ether and matter.[11]

These readings were a good cure for Einstein's earlier espousal of a naive mechanistic concept of the ether. After learning Maxwell's theory from a young admirer of Mach and from the philosophically inclined Hertz, he adopted a philosophico-critical approach to physics in the spirit of the elder masters Helmholtz and Hertz. This happened before he read Mach and Hume. Einstein was ready to develop his own criticism of the most challenging physics of his time.

Toward a Lorentzian electrodynamics

In August 1899, Einstein wrote to his close friend and future wife Mileva Marić:

> I am more and more convinced that the electrodynamics of moving bodies in today's formulation [*Darstellung*] does not agree with the truth, that a simpler formulation will be possible. The introduction of the name "ether" in the electric theories has led to the idea of a medium whose motion can be spoken about, although I do not think we can attach a physical meaning to this way of speaking. I believe that electric forces

---

[10] Drude 1894, p. 9. On optical dispersion, see pp. 110–111, 142, this volume.

[11] Einstein to Marić, August 1899, in *ECP* 1, pp. 225–227; Helmholtz 1881, 1892, 1893a; Hertz 1892. On Helmholtz and Hertz, see pp. 133–137, 142, this volume.

can be directly defined only in empty space. Further, we will have to conceive electric currents not as "the vanishing in time of electric polarization" but as the motion of true electric masses, the reality of which the electrochemical equivalents seem to prove. . . . Electrodynamics would then be the science of the motions in empty space of moving electric and magnetic masses. The radiation experiments should tell us which of the two pictures [of the electric current] must be chosen.

The "electrodynamics of moving bodies" that Einstein was here criticizing referred to Hertz's memoir of 1890 on this topic, reproduced in his *Untersuchungen* of 1892. Einstein shared Heaviside's, Föppl's, and Drude's reluctance to introduce the velocity of the ether, and instead focused on the empirically accessible motion of matter. By calling ether "a name" he may have been alluding to Hertz's reducing electric charge, current, and polarization to "names" given to certain mathematical expressions. Indeed, Einstein proscribed the naive interpretation of the ether as a movable substance just as much as Hertz (and Maxwell) rejected the concept of electricity as a substance. Einstein further rejected Hertz's operational definition of the electric force $\mathbf{E}$ in a material dielectric, which involved a test point charge in an imaginary cavity in the medium. He did not accept the Maxwellian interpretation of the electric current as a varying polarization, to which Hertz and Drude repeatedly alluded. He instead suggested a return to the basic continental intuition of electricity and magnetism as moving "masses," in conformity with Helmholtz's interpretation of the electrolytic current.[12]

As is clear from subsequent letters, Einstein did not mean, at this point, to completely eliminate the ether. Rather he imagined a theory in which particles of matter freely moved through a stationary ether, as Lorentz, Larmor, and Wiechert had already done. Although he was probably unaware of the latter theories, he had read Hertz's suggestion that "The true theory should rather be one that distinguishes the states of the ether from those of the imbedded matter at every point." He also had seen Helmholtz's theory of anomalous dispersion, which implied a similar separation of ethereal and material vibrations. At the end of the former citation, we read that he hoped "radiation experiments" to demonstrate his concept of electric motion. The following month he wrote to Marić that he had "a good idea to investigate which effect the relative motion of bodies with respect to the ether has on the velocity of propagation of light in transparent bodies."[13]

Einstein then wrote a piece on "the relative motion of the ether with respect to ponderable matter," and showed it to his principal physics professor at the ETH, Heinrich Weber, who reacted "like a stepmother" (*stiefmütterlich*). Weber was probably aware of Lorentz's theory, which had recently received much publicity at the *Naturforscherversammlung* of 1898 in Düsseldorf, and he was therefore unimpressed by Einstein's ill-informed proposal. Perhaps on Weber's recommendation, Einstein read Wien's Düsseldorf address on ether motion and found it "very interesting." Indeed, as we saw in Chapter 5, this address summarized the results of numerous ether-drift experiments, including Fizeau's experiment of 1851 and the Michelson–Morley experiment of 1887. About the latter experiment, Wien

[12] Einstein to Marić, August 1899, in *ECP* 1, pp. 226–227.

[13] Hertz 1890b, p. 285; Einstein to Marić, September 10, 1899, in *ECP* 1, pp. 229–230. This sounds like a variant of Fizeau's experiment of 1851.

wrote: "If the ether is at rest, the time that a light ray needs to travel back and forth between two glass plates must change when the plates move. This change is of the order u^2/c^2, but should be observable by interference." Wien briefly characterized Lorentz's theory as based on invariable ions in a stationary ether, stated the correspondence theorem according to which to first order in u/c the laws of propagation of light in a system of bodies moving at the velocity u in the ether are the same as the laws in a system at rest provided the true time is replaced with Lorentz's local time. Wien also mentioned the attempts to explain the negative result of the Michelson–Morley experiment through the Lorentz contraction in a stationary ether.[14]

Einstein kept working on the electrodynamics of moving bodies, although thermal and kinetic-molecular physics then occupied most of his time. In March 1901, he wrote to Marić about the near completion of "our work on relative motion." In April, he discussed "the principal separation of ether and matter and the definition of absolute rest" with his friend Michele Besso. In September he told Grossmann he now had "a much simpler method for investigating the relative motion of matter with respect to the ether . . . based on ordinary interference experiments." This might have been a variant of Babinet's experiment of 1839, or of the Michelson–Morley experiment of 1887. In December he told Marić that he was "working hard on an electrodynamics of moving bodies that promises to yield a capital memoir" and that his doctoral adviser Alfred Kleiner had encouraged him to publish the theory and the attached experimental method. Lastly, Einstein mentioned his intention to read what Lorentz and Drude had written on this topic. The promised memoir never came out.[15]

There is no evidence that in this attempt Einstein adopted the relativity principle. On the contrary, we see that in 1901 he was still interested in testing the effects of motion through the ether. Most likely, he was nearing an optics of moving bodies similar to Lorentz's. When, in 1902, he finally read Lorentz (the *Versuch* of 1895), he may have realized that most of his considerations had been either anticipated or refuted, and that he had no better alternative. This would explain why he gave up on publication. Three years elapsed before he arrived at his theory of relativity. Unfortunately, there are very few traces of his intermediate reflections. All we know, from a letter to Besso, is that in January 1903 he was planning "comprehensive studies in electron theory." For the rest, we can safely assume that sometime in 1902–1904 Einstein became convinced of the absolute impossibility of detecting effects of motion with respect to the ether. From credible, convergent recollections of his, we gather that in this period he explored the possibility of an emission theory in which the velocity of the source adds to the velocity of the emitted waves.[16]

[14]Einstein to Marić, September 28?, 1899, *ECP* 1, pp. 233–235; Wien 1898, p. XV.

[15]Einstein to Marić, March 27, 1901, *ECP* 1, p. 280; April 4, 1901, p. 285 (Besso); December 17, 1901, p. 325 (capital study); December 19, 1901, p. 328 (Kleiner); December 28, 1901, p. 330 (Lorentz and Drude); Einstein to Grossmann, September 6?, 1901, p. 316. Drude's *Lehrbuch der Optik* (1900) had a chapter (II: 8) on Lorentz's optics of moving bodies.

[16]Einstein to Besso, January 22?, 1903, *ECP* 5, p. 10. Cf. Stachel et al., *ECP* 2, pp. 258–261.

7.2 Alternatives to Lorentz's theory

Instead of attempting a detailed reconstruction of Einstein's path to relativity from such meager data, we will consider the theories developed in 1900–1908 by theorists who, like Einstein, judged Lorentz's theory insufficient. This will relativize the singularity of Einstein's dissidence, and, by comparison, this may give us some hints about his motivations. In the previous chapter, we saw how Poincaré posited the exact validity of the relativity principle, reproached Lorentz with an incremental, ad hoc, and approximative compliance with this principle, reinterpreted Lorentz's corresponded states as states measured under adequate conventions, and lastly obtained the exact Lorentz-group symmetry in conformity with this principle. We will now consider three other approaches to the electrodynamic of moving bodies by Emil Cohn, Alfred Bucherer, and Walther Ritz.

Cohn's theory

The Strasburg professor Emil Cohn significantly contributed to the German reception of Maxwell's theory after Hertz's experiments. In 1900, he published one major treatise on electromagnetic theory. Like Drude, he approved of Mach's economy of thought and tried to formulate the electromagnetic theory directly in terms of observable quantities introduced inductively from simple experiments. As announced in his title, *Das elektromagnetische Feld*, the basic quantities of his version of Maxwell's theories were the fields **E** and **H** and the associated polarizations. He completely avoided the name and concept of ether, and regarded the fields and their lines of force as self-sufficient physical entities endowed with energy. Remember that for all of Cohn's predecessors since Faraday, the field was the portion of space in which electric or magnetic actions occur. As an adviser for Klein's *Encyclopädie der mathematischen Wissenschaften*, Cohn was responsible for the modern definition of a field as the continuous distribution of a physical quantity in space.[17]

   Although Cohn admitted that optical dispersion and magneto-optical phenomena involved the molecular structure of matter, he believed that most electrodynamic and optical phenomena should be describable at the macroscopic, phenomenological level. This is why in his contribution to the Lorentz jubilee in 1900 he proposed a new set of macroscopic field equations for the electrodynamics of moving bodies. Without any of Lorentz's microphysical assumptions, this theory agreed with all known phenomena in the optics of moving bodies, including Fizeau's experiment of 1851 and the Michelson–Morley experiment of 1887.[18]

   Let us first consider the macroscopic field equations that derive from Lorentz's microscopic equations by averaging over all electrons contained in a volume element of a dielectric at rest on earth. The microscopic field equations in the ether frame read

$$\nabla \times \mathbf{e} = -\frac{1}{c}\frac{\partial \mathbf{b}}{\partial t}, \quad \nabla \times \mathbf{b} = \frac{1}{c}\left(\rho \mathbf{v} + \frac{\partial \mathbf{e}}{\partial t}\right), \quad \nabla \cdot \mathbf{e} = \rho, \quad \nabla \cdot \mathbf{b} = 0. \tag{7.1}$$

[17] Cohn 1900a. Cf. Darrigol 1995b.

[18] Cohn 1900b.

In the earthbound frame moving at the velocity $\mathbf{u}$ in the ether, the substitutions

$$\partial/\partial t \rightarrow \partial/\partial t - \mathbf{u} \cdot \nabla \quad \text{and} \quad \mathbf{v} \rightarrow \mathbf{v} - \mathbf{u} \tag{7.2}$$

lead to the equations

$$\nabla \times \left(\mathbf{e} + \frac{\mathbf{u}}{c} \times \mathbf{b}\right) = -\frac{1}{c}\frac{\partial \mathbf{b}}{\partial t}, \quad \nabla \times \left(\mathbf{b} - \frac{\mathbf{u}}{c} \times \mathbf{e}\right) = \frac{1}{c}\left(\rho\mathbf{v} + \frac{\partial \mathbf{e}}{\partial t}\right), \quad \nabla \cdot \mathbf{e} = \rho, \quad \nabla \cdot \mathbf{b} = 0. \tag{7.3}$$

Call $\mathbf{s}$ the shift of the electrons of the dielectric owing to the Lorentz force

$$\mathbf{f} = \rho[\mathbf{e} + c^{-1}(\mathbf{v} + \mathbf{u}) \times \mathbf{b}] \approx \rho[\mathbf{e} + c^{-1}\mathbf{u} \times \mathbf{b}]. \tag{7.4}$$

For the average macro-fields

$$\mathbf{E} \equiv \left\langle \mathbf{e} + c^{-1}\mathbf{u} \times \mathbf{b} \right\rangle, \quad \mathbf{H} \equiv \left\langle \mathbf{b} - c^{-1}\mathbf{u} \times \mathbf{e} \right\rangle, \quad \mathbf{D} \equiv \langle \mathbf{e} \rangle + \langle \rho\mathbf{s} \rangle, \quad \mathbf{B} \equiv \langle \mathbf{b} \rangle, \tag{7.5}$$

we have

$$\nabla \times \mathbf{E} = -\frac{1}{c}\frac{\partial \mathbf{B}}{\partial t}, \quad \nabla \times \mathbf{H} = \frac{1}{c}\frac{\partial \mathbf{D}}{\partial t}, \quad \nabla \cdot \mathbf{D} = 0, \quad \nabla \cdot \mathbf{B} = 0. \tag{7.6}$$

The electronic polarization $\langle \rho\mathbf{s} \rangle$ being proportional to the average Lorentz force $\mathbf{E}$ and magnetic polarization being ignored, we also have

$$\mathbf{D} = \varepsilon\mathbf{E} - c^{-1}\mathbf{u} \times \mathbf{B}, \quad \mathbf{B} = \mathbf{H} + c^{-1}\mathbf{u} \times \mathbf{E}. \tag{7.7}$$

In terms of the local time $t - \mathbf{u} \cdot \mathbf{r}/c^2$, the substitution $\nabla \rightarrow \nabla - c^{-2}\mathbf{u}\,\partial/\partial t$ would yield

$$\nabla \times \mathbf{E} = -\frac{1}{c}\frac{\partial \mathbf{H}}{\partial t}, \quad \nabla \times \mathbf{H} = \frac{1}{c}\frac{\partial \varepsilon\mathbf{E}}{\partial t}, \quad \nabla \cdot (\varepsilon\mathbf{E}) = 0, \quad \nabla \cdot \mathbf{H} = 0 \tag{7.8}$$

if only the relation $\mathbf{D} = \varepsilon\mathbf{E} - c^{-1}\mathbf{u} \times \mathbf{B}$ were replaced with $\mathbf{D} = \varepsilon\mathbf{E} - c^{-1}\mathbf{u} \times \mathbf{H}$. This correction being of second order in u/c, to every state of an optical system at rest on earth there corresponds a state of an optical system at rest in the ether in the first-order approximation. This is a special case of Lorentz's theorem of corresponding states, which implies the lack of effects of the ether wind on optical experiments on earth to first order in u/c.[19]

As Cohn remarked, the correspondence becomes exact if we correct the expression of $\mathbf{D}$ as was just indicated. Returning to the ether frame and generalizing to a continuous,

[19] Lorentz 1895.

space-pervading material medium of permittivity ε, permeability μ, local velocity $\mathbf{v}$, charge density ρ, and current density $\mathbf{j}$, Cohn posits the equations

$$\nabla \times \mathbf{E} = -\frac{1}{c}\frac{D\mathbf{B}}{Dt}, \quad \nabla \times \mathbf{H} = \frac{1}{c}\left(\frac{D\mathbf{D}}{Dt} + \mathbf{j}\right), \quad \nabla \cdot \mathbf{D} = \rho, \quad \nabla \cdot \mathbf{B} = 0, \qquad (7.9)$$

$$\mathbf{D} = \varepsilon\mathbf{E} - c^{-1}\mathbf{v} \times \mathbf{H}, \quad \mathbf{B} = \mathbf{H} + c^{-1}\mathbf{v} \times \mathbf{E}, \qquad (7.10)$$

and shows that they imply the lack of any observable effect of the motion of the earth on optical and electrodynamic experiments on earth. Yet, strictly speaking his equations are valid in one frame only, since the constitutive relations (Equation (7.10)) depend on the absolute velocity $\mathbf{v}$ of the medium. Cohn thereby admits an absolute space referred to the fixed stars.[20]

In 1904, Cohn became aware of Lorentz's latest improvement of the technique of corresponding states, now working at any order in u/c. On the one hand, the transformation

$$\mathbf{r}' = (\gamma, 1)\mathbf{r}, \quad t' = \gamma^{-1}t - \gamma\mathbf{u}\cdot\mathbf{r}/c^2, \qquad (7.11)$$

with the concomitant field transformation

$$\mathbf{e}' = (1, \gamma)(\mathbf{e} + c^{-1}\mathbf{u} \times \mathbf{b}), \quad \mathbf{b}' = (1, \gamma)(\mathbf{b} - c^{-1}\mathbf{u} \times \mathbf{e}), \qquad (7.12)$$

turns the Lorentz field equations on earth into the field equations for a system at rest. On the other hand, the transformation

$$\bar{\mathbf{r}} = \mathbf{r}, \quad \bar{t} = t - \mathbf{u}\cdot\mathbf{r}/c^2 \qquad (7.13)$$

turns Cohn's field equations on earth into the same field equations for a system at rest. Consequently, the product of the first transformation by the inverse of the second (with a concomitant field transformation), namely,

$$\mathbf{r}'' = (\gamma, 1)\,\mathbf{r}, \quad t'' = \gamma^{-1}t, \qquad (7.14)$$

turns the Lorentz field equations into the Cohn field equations. As Cohn remarked, in Lorentz's theory moving rods are contracted and moving clocks slow down, so that the coordinates γx and $\gamma^{-1}t$ represent the measured abscissa (along $\mathbf{u}$) and time rate. As Cohn further remarked, in his theory the propagation of light on earth is exactly isotropic with respect to the local time $\bar{t} = t - \mathbf{u}\cdot\mathbf{r}/c^2$. "In optics," Cohn went on, "we define identical moments of time by assuming a spherical propagation in any isotropic medium that is relatively [with respect to the earth] at rest." Therefore, the local time is the one given by optically synchronized clocks, as Poincaré had earlier remarked in the Lorentz-jubilee

---

[20]Cohn 1900b, 1902. Cohn completed his field equations with energy and force considerations. D/Dt here denotes the convective derivative of a flux, as given on p. 128, Equation (5.25), this volume.

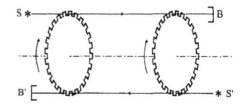

Fig. 7.1. Wien's suggested ether–wind experiment of 1904. The light from the sources S and S′ travels through the two synchronically rotating toothed wheels in opposite directions. Its intensity is measured by the bolometers B and B′.

volume in which Cohn's theory first appeared. In Lorentz's theory, the propagation of light is isotropic with respect to the primed coordinates of Equation (7.10), so that $\gamma^{-1}t - \gamma\mathbf{u}\cdot\mathbf{r}/c^2$ is the measured time. Whereas in Cohn's theory the absolute time t can be measured by mechanical means, in Lorentz's theory the electromagnetic behavior of any conceivable measuring agency prevents such measurement. Cohn objected to this inaccessibility of the true time: "Lorentz's conception requires that we distinguish between measured length and time, and true length and time. However, it fails to provide the experimental means to solve the problem even by assuming ideal measuring instruments." When, early in 1904, Wien proposed an ether-wind experiment in which light was sent through two synchronically rotating toothed wheels (see Figure 7.1), Cohn responded that in Lorentz's theory the experiment should give a negative result because the synchronization of the two wheels necessarily relies on the local time with respect to which the propagation of light is isotropic, whereas in Cohn's theory mechanically synchronized wheels would lead to a positive result.[21]

Cohn's phenomenological theory elegantly explained all known phenomena of the optics of moving bodies through a few simple field equations. As Cohn further showed, it was compatible with the energy principle and with the empirically known forces on currents, charges, and magnets. It implied irremediable though undetectable violations of the relativity principle and the reaction principle. This would not deter anyone who, in Mach's spirit, refused to accord special privileges to the principle of mechanics. More problematic was the purely macroscopic character of Cohn's theory at a time when the theory of electrons attracted most attention. In particular, his theory required a continuous material medium filling all space in optical systems, for instance the air in the Michelson–Morley experiment, and it failed in cases of extreme dilution. Cohn still defended his theory as a simple valid description at the macro level. His competitors Lorentz, Wien, and Abraham took him seriously despite their being engaged in the competing program of molecular, electromagnetic reduction. When, in 1907, Einstein reviewed early literature on the electrodynamics of moving bodies, he referred to Cohn's "relevant" contributions on which "he however did not rely." Before 1905, he probably saw Cohn's article of 1902 in *Annalen der Physik*, for he was regularly accessing this journal. With Cohn he shared the elimination of the ether, the focus on empirically determinable quantities, and the attention to the measurement of space and time. Yet he would have disapproved Cohn's liberties with

---

[21] Cohn 1904, pp. 1299–1300, 1408–1409. Wien 1904a.

the principles of mechanics and his complete neglect of the microphysical foundations of electrodynamics.[22]

Bucherer's theory

In 1902, the Bonn *Privatdocent* Alfred Bucherer conceived an ether-wind experiment in which the intensity of the light emitted from the same source was compared in two different directions. Fizeau had already imagined a similar test in 1852. So did too Einstein around 1900 according to one of his reminiscences. After exchanging a few letters with Lorentz, Bucherer admitted that the compared intensities should be equal in Lorentz's theory. His student Paul Nordmeyer performed the experiment and verified this prediction. Bucherer still disliked the fact that in Lorentz's theory, an obviously asymmetric situation (in which one beam travels along and the other against the ether wind) should be leading to a symmetric result. He knew that, despite contrary theoretical expectations, all trustable ether-wind experiments had given negative results. This prompted him to posit that optical and electromagnetic phenomena were affected only when matter moves with respect to matter. He pursued:

> One who would consistently adopt this point of view would have to renounce the ether-based picture of a temporal propagation of electromagnetic disturbances. But does this renunciation weigh much against the fact that the hypothesis of an ether at rest contradicts both the experiment—I mean Michelson-Morley's—and an important principle of mechanics, the conservation of the center of mass? With the principle: There are only actions from matter to matter, one would return to matter the properties artificially lent to the ether, and thus move from a dualistic to a monistic conception of nature.

In Bucherer's eyes, the ether was just a temporary "scaffolding" in the construction of a unified electromagnetic theory. Time was ripe "to bring down the scaffolding and to show the greatness and beauty of the monument."[23]

Bucherer's idea was to redefine electrodynamics as a theory of the interaction of electrons through fields regarded as mere computational aids. Somehow, he believed he could maintain Lorentz's field equations and nonetheless satisfy the principle of relativity. In 1903, he vaguely indicated that the velocities occurring in the solutions of these equations would have to be relative velocities (of the emitting and receiving electrons). After reading Lorentz's memoir of 1904 with the contractible electron, he more conservatively tried a constant-volume electron in order to avoid cohesive forces. He returned to his etherless theory in 1906, after judging that the Rayleigh–Brace experiments excluded the constant-volume electron and Kaufmann's experiments Lorentz's contractible electron. He at last made clear how he could conciliate the Lorentz field equations with the relativity principle:

---

[22] Einstein 1907c, p. 413n.

[23] Bucherer to Lorentz, February 15, August 6, December 8, 1902, Archive for the History of Quantum Physics (AHQP); Nordmeyer 1903; Bucherer 1903, p. 282; 1904, p. 131. Cf. Goldberg 1970c; Darrigol 2000a, pp. 369–372. For the anticipations, see Fizeau 1852 and Einstein 1922, p. 629.

> Whenever we speak of the dynamical interaction of systems we stipulate that the sys-
> tem acted upon . . . experiences the same force as it would in the Maxwellian theory on
> the assumption that it is at rest in the ether and the other system is moving relatively
> to it.

For two electrons moving at a constant velocity, this stipulation evidently implies a force depending on the relative velocity and satisfying the equality of action and reaction. Bucherer believed this property to remain true for any motion of the electrons, and he also believed that his stipulation was compatible with all known electrodynamic phenomena.[24]

Unlike Cohn's theory, Bucherer's suggestion was poorly received. In its earliest formulation, it was too vague to be even worth a comment. In the more precise formulation of 1906, Planck refused publication in the *Annalen*, and the early British relativist Ebenezer Cunningham attacked Bucherer in the *Philosophical Magazine*. Neither of them could see how Bucherer could apply the Lorentz–Maxwell equations in two different inertial frames without adopting Einstein's new kinematics, for they overlooked the merely formal character of the fields in Bucherer's theory. The fatal blow came in 1908, when Walther Ritz proved that Bucherer's stipulation was incompatible with Ampère's expression for the force between two linear currents. In the same year, Bucherer vindicated the Lorentz–Einstein theory through new measurements of the deflection of the beta rays from radium. Einstein was probably aware of Bucherer's proposal of 1903, published in *Annalen der Physik* in the context of Nordmeyer's experiment. Just like Bucherer, Einstein condemned artificial asymmetries in Lorentz's theory, and rejected the ether responsible for these asymmetries. But he did not share Bucherer's purely formal understanding of the mediating fields.[25]

Ritz's theory

The third theoretical alternative worth considering at this point is Walther Ritz's emission theory, even though it appeared well after Einstein's theory of relativity, in 1908. Ritz's motivations were indeed similar to Bucherer's, his theory could conceivably have appeared at an earlier date, and Einstein himself entertained an emission theory sometime between 1902 and 1904.[26]

In the name of "the economy of thought," Ritz praised past attempts at eliminating force from mechanics and narrowly defined the aim of electrodynamics as the determination of the motion of a system of charges through differential equations depending only on the space–time relations of these charges. In Lorentz's theory, the Maxwell–Lorentz equations yield the retarded potentials, through which the Lorentz force acting on a given charge can be expressed as a function of the (retarded) position and velocity of the other charges. Together with Newton's second law, this procedure yields equations of motion that no longer contain the fields. The field may thus be regarded as a mathematical, eliminable intermediate without physical import. This would no longer be permitted, Ritz conceded,

---

[24] Bucherer 1904; 1905; 1906; 1907, p. 414.

[25] Planck to Wien, November 29, 1906, AHQP, discussed in Pyenson 1985, p. 201; Cunningham 1907, 1908; Bucherer 1908a; Ritz 1908a, p. 204; Bucherer 1908b.

[26] Ritz 1908a. Cf. Martínez 2004; Darrigol 2012c.

"if it *were possible to perceive [the field's] existence at a point of the ether without placing any matter at this point.*" But any such attempt, for instance Lodge's interferential probing of a bulk motion of the ether, had failed. Ritz drew a radical conclusion: every conceptual vestige of the ether, including the fields and the constant velocity of light, had to be eradicated. In his words and with his emphasis:[27]

> *The ether does not exist, or more exactly; one must refrain from using this picture; the motion of light is a relative motion as any other; relative velocities alone play a role in the laws of nature; lastly, we must renounce the partial differential equations and the notion of field as this notion introduces absolute motion.*

Having excluded the Bucherer option, Ritz directly sought an elementary force law that would satisfy the relativity principle and correctly represent the known electromagnetic phenomena. For this purpose, he relied on the emissionist intuition according to which light is a flux of particles emitted by a source at constant radial velocity *with respect to the source.* In this picture, it is evident that a uniform translation of the optical system does not affect the relative progression and interaction of light within the system. Although Ritz regarded the light particles as purely fictitious, he required that the force between two charges should only depend on the disposition and relative velocity of the particles emitted by the first charge around the second charge. Specifically, he assumed that this force depended on the distance traveled by the particles between the two charges, on the velocity of the particles with respect to the second charge, and on their density in the vicinity of this charge (assuming that the first charge emits particles at a constant rate proportional to its charge), and he also relied on partial analogy with the formula Schwarzschild had obtained for the retarded interaction between two charged particles.[28]

The resulting force formula involved three parameters that Ritz could adjust to reproduce quasi-stationary electrodynamics, Hertzian oscillations, Kaufmann's deflection experiments, and even the advance of Mercury's perihelion by transposing the velocity dependence of electromagnetic forces to gravitational forces. The only phenomenon that did not easily fit into this theory was the Fresnel–Fizeau drag of waves by moving transparent bodies. In order to retrieve Lorentz's explanation of the partial drag, Ritz artificially assumed that the light reemitted by the electrons of the dielectric retained the velocity it had with respect to the primary source. Ritz also deplored the complexity of his formulas when compared with Lorentz's, although he insisted that his theory most directly accounted for the relativistic invariance of electrodynamic phenomena. He regarded his theory as a still imperfect and incomplete sketch of a future electrodynamics.[29]

[27] Ritz 1908a, pp. 162, 207–208.

[28] Ritz 1908a, pp. 212–223; Schwarzschild 1903b. That Ritz did not take the light-particle model seriously is confirmed by the disappearance of this model in his later expositions of the theory. Ritz 1908b is strictly limited to the elementary actions between electric particles; Ritz 1908c and Ritz 1909 rely on the notion of a "projected energy" from the luminous source.

[29] Ritz 1908a, pp. 223–252 (quasi-stationary case), 252–254 (oscillations), 260–267 (Kaufmann); 267–271 (Mercury).

7.3 Einstein's relativity theory

Adopting the relativity principle

Let us return to the young Einstein. Sometime in 1902–1903, he ceased to plan ether-drift experiments and accepted that the global motion of a system through the ether could not have any detectable effect. In this period, there were only two theorists who assumed the relativity principle in this strict sense: Poincaré and Bucherer. It is possible that Einstein read Poincaré's relevant arguments, either in *La science et l'hypothèse*, published in 1902, or in the German translation of his address at the International Physics Congress of 1900, published in 1900 in *Physikalische Zeitschrift*. He more likely read Bucherer's article of 1903. He may also have reached this conclusion by similar though independent arguments. Having read Lorentz's *Versuch* in early 1902, he was thoroughly familiar with the numerous failures to detect effects of the ether wind. Having read Drude in 1898, Mach in 1899, and plausibly Cohn in 1902, he understood how Mach's anti-mechanist economy of thought could lead one to question the existence of the ether (although Mach was himself willing to accept the ether as the medium responsible for inertial behavior). If there was no ether, then there was no ether wind and the motion of the earth had no more reason to affect optical experiments than it had to affect mechanical experiments on earth.[30]

If we believe some of Einstein's accounts of the genesis of relativity theory and if we take the introductory argument of his relativity article of 1905 seriously, there is another way in which he may have reached the relativity principle. Just like Bucherer, Einstein deplored that Lorentz's theory implied asymmetries with no observational counterpart. His decisive example was the electromagnetic induction observed in a circuit moving with respect to a magnet. In Lorentz's theory, when the circuit moves in the ether while the magnet remains at rest, the induced current is caused by the Lorentz force $e\mathbf{v} \times \mathbf{B}$ on the electrons, whereas in the symmetric case it is caused by the electric field $-\partial \mathbf{A}/\partial t$. Two different explanations seem necessary even though the observable circumstances of the phenomenon are exactly the same. The invisible, stationary ether is the culprit. As we saw in Chapter 5, in Faraday's original representation of induction phenomena, all that matters is the cutting of the lines of force of the magnet by the conducting wire, which depends on their relative motion only. In the continental theories of induction by Weber and Neumann, the relevant elementary forces or potential variations again depend on relative motion only. In Maxwell's, Heaviside's, and Hertz's theory, the theoretical symmetry of the two cases is less perfect, since the ether outside the magnet and the circuit may be largely undisturbed by their motion. Yet in both cases of motion, a true electric field, not a Lorentz force, is responsible for the induced current: this is so because the ether, when dragged by the moving circuit in the field of the magnet, acquires an electric polarization. The asymmetry denounced by Einstein entered Lorentz's theory surreptitiously when he replaced Maxwell's fully dragged ether with a stationary ether.[31]

[30] Poincaré 1900c, 1902a; Bucherer 1903; Lorentz 1895; Drude 1894; Mach 1897; Cohn 1902.

[31] Einstein 1905a, p. 891. See, e.g., Einstein 1920, p. 20: "The phenomenon of magneto-electric induction compelled me to postulate the (special) principle of relativity." Cf. Rynasiewicz 2000, pp. 169–171. On earlier takes on the relativity of electromagnetic induction, see pp. 154–155, this volume. For a reconstruction of Einstein's path to

To this consideration we may add that Einstein was aware of Violle's constructive use of the "principle of relative motion" in mechanics, since he had read the relevant section of Violle's physics textbook while preparing for the ETH entrance exam. He also knew that Drude, in his *Physik des Aethers*, had directly integrated the empirical relativity of induction phenomena in the field equations without appealing to Hertz's concept of ether motion. He may have read Föppl, who did roughly the same and wondered how the force between two charges could depend on absolute motion while the force between a charge and a magnet did not. Einstein decided, like Poincaré and Bucherer, that electrodynamic phenomena should depend only on the motion of matter with respect to matter. Unlike Poincaré and more like Bucherer, he believed that any ether theory would conflict with this principle, because the theoretical description then depends on the motion of matter with respect to the ether. As an admiring reader of Hertz and Mach, he rejected the resulting redundancy in the representation of phenomena.[32]

Attempting an emission theory

Having rejected the ether, Einstein was free to admit that the velocity of light was no longer a constant of the ether. In order to comply with the relativity principle, he assumed, in conformity with the old Newtonian intuition, that this velocity was a constant with respect to the source. Yet Einstein did not attach this assumption to the corpuscular concept of light (even though in 1905 he would defend a new corpuscular heuristics). Instead he wanted to modify the expression of the retarded fields so that their propagation depends on the velocity of source. He thereby differed from Ritz, who threw the fields out with the ether and defended a strictly monistic, matter-based ontology.[33]

If we believe Einstein's recollections, he encountered several difficulties in this project. First, he could not find differential equations for the modified fields. Second, as Ritz later deplored in his discussion of the Fresnel–Fizeau drag, the emission theory made the velocity of scattered light implausibly depend on the initial velocity of the scattering electron, with absurd consequences for the effect of moving screens, mirrors, and glass plates on the propagation of light. Third (as Ritz also noted), the light from an accelerated source would behave in a very odd manner:

> If an appropriately accelerated light source emits light in one direction (*e.g.*, in the direction of acceleration), then planes of equal phase move with different velocities, and thus one can arrange it so that all the surfaces of equal phase come together at a given location, so that the wavelength there becomes infinitely small. From there on the light reverses itself, so that the rear part overtakes the front.

relativity theory giving a central role to the relativity of electromagnetic induction, see Norton 2004, 2014. On the relevant meaning of "symmetry," cf. Hon and Goldstein 2005.

[32] On Violle, Drude, and Föppl, see pp. 39, 191–192, 140–141, this volume.

[33] Einstein to Ehrenfest, April 25, 1912, in *ECP* 5, p. 450. Cf. Norton 2004, pp. 58–66; Martínez 2004, pp. 9–11; Darrigol 2012c, pp. 233–234.

Lastly, Einstein deplored the absurdity of another consequence of the variability of the velocity of light:[34]

> If there is no fixed light velocity at all, then why should it be so, that all light that is emitted by "stationary" bodies has a velocity *completely independent of the color*? This seemed absurd to me. Therefore, I rejected this possibility as a priori improbable.

The final breakthrough

In early 1905, Einstein still did not know how to proceed. He could not make the emission theory work, and he did not yet see how Lorentz's theory could be reconciled with the relativity principle. In the meantime, he had read *La science et l'hypothèse*, at least in the German translation of 1904. There he could see Poincaré's arguments for the strict validity of the "principle of relativity"—so named here for the first time—and the criticism of Lorentz's step-by-step elimination of the effects of the ether wind. He could also appreciate Poincaré's insistence on the conventions that necessarily control the measurement of space, time, and mechanical quantities. In particular, he could read:

> There is no absolute time; to say two durations are equal is an assertion which has by itself no meaning and which can acquire one only by convention. Not only have we no direct intuition of the equality of two durations, but we have not even direct intuition of the simultaneity of two events occurring in different places: this I have explained in an article entitled *La mesure du temps*.

The German editor added a citation from the latter article:[35]

> We do not have a direct intuition of simultaneity, nor of the equality of two durations. If we think we have this intuition, it is an illusion. We replace it by the aid of certain rules which we apply almost always without taking count of them. But what is the nature of these rules? No general rule, no rigorous rule; a multitude of little rules applicable to each particular case. These rules are not imposed upon us and we might amuse ourselves in inventing others; but they could not be cast aside without greatly complicating the enunciation of the laws of physics, mechanics and astronomy. We therefore choose these rules, not because they are true, but because they are the most convenient, and we may recapitulate them as follows:
> "The simultaneity of two events, or the order of their succession, the equality of two durations, are to be so defined that the enunciation of the natural laws may be as simple as possible. In other words, all these rules, all these definitions are only the fruit of an unconscious opportunism."

---

[34] Einstein to Hines, February 1952, Einstein Archive, quoted in Martínez 2004, p. 10; Ritz 1908a, pp. 374, 404–405.

[35] Poincaré 1902, p. 111; 1904b, pp. 92, 286 (English from Poincaré 1913b, pp. 92–93, 306). That Einstein and his friends of the Akademie Olympia read this text before 1905 is attested in Solovine 1956, pp. vii–viii: "This book profoundly impressed us and kept us breathless for weeks on end."

In these extracts, there is no mention of the constancy of the velocity of light as a convention for measuring time and for judging simultaneity. There are several ways, however, in which Einstein could have become aware of this aspect of Poincaré's discussion. First, he could have read Poincaré's contribution to the Lorentz jubilee in 1900, which contained the interpretation of Lorentz's local time through optical synchronization. This article was frequently cited by the electron theorists, for Abraham regarded it as the source of his concept of electromagnetic momentum. Einstein himself cited it and used it in 1906 in a derivation of the mass–energy relation. Second, Einstein might have read Cohn's article of 1904, in which the local time was defined as the time for which the propagation of light is isotropic (without reference to Poincaré). He is less likely to have read Poincaré's Saint-Louis lecture (in which Poincaré repeated his interpretation of the local time), for it did not appear in any German journal.[36]

Although there is evidence that Einstein followed the literature on electron theory, we do not know how deeply and consistently he studied it. He probably disliked its highly formal and mathematical character, and he may already have doubted the possibility of an electromagnetic model of the electron, despite Kaufmann's alleged verification of Abraham's theory. He is not likely to have read Lorentz's important memoir of 1904, even though he could see it mentioned in *Annalen der Physik*. He was certainly familiar with Lorentz's theorem of corresponding states, at least in the approximation given in the *Versuch* of 1895. He may not have read Lorentz's memoir of 1899 or Larmor's *Aether and matter* (1900) in which the exact Lorentz transformations for fields and coordinates first appeared. The exact coordinate transformations already occurred in Lorentz's French memoir of 1892; and in an article of 1904 by Wien in *Annalen der Physik*. In a note in the same journal, Wien mentioned Lorentz's memoir of 1904 with the Lorentz-contracted electron (without giving the corresponding mass formulas). In *Physikalische Zeitschrift*, Wien discussed the field transformations of this memoir, and Abraham gave the mass formulas for the Lorentz-contracted electron.[37]

From a few remarks in Einstein's first relativity article, we may infer that he was aware of the existence of several models of the electron, most likely Abraham's and Lorentz's (through Wien), and possibly Bucherer's if he read Bucherer's well-diffused book of 1904 on electron theory. However, he may have overlooked Lorentz's mass formulas of 1899 and 1904 since he gave them as new in his article of 1905. As for the Lorentz transformations, he also gave them as new, having noted in his introduction that the invariance of the electrodynamic and optical laws had already been proven to first order in u/c.[38]

According to one of Einstein's remembrances, he suddenly realized, during a conversation with a friend in early 1905, that the constancy of the velocity of light could be reconciled with the relativity principle if only space and time were redefined accordingly. Whether or not he had seen Poincaré's relevant considerations, he may have recognized

[36] Poincaré 1900a, p. 272; Einstein 1906, p. 627; Cohn 1904. On Poincaré's and Cohn's arguments, see pp. 167–169, 197–198, this volume.

[37] Wien 1904b, 1904c, 1904d; Abraham 1904, p. 578.

[38] Einstein 1905a, p. 891 (first-order invariance), p. 919 ("comparison of different theories of the motion of the electron").

that Lorentz's local time, in the first-order approximation, was the time given by optically synchronized clocks under the assumption that the velocity of light is a constant for the moving observers.[39] There being in his mind no ether that could serve as reference for the true constancy of light, he treated the times optically defined in different inertial frames on equal footings. Time thus became relative to the frame in which it was defined. Einstein then determined the exact form of the transformations between the space and time coordinates in two different frames. He proved the invariance of the Maxwell–Lorentz equations under these transformations, with concomitant field transformations. The invariance of optical and electrodynamic phenomena thus came to depend on a new kinematics for which the velocity of light was the same constant in any inertial frame. Lastly, Einstein modified Newtonian mechanics to make it compatible with this new kinematics.[40]

On the electrodynamics of moving bodies

Annalen der Physik received Einstein's article on June 30 and published it on September 26, 1905. In the introduction, Einstein deplores the existence, in Lorentz's theory, of asymmetries without phenomenal counterpart and gives the example of electromagnetic induction. This difficulty, together with the failure of most attempts to detect motion with respect to the ether, brings him to the "principle of relativity" according to which the laws of electrodynamics should be the same in any inertial frame. He next introduces the "seemingly incompatible" principle that the velocity of light in empty space is a constant independent of the velocity of the source. He announces a fully coherent extension of the received electrodynamics of bodies at rest to bodies in motion on the basis of these two principles, through a new analysis of space and time measurements.[41]

This analysis, based on rods, clocks, and frames, is the object of Einstein first "kinematic part." It begins with section 1, about the definition of simultaneity through the optical synchronization of clocks attached to the same rigid frame: two identically built clocks at two different locations A and B are synchronized if and only if the time taken by a light signal to travel from A to B, as judged by the difference of the readings of the two clocks, is the same as the time taken by the reverse trip. According to his second principle, Einstein thereby assumes that the time taken for a roundtrip, as measured by one of the clocks, is given by $2AB/c$, where c is a universal constant. In section 2, Einstein gives the precise formal statement of his two principles:

> The laws according to which the states of physical systems evolve do not depend on which of two coordinates systems in uniform, rectilinear translation relative to each other this evolution is referred to.

---

[39] In Einstein's first historical sketch (1907c, p. 413), this was indeed the first crucial step: "Surprisingly, it turned out that in order to escape from this difficult [Lorentz's artificial explanation of Michelson–Morley] one needed only to grasp the concept of time in a sufficiently sharp manner. It was enough to realize that a certain auxiliary quantity introduced by Lorentz and called 'local time' could be simply defined as 'time'."

[40] The remembrance is from Einstein 1922. At the end of his article (1905a, p. 306), Einstein thanked Michele Besso for his faithful presence and for some valuable stimulation.

[41] Einstein 1905a, pp. 891–892.

Every light ray moves in the coordinate system "at rest" with the definite velocity c, whether this ray is emitted from a body at rest or from a moving body. Velocity here means light path divided by elapsed time as given by the former definition [of synchronous clocks].

By system "at rest," Einstein means any arbitrarily selected inertial system. Contrary to a frequent misrepresentation, his second principle is not the constancy and invariance of the velocity of light. It is just the constancy in a given inertial frame, as would be expected in Lorentz's theory if this frame were the ether frame. That the velocity of light is the same in any inertial frame is a consequence of Einstein's two principles jointly.[42]

Einstein next defines the length of a moving rigid rod as measured by an observer at rest: this length is the distance between the two points of the rest frame that coincide with the extremities of the rod at a given time of the rest frame. As Einstein already announces, it differs from the length measured by an observer moving together with the rod. That is, length is a relative notion. So too is simultaneity, because two clocks attached to the extremities A and B of the moving rod and constantly synchronous in the rest frame do not comply with the synchronization criterion in the comoving frame: they would give $l/(c-u)$ and $l/(c+u)$ for the duration of the AB and BA light trips, respectively, if l denotes the length of the rod (as seen from the rest frame) and u the velocity of the rod.

Einstein sharpens these considerations in section 3, in which he derives (what we now call) the Lorentz transformations connecting the space–time coordinates $(\mathbf{r}, t)$ in a given inertial frame to the coordinates $(\mathbf{r}', t')$ in a frame moving at the constant velocity $\mathbf{u}$ with respect to the first frame. He first proves that the synchronicity conditions in the two frames require

$$\mathbf{r}' = \varepsilon(\gamma, 1)(\mathbf{r} - \mathbf{u}t), \quad t' = \varepsilon\gamma(t - \mathbf{u} \cdot \mathbf{r}/c^2), \quad \text{with } \gamma = (1 - u^2/c^2)^{-1/2}, \qquad (7.15)$$

ε being a positive function of $\mathbf{u}$. Like Poincaré, he proves that the equation $c^2t^2 - \mathbf{r}^2 = 0$ of a wave sent at time zero from the origin of coordinates is invariant under this transformation. Again somewhat like Poincaré, he shows that on the one hand the inversion of this transformation requires $\varepsilon(-\mathbf{u}) = \varepsilon^{-1}(\mathbf{u})$ while the invariance of the length of a rod perpendicular to $\mathbf{u}$ requires $\varepsilon(\mathbf{u}) = \varepsilon(-\mathbf{u})$, so that $\varepsilon = 1$.

In section 4, Einstein derives the contraction of a moving rigid body (e.g., a rigid sphere) by γ^{-1} in the direction of $\mathbf{u}$ when measured in the rest system, by simply injecting, into the equation $\Phi(\mathbf{r}') = 0$ of the surface of the body, the expression $\mathbf{r}' = (\gamma, 1)\mathbf{r}$ given by the Lorentz transformation for $t = 0$. For a clock moving at the velocity $\mathbf{u}$ in the rest frame, to the elapsed time t in the rest frame corresponds the elapsed time

$$t' = \gamma(t - \mathbf{u} \cdot \mathbf{r}/c^2) = \gamma(t - u^2t/c^2) = \gamma^{-1}t, \quad \text{since } \mathbf{r} = \mathbf{u}t. \qquad (7.16)$$

Consequently, the moving clock experiences a growing retardation with respect to the clocks of the rest system on its trajectory. This remains true for a succession of rectilinear

[42] Einstein 1905a, p. 895. In this article, Einstein used V for the velocity of light; c was Abraham's choice, and also Lorentz's since 1904.

uniform motions of the clock, and also, Einstein tells us, in the continuously curved limit of such a broken trajectory. If the moving clock returns to its original location on a curved path with constant velocity modulus u, it will retard by $(1 - \gamma^{-1})t \approx (1/2)(u^2/c^2)t$ with respect to a clock that has remained in this location. This is the effect that Langevin later illustrated by replacing the two clocks with travelers.[43]

Section 5 is about the composition of velocities. For a material point moving at the velocity $\mathbf{v}'$ along the x axis of a frame itself moving at the velocity $\mathbf{u}$ in the same direction with respect to the rest frame, Einstein finds

$$\mathbf{v}' = \frac{d\mathbf{r}'}{dt'} = \frac{d\mathbf{r} - \mathbf{u}dt}{dt - \mathbf{u} \cdot d\mathbf{r}/c^2} = \frac{\mathbf{v} - \mathbf{u}}{1 - \mathbf{u} \cdot \mathbf{v}/c^2}, \quad \text{or} \quad \mathbf{v} = \frac{\mathbf{v}' + \mathbf{u}}{1 + \mathbf{u} \cdot \mathbf{v}'/c^2}. \tag{7.17}$$

The velocity of light, Einstein tells us, plays the role of an infinite velocity, because it retains its value when compounded with any subluminal velocity. The product of two Lorentz transformations with the parallel velocities of amplitude u and v is a Lorentz transformation with the velocity

$$w = \frac{u + v}{1 + uv/c^2}. \tag{7.18}$$

Einstein notes that these parallel transformations form a group, while Poincaré noted that the Lorentz transformations and spatial rotations together formed a group.

Now begins the second part of Einstein's memoir, on electrodynamics. In section 6, he establishes the invariance of the Maxwell–Lorentz equation in a vacuum under the transformations

$$\mathbf{r}' = (\gamma, 1)(\mathbf{r} - \mathbf{u}t), \quad t' = \gamma(t - \mathbf{u} \cdot \mathbf{r}/c^2) \tag{7.19}$$

$$\mathbf{e}' = (1, \gamma)(\mathbf{e} + c^{-1}\mathbf{u} \times \mathbf{b}), \quad \mathbf{b}' = (1, \gamma)(\mathbf{b} - c^{-1}\mathbf{u} \times \mathbf{e}), \tag{7.20}$$

where $\mathbf{e}'$ and $\mathbf{b}'$ represent the fields the moving observers would measure by means of a test point charge and a test magnetic pole, respectively. "In the old way of speech" and to first order in u/c the force acting on a moving electric point charge is the sum of the electric force $\mathbf{e}$ and the "electromotive force" $\mathbf{u} \times \mathbf{b}$. "In the new way of speech," it is the electric force acting on the charge in a comoving (tangent) frame. Einstein thus means to solve the asymmetry paradox with which he began his memoir.[44]

In section 7, Einstein derives the Doppler effect and stellar aberration by reexpressing the phase φ of a monochromatic plane wave in a moving frame as a function of the coordinates in the rest frame. This is a generalization and a variant of the first-order reasoning

---

[43] Langevin 1911, pp. 49–52. Langevin did not use twins and he did not call the effect a paradox.

[44] There is an evident similarity to Bucherer's representation of electromagnetic forces, discussed on p. 199–200, this volume.

Lorentz had used in 1895. In modern notation we have

$$\varphi = \mathbf{k} \cdot \mathbf{r} - \omega t = \mathbf{k}' \cdot \mathbf{r}' - \omega' t', \quad \text{with } \mathbf{k}' = (\gamma, 1)(\mathbf{k} - \mathbf{u}\omega/c^2), \quad \omega' = \gamma(\omega - \mathbf{k} \cdot \mathbf{u}), \quad (7.21)$$

from which follow the desired relations for the frequency $\omega/2\pi$ and the direction of propagation $\mathbf{n} = c\mathbf{k}/\omega$. The usual, pre-relativistic aberration and Doppler formulas are first-order approximations of these new formulas. Einstein further determines the energy density of the wave in the moving frame. For this purpose, he exploits the equality of the electric and magnetic energies, and he decomposes the electric field of the wave into a component perpendicular to the $(\mathbf{u}, \mathbf{n})$-plane and a component in this plane.[45] The two similar components being permuted for the magnetic field, it is sufficient to consider the first component, for which

$$\mathbf{e}' = \gamma(1 - \mathbf{n} \cdot \mathbf{u}/c)\mathbf{e}. \quad (7.22)$$

The energy density ε of the waves then transforms according to

$$\varepsilon' = \gamma^2(1 - \mathbf{n} \cdot \mathbf{u}/c)^2\varepsilon. \quad (7.23)$$

In section 8, Einstein considers a "light complex," that is, the light included in a sphere of volume V moving at the velocity $c\mathbf{n}$ in the rest frame. The energy contained in this sphere is evidently constant. When seen from the moving frame, the sphere becomes an ellipsoid of volume

$$V' = V\gamma^{-1}(1 - \mathbf{n} \cdot \mathbf{u}/c)^{-1}. \quad (7.24)$$

so that the energy $E = \varepsilon V$ of the light complex transforms into

$$E' = E\gamma(1 - \mathbf{n} \cdot \mathbf{u}/c). \quad (7.25)$$

Perhaps having the lightquantum in mind, Einstein remarks that the energy transforms exactly like the frequency. Lastly, he determines the wave reflected by a moving mirror by means of the comoving frame, and he computes the radiation pressure on the mirror by equating its work to the difference between the incoming and outgoing energy fluxes. He ends with the remark that all the optics of moving bodies can be treated by similar methods.[46]

---

[45] The full proof is given in Einstein 1907a, p. 431.

[46] It would have been easier to consider the rectangular prismatic complex delimited by the six planes $x = ct\cos\theta$, $x = ct\cos\theta + \Delta x, y = ct\sin\theta, y = ct\sin\theta + \Delta y, z = 0, z = \Delta z$, with the axis Ox parallel to $\mathbf{u}$, the axis Oz perpendicular to the $(\mathbf{u}, \mathbf{n})$-plane, and θ the angle between $\mathbf{u}$ and $\mathbf{n}$. The substitutions $x = \gamma x'$ and $t = \gamma x' \mathbf{u} \cdot \mathbf{n}/c$ for $t' = 0$ then lead to a prism contracted or dilated by $\gamma^{-1}(1 - \mathbf{n} \cdot \mathbf{u}/c)^{-1}$ in the Ox direction. On the interconnectivity of Einstein's interests in 1905, cf. Holton 1960; Rynasiewicz 2000.

In section 9, Einstein proves the invariance of the complete Maxwell–Lorentz equations, including the convection current $\rho\mathbf{v}$ in the second circuital equation and the electric density in the first divergence equation. Like Poincaré, he gives the transformed charge and velocity as

$$\rho' = \gamma\rho(1 - \mathbf{u}\cdot\mathbf{v}/c^2), \quad \mathbf{v}' = (1, \gamma^{-1})(\mathbf{v} - \mathbf{u})(1 - \mathbf{u}\cdot\mathbf{v}/c^2)^{-1}, \qquad (7.26)$$

for which the equation $\nabla\cdot(\rho\mathbf{v}) + \partial\rho/\partial t = 0$ for the conservation of charge is preserved.

In his last section, Einstein applies the Lorentz transformations to derive the equations of motion of a particle of mass m carrying the electric charge q. In the tangent inertial frame moving at the velocity $\mathbf{v}$ of the electron at a given instant, Einstein assumes the validity of Newton's acceleration law

$$m\frac{d^2\mathbf{r}'}{dt'^2} = q\mathbf{e}'. \qquad (7.27)$$

at the corresponding instant (for which $d\mathbf{r}'/dt' = 0$). The Lorentz transformation then gives

$$dt' = \gamma dt\left(1 - \mathbf{v}\cdot\frac{d\mathbf{r}}{dt}\right), \quad \frac{d\mathbf{r}'}{dt'} = \gamma^{-1}\left(1 - \mathbf{v}\cdot\frac{d\mathbf{r}}{dt}\right)^{-1}(\gamma, 1)\left(\frac{d\mathbf{r}}{dt} - \mathbf{v}\right), \quad \frac{d^2\mathbf{r}'}{dt'^2} = \gamma^2(\gamma, 1)\frac{d^2\mathbf{r}}{dt^2} \qquad (7.28)$$

in the infinitesimal neighborhood of the instant for which $d\mathbf{r}/dt = \mathbf{v}$. Together with the expression of $\mathbf{e}'$ in Equation (7.20), this gives the equation of motion

$$m(\gamma^3, \gamma)\frac{d^2\mathbf{r}}{dt^2} = q\left(\mathbf{e} + \frac{\mathbf{u}}{c}\times\mathbf{b}\right), \qquad (7.29)$$

a result already obtained by Lorentz in 1899 through his corresponding states, and again in 1904 through the model of the contractible electron. Einstein gives the corresponding expressions of the longitudinal and transverse masses of the electron under conventions of his own, and he discusses the consequences for magnetic and electric deflections experiments. He thereby gives the expression of the kinetic energy of the electron as

$$T = \int_{\dot{r}=0}^{\dot{r}=\mathbf{v}} q\mathbf{e}\cdot d\mathbf{r} = \int_{\dot{r}=0}^{\dot{r}=\mathbf{v}} m\gamma_r^3\ddot{\mathbf{r}}\cdot\dot{\mathbf{r}}dt = \int_{\dot{r}=0}^{\dot{r}=\mathbf{v}} m\dot{\gamma}_r dt = mc^2(\gamma_\mathbf{v} - 1). \qquad (7.30)$$

He remarks that his reasoning does not depend on the deeper nature of the mass of the electron. Lastly, he thanks Michele Besso for valuable stimulation.[47]

The inertia of energy

Remember that in 1900 Poincaré had considered the emission of radiation from a moving source and found a discrepancy between the recoil force in the rest frame and the recoil

---

[47] On Lorentz's earlier reasoning, see pp. 147, 173–174, this volume.

force in the comoving frame. He then regarded this discrepancy as a violation of the principle of relativity, in harmony with the violation of the principle of reaction (since he did not accept radiation momentum). In the early fall of 1905, Einstein similarly considered radiation from a moving source, both in the rest frame and in the comoving frame. The only difference is the symmetry of the process he considers: for an observer bound to the emitter, the same energy $J/2$ is emitted by the light source in two opposite directions. For an observer moving at the velocity $\mathbf{u}$ with respect to the source on the emission line, Einstein's transformation law (Equation (7.25)) for the energy of a light complex gives $\gamma(1 + u/c)J/2$ for the energy emitted in one direction and $\gamma(1 - u/c)J/2$ for the energy emitted in the other. The sum of these energies exceeds the energy J by $J(\gamma - 1)$. As the kinetic energy of the emitter is the product of its mass by $(\gamma - 1)c^2$, a variation $-J/c^2$ of this mass during the emission restores the energy balance. To his friend Conrad Habicht, Einstein wrote:

> Another consequence of the electrodynamics paper came to my mind. Together with Maxwell's fundamental equations, the relativity principle implies that mass is a measure of the energy content of bodies. Light transports mass. There should be a sensible diminution of mass in the case of radium. The line of thought is amusing and fascinating. But is not the dear Lord laughing about it? Is not he pulling me by the nose? This much I cannot know.

Einstein soon published his reasoning under the interrogative title "Does the inertia of a body depends on its energy content?", including the suggestion that measurements on radium salts might be used to test his theory.[48]

The following year, Einstein argued that mass–energy equivalence was the necessary and sufficient condition for an extension of the theorem of the center of mass to electromagnetic systems. At the beginning of this memoir, he notes that Poincaré's memoir of 1900 contains "the simple formal considerations on which the proof of this assertion is based." Indeed, Poincaré's equation

$$\frac{d}{dt} \int c^{-2} j \mathbf{r} d\tau + \int m \mathbf{v} d\tau + \int (c^{-2} \rho \mathbf{v} \cdot \mathbf{e}) \mathbf{r} d\tau = \text{constant} \tag{7.31}$$

will result from the equation

$$\frac{d}{dt} \left[\int c^{-2} j \mathbf{r} d\tau + \int m \mathbf{r} d\tau \right] = \text{constant} \tag{7.32}$$

for the uniform motion of the center of mass if the mass density of matter follows the increase $\rho \mathbf{v} \cdot \mathbf{e}$ of its energy content according to

$$\mathrm{D}m/\mathrm{D}t \equiv \partial m / \partial t + \nabla \cdot (m\mathbf{v}) = <\rho \mathbf{v} \cdot \mathbf{e}/c^2>, \tag{7.33}$$

---

[48] Einstein to Habicht [June to September 1905], *ECP* 3; Einstein 1905b. For Poincaré's reasoning, see pp. 165–166, this volume. Cf. Darrigol 2000b.

wherein the symbols < > indicate an average at the scale at which the mass density m is defined. As Einstein notes, the reasoning makes sense only if the expression $m\mathbf{v}$ of the momentum density is applicable and if the only relevant energies are the energy of the electromagnetic field and the internal energy of matter. This can be the case if the mass density m is defined at a macroscopic scale for which each volume element includes many molecules and if the macroscopic kinetic energy remains small.[49]

More concretely and more in the spirit of Newton's and Poincaré's connection between the reaction principle and the impossibility of perpetual motion, Einstein considers an emitter and an absorber of radiation that face each other and belong to the same rigid cylinder. He imagines the following cycle:

- The source emits a radiation pulse with the energy J in the direction of the absorber, which implies a recoil momentum $-J/c$ for the cylinder.
- When the pulse reaches the absorber, the cylinder returns to rest.
- A massless carrier then brings the energy J back to the absorber.
- At the end of this cycle, the cylinder has shifted by the amount $-(J/Mc)L/c$ (in a first approximation), where M is the mass of the solid and L the distance between emitter and absorber.

In order to avoid this perpetual motion, Einstein assumes that the return of the energy J to the emitter involves a transfer of mass J/c^2. The center of mass of the global system does not move during this transfer. Therefore, the solid moves by the amount $L(J/c^2)/M$ (in a first approximation), which compensates for the shift in the first step of the cycle.[50]

This was neither the last nor the most convincing of Einstein's derivations of the inertia of energy. The early derivations were confined to cases in which electromagnetic forces or radiation are responsible for the mass variations. Einstein nonetheless believed in the general validity of this astonishing consequence of relativity theory. He must have felt that general principles, not the peculiarities of the system under consideration, were responsible for the equivalence between mass and energy.[51]

7.4 Early reception 1905–1908

When Einstein's theory was published, it was competing with a few other theories, notably Lorentz's, Abraham's, and Cohn's. The evident superiority of Einstein's theory is only a retrospective illusion. In 1905, Abraham's theory could easily be judged superior to Einstein's and Lorentz's, because of its strict compliance with the enticing electromagnetic worldview. Cohn's theory remained an economical, faithful representation of the macroscopic electrodynamics of moving bodies, devoid of any of the oddities introduced by Lorentz and

---

[49] Einstein 1906, pp. 629–633. The reasoning can easily be extended to fast-moving material elements: one only has to replace m with $m/\sqrt{1 - v^2/c^2}$. On Poincaré's reasoning and formulas, see p. 167, this volume.

[50] Einstein 1906, pp. 627–629.

[51] On later derivations by Einstein and others, cf. Fernflores 2018. Laue's derivation is discussed on p. 226, this volume.

Einstein. Kaufmann kept refining his deflection measurements and found, in 1905, better agreement with Abraham's and Bucherer's electron theories than with Einstein's and Lorentz's theories. His announcement shook Lorentz's confidence and delighted the supporters of the electromagnetic worldview, while Einstein and Poincaré remained confident that future experiments would confirm the relativity principle.[52]

The pros and cons of a principles-based theory

For Göttingen-style electron theorists, Einstein's theory, even if it were to pass the test of beta-ray deflection experiments, would still be lacking a definite model of the electron. Sommerfeld indeed complained about the excessive abstraction of Einstein's physics. In December 1907, he wrote to Lorentz:

> We are all waiting very eagerly for your opinion on the whole bunch of Einstein's papers. Genial though they are, I feel there is something almost unhealthy in this non-constructible and non-intuitive dogmatic. An Englishman would hardly have produced these theories; perhaps what we see here is the abstract manner of the Semite, as we see it in Cohn. Hopefully you will manage to fill this genial concepts-skeleton with true physical life.

Earlier in 1907, Paul Ehrenfest similarly opined that the relativity theory, as "a closed system," ought to determine the structure of the electron. Einstein famously replied:

> The principle of relativity, more exactly, the relativity principle together with the principle of the constancy of the velocity of light, should not be regarded as "a closed system," not even as a system *tout court*, but only as a heuristic principle which, considered in itself, contains only statements about rigid bodies, clocks, and light signals. The only way the theory of relativity provides more is by requiring relations between regularities that otherwise appear independent.

Einstein went on to explain how the relativity principle, together with what we would now call a correspondence principle, led to the relativistic dynamics of any particle, independently of its inner constitution. He also compared the relativity principle with the second law of thermodynamics (called principle in German and in French), to relate given laws to other laws. This was Einstein's first intimation that he was then favoring "theories of principles" constrained from above by principles of empirical origin over "constructive theories" based on models of the ultimate reality.[53]

Einstein did not yet exclude that a proper dynamics of rigid bodies might permit a relativistic model of the electron. However, he soon remarked that relativity theory excluded any dynamical concept of rigidity, since such a concept would allow the instantaneous

---

[52] Kaufmann 1905; Einstein 1907c, pp. 437–439; Lorentz to Poincaré, March 8, 1906, in Walter 2007a. Cf. Miller 1980, pp. 83–84; Stachel et al., 1989, pp. 270–273; Hentschel 1992; Potters 2019.

[53] Sommerfeld to Lorentz, December 26, 1907, in Kox 2008, p. 236; Ehrenfest 1907; Einstein 1907a, pp. 206–207. For a thorough study of Einstein's distinction, cf. Giovanelli 2020.

transmission of a force from one extremity of a rigid rod to the other. In a letter to Sommerfeld in the following year, he acknowledged that relativity theory was "like thermodynamics before Boltzmann"—that it begged for a deeper, constructive foundation. At the same time, he cited the difficulties in constructing a model for black-body radiation as reason to think that the elements of the future foundation would not be found in the received electromagnetic theory. For the time being at least, the theory of relativity had to remain a theory of principles.[54]

This aspect of Einstein's theory seduced a powerful champion of the thermodynamic way of thinking, Max Planck. Since his early works on thermodynamics, Planck had been favoring a physics based on general, macroscopic principles instead of constructive models. Even before Einstein declared himself a supporter of this approach, Planck praised his derivation of the dynamics of a particle for its being independent of any assumption on the structure of the particle. In the fall of 1905, he discussed Einstein's seminal article at the physics colloquium of Berlin University. In the following months, he published several related articles and he interested his assistant Max Laue and his student Kurd von Mosengeil in the new theory.[55]

Like his former Berlin professor Helmholtz, Planck believed the principle of least action to be the ultimate foundation of all physics. In 1906, he rewrote Einstein's Equation (7.29) for the motion of an electron in an electromagnetic field as

$$\frac{d(m\gamma\mathbf{v})}{dt} = \mathbf{f} \quad \text{with} \quad \mathbf{f} = q\left(\mathbf{e} + \frac{\mathbf{v}}{c} \times \mathbf{b}\right), \tag{7.34}$$

and he exhibited the Lagrangian form

$$\frac{d}{dt}\frac{\partial L}{\partial \mathbf{v}} = \mathbf{f} \quad \text{with} \quad L = m\gamma^{-1} = m\sqrt{1 - \mathbf{v}^2/c^2}. \tag{7.35}$$

As he showed in a subsequent article, this formulation eased comparison with the equations of other theories of the electron, which Abraham had already put in Lagrangian form. Planck judged Kaufmann's results inconclusive and kept supporting the "Einstein–Lorentz" theory, with mention of Poincaré's relevant article in the *Compte rendus*. In 1907, he confirmed Einstein's idea of the inertia of energy while extending the relativity principle and the principle of least action (in Helmholtz's form) to thermal radiation and thermodynamic systems. In the same year, Laue showed that the Fresnel–Fizeau drag simply resulted from Einstein's law for the composition of velocities, a result anticipated in Lorentz's derivation of this drag through the local time. Mosengeil combined Planck's radiation theory and relativity theory in a discussion of thermal radiation in a moving cavity. Einstein soon integrated these contributions by Planck and his disciples into his own expositions of relativity theory.[56]

---

[54] Einstein 1907a, pp. 207–208; 1907b, pp. 422–424; Einstein to Sommerfeld, January 14, 1908, *ECP* 5.

[55] Cf. Goldberg 1976; Liu 1991, 1997; Goenner 2010.

[56] Planck 1906a, 1906b, 1907, 1908a; Laue 1907 (see also Laub 1907); Mosengeil 1906; Einstein 1907c, pp. 414, 426, 434–436, 439, 451–453. On the history of relativistic thermodynamics, cf. Liu 1991.

Minkowski's world

In Göttingen, Einstein's former mathematics professor Hermann Minkowski, who had there codirected with Hilbert a seminar on electron theory, became the most enthusiastic and creative interpreter of Einstein and Poincaré. In an address of November 5, 1907 to the Göttingen Mathematical Society, he argued that pure mathematicians like himself were best equipped to anticipate, interpret, and develop the conceptual revolution brought by the relativity principle. He also declared that "the new approaches, provided they truly agree with phenomena, would mean the greatest triumph ever accomplished in the application of mathematics." He then showed that the Lorentz invariance of the Lorentz–Maxwell equations, painstakingly proven by Poincaré and Einstein, was made manifest by means of a four-dimensional pseudo-Euclidean formulation. For this purpose, he broadly extended Poincaré's modest use of four-dimensional vectors in the construction of relativistic invariants. He introduced the four-vector potential $(\mathbf{A}, i\varphi)$ and the four-current $(\rho\mathbf{v}, i\rho)$ transforming like the four-vector $(\mathbf{r}, it)$ under the Lorentz transformations that leave $\mathbf{r}^2 - t^2$ invariant (with $c = 1$). Using the index α running from one to four for the components of a four-vector, he gave the Maxwell–Lorentz equations as

$$\partial_\alpha F_{\alpha\beta} = j_\beta, \quad \text{with} \quad F_{\alpha\beta} = \partial_\alpha A_\beta - \partial_\beta A_\alpha, \quad \partial_\alpha \partial_\alpha A_\beta = 0, \quad \partial_\alpha A_\alpha = 0, \tag{7.36}$$

whose covariance is evident (he wrote the here omitted sums over repeated indices).[57]

In addition, Minkowski offered covariant equations for the macroscopic polarization, based on the covariant generalization $P_{\alpha\beta}$ of Lorentz's macroscopic polarization vectors for dielectrics and magnetized bodies. In analogy with Lorentz's relations $\nabla \cdot \mathbf{E} = \rho - \nabla \cdot \mathbf{P}$ and $\nabla \times \mathbf{H} = \mathbf{j} + \nabla \times \mathbf{M}$ between the macroscopic fields $\mathbf{E}$ and $\mathbf{H}$, the free charge ρ, the conduction current $\mathbf{j}$, the electric polarization $\mathbf{P}$, and the magnetic polarization $\mathbf{M}$, Minkowski had

$$\partial_\alpha F_{\alpha\beta} = j_\beta - \partial_\alpha P_{\alpha\beta} \tag{7.37}$$

for the Maxwell equations in polarizable media. For relativistic extensions of radiation theory, dynamics, and thermodynamics, he summarized Planck's recently published theory.

For the mathematicians in his audience, Minkowski remarked that the Lorentz transformations, being those leaving the hyperboloid $x^2 + y^2 + z^2 - t^2 = 1$, defined a four-dimensional version of the hyperbolic geometry that had been widely discussed in the previous century as a model of Lobachevski's geometry. With an erroneous definition of the four-velocity as $(\mathbf{v}, i\sqrt{1 - \mathbf{v}^2})$ (he should have taken dx_α/ds, with $ds^2 = -dx_\alpha dx_\alpha$), he indicated how the hyperboloid can be used to transform velocities. The remark was meant to confirm the mathematical triumph announced in his introduction:[58]

> In a certain sense, the world in space and time is a four-dimensional non-Euclidean manifold. Clearly it would shed glory onto the mathematicians and it would baffle

[57] Minkowski 1907, p. 927. Cf. Galison 1979; Walter 1996a, 1999a, 1999b, 2008, 2010; Corry 2010.

[58] Minkowski 1907, pp. 927–928.

the rest of mankind, if the mathematicians had created purely in their imagination a large domain that would someday come to completely real existence, even though this would never have been the intension of such idealistic minds.

The following month, Minkowski published a full electrodynamics of moving media based on the new intrinsic geometry. In this memoir he first reexpresses the Lorentz transformations, which he thus names with reference to Poincaré, as rotations in the pseudo-Euclidean four-space with an imaginary time coordinate. For a transformation leaving the y and z coordinates unchanged, this gives

$$x' = x \cos i\psi + it \sin i\psi, \quad it' = -x \sin i\psi + it \cos i\psi, \quad \text{with} \quad \tan i\psi = iu \text{ (and } c = 1). \quad (7.38)$$

He gives again the intrinsic form (Equation (7.36)) of the Maxwell–Lorentz equations, with a now correct definition of the four-velocity as $v_\alpha = dx_\alpha/\sqrt{-dx_\beta dx_\beta}$. For the macroscopic field equations, he starts with the Maxwell–Hertz equations in an isotropic, linear, nonmoving medium of permittivity ε, permeability μ, and conductivity σ:

$$\nabla \cdot \mathbf{D} = \rho, \quad \nabla \times \mathbf{H} - \frac{\partial \mathbf{D}}{\partial t} = \mathbf{j}, \quad \nabla \cdot \mathbf{B} = 0, \quad \nabla \times \mathbf{E} + \frac{\partial \mathbf{B}}{\partial t} = 0, \quad \mathbf{D} = \varepsilon \mathbf{E}, \quad \mathbf{B} = \mu \mathbf{H}, \quad \mathbf{j} = \sigma \mathbf{E}.$$
$$(7.39)$$

He then imagines the simplest covariant extension of these equations to a moving medium thanks to the four-vector calculus.[59]

Allow me a small digression on matters of notation. Minkowski and his followers all used Minkowski's imaginary four-vectors. For the tensors and operations on them, Minkowski sometimes used matrices, sometimes intrinsic notation. For instance, he introduced a "lor" operator (for Lor/entz) corresponding to our differential operator $\partial_\mu \equiv \partial / \partial x^\mu$. In the following years, Abraham, Born, and Laue proceeded similarly, with slightly different terminology. In reverence to Woldemar Voigt, they called tensor what Minkowski called "vector of the second kind" (in the antisymmetric case). In 1910, Sommerfeld systematized the intrinsic approach, with operations and names suggested by analogy with ordinary vector and tensor fields. He still used an imaginary fourth coordinate. The modern usage, with real coordinates, covariant/contravariant indices, and the summation rule emerged in the context of general relativity, when Einstein accommodated Minkowski's four-space with Ricci's absolute differential calculus. For the convenience of the reader, modern tensor notation is used in the following. Note, however, that Minkowski and his first followers were coping with more complex and disparate notation.[60]

In modern notation, Minkowski's electrodynamics regroups the fields $\mathbf{D}$ and $\mathbf{H}$ in the antisymmetric tensor $G^{\mu\nu}$ such that $G^{0i} = -D_i$ (with $i = 1, 2, 3$), $G^{12} = H_3$, $G^{23} = H_1$, $G^{31} = H_2$, and the fields $\mathbf{E}$ and $\mathbf{B}$ similarly in the tensor $F^{\mu\nu}$. His extended field equations then read

[59] Minkowski 1908. Cf. Stachel et al. 1989, pp. 502–507.
[60] Cf. Walter 2007b.

$$\partial_\mu G^{\mu\nu} = j^\nu, \quad \partial_\mu \tilde{F}^{\mu\nu} = 0, \quad v_\mu G^{\mu\nu} = \varepsilon v_\mu F^{\mu\nu}, \quad v_\mu \tilde{F}^{\mu\nu} = \mu v_\mu \tilde{G}^{\mu\nu}, \quad j^\mu - (v^\nu j_\nu)u^\mu = \sigma v_\nu F^{\mu\nu},$$
$$(7.40)$$

where $\tilde{F}^{\mu\nu}$ and $\tilde{G}^{\mu\nu}$ are the duals of the tensors $F^{\mu\nu}$ and $G^{\mu\nu}$ (in which electric and magnetic fields are exchanged), and v^μ is the four-velocity of the material medium. These equations are manifestly covariant, and they are easily seen to reduce to Equation (7.39) when $v^0 = 1$ and $v^1 = v^2 = v^3 = 0$ for a medium at rest. To an empirically sufficient approximation, they agree with the less general and not quite covariant field equations that Lorentz had earlier deduced by an averaging process in the theory of electrons.[61]

Minkowski further builds the traceless tensor

$$T^{\mu\nu} = F^{\mu\rho}G_{\rho\nu} - \tfrac{1}{4}\eta^{\mu\nu}F^{\rho\sigma}G_{\rho\sigma}, \qquad (7.41)$$

whose spatial components agree with Maxwell's stress tensor. As Minkowski further shows, the T^{00} component agrees with Maxwell's energy density, the T^{0i} components with the Poynting vector $\mathbf{E} \times \mathbf{H}$, and the T^{0i} components with the momentum density $\mathbf{D} \times \mathbf{B}$. Through a complex variational method involving a field action and the action of a continuous distribution of mass, Minkowski next obtains the equation of motion of an element of the medium:

$$m\frac{du^\mu}{d\tau} = F^\mu \quad \text{with} \quad F^\mu = \omega f^\mu \quad \text{and} \quad f^\nu = -\partial_\mu T^{\mu\nu} + (u_\rho \partial_\mu T^{\mu\rho})u^\nu, \qquad (7.42)$$

where m is the mass of the element, ω its volume at rest, $u^\mu = dx^\mu/d\tau$ its four-velocity, $d\tau = \sqrt{dx^\mu dx_\mu}$ the differential "proper time," and F^μ the four-force generalizing the Newtonian force. The second, corrective term in the expression of the four-force density f^μ is needed for consistency with $u^\mu F_\mu = 0$ (which results from $u^\mu u_\mu = 1$). From the latter relation, Minkowski gets

$$(d\tau/dt)\mathbf{F} \cdot d\mathbf{r} = F^0 d\tau = d[m(1 - \mathbf{v}^2/c^2)^{-1/2}], \qquad (7.43)$$

which he interprets as the balance between the work of the force $(d\tau/dt)\mathbf{F}$ and the variation of Einstein's kinetic energy. Conversely, he remarks that energy conservation, being the time component of a four-vector relation, can generate all the laws of relativistic mechanics when combined with the covariance postulate.

In this study, Minkowski invented much of the formal and conceptual apparatus of modern relativity theory, with the pseudo-Euclidean four-space, and with the attached scalars, vectors, and tensors: proper time, action, four-force, four-momentum, electromagnetic field tensors, and the stress–energy tensor. He also showed, in the case of macroscopic electrodynamics, how the covariance requirement could generate the field equations and

---

[61] Minkowski himself planned a theory based on the electrons' relative displacements in the moving matter. The discrete character of electricity being largely irrelevant, he replaced the set of electrons with a continuum. His assistant Max Born completed the project after his death (Minkowski and Born 2010). The resulting field equations were more general than those of 1908, for they covered anisotropic and dispersive media. Ponderomotive forces were not discussed.

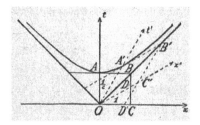

Fig. 7.2.   Minkowski's diagram for constructing a Lorentz transformation. From Minkowski 1909, p. 105.

the equations of motion from their already known expression in a special case. For the time being, he said very little about the nature of space and time in his new theory.

Minkowski saved his more revolutionary claims for his brilliant "Space and time" address of September 21, 1908 at the *Naturforscherversammlung* in Cologne. He there began with a startling announcement:

> The views on space and time I will develop grew on a physical-experimental soil. That is their strength. Their tendency is radical. From now on, space for itself and time for itself should be completely reduced to shadows, and only a kind of union of them should preserve its autonomy.

Minkowski went on to speculate that mathematicians, had they scrutinized the symmetries of Newtonian mechanics, would have been naturally led to the Lorentz group. They would have observed that the group of spatial rotations and the group of Galilean transformations were treated in an oddly disparate manner, and they would have naturally considered the "more intelligible" group G_c of transformations that leaves the form $c^2 t^2 - \mathbf{r}^2$ invariant, since the limit G_∞ of this group contains the two former disjoint symmetries.[62]

Here and elsewhere, Minkowski reasoned geometrically. He defined a *world-point* through the coordinates $(\mathbf{r}, t)$, the *world* as the continuous set of these points filled with a substance, and a *world-line* as the trajectory of a particle of the substance in the world. In the two-dimensional case for which x and t are the only coordinates, he introduced a diagram through which a linear transformation preserving $c^2 t^2 - x^2$ could be geometrically constructed. With $c = 1$, we first draw the upper part of the hyperbola $t^2 - x^2 = 1$ (Figure 7.2). Any linear transformation can be obtained by projecting the world-point on two new axes Ox' and Ot'. The unit vector OA' on Ot' having the coordinates $t' = 1$ and $x' = 0$, and $t^2 - x^2$ being conserved, it must belong to the hyperbola. The simultaneous vanishing of $t^2 - x^2$ and $t'^2 - x'^2$ further implies that the asymptote $x = t$ must bisect the angle x'Ot ' and that the unit vector OC' on Ox' has the same length as OA'.

Minkowski used his space–time diagrams to discuss the G_∞ limit, to derive the Lorentz contraction as a mere projection of a segment parallel to Ox over the new Ox' axis, to ease the expression of the Liénard–Wiechert retarded interaction between two charged particles, and, most importantly, to distinguish between three kinds of vectors: time-like, space-like, and light-like, according to whether their extremity is located within the light cone, outside the light cone, and on the light cone, respectively (Figure 7.3). With his diagrams, he showed

[62] Minkowski 1909, p. 104.

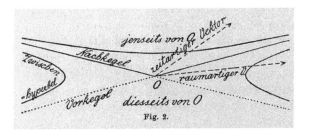

Fig. 7.3.   Minkowski's diagram for distinguishing between different kind of intervals (*Nachkegel* for the future light cone, *Vorkegel* for the past light cone, *zeitartiger Vektor* for time-like vector, and *raumartiger V.* for space-like vector). From Minkowski 1909, p. 107.

that space-like intervals can be made purely spatial and time-like intervals purely temporal (with preserved sign of the time difference) by a Lorentz transformation.

Lastly, Minkowski introduced the "proper time"

$$\tau = \int \sqrt{dt^2 - c^{-2}d\mathbf{r}^2} \qquad (7.44)$$

of a world-line and built the four-vector $\dot{x}^\mu = dx^\mu/d\tau$. He defined the four-force F^μ on a particle of mass m as the product $m\ddot{x}^\mu$, and showed that the usual Lorentz force is the spatial component of $F^\mu d\tau/dt$. He also defined the four-momentum p^μ as $m\dot{x}^\mu$, and used $u^\mu F_\mu = 0$ to derive

$$(d\tau/dt)\mathbf{F} \cdot \mathbf{dr} = F^0 d\tau = dp^0 \qquad (7.45)$$

for the work done by the force on the particle. He showed that $p^0 \equiv m(1 - \mathbf{v}^2/c^2)^{-1/2}$ represents the kinetic energy of the particle, as he had done in his earlier study.

Minkowski emphasized that his world was defined intrinsically and absolutely as a four-dimensional space, just as ordinary Euclidean space exists independently of its representation in various systems of coordinates. He called this statement the *world postulate*. He remarked that not only time but also space is a relative notion depending on the selected reference system. Just as there are many planes in ordinary space, there were many spaces in Minkowski's world. In Minkowski's opinion, Lorentz and Einstein had both failed to appreciate this relativity of space. In his conclusion, Minkowski imbedded the relativity principle and the world postulate in the electromagnetic worldview cherished by his Göttingen colleagues, and he predicted that the ongoing experimental confirmation of the word postulate should reconcile conservative physicists with the "preestablished harmony" between pure mathematics and physics.[63]

The following day at the same Cologne meeting, Bucherer announced he had confirmed the Lorentz–Einstein electron mass formula through new beta-ray deflection experiments,

---

[63] Minkowski 1909, p. 111.

using parallel instead of perpendicular electric and magnetic fields. Minkowski responded enthusiastically:

> I will express my joy in seeing experiments speak in favor of Lorentz's theory and against the rigid electron. From a theoretical point of view, this was to be expected. In my opinion, the rigid electron is a monster in the company of Maxwell's equations, whose deepest harmony lies in the relativity principle. . . . The rigid electron is not a working hypothesis, it is a working hindrance.

Minkowski's pronouncements and Bucherer's announcement efficiently advertised relativity theory at the Cologne meeting. Minkowski died suddenly in January 1909, without seeing the quick spread of his views and methods among a few important physicists and mathematicians. In March of the same year, Wien wrote to Lorentz: "The theory of relativity now seems to be gradually confirmed. It is very enticing in Minkowski form." In January 1910, Sommerfeld wrote to Lorentz: "I have now converted to the theory of relativity. Minkowski's systematic form and conception has helped me understand it." Leading electron theorists were thus abandoning Abraham's electron and endorsing relativity theory. Abraham himself worked to explore the consequences of relativity theory in Minkowski form, even though he still doubted the correctness of its predictions. We will now take a closer look at this rapid change in the perception of relativity theory.[64]

Forces, energy, and momentum

Independently of the Cologne address, Minkowski's electrodynamics of moving media attracted considerable attention. As Einstein deplored that Minkowski's relevant memoir "demanded quite a lot from the reader on the mathematical side," he decided to present, in collaboration with Wien's assistant Jakob Laub, a more elementary derivation of the same equations in ordinary vector form. The idea was to apply a Lorentz transformation to the Maxwell–Hertz equations for bodies at rest, and to assume the resulting equations to remain valid for a variable, nonuniform velocity of the bodies. In a sequel, Einstein and Laub dealt with the forces on the moving matter, and argued on an electron-theoretical basis that Minkowski's expression of the Lorentz force in a magnetized medium was incorrect.[65]

Minkowski's formal determination of the ponderomotive forces turned out to be ambiguous. In 1909, Abraham and the Finnish newcomer Gunnar Nordström debated on their proper expression. Abraham's expression implies a symmetric stress–energy tensor. The symmetry of the spatial components, $T^{ij} = T^{ji}$, reproduces a well-known property of the stress tensor introduced by Augustin Cauchy in the theory of elasticity. The mixed symmetry, $T^{i0} = T^{0i}$, extends the relation $\mathbf{S} = \mathbf{g}c^2$ between the Poynting vector $\mathbf{S}$ and

---

[64] Bucherer 1908b; Minkowski 1909, p. 762; Wien to Lorentz, March 20, 1909, in Kox 2008, p. 264; Sommerfeld to Lorentz, January 9, 2010, ibid., p. 297. In the second edition of his *Theorie der Elektrizität* (1908), Abraham used the Lorentz transformations but rejected the "exceedingly strange" dilation of time as well as the Lorentz contraction of the electron. On Minkowski's role in the electron theorists' conversion to relativity theory, cf. Walter 2018.

[65] Einstein and Laub 1908a, 1908b. Cf. Stachel et al. 1989, pp. 503–507.

the momentum density **g**. Planck had earlier claimed, without a general proof, that this relation was a general consequence of the principle of reaction for any form of energy. As Einstein put it in 1913, it is the "simplest expression of the equivalence between mass and energy," since according to this equivalence an energy flux implies a mass flux, alias momentum.[66]

Nordström and Abraham also wondered how to take the Joule heat into account in a moving conductor. Minkowski's equation of motion may be written as

$$m\frac{\mathrm{d}\,u^{\mu}}{\mathrm{d}\tau} = F^{\mu}_{\mathrm{M}} \quad \text{with} \quad F^{\mu}_{\mathrm{M}} = F^{\mu} - (u_{\nu}F^{\nu})u^{\mu}, \quad F^{\mu} = \omega f^{\mu}, \quad f^{\nu} = -\partial_{\mu}T^{\mu\nu} \quad (\text{with } c = 1) \tag{7.46}$$

The complementary force $-(u_{\nu}F^{\nu})F^{\mu}$ is needed to comply with $u_{\mu}\dot{u}^{\mu} = 0$. For a medium of unit permittivity and unit permeability, $-T^{ij}$ is the usual Maxwell stress tensor τ_{ij}, T^{00} is the Maxwell energy density w, T^{i0} is the Poynting vector **S**, and T^{0i} is the Poincaré–Abraham momentum density **g**. For a unit volume and in a unit time, $\partial w / \partial t$ represents the variation of the included electromagnetic energy, $\nabla \cdot \mathbf{S}$ represents the outgoing energy flux, and the product $(\mathbf{u}/u^{0}) \cdot \mathbf{f}$ represents the work done by the electromagnetic forces. Consequently, the quantity

$$\dot{Q} = u_{\mu}F^{\mu} = -\omega u_{\nu}\partial_{\mu}T^{\mu\nu} = -\omega u^{0}\left(\frac{\partial w}{\partial t} + \nabla \cdot \mathbf{S}\right) - \omega\mathbf{u} \cdot \mathbf{f} \tag{7.47}$$

represents the Joule heat developed in the volume element ω in a unit of proper time. Minkowski's equation of motion may then be rewritten as

$$m\frac{\mathrm{d}u^{\mu}}{\mathrm{d}\tau} = F^{\mu} - \dot{Q}u^{\mu}, \tag{7.48}$$

where F^{μ} is the usual Lorentz four-force. As Abraham notes, we may instead take

$$\frac{\mathrm{d}(mu^{\mu})}{\mathrm{d}\tau} = F^{\mu}, \quad \text{with} \quad \frac{\mathrm{d}m}{\mathrm{d}\tau} = \dot{Q}. \tag{7.49}$$

Then the Lorentz force need not be corrected, but the mass of the element increases with its heat content.[67]

Nordström and Abraham consider the case of an infinite plate perpendicular to the z axis and moving at a constant velocity along this axis. A constant, uniform current runs in this plate in the direction Oz. In Minkowski's theory there is no force acting on this plate because the complementary force $\dot{Q}\mathbf{u}$ exactly compensates for the electromagnetic force **F**. In the modified theory, there is a force **F** on the plate despite the uniform motion. In conformity with Einstein's equivalence between mass and energy, the Joule heating of

[66] Abraham 1909a, 1909b; Nordström 1909, 1910; Planck 1908a; Einstein to Lorentz, August 16, 1913 (in agreement with Laue), *ECP* 5, p. 552.

[67] Abraham 1909b, 1910a. Cf. Liu 1991.

the plate increases its mass, by just the amount needed to balance the Lorentz force. To this analysis, Abraham adds the remark that in the rest frame, some "mechanical" force must counteract the electromagnetic force so that its elements are in equilibrium. In order to comply with the relativity principle, he further assumes that these forces derive from an energy–stress tensor with properties similar to its electromagnetic counterpart, including the aforementioned symmetry. He surmises that this covariant balancing of electromagnetic and nonelectromagnetic forces can serve to construct the electron. As we will see in a moment, this is what Laue managed to do two years later.[68]

7.5 Constructing a relativistic electron

As was mentioned, the electron theorists were not content with Einstein's principle-based derivation of the equation of motion of an electron. They wanted a constructive model, preferably on a purely electromagnetic basis. In 1909–1910, Einstein himself tried and failed to construct both electrons and light quanta as singularities in the solutions of a modified Maxwell field. The modification was too deep to pass for a new implementation of the electromagnetic worldview. In contrast, Minkowski believed his four-dimensional world to be the ultimate realization of this worldview. In his mind, the Lorentz-group symmetry was the essence of electromagnetism, and any theory that complied with this symmetry was therefore electromagnetic in nature.[69]

The Born-rigid electron

Minkowski did not explain how he would construct a relativistic electron out of electromagnetic fields. His assistant Max Born did it for him in his *Habilitationschrift* of 1909. Born's plan was to do what Einstein had already deemed impossible: to graft the electromagnetic field on a rigid frame. The key was a new concept of rigidity allegedly compatible with Minkowskian relativity. While ordinary rigidity requires the distance between any two material points of the body to be invariable, Born rigidity requires the constancy of the distance of infinitesimally close world-lines of the body when reckoned in the common orthogonal three-space. Born developed and employed this notion in true Göttingen style, with elegant but difficult mathematics. Only a few outstanding points and results are retained in the following.[70]

In the Lagrangian picture used by Born, the motion of the body is given by the position $x^\mu(\xi^0, \xi^1, \xi^2, \xi^3)$ of each of its material points as a function of three spatial parameters ξ^1, ξ^2, ξ^3 and the proper time $\xi^0 = \tau$. For two infinitesimally close world-lines, we have $dx^\mu = \partial_\nu x^\mu d\xi^\nu$. The orthogonal three-space at a given point of a given world-line is defined by $u_\mu \partial_\nu x^\mu d\xi^\nu = 0$, with $u^\mu = dx^\mu/d\tau$. Born eliminates $d\xi^0$ through this relation and thus gets

---

[68] Abraham 1909b, pp. 739–740. Nordström (1910) persisted in favoring Minkowski's $md u^\mu/d\tau$ over the Abraham–Einstein $d(m u^\mu)/d\tau$.

[69] Einstein 1909, p. 193; Minkowski 1909, p. 111.

[70] Born 1909a. Cf. Pauli 1921, pp. 689–690; Maltese and Orlando 1995, pp. 272–279.

$$-ds^2 = -dx^\mu dx_\mu = a_{ij} d\xi^i d\xi^j \qquad (7.50)$$

for the spatial quadratic form expressing the local shape of the body. The condition $da_{ij}/d\tau = 0$ defines rigidity. As Paul Ehrenfest soon remarked, this condition can be re-expressed in a more physical manner as the condition that the dimensions of a small portion of the body, when judged in the tangent frame, do not change in time; or, equivalently, that these dimensions, when judged from the rest frame, are subject to the Lorentz contraction corresponding to the velocity of the tangent frame.

As the simplest case of rigid motion, Born gives the "hyperbolic" motion for which the material points move in Minkowski two-space on the world-lines given by

$$X = (x + a^{-1}) \cosh at - a^{-1}, \quad T = (x + a^{-1}) \sinh at, \qquad \text{(with } c = 1\text{)} \qquad (7.51)$$

where X and T are the abscissa and time of one of the material points in a fixed inertial frame of origin O, x the position of the material point at the origin of time ($T = t = 0$), t a global time variable and a a constant. For the interval between two neighboring particles, we have

$$ds^2 = dT^2 - dX^2 = (1 + ax)^2 dt^2 - dx^2. \qquad (7.52)$$

Consequently, t is the proper time along the world-line starting from the origin, and the proper time on another world-line is given by

$$t_x = (1 + ax)t. \qquad (7.53)$$

At a given t-time on any of the world-lines, the velocity is $dX/dT = \tanh at$, so that all the material points share the rapidity at. The Galilean-invariant acceleration at/t_x of one of the material points is a constant in time, but it varies from point to point since the proper time does. Let us define the tangent frame at time t as the inertial frame of rapidity at and of space–time origin coinciding with the material point that started from O. In this frame, the space coordinate of a material point is x, and its time coordinate is zero. In other words, the mutual distances of the particles are invariable when measured in the tangent frame of one of the particles. This is the definition of Born rigidity.[71]

Born then shows that for a hyperbolically moving electron there is no radiation field (in other words, the field accompanies the electron). He determines the equation of motion in a covariant manner, by balancing the external four-force with the volume integral of the internal four-force density *in the tangent frame*. The result is the same as in Lorentz's theory, except for a negligible dependence of the electromagnetic mass on the acceleration. The problems with the deformation work in Lorentz's model seem to disappear because there is no deformation in the tangent frame in which internal and external forces balance each other. There is no sideways force acting on a rectilinearly accelerated electron, as long as the charge distribution or the electron is spherically symmetric. For a slow variation

---

[71] Minkowski (1909, p. 108) had already discussed the hyperbolic motion of a single point, as an especially simple motion. The name rapidity for the additive velocity parameter of a Lorentz transformation belongs to Robb 1911.

of the acceleration, Born assumes the self-field of the electron to be the same as in the hyperbolic case. He ignores any rotation of the electron, since the only rigid motion he knows how to compute is purely translatory.[72]

Born presented his theory at the *Naturforscherversammlung* of September 1910 in Salzburg. In a subsequent discussion with Einstein, he became aware that his definition of rigidity might not allow rotation. Paul Ehrenfest independently published a simple argument to the same effect. According to the Born–Ehrenfest rigidity criterion, an element of the moving rigid body with the instantaneous velocity **v**, when considered by an observer at rest, experiences the same Lorentz contraction as it would if it were moving uniformly and rectilinearly at the velocity **v**. Consequently, when a cylinder (or a disk) is set in rotation around its axis, a radial element of length is seen as unchanged, while an orthoradial element is seen as contracted by the observer at rest. Globally, the circumference appears to become less than 2π times the radius. In order to grasp the paradox here detected by Ehrenfest, we may imagine a radius and the circumference of the disk, before being set in motion, to be covered with tiny equal, adjacent rods, so that the number of circumferential rods is 2π times the number of radial rods. After the disk has been set in motion, the contiguity of the rods is obviously preserved. This contiguity also holds when the rotating disk is seen from the rest frame, and yet the apparent length of a radial rod differs from that of a circumferential rod.[73]

In agreement with this simple objection, Gustav Herglotz and Fritz Noether both proved that a Born-rigid body only had three degrees of freedom. Although the body may rotate *uniformly* around a fixed point, this is only a singular case and in general the motion of one of the points of the body completely determines the motion of all of its other points: it undergoes translation without rotation. Born almost welcomed this result, since after all no known observation required rotating electrons. He also tried a less restrictive definition of rigidity, to little avail. The fatal blow came in January 1911 from Laue, who argued that in relativity theory a continuous body necessarily has an infinite number of degrees of freedom. Laue imagines the body to be at rest for $t \leq 0$ in a given frame, and has n of its points start moving at $t = 0$. In Minkowski space, the sections of the future light cones of these points through the space $t = \varepsilon$ do not overlap for a small enough ε. In plain words, the motions of the n points are completely independent because there cannot be any causal connection between these points. Since n is arbitrarily large, there are infinitely many degrees of freedom. To put it another way, if there were rigid bodies with a finite number n of degrees of freedom, then by fixing $n - 1$ points of this body, and acting on an nth point, we would instantaneously perturb all remaining points and thus contradict the relativistic exclusion of supraluminal communication.[74]

Consequently, for any continuous body the Born-rigid motions are artificially selected from among an infinitely larger variety of possible motions, just as a rigid motion of a

---

[72] Born 1909a, pp. 46–49 (internal work), 53 (spherical electron). On the deformation work, see pp. 175 and 227–229, this volume. The sideways force was one of Ehrenfest's objections to the Lorentz electron: Ehrenfest 1907, based on Abraham's similar objection for a nonspherical Abraham-rigid electron in Abraham 1903, p. 174.

[73] Born 1909b; 1910a, p. 233n (about Einstein's remark); Ehrenfest 1909, 1910. Cf. Stachel 1980, 1989a, 2007.

[74] Herglotz 1910; Noether 1910; Born 1910a, 1910b; Laue 1911a, also Laue 1913, p. 181.

portion of a fluid is only a very special case of motion. There cannot be any strict kine-matic constraint on relativistic motion; there can only be cohesive forces approximately mimicking rigidity. Laue agreed with Planck and Abraham that a relativistic theory of (quasi-)rigid bodies could only be based on a new, Lorentz-covariant theory of elasticity.[75]

Laue's relativistic continuum dynamics

More ambitiously, Laue wanted to base all physics on a relativistic theory of continu-ous media. His first implementation of this program appeared a few months later in *Das Relativitätsprinzip*, the first monograph ever published on relativity theory.[76]

Laue's relativistic continuum theory draws on a wide generalization of Minkowski's energy–momentum tensor. Remember that in Minkowski's electrodynamics, the four-force generalizing Lorentz's force is given as $-\partial_\mu T^{\mu\nu}$, where $\tau_{ij} = -T^{ij}$ agrees with the Maxwell stress tensor, $w = T^{00}$ with the Maxwell energy density, $S_i = T^{i0}$ with the Poynting vector, and $g_i = T^{0i}$ with the Poincaré–Abraham momentum. In vacuum, the vanishing of $\partial_\mu T^{\mu\nu}$ gives

$$\frac{\partial w}{\partial t} + \nabla \cdot \mathbf{S} = 0, \quad \frac{\partial g_j}{\partial t} - \partial_i \tau_{ij} = 0. \tag{7.54}$$

That is to say, the energy flux entering a volume element balances the variation of its energy content, and the pressure on its surface (i.e., momentum flux across this surface) balances the momentum content. Laue decides to regard any energy and momentum in nature as distributed continuously in space–time according to a tensor with components related and interpreted just as in the electromagnetic case. In particular, he requires this tensor to be symmetric, thus integrating Planck's relation between energy flux and momentum density. In this picture, there are no point-like particles, and any body has a finite extension over which its energy and momentum are distributed.[77]

Laue then considers a body with the global velocity $\mathbf{v}$ in a fixed inertial frame and as-sumes the energy flux to vanish in the comoving frame, so that $\bar{T}^{0i} = \bar{T}^{i0} = 0$ (the overline is for quantities in the comoving frame). The Lorentz transformation from the comoving frame to the fixed frame gives:

$$w = \gamma^2(\bar{w} - v_i v_j \bar{\sigma}_{ij}), \quad g_i = v_i \gamma^2(\bar{w} - v^{-2} v_j v_k \bar{\sigma}_{jk}) - \gamma(v_j \bar{\sigma}_{ij})_\perp, \tag{7.55}$$

with $\gamma = (1 - v^2)^{-1/2}$, $\sigma_{ij} = -T^{ij}$, and for units in which $c = 1$. Laue integrates over the volume V of the body and uses $d^3x = \gamma^{-1}d^3\bar{x}$ to get the energy E and the momentum $\mathbf{p}$ of the body. In the simple case in which the stresses in the comoving frame are reduced to the uniform pressure $\bar{P}$, this gives

---

[75] Laue 1911a, p. 86; Abraham 1909b, p. 740, 1910a, p. 531; Planck 1910. Abraham suspended his judgment on relativity until the needed covariant elasticity theory would be available.

[76] Laue 1911b. Cf. Rowe 2008; Miller 1981a, p. 347; Janssen 1995, chap. 2; Janssen 2003.

[77] Laue 1911b, pp. 79–88, 135–138, 147–150; more concisely in Laue 1911c.

$$E = \gamma(\bar{E} + v^2 \bar{P}\bar{V}), \quad \mathbf{p} = \mathbf{v}\gamma(\bar{E} + \bar{P}\bar{V}), \tag{7.56}$$

relations Planck had first obtained in 1907 by thermodynamic means. The momentum of the body thus has the desired Planck–Einstein form, with a mass depending on the energy content and on the pressure in the comoving frame.[78]

Laue next considers a "completely static system" for which the body is by itself in equilibrium in the comoving frame. In this case, the net force $\int\int \bar{\sigma}_{ij} d\bar{S}_j$ on any section of constant $\bar{x}_j$ vanishes. Indeed the section can be closed by a surface situated outside the body (on which $\bar{\sigma}_{ij} = 0$), and the net force on any closed surface vanishes according to Stokes's theorem together with the equilibrium condition $\partial_i \bar{\sigma}_{ij} = 0$. Consequently, we have

$$\int \bar{\sigma}_{i1} d^3x = \int \left(\int\int \bar{\sigma}_{i1} d\bar{S}_1 \right) d\bar{x}_1 = 0, \text{ etc.} \tag{7.57}$$

This is "Laue's theorem": for any closed system in equilibrium, the volume integral of the stresses vanishes in the comoving frame. Together with Equation (7.55) the theorem implies

$$E = \gamma\bar{E}, \quad \mathbf{p} = \mathbf{v}\gamma\bar{E}, \tag{7.58}$$

in conformity with the relativistic dynamics of a material point.[79]

In Laue's eyes, this is a highly important result. First, it justifies his claim that continuum dynamics should or at least can precede the dynamics of material points. Second, it corroborates his assertion that the relation $\mathbf{S} = \mathbf{g}c^2$ between energy flux and momentum density expresses the equivalence between mass and energy. Third, it proves that any static construct of any shape can serve as a model of the electron. No purely electromagnetic model is possible since it cannot produce internal equilibrium; and there are infinitely many possible choices for the equilibrating forces and for the equilibrium shape. In particular, a nonspherical electron would still be able to move rectilinearly and uniformly without a transverse force, *pace* Abraham, Born, and Ehrenfest. Taken separately, the electromagnetic momentum of such an electron would not be parallel to its velocity, but it would become so when combined with the momentum from the cohesive forces.

As Laue notes, a compensation of the same kind occurs in the case of the Trouton–Noble experiment: the electromagnetic torque on the moving charged condenser is balanced by another torque associated with the forces that keep the condenser in internal equilibrium. To be sure, Lorentz and Larmor had already explained the null result of this experience. But they had done so by generalizing the correspondence theorem at the electronic level, thus leaving unresolved the mechanism for the compensation of the electromagnetic torque. As Laue indicates, this deficiency partly motivated his continuum dynamics. Both for the electron and for the Trouton–Noble condenser, two kinds of force are at play: electromagnetic and mechanical. One should not take the relativistic invariance of both mechanical and electromagnetic phenomena as an invitation to reduce the latter forces to the former. Instead one should try to subsume both kinds of phenomena under

---

[78] Laue 1911b, pp. 150–158; Planck 1907, pp. 562–563, Equations 7.43, 7.46; also Einstein 1907c, pp. 445–447.

[79] Laue 1911b, pp. 186–170.

a unified conceptual scheme. This is exactly what Laue meant to do with his relativistic continuum dynamics.[80]

Revisiting the contractible electron

We may now revisit the models of a Lorentz-contracted electron in light of Laue's continuum dynamics. For Lorentz's purely electromagnetic model, Laue's energy–momentum formulas (Equation (7.55)) give

$$E = \gamma(1 + v^2/3)U_s, \quad \mathbf{p} = \gamma\mathbf{v}(4/3)U_s, \tag{7.59}$$

where U_s denotes the electrostatic energy of the electron. Indeed, using the symbol $\langle\rangle$ for the integral over the entire field space of a spherical electron and taking the x axis along $\mathbf{v}$, we have

$$\langle v_i v_j \bar{\sigma}_{ij} \rangle = v_x^2 \langle E_x^2 - \tfrac{1}{2}E^2 \rangle = v_x^2 \langle \tfrac{1}{3}E^2 - \tfrac{1}{2}E^2 \rangle = -(1/3)v^2 U_s. \tag{7.60}$$

The momentum formula is the one Lorentz obtained in 1904 by integrating the electromagnetic momentum, and the energy formula is the one given by Abraham in 1905 for the Lorentz electron. The Poincaré electron of 1905 includes, besides the electrostatic stress system $\bar{\tau}_{ij}$, the cohesive tension $\bar{\kappa}_{ij} = \alpha\delta_{ij}$. According to Laue's theorem, the electron's equilibrium implies $\langle\bar{\tau}_{ij} + \bar{\kappa}_{ij}\rangle = 0$, and Equation (7.55) (with $\bar{\sigma}_{ij} = \bar{\tau}_{ij} + \bar{\kappa}_{ij}$) implies the energy–momentum formulas

$$E = \gamma(U_s + U_p), \quad \mathbf{p} = \gamma\mathbf{v}(U_s + U_p), \quad \text{with} \quad U_p = \alpha V = (1/3)U_s. \tag{7.61}$$

Poincaré's equilibrium condition $\alpha = U_s/3V$ directly results from Laue's relation $\langle\bar{\sigma}_{ij} + \bar{\tau}_{ij}\rangle = 0$.

For the Lorentz electron, the components $(E, \mathbf{p})$ do not form a four-vector. That they need not do so is obvious from their being, according to Laue, the four components of the integral $\int T^{0\mu}d^3x$.[81] This lack of covariance is a true difficulty for the Lorentz electron, in addition to the lack of cohesion. It stands in the way of a coherent generalization of the force–momentum relation $\mathbf{f} = d\mathbf{p}/dt$, and it divorces the inertial mass $(4/3)U_s/c^2$ from the rest energy U_s. The latter defect is easily mended by adding a nonelectrostatic component to the energy of the electron, as Abraham and Poincaré did. There remains a mystery: why is it that the purely electromagnetic momentum $\mathbf{p} = \gamma\mathbf{v}(4/3)U_s$ of the Lorentz electron agrees with the mixed momentum of the Poincaré electron? If we compare the derivations of these momenta through Laue's general relation (Equation (7.55)), we see that their agreement requires $U_p = -\langle\bar{\tau}_{xx}\rangle$ or, equivalently (according to Laue's relation), $U_p = \langle\bar{\kappa}_{xx}\rangle$. The latter

---

[80] Laue 1911c, pp. 525–542 (Trouton–Noble); Laue 1911b, p. 186. Cf. Janssen 1995. On the Trouton–Noble experiment, see p. 173, this volume.

[81] As Enrico Fermi showed in 1922, for a Lorentz electron of four-velocity v^μ, it is possible to redefine the proper electromagnetic energy and momentum through the covariant integral $P^\mu = \int T^{\mu\nu}v_\nu\gamma^{-1}d^3x$, so that they yield the four-vector $U_s v^\mu$ (Fermi 1922; also Rohrlich 1960, 1965).

equality is contingent on Poincaré reducing the cohesive stresses $\bar{\kappa}_{ij}$ to a uniform, isotropic, and constant tension within the electric shell of the electron. Another choice would lead to a different momentum. In particular, for an anisotropic stress as in a miniature Trouton–Noble charged condenser, the transverse term $-\gamma(v_j\bar{\sigma}_{ij})_\perp$ in Laue's momentum density formula (Equation (7.55)) would contribute to the electromagnetic momentum, while the total momentum would still be parallel to the velocity.[82]

Abraham's original objection to the Lorentz electron did not rest on the lack of covariance of $(E, \mathbf{p})$. Indeed, he did not think in Minkowskian terms and he could not worry about the mass–energy relation, which Einstein had not yet proposed. What he condemned was the lack of agreement between two ways of computing the longitudinal mass of the electron: through the electromagnetic momentum, and through the electromagnetic energy. In the latter way, the variation dE/dt of the energy of the electron is equated to the work $\mathbf{v} \cdot d\mathbf{p}/dt$ of the external force acting on it. For the Lorentz–Abraham expressions of the energy E and the momentum $\mathbf{p}$, a simple calculation gives

$$\frac{dE}{dt} - \mathbf{v} \cdot \frac{d\mathbf{p}}{dt} = \frac{1}{3}\gamma v \frac{dv}{dt} U_s, \tag{7.62}$$

which does not vanish. Abraham correctly identified the cause of this discrepancy: for a deformable electron, the work of the external forces is not simply given by $\mathbf{v}\cdot d\mathbf{p}/dt$. Contrary to a popular opinion, the expressions (Equation (7.59)) of the energy and momentum of the Lorentz electron are as coherent as one should expect from their being derived from the same electromagnetic field. This is verified in the following calculation, which belongs to Cunningham.[83]

Any change in the velocity $\mathbf{v}$ of the center of the electron implies a change in its contraction, which in turn implies a nonuniform value for the velocity $\dot{\mathbf{r}}$ of the various parts of the electron. For the latter velocity, we have

$$\dot{\mathbf{r}} = \mathbf{v} + \mathbf{i}\bar{x}\frac{d\gamma^{-1}}{dt} = \mathbf{v} - \mathbf{i}\bar{x}\gamma v\dot{v}, \tag{7.63}$$

where $\mathbf{i}$ denotes the unit vector in the direction of the velocity $\mathbf{v}$, and $\bar{x}$ the abscissa in the comoving frame whose origin is at the center of the electron. In effect, Cunningham is assuming a Born-rigid motion, since at every instant he regards the electron to be contracted by the amount corresponding to the instantaneous velocity $\mathbf{v}$. Taking into account this contraction, the work done by the internal electromagnetic forces on the moving electron is

$$\int \rho\dot{\mathbf{r}} \cdot \mathbf{e}d^3x = \mathbf{v} \cdot \int \rho\mathbf{e}d^3x - \gamma v\dot{v}\int \rho\bar{x}e_x d^3x. \tag{7.64}$$

---

[82]Cf. Janssen and Mecklenburg 2006.

[83]Abraham 1905, pp. 203–208; Cunningham 1907. On Abraham's objection, see p. 175, this volume. Cunningham wrongly believed that Abraham had assumed the validity of $dE = \mathbf{v} \cdot d\mathbf{p}$ for a deformable electron. For this widespread opinion, see, e.g., Pauli 1921, p. 751.

In a purely electromagnetic electron, the internal electromagnetic force is exactly balanced by the external forces. We may therefore rewrite the previous equation as

$$-\frac{\mathrm{d}E}{\mathrm{d}t} = -\mathbf{v} \cdot \frac{\mathrm{d}\mathbf{p}}{\mathrm{d}t} - \gamma v \dot{v} \int \rho \bar{x} e_x \mathrm{d}^3 x. \qquad (7.65)$$

The integral is easily calculated by transforming to the comoving frame in which the electron is spherical and the field is static (in the quasi-stationary approximation), with the result $U_s/3$. Consequently, the true energy–momentum relation (Equation (7.65)) agrees with the Lorentz–Abraham expressions for the energy and the momentum.[84]

7.6 Outside Germany

There is not much to say about the non-German reception of relativity theory in the years preceding general relativity, because, with the exception of Nordström, the only theorists who significantly contributed to the theory were all German. There are a number of reasons for this state of affairs. Einstein's and Poincaré's theories were originally competing with not yet discredited alternative theories. Until 1908, it was commonly believed that Kaufmann's experiments excluded the Lorentz–Einstein electron. Many must have thought it was not worth studying a highly eccentric theory as long as experimental facts did not compel them to do so. These circumstances, which also applied to the Germans, partly explain why the theory had so few followers before 1908. The non-Germans, in addition, had little interest in the problems that relativity theory purported to solve in the electrodynamics of moving bodies and in the electron theories. Lorentz, Larmor, and Poincaré were singular in their abundant contributions to these fields. Although they were well equipped to understand relativity theory, they did not actively promote it in their own country.[85]

In the Netherlands, Lorentz was slow to react to the new relativity theory. When in 1906 Kaufmann's measurements came to contradict the relativistic electron, he quickly gave up:

> Unfortunately, my hypothesis of the flattening of electrons is in contradiction with Kaufmann's results, and I think I have to abandon it. I am therefore *au bout de mon latin*, and it seems to me impossible to establish a theory that requires the complete lack of effects of translation on electromagnetic and optical phenomena.

Lorentz doubted that the principle of relativity applied strictly and generally. Until late 1907 at least he did not understand that Einstein (and Poincaré) had obtained exact Lorentz covariance. Despite his high esteem for the young Einstein and despite his later contribution to general relativity, his first discussions of relativity theory occurred as late as 1909 in

---

[84] We have $\int \rho \bar{x} e_x \mathrm{d}^3 x = \int \bar{\rho}\, \bar{x}\, \bar{e}_x \mathrm{d}^3 \bar{x} = \int (\partial_{\bar{x}} \bar{e}_x) \bar{x}\, \bar{e}_x \mathrm{d}^3 \bar{x} + 2 \int (\partial_{\bar{y}} \bar{e}_y) \bar{x}\, \bar{e}_x \mathrm{d}^3 \bar{x}$. Partial integration, $\nabla \times \mathbf{e} = \mathbf{0}$, and spherical symmetry give $\int (\partial_{\bar{x}} \bar{e}_x) \bar{x}\, \bar{e}_x \mathrm{d}^3 \bar{x} = -\frac{1}{2} \int \bar{e}_x^2 \mathrm{d}^3 x = -\frac{1}{3} U_s$, $\int (\partial_{\bar{y}} \bar{e}_y) \bar{x}\, \bar{e}_x \mathrm{d}^3 \bar{x} = -\int \bar{x}\, \bar{e}_y \partial_{\bar{y}} \bar{e}_x \mathrm{d}^3 \bar{x} = -\int \bar{x}\, \bar{e}_y \partial_{\bar{x}} \bar{e}_y \mathrm{d}^3 \bar{x} = \frac{1}{3} U_s$. This reasoning is based on the assumption (shared by Lorentz, Abraham, and Born that the integral $-\int \rho \mathbf{e} \mathrm{d}^3 x$ (which is also the time derivative of $\int c^{-1} \mathbf{e} \times \mathbf{b} \mathrm{d}^3 x$) truly represents the external force acting on a Born-rigid electron. As explained in Kalckar and Ulfbeck 1982, this assumption does not take into account the fact that different points of a linearly accelerated Born-rigid body have different accelerations in the rest frame. It may be corrected to yield the desired U_s/c^2 for the inertial mass.

[85] For global comparative studies, cf. Goldberg 1969; Glick 1987. For the reasons for acceptance, cf. Brush 1999.

an amplified version of his Columbia lectures of 1906, and in 1910 in his second Wolfskehl lecture at Göttingen. He there acknowledged the higher simplicity and the empirical adequacy of the new theory. But he was still reluctant to follow Einstein and Minkowski in their "bold" and "fascinating" revision of space and time. In his opinion, one could still keep the ether as a privileged reference frame for true space and time, despite the empirical impossibility of deciding between various inertial frames:

> I cannot but regard the ether, which can be the seat of an electromagnetic field with its energy and its vibrations, as endowed with a certain degree of substantiality, however different it may be from all ordinary matter. In this line of thought, it seems natural not to assume at starting that it can never make any difference whether a body moves through the ether or not, and to measure distances and lengths of time by means of rods and clocks having a fixed position relatively to the ether.

That is to say, Lorentz was favoring Poincaré's version of the theory without naming Poincaré.[86] In Britain, Larmor mostly ignored relativity theory, although he later reacted to general relativity. In 1907, his former student Ebenezer Cunningham referred to Einstein's article of 1905 for the exact Lorentz transformations. Like Poincaré, Cunningham associated the invariance of the Maxwell–Lorentz equations with the impossibility of detecting effects of the ether wind. He noted that this impossibility implied the invariance of the velocity of light under the coordinate transformations, and used this property to derive the expression of these transformations. In collaboration with the mathematician Harry Bateman, he later discovered the conformal invariance of the Maxwell–Hertz equations, and used it to extend the relativity principle to accelerated frames. He worked in the received electron-theoretical framework, and ignored the revolutionary side of Einstein's relativity theory, at least until he came to appreciate Laue's treatise of 1911. In contrast, at the Cavendish laboratory in Cambridge, Norman Campbell welcomed Einstein's theory as a new, etherless electrodynamics based on a coherent doctrine for the measurement of space and time. This receptivity may be explained by his general endeavor to found physics on measurement, and by his early belief in the superiority of Faraday's lines of force over an ungraspable ether. His and Cunningham's writings were the main British sources on relativity theory before Eddington.[87]

In France, Poincaré kept lecturing on the *mécanique nouvelle*, that is, his own version of relativity theory with the relativity principle, the Lorentz group, the interpretation of the transformed coordinates and fields as those measured by observers in motion, the ether, and the persisting distinction between apparent states and true states. After the publication of the Palermo study in 1906, he no longer worked on the theory and he did not comment on the German advances in this domain, not even when in April 1909, soon after Minkowski's tragic death, he devoted the last of his Wolfskehl lectures in Göttingen to the *mécanique nouvelle*. He never named Einstein or Minkowski in this context, although

---

[86]Lorentz to Poincaré, March 8, 1906, in Kox 2008, p. 203; Lorentz to Wien, September 22, 1907, in Kox 2008, pp. 221–222 (imperfect covariance of Einstein's theory); Lorentz 1909, pp. 223–230 (cit. p. 230); 1910, p. 1236.

[87]Cunningham 1907, 1910, 1912, 1914; Campbell 1910, 1911, 1913, pp. 354–383. Campbell had earlier imagined an electrodynamics of moving bodies based on the relative motion of Faraday tubes. Cf. Warwick 1992, 1993, 2003. Warwick traces Cunningham's mathematical-transformation approach to the culture of the Cambridge mathematical Tripos, and Campbell's empirico-critical approach to the culture of the Cavendish laboratory.

an implicit reference to Minkowski's approach can be found in his London lecture of May 1912. There is little doubt that Poincaré knew about these German contributions soon after their publication. He may have thought they added nothing essential to his and Lorentz's theory, or he may have resented the German's failure to properly cite his own contributions. His French colleagues, even his former student Paul Langevin, were equally unconcerned with his *dynamique de l'électron*. When at last, in 1910–1911, Langevin lectured on relativity theory at the Collège de France, he favored the Einstein–Minkowski formulation and ignored Poincaré's contribution. The lack of earlier French interest in relativity theory is easy to understand, because France did not have a proper tradition of theoretical physics in the British and German sense.[88]

Conclusions

Whatever Einstein's debts to his predecessors, whatever the exact origin of the flash of insight in which he redefined time and simultaneity through optical synchronization, this redefinition and the concomitant rejection of absolute time were highly original and consequential. Like Hertz, Einstein required a good theory to afford a one-to-one correspondence between representation and phenomenon, with none of the redundancy brought by unnecessary garments. In Cohn's, Bucherer's, Ritz's, and Einstein's eyes, the ether was such a garment. Of this invisible medium, Einstein retained only the constancy of the velocity of light. He otherwise assumed the complete equivalence of all inertial frames, not only from an empirical point of view but also from a representational point of view. Then the velocity of light has to be the same in any inertial frame, and this can be true only if different time coordinates are used in different inertial frames. This new kinematics not only explains the observed relativity of optical and electromagnetic phenomena, it also implies the contraction of length and the dilation of time as simple consequences of the coordination of measurements done in two different frames. What Lorentz and Larmor regarded as a marginal side effect of the motion of rod and clocks through the ether becomes a symmetric, perspectival effect of primary importance.

Einstein's other essential innovation was the inertia of energy. The electromagnetic worldview, or a milder formal analogy between inertial forces and electromagnetic forces had led to the idea of a velocity-dependent mass of the electron. In addition, the mass of Abraham's purely electromagnetic electron was known to be its electrostatic energy divided by c^2. At the electronic scale, the shared transformation properties of the inertial force and the Lorentz force are sufficient to harmonize the electron's dynamics with the relativity principle. As Einstein realized through a thought experiment à la Poincaré, this is no longer the case at the macroscopic scale in which the energy of a body may change without concomitant change of momentum (e.g., by emitting the same amount of radiation in two opposite directions). In order to harmonize the energy–momentum balances in two different frames, Einstein had the mass of the body vary with its energy content. He foresaw

---

[88] Poincaré 1906b, [1906–1907], 1909, 1910a (Wolfskehl), 1912, 1913a; Langevin [1910–1911], 1911, 1912. Cf. Cuvaj 1970; Biezunski 1981; Paty 1987, 1993, 2002; Walter 2011; Bracco and Provost 2013. On French mathematical physics versus British and German theoretical physics, cf. Pestre 1984; Príncipe 2008.

that this relation, originally obtained on a specific electromagnetic system, was a general one, with someday measurable consequences.

By 1906, there were two theories of relativity: Poincaré's and Einstein's. Poincaré's appeared in summary form in the *Comptes rendus* for the Academy session of 5 June 1905, and in full in the Palermo *Rendiconti* for 1906. The German *Annalen* received Einstein's memoir on June 30, 1905 and published it in September 1905. There are important similarities between the two theories: they both introduce the Lorentz group as a fundamental symmetry shared by all physical theories; they both regard this symmetry as the formal expression of the relativity principle; they both interpret the Lorentz-transformed quantities as those measured by moving observers; and their observable predictions are the same for any conceivable experiment in the optics or electrodynamics of moving bodies.

Yet there are multiple differences. Poincaré is the only one to address gravitation and to propose a relativistic model of the electron. He develops and exploits the group structure of the Lorentz transformations much more than Einstein does. He retains the ether as a privileged reference frame in which true space and time are defined, even though he denies any empirical possibility to decide which inertial frame truly is the ether frame. He introduces the Lorentz transformations as those leaving the Maxwell–Lorentz equations invariant, even though he understands that the transformed time is the one measured in a moving frame under the convention of the constant velocity of light. In contrast, Einstein ignores all electron models and completely eliminates the ether. He requires the expression of physical laws (not only their empirical content) to be the same in any inertial frame, and he abolishes the distinction between true and apparent quantities. To the relativity principle thus understood, he adds the principle of the constancy of light in any given inertial frame, and he derives the Lorentz transformations from these two principles. He fully discusses the consequences of these transformations on the measurement of space and time, thus reinterpreting the contraction of lengths and the dilation of time as perspectival effects of motion. He obtains the relativistic equations of motion of an electrically charged particle in the electromagnetic field without any model of the particle, from the validity of Newton's acceleration law at small speeds combined with the Lorentz transformations (Poincaré will do the same in 1906). Later in 1905 he gives his first argument for the inertia of energy.

Poincaré's relativity theory is strikingly modern in two respects: through its higher emphasis on the group structure, and also through its appeal to the principle of least action (in the dynamics of the electron). Yet there is no doubt that Einstein's theory is in other respects closer to special relativity as we now understand it: through the radical redefinition of space and time, through the physical understanding of the resulting kinematical and dynamical consequences, and through the pure principles-based approach. Although Poincaré was first to introduce the distinction between "the physics of principles" and a physics made on constructive models, in his *dynamique de l'électron* he adopted a hybrid approach in which the principles served to select a constructive model. In contrast, Einstein tried to get as much as he could with the principles only. Lastly, there is a major stylistic difference: whereas Einstein's article of 1905 is a model of clear, polished writing, with transparent logic and proper display of the most important results, Poincaré's long, winding memoir seems to have been written in a hurry, with no time to distill the contents for the physicist reader.

That said, the similarities between Poincaré's and Einstein's theories are too important to be merely fortuitous. They may be explained by noting that both men started
with a critical appraisal of Lorentz's theory. Lorentz prepared the ground by removing
from the ether its most crudely mechanical attribute, velocity, and by developing a powerful correspondence between the states of a system at rest and the states of a system
in motion (in the ether). This correspondence contained the Lorentz transformations in
germ and formally anticipated Einstein's and Poincaré's use of these transformations in
the electrodynamics of moving bodies. Einstein and Poincaré also shared a critical attitude
toward physical theories, probing and questioning concepts as basic as space, time, and the
ether, and paying special attention to the necessary conventions of measurement. Einstein
could have gotten this attitude from Hertz, Drude, and Mach, but also from Poincaré's
La science et l'hypothèse. The shared attitude and the shared attention to Lorentz's theory
may explain why Einstein and Poincaré both adopted the relativity principle and why they
both reinterpreted Lorentz's corresponding states as the states measured by observers in
motion, with a very similar appeal to optical synchronization. Or it could be that Einstein
got both the principle and the reinterpretation from Poincaré, while reading his relevant
writings of 1900 and 1902. Or else it could be, as Peter Galison propounded, that Poincaré
and Einstein both found inspiration in contemporary techniques for the distribution of
time, which implied synchronization by telegraphic signals.[89]

As for the differences between Poincaré and Einstein, they too may be explained in a variety of ways. Poincaré, as a mathematician immersed in the theory of Lie groups, naturally
developed the group-theoretical aspects of relativity theory. He could easily assimilate the
mathematical intricacies of electron theory, whereas the young Einstein regarded heavily
mathematized theories with suspicion. Poincaré's interest in the electron theories and the
electromagnetic worldview partly explains why he maintained the ether as the universal
carrier of all forms of energy. In contrast, Einstein's early quantum-theoretical reflections
led him to question both the mechanical and the electromagnetic worldview and to favor
a principles-based approach.

There also were epistemological reasons for Einstein's and Poincaré's different attitudes
toward the ether. Whereas Einstein required a good theory to be devoid of surplus content, Poincaré tolerated models and conventions that eased the intuition of phenomena
and helped preserve ancestral habits of thought. As a mathematician familiar with Klein's
Erlangen program, Poincaré was well aware of the definition of a geometry through the invariance group of its figures, and he must have seen that the Lorentz group could define a
new geometry of space–time. Yet, when toward the end of his life he alluded to Minkowski's
realization of this possibility, he anticipated that physicists would rather keep the ether to
preserve their habitual conception of space and time. In his conventionalist approach to the
foundations of geometry, Poincaré had earlier regarded the choice of Euclidean geometry
as a matter of convenience, because the same phenomena can be represented under a different geometry if only the expression of physical laws is modified accordingly. Similarly,
Poincaré now regarded the choice of Newtonian space and time as a matter of habit and

---

[89] Galison 2003.

convenience, whereas Einstein wanted the conventions of measurement to reflect concrete practice and to shun the undetectable ether.[90]

In the subsequent history of relativity, Minkowski was the only author who significantly drew on Poincaré's relativity theory. From this fellow mathematician, he borrowed the interpretation of the Lorentz transformations as rotations in four-space with a fourth imaginary coordinate, the four-vectors, and the general idea of building invariants thereupon, including the field and particle actions. He must have been impressed by Poincaré's remark that the Lorentz transformations do not by themselves form a group and that the product of two Lorentz transformations generally involves a spatial rotation, whereas Einstein only considered the subgroup of parallel boosts. Minkowski indeed insisted on the nonfactorable character of the Lorentz group as defined by Poincaré, and thereby justified the fusion of space and time into a single entity. The very idea of defining a new space–time geometry through the Lorentz group only makes sense under Poincaré's definition. Altogether, Minkowski's debt to Poincaré seems more considerable than is usually assumed.

References to Poincaré are very scarce in early literature on relativity and electron theory. In his memoir of 1908, Minkowski credited Poincaré with naming the Lorentz transformations and the Lorentz group and for a relativistic generalization of Newton's law of gravitation, but he did not name Poincaré at all in his famous Cologne lecture on space and time. Lorentz, whom Poincaré had directly informed of his progress in May 1905, mentioned Poincaré in print for the first time in 1909 and only for his model of the electron. This is the more surprising because Lorentz's ether-based reinterpretation of Einstein's relativity resembled Poincaré's theory. In 1906, Kaufmann and Planck both referred to Poincaré's note in the *Comptes rendus* for his improving Lorentz's electron model. Laue too referred to Poincaré's electron in 1911, then realizing the similarity to his own solution to Abraham's energy–momentum paradox. Most fairly but quite singly, in the second edition of his *Theorie der Elektrizität*, published in 1908, Abraham credited Poincaré for the optical-metric interpretation of Lorentz's local time. As for Einstein, in 1906 he referred to Poincaré's Lorentz-jubilee article of 1900 in his second derivation of the mass–energy relation, and yet in his relativity review article of 1907 he did not mention that Poincaré had given a very relevant interpretation of Lorentz's local time in this article. Nor did he cite Poincaré's *dynamique de l'électron*, even though Planck's discussion of Kaufmann (which Einstein commented on) should have alerted him to this work. Was it dissimulation or just carelessness? No one knows. In those years there was not much inclination to cite French sources in Germany, and vice versa. As was already said, Poincaré never explicitly mentioned Einstein's and Minkowski's relativity theories.[91]

Minkowski, mathematician though he was, did not have the difficulty Poincaré had in catching the physicists' attention. While Einstein initially undervalued Minkowski's new tensor calculus, Göttingen-trained electron theorists immediately saw the formal simplicity this calculus brought to relativity theory, and they used it to further develop this theory. Minkowski inspired a wide audience with his new world of shadows, and then proposed

[90] Poincaré 1912, p. 170, cited p. 182, this volume.

[91] Minkowski 1908, pp. 54, 109n; Lorentz 1909, p. 213; Kaufmann 1906, p. 494; Planck 1906b, p. 756n; Laue 1911c, p. 542; Abraham 1908, pp. 365, 402; Einstein 1906, p. 627; Einstein 1907c.

his space–time diagrams as an efficient remedy for this abstraction. He died as the herald of a new vision in which mathematical structure, rather than Einstein's operational considerations, became the essence of relativity theory.

Minkowski's new world geometry enabled his disciple Max Born to construct a purely electromagnetic and ideally rigid electron, in conformity with Minkowski's persisting adherence to the electromagnetic worldview. The underlying new concept of rigidity was perhaps more important for its failure than for its success. Criticism by Einstein and others revealed its incompatibility with the setting of a body in rotation, as well as the strange non-Euclidean relations of lengths measured in a rotating frame. Criticism by Laue led to a proof that a continuously extended body, in relativity theory, necessarily had an infinite number of degrees of freedom. This result called for a new relativistic dynamics of continua, which Laue found in a generalization of the energy–momentum tensor Minkowski had introduced as a four-dimensional generalization of Maxwell's stress tensor. Laue associated such a tensor with every kind of force, and he used its symmetry for a general demonstration of the mass–energy equivalence. He proved that any system in equilibrium under balancing stresses globally behaved as a point-like system does in Einstein's relativistic dynamics. When applied to the electron, this theorem opened up infinitely many possible models, with much freedom in the choice of the cohesive stress and the shape of the charge distribution. At the same time, the growing conviction that the quantum of action should affect the constitution of atomic or subatomic entities discouraged classical electron modeling. What mattered most, for the impact of Laue's relativistic dynamics, was the general tools developed to represent force, energy, and momentum in relativistic field theories.

The publication of Laue's *Das Relativitätsprinzip* in 1911 marked the maturity of (special) relativity theory, as well as its acceptance by most experts in the concerned fields, in one form or another. This happened without any strong empirical confirmation of specific quantitative predictions. The only alleged confirmation, Bucherer's electron-deflection experiment of 1908, turned out to be nearly as doubtful as Kaufmann's contrary result of 1906. The next, indirect confirmations occurred in late 1915 with Einstein's explanation of the anomaly in Mercury's perihelion, and in 1916 with the success of Sommerfeld's relativistic theory of the fine structure of the hydrogen atom. Direct confirmation of the mass–energy relation (in nuclear reactions) and of relativistic time dilation (through the Doppler effect) waited until the 1930s. In contrast, the main pillar of the theory, the relativity principle, had been abundantly confirmed through the null result of ether-drift experiments, even in a case, the Rayleigh–Brace experiments, in which most of the competing theories predicted a positive result. In the experts' eyes, what made relativity theory so convincing was the empirical solidity of the relativity principle, the formal simplicity brought to electrodynamics, and the heuristic promise of fitting all physics into a single framework.[92]

---

[92] On doubts about Bucherer's result, cf. Battimelli 1981. On Mercury's perihelion, see p. 317, this volume. On Sommerfeld's theory, cf. Kragh 1985.

8

FROM RIEMANN TO RICCI

> Either the actual foundation of space must be a discrete manifold, or the ground of the metric structure must be sought externally in the binding forces that act on the manifold.[1] (Bernhard Riemann, 1854)
>
> I still have vivid memories of the extraordinary impression that Riemann's trains of thought made on young mathematicians [when his habilitation lecture was published]. Much of it seemed obscure and hard to understand and yet of unfathomable depth. Today's mathematicians, who have integrated all these things in their way of thinking, still admire the clarity and fertility of the analysis.[2] (Felix Klein, 1926)

In 1908, Minkowski argued that relativity theory could have been earlier invented by mathematicians, merely by removing the group-theoretical heterogeneity between spatial rotations and Galilean boosts. Whatever the merit of this speculation, the historical genesis of special relativity did not essentially imply mathematics that was not already familiar to physicists. This is surely the case for Einstein, who relied on elementary algebra and analysis and did not even employ the vector notation. Poincaré's recourse to the theory of Lie groups and Minkowski's four-dimensional calculus do not really count as vital injections of pre-physical mathematics, because they are easily obtained by generalizing familiar three-dimensional notions. All we can say is that Poincaré's and Minkowski's interpretations of relativity theory depended on their familiarity with the new geometries of the nineteenth century: Minkowski's by defining a new space–time through the Lorentz group, Poincaré's by resisting this new space–time in the name of his conventionalist denial of the physical import of new geometries.[3]

In contrast, we will see that Einstein's general relativity crucially depended on "the absolute differential calculus" to which his mathematician friend Marcel Grossmann directed him while he was searching for a gravitational field equation in the 1910s. This was an early version of what we would now call the tensor calculus on differential manifolds. Especially important was Riemann's curvature tensor, on which Einstein based his final equations for the gravitational field. In the history of modern physics, it repeatedly happened that the construction of new theories involved the simultaneous creation of new mathematics.

---

[1] Riemann 1867 [1854], p. 149.

[2] Klein 1926–27, Vol. 2, p. 16.

[3] For Minkowski's argument, see p. 218, this volume. Minkowski's use of hyperbolic geometry to represent relativistic velocities (cf. Walter 1999b) did not interest physicists.

Think for instance of differential calculus in Newton's mechanics, or field-operator calculus in Maxwell's electromagnetic field theory, or tensor calculus in the theory of elasticity. More rarely, an important piece of mathematics that had long remained unemployed suddenly became the centerpiece of a new physical theory. Therefore, the history of this theory cannot properly be told without first attending to the genesis of the employed mathematics. This is why we need a chapter on the origins of the absolute differential calculus.

It all started on June 10, 1854 when Bernhard Riemann delivered his habilitation lecture "On the hypotheses which lie at the basis of geometry" in front of the Göttingen faculty of philosophy. This lecture left much for mathematicians and historians to wonder about. In particular, the concept of curvature for a manifold of any dimension there occurred with no calculations whatsoever and yet with stunning precision. Riemann was addressing a mostly nonmathematical audience who would not have followed the mathematical technicalities. Old Gauss probably was the only attendee to understand the extraordinary profundity of the lecture. Riemann never published his reflections or any relevant calculation, because poor health prevented him from doing so (according to Dedekind) or maybe because his interest in geometry was only occasional.[4]

In 1861, Riemann sent to the French Academy of Sciences a memoir usually called the *Commentatio*, in which the modern reader easily recognizes the precise expression of what is now called the Riemann tensor as well as an argument for the covariance of this tensor. It is tempting to relate the relevant section of this memoir to the geometric problem of finding a criterion of flatness for a manifold, in the wake of the habilitation lecture. Yet the purpose of the *Commentatio* was different: it answered a prize question on the propagation of heat in solid bodies. Its derivations were purely analytic, and its only contact with geometry was in Riemann's remark that some analytical expression could be illustrated as the curvature components of a metric manifold. The prize was not attributed, and Riemann's submission long rested in the archive of the French Academy.[5]

Riemann's habilitation lecture and the *Commentatio* were published posthumously, the former by Richard Dedekind in 1867, and the latter by Heinrich Weber in 1876 in Riemann's *Werke*. Riemann was no longer there to answer the following questions: How did he obtain the results enunciated in the habilitation lecture? Granted he relied on precise calculations, what were these? Do the formulas of the *Commentatio* have anything to do with this lecture? Or could Riemann have conceived them in a purely algebraic or analytic manner?

In the absence of relevant manuscript materials, the answer to these questions can only be conjectural. The best we can do is to examine the resources available to Riemann and to guess how he might have exploited them. Most important in this regard are Gauss's *Disquisitiones* of 1828 on curved surfaces, for Riemann regarded his own concept of curvature as a generalization of the intrinsic curvature invented by Gauss in the two-dimensional case. This is why Section 8.1 of this chapter is devoted to Gauss's theory of surfaces and to a rephrasing of his main results and proofs in a manner that lends itself to higher-dimensional generalization.

---

[4]On these circumstances, cf. Dedekind 1876, p. 517.

[5]On the purpose and fate of the *Commentatio*, cf. Farwell and Knee 1990a.

Section 8.2 exploits this Gaussian background to analyze Riemann's results regarding the curvature of a manifold in his habilitation lecture, his investigation of the transformation properties of quadratic differential forms in the *Commentatio*, and his speculations on a role of curved space in physics. It will be shown that the generalization of Gaussian notions naturally leads to Riemann's concepts of geodetic distortion and sectional curvature. In light of heretofore unexploited manuscripts in the Göttingen University archive, it will appear that the calculations of the *Commentatio* were done by mostly algebraic means, with hints given by partial geometric interpretation.

Riemann's willingness to speculate on a curved physical space is not unrelated to the earlier advent of non-Euclidean geometry, which is briefly recounted in Section 8.3. We will see how János Bolyai and Nikolai Lobachevsky pioneered hyperbolic geometry, how Eugenio Beltrami connected it to Riemannian geometry with constant negative curvature, how Hermann Helmholtz redefined the problem of space through the group of displacements of rigid bodies, how Felix Klein redefined geometry through invariance groups, how Poincaré developed his conventionalist attitude in face of the multiplicity of the new geometries, and how new kinds of space were occasionally invoked to solve astronomical paradoxes or to dream of new worldviews.

Section 8.4 is about the way Riemann's posthumous readers developed the geometric and analytic-algebraic aspects of Riemann's four-index parenthesis $(\iota\iota', \iota''\iota''')$, which corresponds to our Riemann tensor. They did so while filling the gaps in Riemann's texts or while building theories that formally implied some of Riemann's constructs. The Riemann parenthesis has a geometrical aspect as the measure of curvature in a metric manifold, and an algebraic aspect as the crucial quantity in the equivalence problem for quadratic differential forms.[6] From a modern point of view, these two aspects are conflated by interpreting any quadratic differential form as a metric on a differential manifold. Although Riemann clearly perceived the interconnection between the two aspects, he also understood that they could be developed independently: the mostly algebraic nature of the relevant section of the *Commentatio* is evidence for that. Some of his followers, including Dedekind, Tullio Levi-Civita, and Hermann Weyl, combined both aspects. Others, including Elwin Christoffel, Rudolf Lipschitz, and Gregorio Ricci-Curbastro favored the algebraic or analytic aspect. While Christoffel and Ricci connected their calculus with the theory of invariants, Lipschitz connected his own formalism with analytical mechanics. Ricci insisted that the new calculus, which he called the absolute differential calculus, should stand on its own, geometry being only one application among others. This is the conception Einstein inherited when he built general relativity. As we will see at the end of this chapter, in a later attempt to interpret a mysterious differential expression in Riemann's *Commentatio*, Tullio Levi-Civita reintroduced geometric intuition in Ricci's calculus through the concept of parallel displacement.

In the following, modern vector and tensor notation is used for the sake of compactness and legibility. These notations should only be regarded as abbreviations for the Cartesian coordinate or sum-index notations that were actually used by Gauss, Riemann, and their followers, with no intended allusion to invariance or covariance properties. However, the

---

[6] A *quadratic differential form* is an expression of the type $\sum_{\mu\nu} a_{\mu\nu}(x)\mathrm{d}x^\mu\mathrm{d}x^\nu$.

notation for differentials and variations (d and δ symbols) will not be modernized, because the ambiguities inherent in this notation played a significant historical role. The context dependence of these symbols caused misunderstandings for instance between Riemann and Levi-Civita, and it raises interpretive difficulties for modern readers accustomed to more precise notation. But it is part of the historical dynamics we are trying to grasp.

8.1 Gauss's curved surfaces

In the mid-1820s, the Göttingen astronomer and mathematician Carl Friedrich Gauss investigated the geometry of curved surfaces in the triple context of maps, developable surfaces, and geodesy. In an influential memoir of 1779, Joseph Louis Lagrange had given the mathematical condition for a map to conserve angles: the mapping function should satisfy the d'Alembert–Cauchy–Riemann condition, or, equivalently, it should be the real part of a derivable function of a complex variable. A similar problem, on which Gauss had been reflecting for several years, was to find the condition for a map to preserve the area of surface elements. This problem led Gauss to focus on the properties of a surface that were conserved by "development," that is, by bending without stretching. Through his involvement in a geodetic survey of the state of Hannover, Gauss was also concerned with what I shall call *geodetic distortion*, namely, the error committed in geodetic surveys when geodesic triangles on the curved surface of the earth are assimilated with planar triangles. As Adrien-Marie Legendre had noted in 1787, this error depends on the excess of the sum of the angles over a flat angle. As was known since the sixteenth century from Albert Girard, this excess is equal to the area of the triangle divided by the square of the radius of the sphere. Equivalently, the inverse square of the radius is the angular excess divided by the area of the triangle. Gauss presumably exploited this result to define the geodetic curvature at a point of a surface as the ratio of angular excess over area for a small geodesic triangle around this point. As these quantities depend only on intrinsic metric relations, the geodetic curvature is invariant by development of the surface.[7]

The Theorema egregium

Intuitively, curvature also has to do with the way the normal at a point of the surface rotates when the point moves on the surface. Gauss represented the direction of the normal by a point on a sphere of radius one. On this unit sphere he considered the triangle corresponding to a geodesic triangle, and proved by geometrical means that the angular excess of the former triangle was equal to the area of the latter triangle. In the end, he replaced the geodetic definition with the *normal definition* of the curvature as the ratio between two corresponding small areas on the unit sphere and on the surface. As he proved, this ratio

---

[7]Legendre 1787, p. 358. This reconstruction of Gauss's route to intrinsic curvature is a guess based on hints in Gauss 1828 (pp. 43, 46) and on the order of the earlier manuscript, Gauss [1825], in which the intrinsic character of Gaussian curvature is proved on the basis of its geodetic interpretation (whereas in the final memoir it is proved by analytical means). On Gauss's life, cf. Bühler 1981. On the history of differential geometry in Gauss's times and earlier, cf. Reich 1973. On Gauss's and Riemann's differential geometry, cf. Kline 1972, Vol. 3, chapter 37. For a modern assessment of Gauss's contribution, cf. Torretti 1978, pp. 71–82.

has a definite value independent of the shape of the area drawn on the surface: it is given by the inverse of the product of the two principal radii of curvature at the given point.[8]

The normal curvature is not obviously an intrinsic property of the surface, because the direction of the normal obviously is not. Gauss nonetheless favored the normal definition. In a draft of 1825, he obtained the intrinsic character of the normal curvature by proving its equality with the geodetic curvature. In the final *Disquisitiones* of 1828, he followed a more algebraic route in which the normal curvature is first expressed in terms of the derivatives of the parametric representation $\mathbf{r}(p,q)$ of the surface and then shown to depend only on the coefficients of the fundamental form

$$ds^2 = Edp^2 + 2Fdpdq + Gdq^2 \tag{8.1}$$

that gives the element of length as a function of the variations of the parameters p and q. This is the *Theorema egregium*, which Gauss stated as follows:

> The analysis developed in the preceding article shows us that for finding the measure of curvature there is no need of finite formulas that express the coordinates x, y, z as functions of the indeterminates p, q; but that the general expression for the magnitude of any linear element is sufficient.

As the expression of the element of length evidently does not change by development of the surface, Gauss's theorem implies the intrinsic character of the Gaussian curvature.[9]

Toward the Riemann tensor

The expression of the curvature in terms of the E, F, G coefficients is in itself an important result. It may be derived by directly exploiting the geodetic definition. In order to prepare later generalization, I anachronistically use the indicial expression of the metric (8.1),

$$ds^2 = g_{\mu\nu}dx^\mu dx^\nu, \quad \text{with } \mu = 1, 2, \quad \nu = 1, 2. \tag{8.2}$$

Summation over repeated indices is understood. If $\mathbf{r}(x_1, x_2)$ denotes the point of the surface corresponding to the values x_1 and x_2 of the parameters, then $g_{\mu\nu} = \partial_\mu \mathbf{r} \cdot \partial_\nu \mathbf{r}$. The equation of a geodesic line results from

$$0 = \delta \int \sqrt{g_{\mu\nu}dx^\mu dx^\nu} = \int \frac{1}{2}\partial_\rho g_{\mu\nu}\delta x^\rho dx^\mu dx^\nu/ds + \int g_{\mu\nu}(dx^\mu/ds)\delta dx^\nu. \tag{8.3}$$

---

[8] Gauss [1825], §§ 11, 14. The principal curvatures (Euler 1767) are the minimal and maximal curvature of the lines formed by the intersection of the surface with a plane containing the normal at the given point. For concise modern proofs, cf. Darrigol 2015, pp. 50–51.

[9] Gauss [1825], §14; 1828, §12.

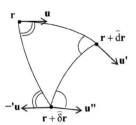

Fig. 8.1.   Angles of a small geodesic triangle.

Introducing $u^\mu = dx^\mu/ds$, and integrating the last integral by parts, we get

$$\frac{du^\rho}{ds} + \Gamma^\rho_{\mu\nu}u^\mu u^\nu = 0, \quad \text{with } g_{\rho\sigma}\Gamma^\rho_{\mu\nu} = \frac{1}{2}(\partial_\mu g_{\sigma\nu} + \partial_\nu g_{\sigma\mu} - \partial_\sigma g_{\mu\nu}). \tag{8.4–5}$$

The couple (u^1, u^2) gives the coordinates of the unit tangent vector **u** in the tangent-plane basis $(\partial_1 \mathbf{r}, \partial_2 \mathbf{r})$. Similarly, any vector **a** in the tangent plane can be expressed by its coordinates a^μ in this local basis. The scalar product of two vectors **a** and **b** is then given by $\mathbf{a} \cdot \mathbf{b} = a^\mu b^\mu \partial_\mu \mathbf{r} \cdot \partial_\nu \mathbf{r} = g_{\mu\nu}a^\mu b^\nu$.

Now consider a small geodesic triangle with summits at $\mathbf{r}$, $\mathbf{r} + \hat{d}\mathbf{r}$, and $\mathbf{r} + \hat{\delta}\mathbf{r}$. Call **u** the tangent unit vector at point **r** of the geodesic joining **r** and $\mathbf{r} + \hat{d}\mathbf{r}$; **u**′ the tangent unit vector at point $\mathbf{r} + \hat{d}\mathbf{r}$ of the same geodesic; **u**″ the unit vector that makes the same angle with the geodesic joining $\mathbf{r} + \hat{d}\mathbf{r}$ and $\mathbf{r} + \hat{\delta}\mathbf{r}$ at point $\mathbf{r} + \hat{\delta}\mathbf{r}$ as the vector **u**′ does with the same geodesic at point $\mathbf{r} + \hat{d}\mathbf{r}$; and ′**u** the unit vector that makes the same angle with the geodesic joining **r** and $\mathbf{r} + \hat{\delta}\mathbf{r}$ at point $\mathbf{r} + \hat{\delta}\mathbf{r}$ as the vector **u** does with the same geodesic at point **r** (see Figure 8.1). By construction, the angle between **u**″ and − ′**u** is equal to the sum of the angles of the geodesic triangle. Equivalently, the angle between **u** and ′**u** is the excess of this sum over a flat angle. These two vectors being unit vectors with very nearly the same direction, their difference **u**″ − ′**u** must be perpendicular to **u**. The vector $\hat{d}\mathbf{r} \times \hat{\delta}\mathbf{r}$ is directed along the normal and its length measures twice the surface of the geodesic triangle. We therefore have

$$\mathbf{u}'' - {}'\mathbf{u} = \tfrac{1}{2}C\mathbf{u} \times (\hat{d}\mathbf{r} \times \hat{\delta}\mathbf{r}) \tag{8.6}$$

where C denotes the geodetic curvature.[10]

For a unit vector **a** carried along a geodesic at constant angle with the tangent vector **u**,[11] the coordinates evolve according to

$$\frac{da^\rho}{ds} + \Gamma^\rho_{\mu\nu}u^\mu a^\nu = 0, \tag{8.7}$$

because this equation is easily seen to preserve $\mathbf{a}^2 = g_{\mu\nu}a^\mu a^\nu$ and $\mathbf{a} \cdot \mathbf{u} = g_{\mu\nu}a^\mu u^\nu$.

---

[10] I use the symbols $\hat{d}$ and $\hat{\delta}$ to indicate *finite* increments of which the zero limit will be taken at a later stage of the reasoning.

[11] The general concept of parallel displacement is not needed here, because along a geodesic of a surface the conditions of constant length and constant angle are sufficient to determine the displacement.

In an alleviated notation, the former equation reads $\mathrm{d}a = -\Gamma_{a,\mathrm{d}x}$. A second differentiation along the geodesic leads to

$$\mathrm{d}^2 a = -\mathrm{d}\Gamma_{a,\mathrm{d}x} - \Gamma_{\mathrm{d}a,\mathrm{d}x} - \Gamma_{a,\mathrm{d}^2x}, \quad \text{with } \mathrm{d}\Gamma = \mathrm{d}x^\mu \partial_\mu \Gamma. \tag{8.8}$$

Hence follows

$$\hat{\mathrm{d}}a = -\Gamma_{a,\hat{\mathrm{d}}x} - \tfrac{1}{2}\Gamma_{\hat{\mathrm{d}}a,\hat{\mathrm{d}}x} - \tfrac{1}{2}\hat{\mathrm{d}}\Gamma_{a,\hat{\mathrm{d}}x} \tag{8.9}$$

for the variation of a on the geodesic joining x and $x + \hat{\mathrm{d}}x$ to second order in $\hat{\mathrm{d}}x$. We may compute the vectors $\mathbf{u}'$, $\mathbf{u}$, and $'\mathbf{u}$ of Figure 8.1 through this formula. Now writing d for $\hat{\mathrm{d}}$, and δ for $\hat{\delta}$, this leads to

$$\mathbf{u}'' - '\mathbf{u} = -\tfrac{1}{2}(\Gamma_{\mathrm{d}u,\delta x} - \Gamma_{\delta u,\mathrm{d}x} + \mathrm{d}\Gamma_{u,\delta x} - \delta\Gamma_{u,\mathrm{d}x}). \tag{8.10}$$

Returning to the tensor notation, this is

$$u''^\mu - '\mathbf{u}^\mu = -\tfrac{1}{2}R^\mu{}_{\nu\rho\sigma} u^\nu \mathrm{d}x^\rho \delta x^\sigma, \quad \text{with } R^\mu{}_{\nu\rho\sigma} = \partial_\rho\Gamma^\mu_{\nu\sigma} - \partial_\sigma\Gamma^\mu_{\nu\rho} + \Gamma^\mu_{\rho\tau}\Gamma^\tau_{\nu\sigma} - \Gamma^\mu_{\sigma\tau}\Gamma^\tau_{\nu\rho}. \tag{8.11–12}$$

Incidentally, this reasoning shows that for any three vectors a, b, and c, the expression $R^\mu{}_{\nu\rho\sigma} a^\nu b^\rho c^\sigma$ transforms like the coordinates of a vector under a change of coordinates.

Using the expression (8.6) of the vector $\mathbf{u}' - '\mathbf{u}$, for any vector $\mathbf{v}$ we have

$$(\mathbf{u}' - '\mathbf{u}) \cdot \mathbf{v} = \tfrac{1}{2}C(\mathbf{u} \times \mathbf{v}) \cdot (\mathrm{d}\mathbf{r} \times \delta\mathbf{r}) = \tfrac{1}{2}Cu^\mu v^\nu \mathrm{d}x^\rho \delta x^\sigma(g_{\mu\rho}g_{\nu\sigma} - g_{\mu\sigma}g_{\nu\rho}), \tag{8.13}$$

to be compared with

$$g_{\mu\nu}(u^\mu - '\mathbf{u}^\mu)v^\nu = -\tfrac{1}{2}R_{\mu\nu\rho\sigma}v^\mu u^\nu \mathrm{d}x^\rho \delta x^\sigma, \quad \text{with } R_{\mu\nu\rho\sigma} = g_{\mu\tau}R^\tau{}_{\nu\rho\sigma}. \tag{8.14}$$

We therefore have

$$R_{\mu\nu\rho\sigma} = C(g_{\mu\rho}g_{\nu\sigma} - g_{\mu\sigma}g_{\nu\rho}). \tag{8.15}$$

This implies the symmetry relations

$$R_{\mu\nu\rho\sigma} = R_{\rho\sigma\mu\nu}, \ R_{\mu\nu\sigma\rho} = -R_{\mu\nu\rho\sigma}, \ R_{\nu\mu\rho\sigma} = -R_{\mu\nu\rho\sigma}, \tag{8.16}$$

and there is only one independent $R_{\mu\nu\rho\sigma}$ coefficient, say

$$R_{1212} = C(g_{11}g_{22} - g_{12}{}^2). \tag{8.17}$$

Through this formula and through Equations (8.5) and (8.12), the curvature C is seen to depend only on the $g_{\mu\nu}$ coefficients and their first and second derivatives, as was to be proved.

Geodetic distortion

So far we have encountered two interpretations of curvature in Gauss's theory of surfaces: through the angular excess of geodesic triangles, and through the surface swept by the normal unit vector on the unit sphere. The last section of Gauss's memoir provides a third interpretation through geodetic distortion, which is the surveying error committed in assimilating geodesic triangles with planar triangles. As was already hinted at, in 1787 Legendre proved that the ratios between the sides of a small spherical triangle of angles α, β, and γ were nearly the same as for a planar triangle of angles $\alpha - e/3$, $\beta - e/3$, $\gamma - e/3$, wherein e is the excess of the sum of the angles of the spherical triangle over the flat angle π. Gauss generalized this result to surfaces of variable curvature, and he also obtained a higher-order correction for the large triangles he used in his geodetic mission.[12]

Following Gauss, we will use geodesic polar coordinates generalizing the latitude and longitude on a sphere. For a given point of the surface (not too far from the pole), the first of these coordinates is the geodesic distance s from the pole, and the second is the angle θ that the connecting geodesic makes with a fixed direction at the pole. From s and θ, we build the *normal coordinates* $x^1 = s \cos \theta$ and $x^2 = s \sin \theta$, alias $x^\mu = s\xi^\mu$. Along a geodesic from the pole,

$$\frac{dx^\mu}{ds} = \xi^\mu, \quad \Gamma^\mu_{\nu\rho}\xi^\nu\xi^\rho = -\frac{d^2x^\mu}{ds^2} = 0, \quad \text{and} \quad \xi^\sigma\partial_\sigma\Gamma^\mu_{\nu\rho}\xi^\nu\xi^\rho = -\frac{d^3x^\mu}{ds^3} = 0. \tag{8.18}$$

The direction of the ξ vector being arbitrary, the second identity requires the Γ coefficients to vanish. The third identity requires

$$\partial_\sigma\Gamma^\mu_{\nu\rho} + \partial_\nu\Gamma^\mu_{\rho\sigma} + \partial_\rho\Gamma^\mu_{\sigma\nu} = 0. \tag{8.19}$$

Now consider the point of coordinates δx, draw the vector dx from this point, and move this vector along the geodesic joining this point to the pole in the length-preserving manner given by Equation (8.7), with the result dx'. Since in the normal coordinates the Γ coefficients vanish, we have

$$dx' = dx + \tfrac{1}{2}\delta\Gamma_{dx,\delta x} \tag{8.20}$$

to second order in δx. The length of this vector being the same as the length dx of the vector dx, we have (to fourth order in dx and δx)

$$ds^2 = \underline{ds}^2 + dx \cdot \delta\Gamma_{dx,\delta x}, \tag{8.21}$$

wherein $\underline{ds}^2$ is the expression that the length would have in the planar case, and the dot product is taken with respect to the $g_{\mu\nu}$ at the pole. Identity (8.19) and the expression (8.12) of $R^\mu_{\nu\rho\sigma}$ imply

[12] Gauss 1828, §§23–29; Legendre 1787, p. 358.

$$2\delta\Gamma_{dx,\delta x} + d\Gamma_{\delta x,\delta x} = 0, \quad \text{and} \quad \delta\Gamma_{dx,\delta x} = \tfrac{1}{3}(\delta\Gamma_{dx,\delta x} - d\Gamma_{\delta x,\delta x}) = -\tfrac{1}{3}\delta x^{\nu}R^{\mu}_{\nu\rho\sigma}dx^{\rho}\delta x^{\sigma}, \qquad (8.22)$$

so that the correction to the metric is

$$\delta ds^2 = -\tfrac{1}{3}R_{\mu\nu\rho\sigma}dx^{\mu}\delta x^{\nu}dx^{\rho}\delta x^{\sigma}. \qquad (8.23)$$

Owing to the symmetries (8.16) of $R_{\mu\nu\rho\sigma}$, this expression is a quadratic form of the antisymmetric combinations $dx^{\mu}\delta x^{\nu} - \delta x^{\mu}dx^{\nu}$ (and therefore vanishes when dx and δx are parallel, as expected on geometrical grounds). Using the two-dimensional expression (8.15) of $R_{\mu\nu\rho\sigma}$, we finally get the distortion

$$\delta ds^2 = -\tfrac{1}{3}C[(d\mathbf{r})^2(\delta\mathbf{r})^2 - (d\mathbf{r}\cdot\delta\mathbf{r})^2] = -(4/3)CS^2, \qquad (8.24)$$

where S is the surface of the triangle defined by the pole and the points of coordinates δx and $\delta x + dx$.

8.2 Riemann's curvature

In an abstract of his *Disquisitiones*, Gauss wrote:

> These theorems lead us to consider the theory of curved surfaces from a new point of view, so that a wide and still wholly uncultivated field is open to investigation. If we regard surfaces not as boundaries of bodies, but as bodies of which one dimension vanishes, and if we thereby conceive them as flexible but not extensible, we see that two essentially different kinds of relations must be distinguished: those presupposing a definite form of the surface in space, and those independent of the various shapes that the surface may assume. This discussion is concerned with the latter kind. As we saw, the measure of curvature belongs to this kind. But it is easily seen that the properties of figures constructed upon the surface, their angles, their areas, their integral curvature, the shortest line between two points, and so forth, also belong to this kind. Their investigation must start from the fact that the expression of the linear elements in the form $\sqrt{Edp^2 + 2Fdpdq + Gdq^2}$ fully determines the nature of the curved surface.

Gauss's interpretation of surfaces as "bodies of which one dimension vanishes" and his focus on the intrinsic geometry of such surfaces invite generalization to manifolds of any dimension.[13]

The habilitation lecture

Gauss's student Bernhard Riemann got the hint and developed it in his habilitation lecture of 1854, in a manner that Gauss himself judged highly impressive. Riemann first defined a continuous, differentiable manifold of n dimensions in an intuitive manner that need

---

[13] Gauss 1828, p. 45. It is not clear whether in 1828 Gauss related this remark with his private anticipation of non-Euclidean geometry: cf. Gray 2006. He might have done so after reading Bolyai in 1832: cf. Abardia, Reventós, and Rodríguez 2012.

not concern us here. For the purpose of length measurement, he defined a homogeneous differential form[14] on this manifold and focused on the Gaussian quadratic case we would now write as $ds^2 = g_{\mu\nu}dx^{\mu}dx^{\nu}$. Before analyzing this metric structure, Riemann made clear that Gauss's memoir on surfaces was his guide:

> These metric relations can only be studied through abstract quantity concepts, and their interdependence can only be represented by formulae. Under certain assumptions, however, they can be decomposed into relations which, taken separately, are capable of geometric representation. It thus becomes possible to express the results of calculation geometrically. To come to solid ground, we truly cannot avoid abstract formulas, but at least the results of calculation may subsequently be presented in geometric form. The foundations of these two parts of the question are contained in the celebrated memoir of Privy Councillor Gauss on curved surfaces.

From this extract, it is clear that Riemann performed detailed calculations generalizing Gauss's calculation of intrinsic surface properties, and that he also found a way to extend Gauss's geometric interpretation of these properties.[15]

The coefficients $g_{\mu\nu}$ depend both on the metric structure of the manifold and on the arbitrary choice of coordinates. In order to avoid this indeterminacy, Riemann introduces special coordinates defined through intrinsic metric operations. If O is a given point (pole) of the manifold, the position of another point M (in the neighborhood of O) can be determined by drawing the geodesic from O to M and recording the length s of this geodesic as well as the rectangular coordinates ξ^{μ} of the unit tangent vector at the origin of this geodesic. This is a straightforward generalization of Gauss's geodesic polar coordinates. Riemann next defines the so-called *geodesic normal coordinates* through $x^{\mu} = s\xi^{\mu}$, a natural generalization of Cartesian coordinates to curved spaces.[16] In the flat case, all geodesics are straight lines and the metric element is simply given by $ds^2 = (dx^1)^2 + (dx^2)^2 + ... + (dx^n)^2$ throughout space. In the general case, Riemann asserts that at the lowest order in x the metric element differs from its Euclidean value by a quadratic form in the combinations $x^{\mu}dx^{\nu} - x^{\nu}dx^{\mu}$. Formally, this statement may be rendered as

$$\delta ds^2 = \tfrac{1}{4}S_{\mu\nu\rho\sigma}(x^{\mu}dx^{\nu} - x^{\nu}dx^{\mu})(x^{\rho}dx^{\sigma} - x^{\sigma}dx^{\rho}), \tag{8.25}$$

with $S_{\mu\nu\rho\sigma} = S_{\rho\sigma\mu\nu}$, $S_{\mu\nu\sigma\rho} = -S_{\mu\nu\rho\sigma}$, $S_{\nu\mu\rho\sigma} = -S_{\mu\nu\rho\sigma}$.[17]

Riemann further asserts that the ratio of this quantity to the square of the area of the triangle of vertices O, x, and $x+dx$ (which is the fourth of $x^2(dx)^2 - (x \cdot dx)^2$ in a first approximation) is a finite quantity independent of the choices of x and dx as long as the associated

---

[14]A homogeneous differential form of degree k is an expression of the kind $\sum_{\mu_1\mu_2...\mu_k} a_{\mu_1\mu_2...\mu_k}(x)dx^{\mu_1}dx^{\mu_2}...dx^{\mu_k}$, and it corresponds to what we would now call a completely symmetric tensor of order k.

[15]Riemann 1867, p. 138. On Riemann's lecture and its broader significance in the history of geometry, cf. Scholz 1980, chap. 2; Portnoy 1982; Scholz 1992; Reich 1994, pp. 26–28. For modern readings, cf. Spivak 1970–75, chap. 4, Part B; Torretti 1978, pp. 90–101. On Riemann's biography, cf. Dedekind 1876; Laugwitz 1996.

[16]Nowadays, these coordinates are defined through the "exponential map" in the vicinity of the pole.

[17]Riemann 1867, p. 141.

triangle (more exactly the O edge of this triangle) remains in the same plane. Lastly, Riemann tells us that the latter quantity multiplied by $-3/4$ gives the Gaussian curvature of the surface defined by the set of geodesics whose initial tangent vector belongs to the tangent plane associated with a given value of x and dx. In symbols, and writing δx instead of x (for the sake of conformity with our earlier notation), this curvature is[18]

$$C(dx, \delta x) = -3 \frac{S_{\mu\nu\rho\sigma}dx^{\mu}\delta x^{\nu}dx^{\rho}\delta x^{\sigma}}{(g_{\mu\nu}dx^{\mu}dx^{\nu})(g_{\rho\sigma}\delta x^{\rho}\delta x^{\sigma}) - (g_{\mu\nu}dx^{\mu}\delta x^{\nu})^2}. \tag{8.26}$$

Riemann never published a proof of these statements. Plausibly, he reasoned in the manner given above in the two-dimensional case, or according to the more complex variants proposed by Dedekind (and Weber) in 1876 and by Hermann Weyl in 1919. Such reasoning leads, for any dimension, to Equation (8.23)

$$\delta ds^2 = -\tfrac{1}{3}R_{\mu\nu\rho\sigma}dx^{\mu}\delta x^{\nu}dx^{\rho}\delta x^{\sigma}$$

with the expression (8.12) of $R_{\mu\nu\rho\sigma}$. In order to reach Riemann's quadratic expression (8.25) of the geodetic distortion, one further needs the symmetry relations (8.16):

$$R_{\mu\nu\rho\sigma} = R_{\rho\sigma\mu\nu}, \; R_{\mu\nu\sigma\rho} = -R_{\mu\nu\rho\sigma}, \; R_{\nu\mu\rho\sigma} = -R_{\mu\nu\rho\sigma}.$$

The second relation is the only one that can immediately be read in the expression of $R_{\mu\nu\rho\sigma}$. The others do not strike the eye. Plausibly, Riemann first noted them in the two-dimensional case in the manner given earlier, and then checked their validity for any dimension.[19]

By definition, geodetic distortion derives from the metric relations of the manifold. Reciprocally, the curvature coefficients expressing this distortion completely determine the metric relations provided these coefficients are independent from each other. Riemann justified this reciprocal proposition by arguing that there were $n(n - 1)/2$ independent curvature coefficients in general (as many as there are independent surface directions $\delta x^{\mu}dx^{\nu} - \delta x^{\nu}dx^{\mu}$) and just as many independent changes of the $g_{\mu\nu}$ coefficients that cannot be absorbed by a change of coordinates (since there are $n(n + 1)/2$ metric coefficients and n coordinates). He may also have relied on the intuitive remark that the knowledge of the distortion of small geodesic triangles is sufficient to correctly survey the entire manifold. Consequently, a metric manifold with zero curvature at all points is a Euclidean space.[20]

As for the relation (8.23) between the geodetic distortion and the Gaussian curvature of the geodetic surface tangent to the plane of the vectors dx and δx, Riemann is likely to have obtained it by direct calculation in normal geodesic coordinates for which this surface is defined by $x_3 = x_4 = \ldots = x_n = 0$.[21] In these coordinates, one can rely on the above-given

[18] Riemann 1867, pp. 142–143.

[19] Dedekind and Weber 1876; Weyl 1919. For the two-dimensional case, see pp. 243–244, this volume.

[20] Riemann 1867, pp. 140, 142, 144. The first rigorous proofs of the latter result were given by Lipschitz and Ricci. For a simple proof based on the integrability of parallel displacement, cf. Eddington 1923, pp. 76–77.

[21] This is in substance what Dedekind and Weber do in their reconstruction (1876, pp. 389–391) of Riemann's reasoning, as they introduce Gaussian polar coordinates on the geodetic surface.

two-dimensional derivation of the relation between the angular excess of geodesic triangles and geodetic distortion.

The Commentatio

In 1861 Riemann sent a manuscript to the French Academy of Sciences in answer to a prize question on the propagation of heat in a homogeneous (and isotropic) solid: find the conditions under which the temperature becomes a function of time and of two independent (spatial) variables only. Riemann's answer to this question requires knowing the answer of another question: under what conditions on the $g_{\mu\nu}$ coefficients can the form $g_{\mu\nu}(x)dx^{\mu}dx^{\nu}$ be brought to the simple form $d\mathbf{r}^2$ through the change of variables $x \rightarrow \mathbf{r}(x)$? Toward the beginning of his memoir, Riemann announces: "We will see below that this question can be handled by nearly the same method that Gauss used in the theory of curved surfaces." This handling is found in the second part of the memoir.[22]

Riemann there proves, by purely algebraic means, that $R_{\mu\nu\rho\sigma} = 0$ is a necessary condition for the form $g_{\mu\nu}(x)dx^{\mu}dx^{\nu}$ to be reducible to $d\mathbf{r}^2$, with

$$R^{\mu}{}_{\nu\rho\sigma} = \partial_{\rho}\Gamma^{\mu}_{\nu\sigma} - \partial_{\sigma}\Gamma^{\mu}_{\nu\rho} + \Gamma^{\mu}_{\rho\tau}\Gamma^{\tau}_{\nu\sigma} - \Gamma^{\mu}_{\sigma\tau}\Gamma^{\tau}_{\nu\rho}, \text{ and } g_{\rho\sigma}\Gamma^{\rho}_{\mu\nu} = \frac{1}{2}(\partial_{\mu}g_{\sigma\nu} + \partial_{\nu}g_{\sigma\mu} - \partial_{\sigma}g_{\mu\nu})$$

as in Equations (8.12) and (8.5). The next paragraph of Riemann's *Commentatio* is most mysterious. There, "in order to reveal the character of the equations $[R_{\mu\nu\rho\sigma} = 0]$"[23] he forms the expression

$$X = \delta\delta(g_{\mu\nu}dx^{\mu}dx^{\nu}) - 2d\delta(g_{\mu\nu}dx^{\mu}\delta x^{\nu}) + dd(g_{\mu\nu}\delta x^{\mu}\delta x^{\nu}), \tag{8.27}$$

in which the second-order variations d^2, δ^2, and $d\delta$ are constrained by

$$\delta'(g_{\mu\nu}dx^{\mu}dx^{\nu}) - 2d(g_{\mu\nu}dx^{\mu}\delta'x^{\nu}) = 0, \quad \delta'(g_{\mu\nu}\delta x^{\mu}\delta x^{\nu}) - 2\delta(g_{\mu\nu}\delta x^{\mu}\delta'x^{\nu}) = 0,$$
$$\delta'(g_{\mu\nu}dx^{\mu}\delta x^{\nu}) - \delta(g_{\mu\nu}dx^{\mu}\delta'x^{\nu}) - d(g_{\mu\nu}\delta x^{\mu}\delta'x^{\nu}) = 0 \tag{8.28}$$

for any variation δ'. Without proof, Riemann announces

$$X = -\frac{1}{2}R_{\mu\nu\rho\sigma}(dx^{\mu}\delta x^{\nu} - dx^{\nu}\delta x^{\mu})(dx^{\rho}\delta x^{\sigma} - dx^{\sigma}\delta x^{\rho}), \tag{8.29}$$

with the comment: "From the buildup of this expression, it is clear that through a change of the independent variables this expression is turned into an expression that depends on the new form of $g_{\mu\nu}dx^{\mu}dx^{\nu}$ by the same law." Riemann probably means that the quantity X, which is an invariant by construction, is the same combination of $R_{\mu\nu\rho\sigma}$, dx^{μ} and δx^{μ} whatever the choices of the coordinates x^{μ}. Taking into account the vector character of dx^{μ} and

---

[22] Riemann 1876, pp. 372 (citation), 380–383. On the *Commentatio*, its heat propagation context, and its reception, cf. Farwell and Knee 1990a; Darrigol 2015, pp. 58–67. For a modern reading, cf. Spivak 1970–75, chap, 4, part C.

[23] "Ut indoles harum aequationum melius perspiciatur . . ."

δx^μ and the symmetries of $R_{\mu\nu\rho\sigma}$ by permutation of the indices, this remark is tantamount to recognizing the tensor character of $R_{\mu\nu\rho\sigma}$.[24]

Lastly, Riemann builds the quantity

$$C = \frac{R_{\mu\nu\rho\sigma}(dx^\mu\delta x^\nu - dx^\nu\delta x^\mu)(dx^\rho\delta x^\sigma - dx^\sigma\delta x^\rho)}{4[(g_{\mu\nu}dx^\mu dx^\nu)(g_{\rho\sigma}\delta x^\rho\delta x^\sigma) - (g_{\mu\nu}dx^\mu\delta x^\nu)^2]}, \tag{8.30}$$

which he knows to be the sectional curvature of a manifold of metric $ds^2 = g_{\mu\nu}dx^\mu dx^\nu$ in normal geodesic coordinates. If X vanishes in a given system of coordinates, then its invariance implies that it should also vanish in normal geodesic coordinates. Therefore, the curvature C vanishes at every point, and the manifold must be a flat, Euclidean space. This is how Riemann proves, or rather alludes to a proof that $R_{\mu\nu\rho\sigma} = 0$ is also a sufficient condition of the reducibility of $g_{\mu\nu}(x)dx^\mu dx^\nu$ to $d\mathbf{r}^2$.

Thus we see that Riemann had the curvature tensor that now bears his name, and that he knew how to express it by means of the Christoffel symbols. His notation was $a_{i,i'}$ for $g_{\mu\nu}$, $p_{i,i',i''}$ for $2\Gamma_{\mu\nu\rho}$, and $(\iota\iota', \iota''\iota''')$ for $-2R_{\mu\nu\rho\sigma}$. From contemporary manuscripts we also know that he used both contravariant and covariant coordinates, and that he was even aware of the Bianchi identity (see Figure 8.2). We also learn that his way of reasoning was a mixture of algebraic and geometric reasoning, with frequent appeal to the equation of geodesics (8.4) on a metric manifold.

A puzzling component of Riemann's reasoning is the fourth-order differential invariant X. What is the meaning of this invariant? How did Riemann arrive at it? As we will see in a moment, Lipschitz and Levi-Civita later tried to answer these questions, with important benefits. Based on the fragmentary calculations in Riemann's manuscripts, he most probably reasoned as follows. First, he could not fail to see that the expressions $R_{\mu\nu\rho\sigma}$ in the reducibility condition $R_{\mu\nu\rho\sigma} = 0$ were proportional to the coefficients in the expression

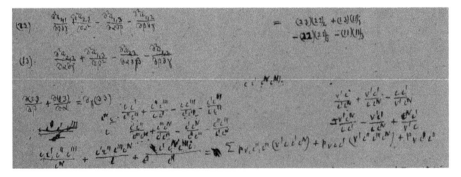

Fig. 8.2.   Extract of folio 4 of Cod. Ms. B. Riemann 9 (courtesy of the Göttingen University Archive). The bottom line is a truncated version of the Bianchi identity.

[24]Riemann 1876, p. 382. The X name is mine. Riemann's expression of X has an additional factor 4 because his sums are restricted to $\mu < \nu$ and $\sigma < \rho$.

(8.25) he had reached in 1854 for geodetic distortion in normal geodesic coordinates. This awareness plausibly led him to consider the expression

$$Y = R_{\mu\nu\rho\sigma}(dx^\mu \delta x^\nu - dx^\nu \delta x^\mu)(dx^\rho \delta x^\sigma - dx^\sigma \delta x^\rho), \tag{8.31}$$

which enjoys the following properties: it is of second order with respect to both d and δ, it is symmetric by permutation of d and δ, and it vanishes identically when dx = δx. In order to prove the covariant character of $R_{\mu\nu\rho\sigma}$ in arbitrary coordinates, he would then have tried to generate this differential expression by combining double variations of the invariants $g_{\mu\nu}dx^\mu dx^\nu$, $g_{\mu\nu}dx^\mu \delta x^\nu$, and $g_{\mu\nu}\delta x^\mu \delta x^\nu$. The only combination that shares the given properties of Y is the expression X (up to a global coefficient). Riemann could then observe that the third-order differentials canceled out, and could seek to eliminate the second-order differentials in a covariant manner. From a geometric point of view, the obvious means to do so is to draw the geodesics beginning at x in the directions of dx, δx, and dx + δx. Riemann's conditions (8.28) indeed result from the vanishing of the variation $\delta' \int ds$ for these geodesics. Riemann could then verify that under these conditions $X = -\frac{1}{2}Y$, as needed for proving the invariance of Y.[25]

Physical geometry

Let us return to the habilitation lecture. Despite heuristic roots in Gauss's theory of surfaces, most of Riemann's lecture is concerned with metric manifolds *in abstracto*. Only in the last section does Riemann come to the application to physical space. Assuming space to be a three-dimensional metric manifold, he conceives three ways of further determining its geometry. First, one may assume the curvature (as given by the angular excess of geodesic triangles) to vanish everywhere, which leads to the usual Euclidean space. Second, one may assume the free mobility of rigid bodies through space, which requires constant curvature. Third, one might require the direction of a line to be defined independently of its position (perhaps Riemann has in mind what Einstein later called teleparallelism).[26]

Riemann then wonders whether the Euclidean structure, well known to cover our spatial experience at medium scale, would extend to very large and very small scales. At large scale, space could very well be finite, though unbounded, if any small amount of constant curvature were admitted. Space would then be the three-dimensional version of the ordinary sphere. At small scale, Riemann speculates that the curvature might fluctuate without affecting its constancy or its vanishing at larger scale. He even conceives the possibility that the metric would no longer be quadratic (Finsler geometry). Concretely, the concept of a freely mobile rigid body and the concept of a light ray would no longer apply at small scale: "It is quite thinkable that the metric structure of space in the infinitely small would not meet the requirements of [ordinary] geometry, and we should indeed assume so much if this allows us to explain phenomena in a simpler manner."[27]

---

[25] For details, cf. Darrigol 2015, pp. 61–66.

[26] Riemann 1867, pp. 146–147.

[27] Riemann 1867, p. 149. Whereas in his opinion the finite or infinite character of space is inconsequential, its metric behavior at small scale concerns all Newtonian physics based on infinitesimal causality.

Riemann further remarks that a continuous manifold, unlike discrete manifolds in which the points can be counted, does not have an inherent metric structure. This structure must be imported: "Either the actual foundation of space must be a discrete manifold, or the ground of the metric structure must be sought externally in the binding forces that act on the manifold." He concludes his lecture with the words:[28]

> These questions can be decided only by departing from the heretofore confirmed conceptions of Newtonian physics, and by reworking them under the pressure of recalcitrant facts. The only purpose of investigations based, like mine, on general concepts, is to avoid that narrow concepts hamper this reworking and that excessive prejudices block progress in our knowledge of the connection of things.
>
> This drives me into the domain of another science, physics, which the nature of today's occasion may not allow me to tread.

8.3 Non-Euclidean geometries

Negating the axiom of parallels

Riemann introduced his habilitation lecture with the words:

> As is well known, geometry presupposes the concept of space as well as basic concepts for constructions in space. Of these concepts it gives only nominal definitions, while the essential determinations occur under the form of axioms. The mutual relationships of these presuppositions thereby remain in the dark. We cannot recognize to which extent their connection is necessary, nor whether they are a priori possible.

Riemann was surely aware of earlier debates on the foundations of geometry, to which his mentor Gauss had contributed. After a long history of failures to prove Euclid's axiom of parallels from his other concepts, the prospects of consistently denying this axiom began to open up in the eighteenth century. A violation of the axiom of parallels is easily seen to imply that the sum of the angles of a triangle differs from a flat angle. In 1733, Girolamo Saccheri proved, using Euclid's other axioms, that the sign and the value of the angular excess are the same for any triangle. The positive case corresponds, in two dimensions, to spherical geometry. It could be outright rejected for being incompatible with Euclid's assumption of infinite straight lines. The negative case was more resistant. Later in the century, Johann Lambert could not find any absurdity in this case and even suggested a correspondence with the geometry of a sphere of imaginary radius. Adrien-Marie Legendre tried hard to prove the necessity of the axiom of parallels and failed repeatedly. In this process he found that the consequences of a negation of this axiom could be widely developed without encountering any contradiction.[29]

In the 1830s, the Hungarian mathematician János Bolyai and the Russian mathematician Nikolai Ivanovich Lobachevsky gave the first fully developed non-Euclidean geometries in synthetic and trigonometric forms. They did not doubt the consistency of

---

[28] Riemann 1867, pp. 149–150.

[29] Riemann 1867, p. 133; Saccheri 1733; Lambert 1786; Legendre 1794. Cf. Gray 2007, chap. 7.

their system, and they regarded it as a generalization of Euclid's geometry in which a certain parameter (the angular defect of a triangle divided by its surface) could take any positive value to be determined astronomically by triangulation. They both communicated their results to Gauss, who privately approved of them as he had independently convinced himself of the possibility of non-Euclidean geometry. However, Gauss was too concerned with the possibility of a hostile reaction from his German colleagues to publicize the new geometry.[30]

Bolyai's and Lobachevsky's theories were largely ignored until the posthumous publication of Gauss's supporting letters in the 1860s. In 1868, the Italian mathematician Eugenio Beltrami read Lobachevsky and proved that in two dimensions the new geometry could be modeled through the geodesics of a surface of constant negative curvature. He also showed that the geodesics became straight lines after projection onto a disk. Implicitly, he thus proved the consistency of the new geometry in two different ways: through an analytic representation and through a correspondence with Euclidean geometry. After studying Riemann's freshly published lecture, he extended his reasoning to higher spatial dimensions. It thus became clear that Riemann's metric manifolds contained hyperbolic geometry as a particular case.[31]

Helmholtz's foundations

In the same year, 1868, Helmholtz published a first version of his influential reflections on the foundations of geometry. His original intention was to prove that Euclidean geometry derived from a single empirical fact (*Thatsache*): the existence of freely mobile rigid bodies. Geometry, being the science of the measurement of space, presupposes our ability to measure lengths and angles through invariable rods and compasses, respectively. What makes such measurements possible is the existence of material bodies that can be displaced without any deformation. In modern terms, Helmholtz assumes a three-dimensional differential manifold, and has a six-parameter, continuous group of displacements (translations plus rotations) act on this manifold. He analyzes the infinitesimal generators of this group and finds that in a sufficiently small region of space the only physically acceptable transformations conserve a positive quadratic form, to be interpreted as the squared distance between two points. This means that the manifold is locally Euclidean. For free mobility at large scale, he further requires the curvature of the manifold to be uniform. As he is a priori excluding the finite space of spherical geometry, he believes to have proved the physico-mathematical necessity of Euclidean geometry.[32]

Helmholtz was originally unaware of Riemann's and Beltrami's relevant works. After reading Riemann, he realized that a good deal of his reflections had been anticipated, save for the derivation of the locally Euclidean character of space. After reading Beltrami, he understood that constant curvature was compatible with an infinite space, if

---

[30] Bolyai 1832; Lobachevsky 1835–1838, 1837,1840. Cf. Gray 2007, chaps. 9–10.

[31] Betrami 1868a, 1868b. Cf. Torretti 1978, chap. 2.1; Gray 2007, chap. 19.

[32] Helmholtz 1868a. Cf. Torretti 1978, chap. 3.1; Hyder 2009; Darrigol 2014, chap. 4. Here and elsewhere, I use "local" and "locally" in the physicists' loose sense of a sufficiently small neighborhood.

only the curvature was negative. He henceforth decided that there were infinitely many possible geometries, depending on the (negative or zero) value of the curvature. In a revised version of Kant's synthetic a priori, he believed the general notion of space as a metric (differentiable) manifold to be a priori necessary while only experience could decide the value of the curvature. Altogether, Helmholtz's geometry was a lot less free than Riemann's since it had only one free parameter. For all practical purposes, it was comparable with the hyperbolic geometry invented by Bolyai and Lobachevski, implicitly contained in Riemann's subcase of constant negative curvature and thus modeled by Beltrami.[33]

Helmholtz nonetheless differed from his predecessors in the status he accorded to the axioms of the new geometry. For Bolyai, they were the legitimate self-evident truths of a purified Euclidean geometry; for Riemann, constant curvature was only a special way to restrict the geometry, and the only reason to favor the quadratic form of the metric was its implying "flatness in the smallest parts" (in contrast to the Finsler geometries); for Helmholtz, hyperbolic geometry (including the Euclidean subcase) was the only geometry complying with the existence of freely mobile rigid bodies. Without altering the core of his reasoning, Helmholtz could have restricted this demand to small regions of space and he would thus have allowed for variable curvature. He did not, presumably because he belonged to a Kantian tradition in which perfect homogeneity was an essential attribute of space. As Riemann saw, once homogeneity has been given up, space can no longer be a purely passive form of sensibility: the variations of curvature must have a cause, to be found in the distribution of matter.[34]

Groups in geometry

From a purely mathematical point of view, an important aspect of Helmholtz's deduction was his focus on what we would now call the Lie algebra of a group of transformations: he first determined this algebra and then obtained finite transformations by exponentiation. This is the more remarkable because Sophus Lie had not yet invented the theory that now bears his name. In the years 1869–1872, the Norwegian mathematician thoroughly analyzed the structure of continuous groups of transformations with a double interest in the transformations occurring in projective geometry and in the theory of systems of differential equations. He shared the former interest with his friend Felix Klein, who advised him to apply the Lie group technique to Helmholtz's space problem. Lie confirmed Helmholtz's chief conclusions while filling some gaps in the deductions.[35]

In the 1860s, group theory was still in its infancy, and it was mostly used in Galois's original context of algebraic equations. Yet there was another domain of mathematics in which groups of transformations played a central role: the projective geometry of Jean-Victor Poncelet, August Möbius, Arthur Cayley, Julius Plücker, Karl von Staudt, and a

[33] Helmholtz 1868b, [1870b].

[34] On Helmholtz's philosophy of geometry, cf. DiSalle 1993; Carrier 1994.

[35] Lie 1886, 1890. On the history of Lie groups, cf. Hawkins 2000.

few others who focused on the properties of figures that are invariant by conical projection. In the early 1870s, young Klein ambitioned to unify all existing geometries. He found that projective geometry, once separated from its Euclidean substratum and abstractly defined through the group of transformations leaving the geometrical objects invariant, offered the desired unifying framework. In the famous *Programmschrift* he wrote in 1872 for his new professorship at Erlangen, he showed how all the known geometries of his time, metric or nonmetric, could be defined by a subgroup of the projective group.[36]

The definition of a geometry through a characteristic group of transformations also became a central component of Poincaré's philosophy of geometry. One of his early breakthroughs as a mathematician, in 1880, depended on an astute reliance on hyperbolic geometry in studying the transformation properties of the functions he called Fuchsian. Later, after reading Helmholtz on geometry, he introduced the geometry of visual space through the group of compensatory motions of the eye and of a perceived body. More broadly, Poincaré considered the group structure as the expression of the mind's ability to conceive the indefinite repetition or combination of identical operations. He regarded this ability as an expression of Kant's synthetic a priori, for it plays an essential role in our analysis of complex physical phenomena and evolutions into elementary, homogeneous phenomena.[37]

Like Beltrami and Helmholtz, Poincaré spent much time displaying the inner consistency of non-Euclidean geometry through adequate models or dictionaries. Yet he did not believe in an empirical decision between the various possible geometries. In his view any geometry could be applied to the physical world if only the physical laws ruling the behavior of rigid bodies and light were properly adjusted to this geometry. Helmholtz had similarly noted that the axioms of geometry needed to be combined with mechanical principles in order to gain empirical content. But in the end he believed the latter principles to be sufficiently settled to allow for an empirical determination of spatial curvature. In contrast Poincaré maintained that in the face of experimental conflict, physicists would rather keep Euclidean geometry as a simple, convenient convention and modify the nongeometrical laws.[38]

Geometry and physics

In mathematics narrowly considered, the new geometries were important as offering new tools for mathematical demonstrations, as a springboard to a group-theoretical definition of geometry, and more broadly as bringing new insights into the nature of mathematics. Mathematics now had to be regarded as a free creation of the mind, at least in part. The axioms of geometry were no longer to be regarded as self-evident truths. Nor could they be entirely determined as the necessary form of external sensibility, against the received

[36] Klein 1871, 1872 (Erlangen program), 1873. Cf. Torretti 1978, chap. 2.3; Gray 2007, chaps. 4–6, 13–15, 20.

[37] Poincaré 1881, 1887, 1891, 1892 (on geometry); 1902a, pp. 83, 90–91 (on groups). On Fuchsian functions, cf. Gray 2013, chap. 3. On Poincaré's philosophy of geometry, cf. Torretti 1978, chap. 4.4.

[38] Helmholtz 1870, p. 30; Poincaré 1902, chap. 5. Cf. Nabonnand 2010.

Kantian view. At the close of the century, David Hilbert epitomized the acquired freedom by founding geometry on axioms in the modern sense. In this new axiomatics, the names given to geometrical objects (points, lines, and planes) were completely immaterial and the meaning of these objects was implicitly defined by the axioms in which they occurred. The axioms being thus divorced from geometric intuition, Hilbert proved their independence and consistency through simple models.[39]

Opinion varied on whether this modern freedom of mathematical construction mattered in physics. Kantian philosophers had long rejected non-Euclidean geometries qua physical geometries, even when they accepted them as legitimate mathematical constructs. A few important mathematicians including Gauss, Riemann, and William Clifford speculated on a non-Euclidean structure of physical space. The cases of Riemann and Clifford are especially interesting, because they both considered variable curvature as a manifestation of matter. After reading Riemann in 1870, Clifford wrote:

I hold in fact
(1) That small portions of space are in fact of a nature analogous to little hills on a surface which is on the average flat; namely, that the ordinary laws of geometry are not valid in them.

(2) That this property of being curved or distorted is continually being passed on from one portion of space to another after the manner of a wave.

(3) That this variation of the curvature of space is what really happens in that phenomenon which we call the *motion of matter*, whether ponderable or etherial.

(4) That in the physical world nothing else takes place but this variation, subject (possibly) to the law of continuity.

According to his posthumous editor Karl Pearson, Clifford also contemplated the possibility of a global, uniform curvature, and he reduced light to an elastically propagated deformation of space. Space would thus replace the ether, and all physics would be reduced to a space dynamics.[40]

The possibility of a constant uniform curvature was most commonly admitted, first by Gauss, Bolyai, and Lobachevsky, then by a few famous astronomers: Friedrich Bessel in 1839 in answer to Gauss, Friedrich Zöllner in 1872 to solve Olbers' dark-sky paradox by means of a closed universe, and Karl Schwarzschild in 1900 in defense of elliptic space. At least one major physicist, Ludwig Boltzmann, judged space curvature possible, in 1904 in his lectures on mechanics. A few mathematicians discussed the effect of space curvature on Mercury's precession: Wilhelm Killing and Carl Neumann in 1885–1886 for positive curvature, Heinrich Liebmann in 1902 for negative curvature. There was occasional speculation on a fourth dimension of space, sometimes for the sake of immersing a curved three-space in a Euclidean four-space, sometimes to leave room for spiritual phenomena, or more soberly to represent the time coordinate on the same footing as the spatial coordinates.

---

[39] Hilbert 1899. Cf. Torretti 1978, chap. 3.4; Gray 2007, chap. 23.

[40] Clifford 1876 (read in 1870); 1885 (through Pearson), pp. 224–226. Cf. Farwell and Knee 1990b.

D'Alembert and Lagrange pioneered the latter concept of the fourth dimension. No one seriously developed this option before Minkowski.[41]

To summarize, there were three conceptions of the relation between geometry and physics: the (strictly) Kantian view according to which Euclidean geometry is the only possible physical geometry, the empiricist view according to which at least some features of the geometry are to be determined empirically, and the conventionalist view according to which all (homogeneous, metric, three-dimensional) geometries can in principle be used, although for convenience only Euclidean geometry is retained. Within the empiricist view, we may distinguish two varieties: the dominant variety in which space remains a fixed, homogeneous receptacle of matter, and the Riemann–Clifford variety in which the structure of space depends on the presence of matter. The dominant variety is itself compatible with two attitudes: the conservative attitude in which Euclidean geometry is regarded as good for all purposes, and the progressive attitude in which non-Euclidean geometry is regarded as an opportunity to solve paradoxes such as the darkness of the night sky. Toward the end of the nineteenth century, philosophers and mathematicians tended to favor the empiricist view. Even though Helmholtz and a few astronomers also believed in a possible relevance of non-Euclidean geometry, most physicists quietly accepted Euclidean geometry either as a commonsense truth (Duhem) or as a solidly established empirical truth (Stallo and Mach).[42]

As we have already observed in the two preceding chapters, the nineteenth-century history of geometry was an important background to Poincaré's and Minkowski's contributions to relativity theory. Poincaré's reluctance to associate the Lorentz group with a new geometry of space–time is easily explained by his geometric conventionalism. Yet another aspect of his conventionalism, the elucidation of conventions of measurement, contributed to his ground-breaking reinterpretation of Lorentz's local time. As for Minkowski, his familiarity with non-Euclidean geometries and with the Erlangen program predisposed him to construct a new geometry of space–time based on the Lorentz group and related to hyperbolic geometry. The same familiarity also explains his success in convincing a few mathematicians to develop his views.[43]

In 1905, Einstein was well aware of the debates about non-Euclidean geometry, if only because he had read Poincaré's *La science et l'hypothèse*. Yet he did not regard his new kinematics as a step toward a new geometry. In Minkowski's space–time he initially saw a mathematical perversion hiding the true physical content behind opaque formalism. As we will see in Chapter 9, his later adoption of Minkowski's tensor formalism was not geometrically motivated. The same is true for his reliance on Ricci's absolute differential calculus

---

[41] Bessel to Gauss, February 10, 1829, in Gauss and Bessel 1880, p. 493; Zöllner 1872, p. 312; Schwarzschild 1900; Boltzmann 1904, p. 334; Killing 1885; Neumann 1886; Liebmann 1902; d'Alembert 1754, p. 1010; Lagrange 1797, p. 223. On non-Euclidean geometry in astronomy, cf. Kragh 2012; on Mercury's perihelion, cf. Roseveare 1982, pp. 162–165; on the fourth dimension, cf. Cajori 1926; Bork 1964; Beichler 1988; Goenner 2008, §1.

[42] On the mathematicians' preference for empiricism over conventionalism, cf. Walter 1997; on Duhem and math, cf. Crowe 1990; on Stallo, Mach, and geometry, cf. Kragh 2012, pp. 35–36.

[43] Cf. Walter 1999a.

in the construction of general relativity. We will now see how this new calculus emerged in connection with Riemann's seminal works.[44]

8.4 The absolute differential calculus

As the calculations behind Riemann's theory of metric manifolds remained largely unpublished, a few mathematicians struggled to reconstruct them after his death. They developed this algebraic aspect of Riemann's allusive theory in various contexts, thus building "the absolute differential calculus," which was Ricci-Curbastro's name for what we would now call the tensor calculus on differential manifolds.

Dedekind

The earliest reader of Riemann's unpublished writings on geometry and quadratic differential forms was his friend Richard Dedekind, who was in charge of his *Nachlass* after his untimely death in 1866. In private notes probably written while he was preparing the publication of Riemann's habilitation lecture, Dedekind tried to restore the missing calculations in the light of the *Commentatio*. He thereby introduced invariant measures such as $\sqrt{g_{\mu\nu}dx^\mu dx^\nu}$ and $\sqrt{g}d^n x$ (g being the determinant of $g_{\mu\nu}$); he explained Riemann's concept of immersed plane directions $dx^\mu \delta x^\nu - \delta x^\mu dx^\nu$; and he discussed immersed lines, geodesics, and the more general notion of a submanifold of extremal content. Presumably having in mind the heat-propagation problem of the *Commentatio*, he also gave a covariant generalization of Green's theorem together with the invariant Laplacian

$$\Delta u = \frac{1}{\sqrt{g}}\partial_\mu \left(g^{\mu\nu}\partial_\nu u\sqrt{g}\right). \tag{8.32}$$

As we will later see, Einstein used this construct on his way to general relativity. He borrowed it from Beltrami, who got it independently of Dedekind in 1869 through a variational procedure.[45]

Dedekind's main purpose was to explicate Riemann's expressions for the sectional curvatures and for geodetic distortion, and to reconstruct the proof of their invariance under any change of coordinates. Heinrich Weber, the chief editor of Riemann's papers, published the relevant considerations in 1876 to accompany the *Commentatio*. On the one hand, Dedekind and Weber's commentary was mostly analytical, in conformity with the title of Dedekind's manuscript: the main purpose was to explicate the calculations that Riemann did not find time to write down or publish. On the other hand, this commentary emphasized the geometrical aspects of the second part of the *Commentatio* by relating it both to Gaussian curvature (as Riemann succinctly did) and to geodetic distortion (as Riemann did not). Their reconstruction of Riemann's X does not match the fragmentary calculations found in the Riemann *Nachlass*, and it is slightly inconsistent. Their derivation

---

[44]Sommerfeld (1949, p. 102) reported Einstein to have said: "Since the mathematicians have invaded the theory of relativity, I do not understand it myself anymore."

[45]Dedekind [c. 1867]; Beltrami 1869. On Beltrami's derivation, cf. Dell'Aglio 1996, pp. 231–235.

of geodetic distortion in normal geodesic coordinates is nonetheless likely to reproduce features of Riemann's unknown derivation of the same thing at the time of his habilitation lecture. As Dedekind and Weber were not pretending to be historians, they cannot be blamed for having creatively mixed ideas that Riemann conceived separately.[46]

Dedekind and Weber thus inaugurated a tradition of regarding the *Commentatio* as an opportunity for Riemann to summarize some of the calculations he must have done to prepare his habilitation lecture. In his introduction, Weber announced this text as being "of especially high interest because therein Riemann's investigations about the general properties of multiply extended manifolds are recorded in their main features and find a remarkable application."[47]

Christoffel

In 1869, two years after the publication of Riemann's habilitation lecture, Dedekind's successor at the Zürich Polytechnicum, Elwin Bruno Christoffel, published a powerful memoir on the mutual transformability of quadratic differential forms. As Christoffel explains in his introduction, "[My] researches originated in the *n*-dimensional extension of the problem of surfaces that can be developed onto one another." The latter problem had motivated Gauss's *Disquisitiones*, and its *n*-dimensional generalization was easily at hand through Riemann's concept of a metric manifold. At the end of his memoir, Christoffel mentions the recent publication of Riemann's lecture of 1854 as well as the existence of a manuscript in which Dedekind gave the relevant calculations. If Christoffel saw this manuscript, then he was aware of Riemann's handling of the transformability problem in the *Commentatio*. If he did not, he was at least aware of the geometrical handling of a particular case of this problem in the lecture of 1854. At any rate, his subsequent considerations were formulated in a purely algebraic manner.[48]

Christoffel's problem is to find under what conditions the form $g_{\mu\nu}(x)dx^\mu dx^\nu$ can be obtained from a given form $g'_{\mu\nu}(x')dx'^\mu dx'^\nu$ through the change $x' \to x$ of the variables. In symbols the effect of this change is

$$g_{\mu'\nu'}[x(x')]\frac{\partial x^\mu}{\partial x'^{\mu'}}\frac{\partial x^\nu}{\partial x'^{\nu'}} = g'_{\mu\nu}(x'). \tag{8.33}$$

Being an expert on the algebraic theory of invariants, Christoffel tries to replace this partial differential equation with purely algebraic conditions for the covariance of certain multilinear forms. A first algebraic condition is the existence, for any given value of x', of n numbers x^μ and n^2 numbers $u^\mu_{\mu'}$ such that

$$g_{\mu\nu}(x)u^\mu_{\mu'}u^\nu_{\nu'} = g'_{\mu'\nu'}(x'). \tag{8.34}$$

[46] Dedekind and Weber, in Riemann 1876, pp. 384–399. For details, cf. Darrigol 2015, pp. 68–70.

[47] Weber, in Riemann 1876, p. 4.

[48] Christoffel 1869, pp. 47 (citation), 69. On Christoffel's biography, cf. Butzer 1981.

If there were only one choice of these numbers, say $x^\mu(x')$ and $u^\mu_{\mu'}(x')$, for which the condition is satisfied and if it could be shown that $u^\mu_{\mu'} = \partial x^\mu / \partial x'^{\mu'}$, then the problem would be solved. This is unfortunately not the case, because any quadratic form can be turned into any other quadratic form with the same signature by linear transformations of the kind (8.34). In order to remedy this indeterminacy, Christoffel constructs other algebraic conditions of the form

$$T_{\mu\nu\rho\sigma...}(x)\, u^\mu_{\mu'}\, u^\nu_{\nu'}\, u^\rho_{\rho'}\, u^\sigma_{\sigma'}\, ... = T'_{\mu'\nu'\rho'\sigma'...}(x') \tag{8.35}$$

through successive differentiations of Equation (8.34). In modern terms, he derives tensors of higher order by repeated differentiation of the metric tensor.[49]

A first differentiation of Equation (8.34) and laborious calculations lead him to

$$\frac{\partial^2 x^\mu}{\partial x'^{\nu'}\, \partial x'^{\rho'}} + \Gamma^\mu_{\nu\rho}(x)\frac{\partial x^\nu}{\partial x'^{\nu'}}\frac{\partial x^\rho}{\partial x'^{\rho'}} = \Gamma^{\mu'}_{\nu'\rho'}(x')\frac{\partial x^\mu}{\partial x'^{\mu'}} \tag{8.36}$$

with the definition (8.5) of the Γ coefficients. A second derivation and a series of complex manipulations bring him to

$$R_{\mu\nu\rho\sigma}(x) u^\mu_{\mu'}\, u^\nu_{\nu'}\, u^\rho_{\rho'}\, u^\sigma_{\sigma'} = R'_{\mu'\nu'\rho'\sigma'}(x'), \tag{8.37}$$

with the definition (8.12) of $R_{\mu\nu\rho\sigma}$.[50] Christoffel does not relate the $R_{\mu\nu\rho\sigma}$ to Riemann's curvature, even though he builds the symmetric quadrilinear form $R_{\mu\nu\rho\sigma}(dx^\mu\delta x^\nu - dx^\nu\delta x^\mu)(Dx^\rho\Delta x^\sigma - Dx^\sigma\Delta x^\rho)$, which is the polar of Riemann's form (8.25). His notation for $R_{\mu\nu\rho\sigma}$ is $(gkhi)$ or $(ii_1i_2i_3)$, not too different from the $(u', \iota''\iota''')$ of Riemann's *Commentatio*, although this may be a coincidence.

Christoffel next shows that for any $T_{\mu\nu\rho\sigma...}$ and $T'_{\mu'\nu'\rho'\sigma'...}$ such that

$$T_{\mu\nu\rho\sigma...}(x)u^\mu_{\mu'}\, u^\nu_{\nu'}\, u^\rho_{\rho'}\, u^\sigma_{\sigma'}\, ... = T'_{\mu'\nu'\rho'\sigma'...}(x'), \tag{8.38}$$

the form

$$D_\lambda T_{\mu\nu\rho\sigma...} = \partial_\lambda T_{\mu\nu\rho\sigma...} - \Gamma^\kappa_{\lambda\mu} T_{\kappa\nu\rho\sigma...} - \Gamma^\kappa_{\lambda\nu} T_{\mu\kappa\rho\sigma...} - \Gamma^\kappa_{\lambda\rho} T_{\mu\nu\kappa\sigma...} - \Gamma^\kappa_{\lambda\sigma} T_{\mu\nu\rho\kappa...} - .... \tag{8.39}$$

transforms according to the same rule. In modern terms, he builds the covariant derivative of a covariant tensor of any order. In addition, he proves that the covariant derivative of $g_{\mu\nu}$ vanishes, a result often called "Ricci's lemma."[51]

---

[49] For a lucid interpretation of Christoffel's difficult memoir, cf. Ehlers 1981. On the invariant-theoretical context, cf. Klein 1926–27, Vol. 2, pp. 195–198; Burau 1981; Hunger Parshall 1994.

[50] Christoffel 1869, pp. 49, 54. The Γ notation belongs to Einstein (with the opposite sign) and Weyl (with the same sign); Christoffel used the symbol $\{^{\nu\rho}_{\mu}\}$ for $\Gamma^\mu_{\nu\rho}$ and $[^{\nu\rho}_{\mu}]$ for $g_{\mu\sigma}\Gamma^\sigma_{\nu\rho}$.

[51] Christoffel 1869, §6. On this aspect of Christoffel's memoir, cf. Reich 1994, pp. 59–65. The notation $D_\lambda T_{\mu\nu\rho\sigma...}$ for the new tensor is mine. The interpretation of this operation as a generalized derivation belongs to Ricci: cf. Dell'Aglio 1996.

Whereas we now regard these results as the formal basis of tensor calculus on a Riemannian manifold, for Christoffel they are only a means to transform the initial problem of the equivalence of two quadratic differential forms into the algebraic problem of the equivalence of a sequence of multilinear forms. The heart of his memoir is the proof that in most cases (when the invariance group of the quadratic differential form is trivial), the algebraic conditions

$$g_{\mu\nu}(x)u^{\mu}_{\mu'}u^{\nu}_{\nu'} = g'_{\mu'\nu'}(x') \quad \text{and} \quad T_{\mu\nu\rho\sigma...}(x)u^{\mu}_{\mu'}u^{\nu}_{\nu'}u^{\rho}_{\rho'}u^{\sigma}_{\sigma'}\, ... = T'_{\mu'\nu'\rho'\sigma'...}(x') \qquad (8.40)$$

for $T_{\mu\nu\rho\sigma...} = R_{\mu\nu\rho\sigma}$, $D_{\tau}R_{\mu\nu\rho\sigma}$, $D_{\upsilon}D_{\tau}R_{\mu\nu\rho\sigma}$, etc. completely determine the values of the numbers x^{μ} and $u^{\mu}_{\mu'}$ for any given x' if the number of repetitions of the covariant derivation is large enough.[52]

Christoffel's initial problem was thus solved through a purely algebraic set of conditions. As Felix Klein later noted, the benefit is not as large as it would seem, because the algebraic elimination problem remains difficult. Later mathematicians, especially Elie Cartan, gave more elegant and more general solutions to Christoffel's problem. The tensor-calculus by products of Christoffel's investigation were nonetheless considerable.[53]

Since these by products later received a geometric interpretation, one may wonder whether Christoffel privately relied on geometric intuition in their derivation. First, he may have reached Equation (8.36) by requiring that the geodesics of the $g_{\mu\nu}$ metric should correspond to the geodesics of the $g'_{\mu\nu}$ metric. This is quite plausible because the notation he used for the Γ coefficients, $\{^{\nu\rho}_{\mu}\}$, is the one he had earlier used to denote the coefficients of the equation of geodesics on a surface.[54] Second, in his discovery of Equation (8.37), $R_{\mu\nu\rho\sigma}(x)u^{\mu}_{\mu'}u^{\nu}_{\nu'}u^{\rho}_{\rho'}u^{\sigma}_{\sigma'} = R'_{\mu'\nu'\rho'\sigma'}(x')$, Christoffel may have been inspired by his awareness of the quadratic form that Riemann associated with the $R_{\mu\nu\rho\sigma}$ coefficients in his lecture of 1854. Some knowledge of the contents of the *Commentatio*, even indirect, could also have helped him. Assuming that Christoffel privately relied on geometric intuitions and derivations, why would he suppress them from the published memoir? Perhaps he wanted to emphasize continuity with his earlier work on the algebraic theory of invariants. Perhaps he was following the growing trend of arithmetizing or algebraizing geometry and analysis. Perhaps, he did not believe that the study of n-dimensional manifolds truly belonged to geometry. Indeed his few mentions of geometrical ideas and problems in his memoir concerned ordinary, two-dimensional surfaces.[55]

---

[52] On this proof, cf. Darrigol 2015, pp. 71–72.

[53] Klein 1926–27, Vol. 2, p. 197. On later solutions, cf. Ehlers 1981.

[54] Christoffel 1868, p. 126.

[55] One of these mentions (Christoffel 1869, p. 47) concerns the developability of surfaces; another (ibid., p. 64) concerns the invalidity of Christoffel's theorem for a surface that can be shifted [*verschiebt*] into itself (e.g., a surface of constant curvature).

Lipschitz

In the same year, 1869, and in the same journal, the Bonn mathematician Rudolf Lipschitz published a memoir on the condition for a given homogeneous differential form[56] to be equivalent to a form with constant coefficients through a change of variables. Despite evident similarities, this problem differs from Christoffel's in two ways. First, Lipschitz considers homogeneous forms of any degree (including *one*), whereas Christoffel regards the degree two (bilinear forms) as the only interesting case. Indeed, for the higher degrees, the first algebraic condition of equivalence (the counterpart of condition (8.34)) completely determines the functions $x^\mu(x')$ and $u^\mu_{\mu'}(x')$, and there is no need for deriving conditions of higher order. Second, Lipschitz treats the equivalence of a differential form with a form with constant coefficients, whereas Christoffel's method excludes this case as well as any case for which the differential form admits nontrivial automorphisms.[57]

Lipschitz acknowledged the stimulus provided by the publication of Riemann's habilitation lecture. In the quadratic case, Riemann had indeed given the condition sought by Lipschitz: the curvature of the metric manifold associated with the quadratic field must vanish. As the full expression of Riemann's curvature appeared only in the unpublished *Commentatio*, Lipschitz was not quite sure that his own equivalence condition agreed with Riemann's.[58] But he surely understood the importance that Gauss and Riemann had given to such a condition and to its geometric interpretation. We will see that analogy with Gauss's intrinsic expression of curvature guided Lipschitz's construction of the general equivalence condition. At the same time, by generalizing to forms of any degree, Lipschitz distanced himself from Gaussian and Riemannian geometry; and his heuristics involved an important nongeometrical component: Lagrange's calculus of variations in the mechanical tradition continued by William Rowan Hamilton and Carl Jacobi.[59] Whereas Christoffel used ordinary differentiation only (affecting only the coefficients of a differential form), Lipchitz used variations that increase the differential order of differential forms.

In his *Mécanique analytique* Lagrange had shown how to derive a higher-order invariant differential expression from a lower-order one.[60] For the invariant form $f(x, \mathrm{d}x)$ (with $x = (x_1, x_2, ..., x_n)$, etc.), the identity

---

[56] In conformity with nineteenth-century terminology, *differential form* of degree k on a manifold here refers to any k-linear form $\sum_{\mu\nu...} a_{\mu\nu...}(x)\mathrm{d}_1x^\mu\mathrm{d}_2x^\nu...$ of the differentials $\mathrm{d}_1x, \mathrm{d}_2x, ..., \mathrm{d}_kx$ (in modern terms this expression is a covariant tensor of order k, which would further need to be completely antisymmetric in order to define a differential form in the modern sense).

[57] Lipschitz 1869. Cf. Reich 1994, pp. 29–32 and Tazzioli 1994.

[58] Lipschitz 1869, pp. 71–72. In a footnote (p. 74n) Lipschitz writes: "I have doubts on whether the results of Riemann's memoir can be interpreted in a manner agreeing with [my] theorem."

[59] On the Lagrangian, mechanical, and variational context, cf. Tazzioli 1994. According to Felix Klein (1926–27, Vol. 2, p. 189), because of its mechanical underpinning Lipschitz's contribution was less appreciated by mathematicians than Christoffel's.

[60] *Differential expression* of order k here refers to any polynomial combination of the differentials $\mathrm{d}^j_i x^\mu$ such that the sum of all j's in each term is equal to k (the index i labels the differentials and the index j gives their order). For instance, $(a\mathrm{d}^2x + b\mathrm{d}x\mathrm{d}y)\delta y$ is a differential expression of the third order. Lagrange's theorem is more general than the sample given here, for it includes f functions depending on the higher differentials $\mathrm{d}^2x, \mathrm{d}^3x$, etc.

$$\delta f(x, dx) = \delta x \cdot \partial f / \partial x + d(\delta x \cdot \partial f / \partial dx) - \delta x \cdot d(\partial f / \partial dx) \tag{8.41}$$

implies the invariance of the new form

$$\omega(x, dx, d^2x, \delta x) = \delta x \cdot [d(\partial f / \partial dx) - \partial f / \partial x]. \tag{8.42}$$

As a first application of this rule, Lipschitz considers the linear form $f = a_\mu dx^\mu$, whose derived invariant form reads

$$\omega = \partial_\mu a_\nu (dx^\mu \delta x^\nu - dx^\nu \delta x^\mu). \tag{8.43}$$

This form vanishes if and only if $\partial_\nu a_\mu - \partial_\mu a_\nu = 0$, which is the condition for the form f to be a differential. Evidently, this condition holds whenever there exists a system of coordinates for which the coefficients a_μ are constant. Reciprocally, if this condition holds, the form f can be integrated according to $f = dF(x)$; in a new system of coordinates in which F is the first coordinate, the differential form has constant coefficients. This solves Lipschitz's problem for forms of degree one.[61]
  In the general case, the derived form is

$$\omega = \left(\frac{\partial^2 f}{\partial \, dx^\mu \, \partial \, dx^\nu} d^2 x^\nu + \frac{\partial^2 f}{\partial \, dx^\mu \, \partial x^\nu} dx^\nu - \frac{\partial f}{\partial x^\mu} \right) \delta x^\mu. \tag{8.44}$$

For the quadratic form $f = g_{\mu\nu} dx^\mu dx^\nu$, this gives

$$\omega = g_{\rho\sigma}(d^2 x^\rho + \Gamma^\rho_{\mu\nu} dx^\mu dx^\nu) \delta x^\sigma, \quad \text{with } g_{\rho\sigma} \Gamma^\rho_{\mu\nu} = \frac{1}{2}(\partial_\mu g_{\sigma\nu} + \partial_\nu g_{\sigma\mu} - \partial_\sigma g_{\mu\nu}). \tag{8.45}$$

Lipschitz's next step is to eliminate the second-order differentials $d^2 x^\mu$ in the invariance relation $\omega = \omega'$. After complicated calculations involving the transformation rule for $d^2 x^\mu$, he obtains the invariance of a fourth-order differential form built from the first and second partial derivatives of the form f. In the quadratic case, this new form reads

$$\Omega = R_{\mu\nu\rho\sigma} dx^\mu \delta x^\nu d'x^\rho \delta'x^\sigma, \quad \text{with } R_{\mu\nu\rho\sigma} = g_{\mu\nu}(\partial_\rho \Gamma^\nu_{\nu\sigma} - \partial_\sigma \Gamma^\nu_{\nu\rho} + \Gamma^\nu_{\rho\tau} \Gamma^\tau_{\nu\sigma} - \Gamma^\nu_{\sigma\tau} \Gamma^\tau_{\nu\rho}), \tag{8.46}$$

in conformity with the expression (8.12) of Riemann's curvature. The invariance of this form implies that it should vanish if there exists a system of coordinates for which the coefficients of the form f are constants. Lipschitz's most impressive achievement is his proof that reciprocally the vanishing of the form Ω implies the existence of a system of coordinates for which the coefficients of the form are constants. For this purpose, he relies on uniform geodetic motion to construct the geodetic coordinates for which the form f becomes uniform.[62]

[61] Lagrange 1811–1815, Vol. 1, pp. 307–310; Lipschitz 1869, p. 77.
[62] Lipschitz 1869, p. 84, §6.

In accordance with the mechanical nature of the latter reasoning, Lipschitz developed mechanical aspects of the study of homogeneous forms in subsequent memoirs. The basic idea was to relate curvature to the inertial force experienced by a particle constrained to move on a curved manifold. He developed the notion of geodesic normal coordinates—which he called central coordinates—again in a mechanical manner associated with geodetic motion. In this context, he was able to prove Riemann's relation between geodetic distortion and curvature. That is to say, he derived the expression $-\frac{1}{3}R_{\mu\nu\rho\sigma}dx^{\mu}\delta x^{\nu}dx^{\rho}\delta x^{\sigma}$ of the correction to the length of the element dx at the point $x + \delta x$ in normal coordinates centered on x.[63]

Another of Lipschitz' achievements was the manifestly covariant expression he gave to the form Ω:

$$\Omega = \delta'x \cdot (D\Delta - \Delta D)d'x, \tag{8.47}$$

where D and Δ are the covariant variations defined by[64]

$$Da^{\mu} = da^{\mu} + \Gamma^{\mu}_{\nu\rho}dx^{\nu}a^{\rho} \text{ and } \Delta a^{\mu} = \delta a^{\mu} + \Gamma^{\mu}_{\nu\rho}\delta x^{\nu}a^{\rho}. \tag{8.48}$$

In 1877, after reading the *Commentatio* in Riemann's collected works, he split Ω into two invariant parts:

$$\Omega = \frac{1}{2}\hat{X} + \Xi, \text{ with} \tag{8.49}$$

$$\frac{1}{2}\hat{X} = D(\delta'x \cdot \Delta d'x) - \Delta(\delta'x \cdot Dd'x) \text{ and } \Xi = -D\delta'x \cdot \Delta d'x + \Delta\delta'x \cdot Dd'x.$$

Using the identities

$$d(a \cdot b) = D(a \cdot b) = (Da) \cdot b + b \cdot (Da), \text{ } Dd'x = D'dx, \text{ } D\delta'x = \Delta'dx, \text{ etc.,} \tag{8.50}$$

we have

$$\hat{X} = dd'(\delta x \cdot \delta'x) + \delta\delta'(dx \cdot d'x) - \delta d'(dx \cdot \delta'x) - d\delta'(\delta x \cdot d'x). \tag{8.51}$$

For $d = d'$ and $\delta = \delta'$ this expression becomes identical to Riemann's X. As for the invariant Ξ, Lipschitz notes that it vanishes under Riemann's conditions (8.28), which are equivalent to $D\delta x = 0$, $Ddx = 0$, and $\Delta\delta x = 0$. Furthermore, he observes that an invariant of the same type as this one occurs in rational mechanics when Gauss's principle of least constraint is expressed in arbitrary coordinates q^{i} by means of the quadratic differential form $g_{ij}dq^{i}dq^{j}$ occurring in the expression $T = (1/2)g_{ij}\dot{q}^{i}\dot{q}^{k}$ of the kinetic energy. Lipschitz thus solved

---

[63] Lipschitz 1870a; 1870b, p. 23.

[64] The notation and the naming are mine. Lipschitz instead uses the covariant ψ vectors such that $\psi_{\mu}(dx, a) = g_{\mu\nu}Da^{\nu}$.

the mystery of Riemann's X in his own manner, that is, through a theory of differential invariants based on variational calculus in a mechanical context.[65]

Ricci

Lipschitz was not alone in connecting the theory of quadratic differential forms to analytical mechanics. So too did Eugenio Beltrami in 1868, Ernst Schering in 1873, Wilhelm Killing in 1885, and Heinrich Hertz in 1894.[66] For others including Dedekind, Aurel Voss, Richard Beez, Klein, Helmholtz, and Lie, the geometrical interpretation was more important. A third tendency, inaugurated by Christoffel, was to develop the theory of differential invariants for its own sake, independently of any application. The Padova mathematician Gregorio Ricci-Curbastro was aware of these three approaches when he began working on the theory of quadratic differential forms in the mid-1880s. Although he was himself interested in geometrical, mechanical, and physical applications of this theory, he believed that for the sake of rigor and generality foundations should precede applications:[67]

> The aim of this writing is to initiate a series of researches on the theory of quadratic differential forms. By founding these researches on purely analytical concepts, we will better penetrate the nature of these forms and we will avoid idle discussions regarding the existence and nature of spaces of more than three dimensions. Interpretations dictated by geometrical or mechanical analogies—such interpretations can indeed be given—will only be illustrations of the theory thus conceived.

In his first memoir on this topic, published in Italian in 1884, Ricci defines the *class* of a quadratic differential form of n variables as the number h such that $n + h$ is the minimal number of variables of a form with constant coefficients from which the original form can be derived by constraining these variables through h relations. Ricci indeed knows from Ludwig Schlaefli that there always is a Euclidean space of dimension $n + h$ of which a given metric manifold of dimension n is a submanifold, with $0 \leq h \leq n(n - 1)/2$. Ricci then proves that the vanishing of Riemann's $R_{\mu\nu\rho\sigma}$ is a necessary and sufficient condition for a form to be of class zero, and he finds similar conditions for forms of class one.[68]

In a series of subsequent memoirs, Ricci thoroughly developed an "absolute differential calculus" (*calcolo differenziale assoluto*) containing all the basic definitions and operations of modern tensor analysis on differential manifolds: covariant and contravariant "systems of functions" (*sistemi di funzioni*, our tensors), tensor products, contraction, raising and lowering of indices through the fundamental (metric) tensor, covariant and contravariant derivation, and the Riemann curvature tensor and its relation to the commutator of two

---

[65] Lipschitz, 1870b, p. 16; 1877, pp. 317–321, §§2–3 (Gauss's principle). On the mechanical considerations, cf. Darrigol 2015, p. 75. Lipschitz was not the first or the last to geometrize analytical mechanics. On Liouville's earlier attempt, Hertz's later attempts, and the relative independence of Lipschitz's, cf. Lützen 1994, 1999, 2005.

[66] Cf. Tazzioli 1994.

[67] Ricci 1884, p. 140. On Ricci's approach, cf. Tonolo 1961; Struik 1993; Dell'Aglio 1996, pp. 235–256; Bottazzini 1999, pp. 243–253. On his calculus and its influence, cf. Reich 1994, pp. 63–110; Giovanelli 2013, §2.2. The first definition of *derivazione covarianti* is in Ricci 1887, p. 202.

[68] Ricci 1884.

covariant derivations. He also offered a number of geometrical and physical applications of his new calculus, including systems of geodesics, congruences, elasticity, heat propagation in curved hyperspaces, and continuous groups of motion. The most systematic exposition of his theory was the one he published in French in the *Mathematische Annalen* in 1901 together with his brilliant disciple Tullio Levi-Civita.[69]

Levi-Civita, parallel transport, and Riemann

In the 1890s, young Levi-Civita abundantly used Ricci's calculus in contributions to the theory of differential invariants, to its application to mechanics, and to its connection with Lie's theory of continuous groups. He returned to this topic in 1917, after Einstein's publication of the general theory of relativity. This impressive application of Riemann's concept of metric manifold and of Ricci's absolute differential calculus induced Levi-Civita to "somewhat reduce the formal apparatus that usually serves to introduce and establish the covariant behavior." For this purpose, he sought inspiration in Riemann's writings.[70]

As is easily seen, the conditions (8.28) of Riemann's *Commentatio* include the covariant relation

$$\delta dx^\mu = d\delta x^\mu = -\Gamma^\mu_{\nu\rho}\delta x^\nu dx^\rho \tag{8.52}$$

between the second-order differential δdx^μ and the first-order differentials δx^μ and dx^μ. After emphasizing the covariant character of this relation, Levi-Civita considers the third-order differential

$$\delta'\delta dx^\mu = -\delta'(\Gamma^\mu_{\nu\rho}\delta x^\nu dx^\rho) = -\delta'\Gamma^\mu_{\nu\rho}\delta x^\nu dx^\rho - 2\Gamma^\mu_{\nu\rho}\delta'\delta x^\nu dx^\rho$$
$$= -\partial_\sigma\Gamma^\mu_{\nu\rho}\delta x^\nu dx^\rho\delta'x^\sigma + 2\Gamma^\mu_{\nu\rho}\Gamma^\nu_{\sigma\tau}dx^\rho\delta x^\sigma\delta'x^\tau, \tag{8.53}$$

the similar differential $\delta\delta'dx^\mu$, and he forms the difference

$$u^\mu = (\delta'\delta - \delta\delta')dx^\mu, \tag{8.54}$$

which he proves to be a contravariant vector. The effective calculation of u^μ yields

$$u^\mu = R^\mu_{\nu\rho\sigma}dx^\nu\delta x^\rho\delta'x^\sigma. \tag{8.55}$$

Consequently, the expression

$$I = g_{\mu\tau}d'x^\tau(\delta'\delta - \delta\delta')dx^\mu = R_{\mu\nu\rho\sigma}d'x^\mu dx^\nu\delta x^\rho\delta'x^\sigma \text{ with } R_{\mu\nu\rho\sigma} = g_{\mu\tau}R^\tau_{\nu\rho\sigma} \tag{8.56}$$

is an invariant, and $R_{\mu\nu\rho\sigma}$ must be a completely contravariant four-tensor. "If I am not mistaken," Levi-Civita notes, "this is the fastest way to arrive at this result."[71]

---

[69] Ricci 1886, 1887, 1888, 1892; Ricci and Levi-Civita 1901.

[70] Levi-Civita 1917a, p. 173. On Levi-Civita and parallel transport, cf. Struik 1989; Reich 1992; Bottazzini 1999, pp. 254–256.

[71] Levi-Civita 1917a, p. 198.

This simple derivation of the Riemann tensor occurs toward the end of Levi-Civita's memoir after an extensive development of the notion of parallel transport. As Levi-Civita reveals at the beginning of his memoir, the order of discovery was different: he first obtained the new derivation of the Riemann tensor in a purely algebraic manner, and then sought a geometric interpretation of this derivation. As he observes in his memoir, for $\delta = \mathrm{d}$ the relation (8.56) has a simple geometric interpretation through the equation of geodesics: for the geodesic passing through x and $x + \mathrm{d}x$, $\mathrm{d}x$ is a tangent vector at x, and $\mathrm{d}x^\mu + \mathrm{dd}x^\mu = \mathrm{d}x^\mu - \Gamma^\mu_{\nu\rho}\mathrm{d}x^\nu\mathrm{d}x^\rho$ is the tangent vector at $x + \mathrm{d}x$ that has the same length as the vector dx. In the general case $\delta \neq \mathrm{d}$, Levi-Civita is aware of the identities

$$d(g_{\mu\nu}\delta x^\mu \delta x^\nu) = 0 \quad \text{and} \quad d(g_{\mu\nu}\mathrm{d}x^\mu \delta x^\nu) = 0, \tag{8.57}$$

whose geometric interpretation is the conservation of the length of the vector δx and of its angle with the vector $\mathrm{d}x$ when the vector δx changes by $-\Gamma^\mu_{\nu\rho}\delta x^\nu \mathrm{d}x^\rho$ along the geodesic line passing through x and $x + \mathrm{d}x$. Intuitively, geodesics are the straightest possible lines on the manifold; therefore, on a two-dimensional manifold for which the direction of a vector is completely determined by its angle with a given line, the direction of the vector $\delta x + \mathrm{d}(\delta x)$ at point $x + \mathrm{d}x$ may be said to differ as little as possible from the direction of vector δx at point x.[72]

Possibly guided by this intuition, Levi-Civita looks for a notion of parallel displacement on a manifold of any dimension. He finds one by regarding the manifold as a hypersurface in a Euclidean space of higher dimension. This immersion provides an extrinsic notion of parallelism for two vectors situated at two different points of the manifold: they have to make the same angle with any given vector of the Euclidean space. For intrinsic parallelism, Levi-Civita tries the next best thing: two vectors **a** and **b** situated at two neighboring points x and $x + \mathrm{d}x$ of the manifold are said to be parallel if and only if they make the same angle with any vector belonging to the tangent hyperplane at point x. If $\mathbf{r}(x)$ denotes the immersion of the point of coordinates x^μ in the larger Euclidean space, this condition gives

$$\mathbf{a} \cdot \partial_\mu \mathbf{r} = \mathbf{b} \cdot \partial_\mu \mathbf{r} \quad \text{to first order in } \mathrm{d}x \text{ for } \mu = 1, ..., n. \tag{8.58}$$

Remembering that $\mathbf{a} = a^\mu \partial_\mu \mathbf{r}(x)$, $\mathbf{b} = b^\mu \partial_\mu \mathbf{r}(x + \mathrm{d}x)$, and $g_{\mu\nu} = \partial_\mu \mathbf{r} \cdot \partial_\nu \mathbf{r}$, this is equivalent to

$$a^\mu g_{\mu\nu} = b^\mu g_{\mu\nu} + b^\mu(\partial_\mu \partial_\rho \mathbf{r}) \cdot \partial_\nu \mathbf{r}\, \mathrm{d}x^\rho. \tag{8.59}$$

Using $(\partial_\mu \partial_\rho \mathbf{r}) \cdot \partial_\nu \mathbf{r} = g_{\nu\sigma}\Gamma^\sigma_{\mu\rho}$, we get $a^\mu = b^\mu + \Gamma^\mu_{\nu\rho}b^\nu \mathrm{d}x^\rho$, or, at the same infinitesimal order,

$$b^\mu = a^\mu - \Gamma^\mu_{\nu\rho}a^\nu \mathrm{d}x^\rho. \tag{8.60}$$

Levi-Civita thus obtains the intrinsic expression of his newly defined parallel displacement and finds it to be identical to the variation he has seen in Riemann's *Commentatio*.[73]

---

[72] Levi-Civita 1917a, §§15–16, pp. 173 (order of discovery), 196 (identities).

[73] Levi-Civita 1917a, pp. 174–175, §§2–3.

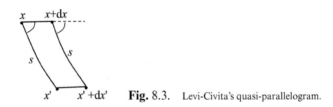

Fig. 8.3.   Levi-Civita's quasi-parallelogram.

Now equipped with this new geometrical concept, Levi-Civita uses it to interpret his expression (8.56) of the Riemann curvature geometrically. For this purpose, he considers the quasi-parallelogram obtained by drawing two geodesics from the points x and $x + dx$ at a constant angle with dx and by marking on these geodesic the points x' and $x' + dx'$ situated at the same distance s from their respective origins (see Figure 8.3). Interpreting d as the first-order parallel displacement along dx when applied to a vector, introducing the first-order differential δ for changes occurring on the xx' line and interpreting this change as the first-order parallel displacement along this line when needed, Levi-Civita obtains

$$\delta ds^2 \equiv g_{\mu\nu}(x')dx'^\mu dx'^\nu - g_{\mu\nu}(x)dx^\mu dx^\nu = g_{\mu\nu}(x)dx^\mu(d\delta - \delta d)\delta x^\nu. \qquad (8.61)$$

Using Equation (8.56), this gives

$$\delta ds^2 = g_{\mu\tau}dx^\tau R^\mu{}_{\nu\rho\sigma}\delta x^\nu \delta x^\rho dx^\sigma = -R_{\mu\nu\rho\sigma}dx^\mu \delta x^\nu dx^\rho \delta x^\sigma \qquad (8.62)$$

for the difference in length between the two small sides of the quasi-parallelogram.[74]

Levi-Civita thus arrived at a new geometric interpretation of Riemann's curvature, based on the construction of a quasi-parallelogram by parallel transport. As he tells us in his introduction, for a while he believed that Riemann had the same interpretation in mind when he gave the calculation based on the X formula of the *Commentatio*. After closer inspection, Levi-Civita realized that he could not find any trace of the quasi-parallelogram in Riemann's text or in Dedekind's reconstruction. Worse, he judged that Riemann's X should identically vanish because of the identities (8.57) that express the conservation of length and angles during parallel displacement. He commented:

> Probably, in Riemann's X there is some writing defect that conceals the underlying concept. I flatter myself I have in substance reconstructed this concept, although I could not adjust the symbols. Provided the adjustment can be done, it will be good to give full justice to Riemann's genius in this respect too.

---

[74]Levi-Civita 1917a, §§17–18. Equation (73) is akin to the theorem of geodesic deviation (Levi-Civita 1927), which relates the covariant variation of the vector connecting two neighboring geodesics to the curvature tensor. For details, cf. Darrigol 2015, pp. 78–79.

As we saw, Riemann probably reached his X through mostly algebraic considerations. At any rate, his d and δ symbols cannot be interpreted in Levi-Civita's manner. Ironically, this misinterpretation led to the all-important concept of parallel displacement.[75]

Conclusions

Through his new concept of manifold, Riemann completed the transition from an older view of geometry as the science of the construction and properties of geometrical figures to the science of space as an abstractly defined background. His manifolds are extremely general: they can be discrete and measured by counting points, or they can be continuous with a not necessarily quadratic measure and with a nonvanishing, even variable curvature in the quadratic case. In this view, space is only one particular three-dimensional manifold, to be selected by empirical means in an infinite variety of possibilities. These considerations mattered in at least three different ways: they contributed to the mathematicians' emancipation from Euclidean geometry by providing models for the new non-Euclidean geometries, they permitted an analytic-algebraic theory of quadratic differential forms that could be used in nongeometrical contexts, and they opened new vistas on physical space.

Although this third aspect includes Riemann's stunning anticipation of a space structure depending on the presence of matter, and although a few astronomers in the nineteenth century invoked a space of nonvanishing (constant) curvature, it is the second aspect that truly served the genesis of general relativity. As we will see in the next two chapters, Einstein relied on Ricci's nongeometric calculus, and geometric insights into the space–time of general relativity came mostly through the efforts of Levi-Civita, Weyl, and Eddington. It is therefore important to understand how the theory of differentiable manifold, originally a mix of geometric and algebraic methods in the hands of Riemann, Dedekind, and Weber, became a new addition to the algebraic theory of invariants in Christoffel's and Ricci's hands.

For the metric manifolds now called Riemannian, Riemann plausibly obtained his powerful concept of sectional curvature by generalizing Gauss's notion of geodetic distortion with the benefit of normal geodesic coordinates, which are themselves a generalization of Gauss's geodesic polar coordinates. The calculations found in the last section of the *Commentatio*, though not dictated by geometrical considerations, were informed by partial geometrical interpretations of intermediate steps and equations. This characterization of Riemann's efforts is based on close reading of Riemann's allusive, fragmentary writings and on some archival material. With even fewer clues, Riemann's posthumous readers faced the challenge of reconstructing his main results. They had a lot of latitude to do so, depending on their own interests.

In their commentary to the *Commentatio*, Dedekind and Weber thoroughly mixed the geometrical and analytical aspects by applying the geometric notion of normal geodesic coordinates to the analytical problem treated in the last section of the *Commentatio*. In

---

[75] Levi-Civita 1917a, pp. 74, 102. As I have shown elsewhere (Darrigol 2015, pp. 62–63), Riemann's X admits a geometrical interpretation. But the relevant geometric figure differs from Levi-Civita's parallelogram and the associated meaning of d and δ is different.

contrast, Christoffel and Lipschitz focused on the algebraic-analytic equivalence problem, and they avoided geometric illustrations in their writings. The former nonetheless admitted to being motivated by the geometric problem of development of a (hyper)surface on another, and both are likely to have relied on analogies with Riemann's or Gauss's more geometric considerations. That said, their analytical developments mostly depended on the principal context in which they inscribed them: the theory of invariants for Christoffel, and the variational calculus of Lagrangian analytical mechanics for Lipschitz. In these contexts, with or without hidden recourse to geometric analogies, they significantly extended the proto-tensor calculus of the *Commentatio*, and Lipschitz succeeded in interpreting the mysterious double-variation I called X in Riemann's calculations. Ricci's absolute differential calculus most drastically eliminated extra-analytical considerations, his aim being to provide a universal, absolute differential calculus that would serve many domains of mathematics and physics.

General relativity and Levi-Civita brought physics and geometry back into the interpretation of the Riemann parenthesis or tensor. More than his algebraically inclined predecessors, Levi-Civita sensed the artificiality and heaviness of some of the new calculus. He suspected the formulas of Riemann's *Commentatio* to depend on hidden geometrical ideas, and he tried to elucidate Riemann's X by recreating these ideas. Some of Riemann's formulas involved a peculiar infinitesimal variation of vectors that Levi-Civita knew to preserve angles and length. His interpretation of this variation led him to the concept of parallel transport. Although this concept does not quite solve the X mystery, it significantly simplifies the apparatus of tensor calculus and differential geometry. It soon opened the door to important and vast generalizations of Riemannian geometry.

To sum up, the ambivalent and elliptical character of Riemann's writings on curvature and on the equivalence problem of quadratic differential forms allowed for diverging developments in contexts including geometry, the theory of invariants, and Lagrangian mechanics. Reacting to this diversity, Ricci propounded a decontextualized, universalist tensor analysis, which he called absolute differential calculus. In the end, general relativity and the Riemann mystery stimulated Levi-Civita's geometrization of Ricci's calculus and thus eased the cross-fertilization of analysis, geometry, and physics.

9

MOSTLY EINSTEIN: TO GENERAL RELATIVITY

> With this, the general theory of relativity is finally made a closed logical construct. The relativity principle in its most general conception, according to which the space and time coordinates become physically meaningless, inevitably leads to a completely definite theory of gravitation explaining the motion of Mercury's perihelion.[1] (Albert Einstein, November 1915)

Newton's law of gravitation is evidently incompatible with special relativity,[2] since it involves instantaneous action at a distance. The founders of relativity theory, Lorentz, Poincaré, Einstein, and Minkowski, immediately saw the necessity of a new theory of gravitation in which the gravitational action would propagate at the velocity of light and the Newtonian predictions of the movement of planets would receive relativistic corrections. Einstein distinguished himself, however, by basing the new theory on a generalization of the relativity principle. The crucial insight occurred in 1907, when he posited that the effects of a homogeneous gravitational field in a given system were equivalent to the effects of a common acceleration impressed on this system. This *equivalence principle* made the gravitational field a frame-dependent notion, as the electromagnetic field already was in special relativity. There were two essential differences, however: any reference frame now became acceptable, and the notion of frame became a local affair since, in a nonhomogeneous gravitational field, the equivalence principle was necessarily restricted to small regions of space.

By 1912 Einstein understood that the new theory should be based on geodesic motion in a four-dimensional metric manifold, with as much covariance as possible for the equations ruling the metric coefficients. In retrospect, this idea, together with a few natural presuppositions, should have quickly led him to the goal.[3] In reality, it took him three additional years of intense struggle to arrive at the final "Einstein equations." The purpose of this chapter is to explain how Einstein reached the basic metric manifold setting of 1912, and why it took him so long to complete the theory in this setting. As there is already much excellent literature on these difficult developments, it may be good to indicate briefly three ways in which my account differs from earlier accounts.[4]

---

[1] Einstein 1915c, p. 847.

[2] Of course, special relativity could only be called so after a generalized relativity existed. "Speziellere Theorie" first occurred in Einstein 1914c; "spezielle Relativätstheorie" in Einstein 1915a, 1916a.

[3] On principles that forcefully lead to general relativity, cf. Ehlers 1973; Darrigol 2014a, chaps. 5 and 7.

[4] Here too I have avoided relying on Einstein's often misleading recollections. The best early accounts are Earman and Glymour 1978a; Pais 1982; Stachel 1989b [1980]. John Norton gave a first penetrating reading of the Zürich

Relativity Principles and Theories from Galileo to Einstein. Olivier Darrigol, Oxford University Press.
© Olivier Darrigol (2022). DOI: 10.1093/oso/9780192849533.003.0009

First, using the long-term perspective adopted in this book I will show that important singularities of Einstein's efforts, for instance his appeal to the equivalence principle or his neglect of geometric considerations, depended on much earlier developments in the history of physics and mathematics. Second, I will pay special attention to his concern with measurement in accelerated frames in 1907–1911 and relate this concern to his later prejudice regarding the physical meaning of coordinates. Third, I will argue that in the setting of 1912, Einstein's chief heuristic principle was the requirement that the four-force on a particle of matter should be the four-divergence of a stress–momentum tensor for the gravitational field, in analogy with the Maxwell stresses in an electromagnetic field. Einstein's principal reason for long rejecting maximally covariant candidates for the gravitational field equation was a (wrongly) suspected incompatibility with this *stress principle*. His belief in the nongenerally covariant *Entwurf* theory of 1913 depended on its being based on this principle. His final return to fully covariant options happened when he realized that after all the covariant options were compatible with the stress principle. This is not to deny that other heuristic principles and other kinds of considerations, for instance correspondence principles, rotating frames, coordinate conditions or restrictions, electromagnetic analogy, and Mach's principle, did not play any role—they will all be discussed in the appropriate places. It is only a matter of relative emphasis: I believe higher clarity is achieved by focusing on the stress principle.[5]

In the introduction to Chapter 7, the reader was warned about two common prejudices regarding the genesis of special relativity: that it was a single man's achievement, and that Einstein worked in isolation. In the case of general relativity, one cannot deny that this theory is mostly Einstein's. Although there were other attempts at a relativistic theory of gravitation in the years of 1906–1915, only Einstein's rested on the equivalence principle and a generalized relativity principle. Einstein was unique in his faith in these principles, and no one competed to develop their consequences, except David Hilbert in late 1915. That said, Einstein did not work in isolation. He greatly benefited from Marcel Grossmann's collaboration in developing the needed tensor calculus. Michele Besso did most of the work regarding the relativistic precession of a planet's orbit in the *Entwurf* theory of 1913, and he is the likely source of the hole argument through which Einstein long denied general covariance. In 1914, young Adriaan Fokker helped Einstein rephrase the scalar theory as a conformal metric theory, with consequences for Einstein's understanding of the role of covariance. In 1911, Abraham's Minkowskian theory of gravitation may have prepared Einstein for the metric formulation of his own theory, even though he mocked his competitor. The mathematician Paul Bernays invited Einstein and Grossmann to base their theory on the principle of least action, a move that proved crucial in the transition

---

notebook in Norton 1984 and in *ECP* 4. John Stachel, John Norton, Jürgen Renn, Tilman Sauer, Michel Janssen, and a few other scholars joined forces to produce the monumental multi-volume *Genesis* (*GGR*). For lucid, short histories, see Stachel 1995; Janssen 2005, 2014; Sauer 2005b, 2013; Janssen and Renn 2015; and Norton 2020.

[5] Norton 2020 came closest to this perspective by arguing for the prevalence of energy–momentum considerations over the equivalence principle in Einstein's "two-tiered" heuristics. As John Norton told me, there is a residual difference: whereas I regard the stress–energy tensor of the gravitational field as a crucial conceptual component of Einstein's developing theory, he tends to see it as an "intermediary."

toward the final theory. Einstein also benefited from epistolary exchanges with friends and colleagues including Ehrenfest, Lorentz, and Hilbert. Although Hilbert did not arrive at the final field equation before Einstein, he revealed the formal advantages of basing the theory on an invariant action.[6]

Section 9.1 of this chapter is about Einstein's and others' heuristic considerations of gravitation and relativity in the years of 1906–1911. These include Poincaré's, Minkowski's, and Sommerfeld's relativistic generalizations of Newton's gravitation law, Einstein's introduction of the equivalence principle, his concepts of measurement in accelerated frames, and his ingenious derivation of three cardinal effects: the gravitational redshift, the curving of light rays by large masses, and the equivalence between gravitational mass and energy. Section 9.2 is devoted to the static theory of 1912, in which Einstein applied the equivalence principle in tangent frames to arrive at a theory based on a nonuniform Minkowskian metric of a special kind. Section 9.3 shows how Einstein, in the so-called Zürich notebook of 1912–1913, struggled to extend this metric approach to arbitrary gravitational fields and how he at some point tried field equations based on the Riemann tensor, only to renounce them for a variety of (retrospectively) bad reasons. At the end of this period, Einstein gave up general covariance and managed to construct a theory based on the stress principle and a correspondence principle. This is the *Entwurf* theory published in 1913 with Grossmann and further worked out with Besso's help, as explained in Section 9.4. After reading Gunnar Nordström, Einstein soon realized that the scalar theory was a viable alternative to the *Entwurf* theory, equally compatible with the equivalence principle. Section 9.5 analyzes the consequences of this new awareness. As is shown in Section 9.6, Einstein's confidence in the *Entwurf* theory nevertheless increased as he discovered arguments justifying restricted covariance, including the (in)famous hole argument. In Section 9.7, we will see that in late 1914, after working out with Grossmann the restricted covariance properties of this theory, Einstein thought he could prove it was the only consistent Lagrangian-based theory for which the field equations do not determine the metric field more than is permitted by the restricted covariance. As is recounted in Section 9.8, Einstein arrived at his final equations in November 1915, after realizing that this proof, as well as all his past arguments against general covariance, was based on errors and misconceptions. In the conclusion, the reader will find an overview of Einstein's efforts toward general relativity, followed by a critical assessment of his aims and methods. Sections 9.3–8 are mathematically much more demanding than the rest of this chapter. Readers who do not wish to enter into all the technicalities may jump directly from Section 9.2 to the conclusion, which includes a summary of the main contents of these sections.

Notation

Now standard notation and conventions (largely inspired by Einstein's *Grundlage* of 1916) are used through this chapter (except for $\Gamma^{\rho}_{\mu\nu}$, which Einstein defined differently). Greek

---

[6]Cf. Weinstein 2015b for the stimulating role of Einstein's conflicts with Abraham, Nordström, Hilbert, Kretschmann, and others.

indices run from zero to three, Latin indices from one to three. The Minkowski metric has the signature $(+,-,-,-)$. Summation over repeated indices is understood. Upper indices are used for contravariant components, lower indices for covariant components (Einstein originally used Latin letters for covariant two-tensors and Greek letters for contravariant two-tensors). The product of a tensor by $\sqrt{-g}$ (giving the associated tensor density) is indicated by bold face; for instance, $\mathbf{T}_\mu^\nu = \sqrt{-g}T_\mu^\nu$ (Einstein used gothic letters).

c	velocity of light as defined in special relativity
$\hat{c}(\mathbf{r})$	location-dependent velocity of light in Einstein's theory of the static gravitational field
G	gravitational constant
x^μ	coordinates of a point of the space–time manifold
ds^2	square of the interval between the point-events of coordinates x^μ and $x^\mu + dx^\mu$
$\eta_{\mu\nu}$	Minkowski metric ($\eta_{00} = c^2$, $\eta_{0i} = \eta_{i0} = 0$, $\eta_{ij} = -\delta_{ij}$)
$g_{\mu\nu}$	metric tensor such that $ds^2 = g_{\mu\nu}dx^\mu dx^\nu$
g	determinant of the metric $g_{\mu\nu}$
∂_μ	partial derivative with respect to x^μ
D_μ	covariant derivative
$\{^\sigma_{\nu\rho}\}$	Christoffel symbol defined by $g_{\mu\sigma}\{^\sigma_{\nu\rho}\} = \frac{1}{2}(\partial_\nu g_{\mu\rho} + \partial_\rho g_{\mu\nu} - \partial_\mu g_{\nu\rho})$
$R_{\mu\nu\rho\sigma}$	Riemann curvature tensor
$R_{\mu\nu}$	Ricci tensor, here defined as $R_{\mu\nu} = R^\rho_{\mu\nu\rho}$
$t^{\mu\nu}$	stress–energy (pseudo-)tensor of the gravitational field
$T^{\mu\nu}$	stress–energy tensor of matter (including the electromagnetic field)
$\theta^{\mu\nu}$	stress–energy tensor of a dust

Frequently used identities are: $g_{\mu\nu}g^{\nu\rho} = \delta_\mu^\rho$, $g_{\mu\nu}\delta g^{\nu\rho} = -g^{\nu\rho}\delta g_{\mu\nu}$, $\delta g = gg^{\mu\nu}\delta g_{\mu\nu}$, $D_\nu g^{\mu\nu} = 0$, and $g_{\mu\rho}D_\nu S^{\mu\nu} = (1/\sqrt{-g})\partial_\mu(\sqrt{-g}S_\rho^\mu) - \frac{1}{2}(\partial_\rho g_{\mu\nu})S^{\mu\nu}$ for any symmetric tensor $S^{\mu\nu}$.

9.1 Heuristic arguments (1906–1911)

Lorentz-invariant gravity

Poincaré gave the first attempt at a Lorentz-covariant theory of gravitation in his Palermo memoir of 1906. In order to satisfy the relativity postulate, he required every force in nature to obey the same transformation laws as electromagnetic forces during a change of inertial frame. In analogy with the pseudo-Euclidean four-vector $X = (\mathbf{r}, ict)$ of squared length $\|X\|^2 = |\mathbf{r}^2 - c^2t^2|$, he introduced the four-vectors x and y that give the position and time of two moving mass points, their four-difference $\rho = x - y$, the four-velocities $u = dx/\|dx\|$ and $v = dy/\|dy\|$, and the four-force $F = md^2x/\|dx\|^2$. He then formed the four invariants $\rho^2, \rho \cdot u, \rho \cdot v, u \cdot v$. The gravitational action of the second particle on the first should be a linear combination of the available four-vectors

$$F = \alpha\rho + \beta u + \gamma v, \tag{9.1}$$

with coefficients depending on the invariants only. The interaction being propagated at the velocity of light, we must have $\rho^2 = 0$. In addition, the definition of the four-force implies $F \cdot u = 0$, which gives the relation $\alpha\rho \cdot u + \beta + \gamma u \cdot v = 0$ between the three coefficients.[7]

Poincaré admits any force formula compatible with these constraints and with Newton's law in the limit of small velocities. As the simplest subcases, he gives

$$F \propto \frac{(u \cdot v)\rho - (\rho \cdot u)v}{(u \cdot v)(\rho \cdot v)^3} \quad \text{and} \quad F \propto \frac{(u \cdot v)\rho - (\rho \cdot u)v}{(\rho \cdot v)^3}, \tag{9.2–3}$$

which depart from Newton's law to second order in v/c only. As Poincaré reminds his reader, Laplace had shown that a first-order modification of Newton's attraction—Laplace meant an aberration in the direction of motion of the "gravitational fluid"—would be incompatible with the observed secular motion of the moon unless the propagation velocity was at least a hundred million times the velocity of light. This argument does not concern Poincaré's theory or Lorentz's earlier electromagnetic theory of gravitation, which both involve a modified force formula and a modified dynamics in addition to Laplace's purely kinematical effect.[8]

After reading Poincaré, Minkowski similarly gave relativistic generalizations of Newton's law. Although his reasoning was more geometrical and less general than Poincaré, his results are easily compared with Poincaré's. The formula he obtained in 1907 through a natural construction in Minkowski space agrees with Poincaré's Equation (9.3), and the one he gave in 1908 in *Raum und Zeit* by analogy with the electromagnetic interaction agrees with Poincaré's Equation (9.2). In his four-vector calculus of 1910, Sommerfeld noted this agreement as well as the higher generality of Poincaré's reasoning. He also suggested a permutation-symmetric variant of Minkowski's formulas.[9]

Poincaré and Minkowski both drew on analogy between electromagnetism and gravity, the former because he transposed Lorentz covariance from electromagnetism to gravity, the latter because he conceived one of his gravitational force formulas in analogy with the Liénard–Wiechert retarded interaction. They both insisted that gravitational interactions, in order to be compatible with the relativity principle, had to be propagated at the velocity of light. Yet, neither of them contemplated a relativistic field theory. They reasoned in terms of retarded action at a distance, to be expressed through a covariant formula for a velocity-dependent force between two mass points.

If we believe Einstein's memories, before considering a generalized relativity principle, he considered the possibility of a scalar field theory of gravitation. He may then have reached a theory similar to Nordström's theory of 1912, on which more in a moment. However,

---

[7] Poincaré 1905, pp. 1505–1506; 1906, pp. 166–175. I have slightly modernized Poincaré's notation. On the history of Lorentz-covariant generalizations of Newton's gravitation law, cf. Walter 2007b.

[8] Poincaré 1906, pp. 174–175; Laplace 1805, Vol. 4, pp. 325–326. On Lorentz's theory, see p. 170, this volume. On the history of attempts to modify Newton's law, cf. Poincaré [1906–1907].

[9] Minkowski 1908, pp. 109–111; 1909, pp. 110–111; Sommerfeld 1910b, pp. 684–689. For Minkowski 1908, the correspondence between his vectors and mine reads: $B = x$, $B^* = y$, $BC = dx$, $B^*C^* = dy$, $OA' = v$, $D^* = (\rho \cdot v)v + y$, $BD^* = (\rho \cdot v)v - \rho$, $B^*D^* = (\rho \cdot v)v$.

he soon realized that in any such theory, a body falling with an initial horizontal velocity would fall more slowly than when released from rest. This is easily seen by considering the same fall from a frame moving uniformly at the initial horizontal velocity of the body: according to relativistic time dilation, the falling time in the earth-frame is longer than the falling time in the former frame, which is the same as the falling time of a body initially at rest owing to the invariance of the scalar field. Einstein disliked this consequence of the scalar theory, for it seemed to imply that the falling time of a body would depend on its internal kinetic energy—and therefore on its mass, contradicting Galileo's observation that all bodies fall with an acceleration independent of their mass. In Einstein's opinion, any future theory of gravitation would have to comply with this universality of free fall.[10]

The principle of equivalence and two kinds of time (1907)

Einstein's first discussion of relativity and gravitation appeared in 1907, at the end of a long review of relativity theory for the *Jahrbuch für Radioaktivität und Elektronik*, at Johannes Stark's request. Ignoring Lorentz-covariant approaches to gravitation, Einstein directly introduced what he later dubbed "the happiest thought in my life": the *principle of equivalence*. Namely, *the laws of physics should be the same in a uniformly accelerated frame without gravity and in an inertial frame with gravity equal and opposed to the acceleration of the former frame.* For the laws of free fall, this equivalence results from Galileo's observation that all bodies fall with the same acceleration independently of their mass. Einstein assumes the equivalence to extend to all laws of physics, thus generalizing the principle of relativity to accelerated frames.[11]

As we saw in Chapter 2, in his *Principia* Newton underlined the proportionality of gravitational and inertial mass, and he understood that in his theory of gravitation the relative motion of a system of bodies was unaffected when they were all subjected to a common acceleration. Concretely, the motion of the moon around the earth is unaffected by their common gravitational motion around the sun. In the nineteenth century, Bélanger defined the "general principle of relative motions" as the principle according to which the relative motion of a system of bodies under given forces is unaffected by a common rectilinear acceleration, and he used this principle to derive Newton's second law of motion. Violle did the same in the physics textbook that Einstein used to prepare for the ETH entrance examination. Although Bélanger and Violle did not explicitly refer to gravitation in their statements of a relativity principle, they understood that gravitation, being the cause of a common acceleration of bodies of any mass, gives us a constant and frequent opportunity to verify the truth of the principle. The proportionality of inertial and gravitational mass is a natural consequence of the perfect homogeneity of the primitive matter of which Newton's atoms were made. As Wien noted in 1900, it is also a natural consequence of the electromagnetic worldview if inertial and gravitational mass both derive from electric charge.[12]

---

[10] See Einstein [1933], p. 177.

[11] Einstein 1907c, p. 454; [1920], p. 265. Cf., e.g., Janssen 2012, p. 160.

[12] See pp. 17–18 (Newton), 37–39 (Bélanger), 39 (Violle), 170 (Wien), this volume.

We thus see that Einstein's attention to the proportionality of inertial and gravitational mass was far from being unprecedented. In particular, the connection with a relativity principle extended to accelerated frames appeared in a textbook Einstein had read. His awareness of these considerations may have eased his invention of the equivalence principle. That said, there is no doubt that Einstein's principle went much further than relativity à la Bélanger. The essential difference hinges on the nature of the compared frames. Whereas for Newton, Laplace, and Bélanger, the acceleration of a frame retained an objective meaning with respect to absolute space (or at least with respect to the privileged class of inertial frames), for Einstein, acceleration was just as relative as velocity was. There is indeed an evident similarity between the equivalence principle and the reasoning through which Einstein introduced the principle of relativity in 1905. In the received electromagnetic theory, Einstein then remarked, we express the laws of electromagnetism differently in the ether frame and in another inertial frame. Similarly, in the received theory of gravitation we regard the free motion of a body in a uniform gravity and the motion of a body in an accelerated frame as fundamentally different processes. In the first case, the asymmetry is removed by making the electric and magnetic fields frame-dependent concepts. In the second case, the asymmetry is removed by making the gravitational field a frame-dependent concept. This explains why Einstein regarded the principle of equivalence as an extension of the relativity principle to accelerated frames. This also explains why he applied it to any physical law, not only to the laws of gravitational motion.[13]

Einstein's next step in his *Jahrbuch* article is kinematical: he relates the measurement of space and time in an accelerated frame to their measurement in an inertial frame. For this purpose, he considers a frame Σ moving with the constant acceleration γ along the x axis of the inertial frame S. At every instant of this motion (with respect to S), there is an inertial frame S′ whose axes and velocity coincide with those of Σ. This frame, now called the *tangent frame*, plays an essential role in the rest of the reasoning.[14]

A priori, Einstein tells us, the lengths of identically built rods and the rate of identically built clocks might differ in the frames Σ and S′ owing to the acceleration of Σ. This effect can only be of second order in γ since opposite accelerations obviously produce the same effect. Einstein therefore decides to ignore it. Consequently, within a small enough time lapse, rod-based surveying and optical synchronization of clocks should yield the same space and time measurements in both frames.

Suppose that at $t = 0$ in S, the origin of Σ coincides with the origin of S and the velocity of Σ with respect to S vanishes. Einstein conceives two ways of defining time in Σ. Identically built clocks being placed at every point of this frame, the first option is to synchronize these clocks at $t = 0$; call σ the time given by these clocks. The second option is to constantly resynchronize the clocks of Σ with its central clock; call τ the time then given by these clocks. According to these definitions, equal σ times in Σ correspond to equal t times in S, and equal τ times in Σ correspond to equal t' times in the tangent frame S′. Setting the origin of the time t' so that it coincides with the time t of S at the common origin of S′ and Σ, we therefore have

[13] Einstein 1905a, p. 891; Einstein [1920], pp. 264–265: "The gravitational field, like the electric field produced by electromagnetic induction, has only a relative existence with respect to an observer."

[14] Cf. Pais 1982, pp. 180–182.

$$\tau - \sigma = t' - t. \tag{9.4}$$

Call v the velocity of S' with respect to S at time t', x' the abscissa in S', and ξ the abscissa in Σ. Neglecting second-order terms in v, we have

$$t' - t = -vx'/c^2 = -v\xi/c^2. \tag{9.5}$$

The σ time and the τ time therefore differ according to

$$\sigma = \tau(1 + \gamma\xi/c^2). \tag{9.6}$$

As Einstein explains, the σ time is the time we should naturally use when defining physical quantities at a given location in σ, because these quantities should be measured with standard devices brought to the location. In contrast, the τ time is the one with respect to which the invariance of physical laws is to be stated. For this reason, Einstein calls τ the "time" *tout court*, whereas he calls σ the "local time."[15]

According to the equivalence principle, the accelerated frame Σ should be equivalent to an inertial frame in which the constant gravity $-\gamma$ acts. The gravitational potential in this field being $\Phi = \gamma\xi/c^2$, we may rewrite Equation (9.6) as

$$\sigma = \tau(1 + \Phi/c^2). \tag{9.7}$$

Now suppose we bring two exemplars of the same clock to two different locations. According to the former argument, they yield the local time σ. Einstein imagines an observer who compares light signals from the two clocks at a third location. The τ time that a signal takes to travel from one clock to the observer is a constant. Accordingly, the period of the clock immersed in the smaller gravitational potential will appear to be larger to the observer. In other words, this clock runs slower. Extending this law to a nonuniform gravitational field and applying it to spectral sources, a spectral line from a source situated at the surface of the sun should appear to us redshifted by about 2×10^{-6} in relative value.[16]

Readers familiar with modern general relativity can easily see that Einstein's local time σ corresponds to the proper time given by the invariant metric element ds, whereas the τ time corresponds to the time coordinate with respect to which the metric coefficients are constants in a static gravitational field. Whereas we now regard the proper time as the only fundamental time, Einstein favors the τ time and consequently talks about the gravitational slowing down of clocks where we would now see the nonconservation of the interval ds between the light signals emitted at two successive ticks of a clock and traveling from this clock to a remote observer (the conserved quantity being d$t = (1 + \Phi/c^2)^{-1}$ds).

Imitating special relativity, Einstein proceeds to show that the form of the Maxwell–Lorentz equations is preserved in the accelerated frame Σ if the constant velocity of light c

---

[15] Einstein 1907c, pp. 454–457.

[16] Einstein 1907, pp. 457–459.

is replaced with $c_\tau = c(1 + \gamma\xi/c^2)$. Consequently, in a gravitational field of acceleration $-\gamma$, light travels with the location-dependent velocity

$$c_\tau = c(1 + \Phi/c^2). \tag{9.8}$$

By analogy with the optics of heterogeneous transparent media, this implies a deviation of the light rays at a rate proportional to γ/c^2. At that time, Einstein considered only the deviation by the gravitational field of the earth, which he judged too small to be detectable.[17]

Lastly, from the electromagnetic field equations in the accelerated frame, Einstein derives the integral equation

$$\int \rho_e \mathbf{u}_\tau \cdot \mathbf{E}(1 + \Phi/c^2)\mathrm{d}\xi\mathrm{d}\eta\mathrm{d}\zeta = \frac{\mathrm{d}}{\mathrm{d}\tau} \int \tfrac{1}{2}(E^2 + H^2)(1 + \Phi/c^2)\mathrm{d}\xi\mathrm{d}\eta\mathrm{d}\zeta, \tag{9.9}$$

where $\mathbf{E}$ denotes the electric field, $\mathbf{H}$ the magnetic field, ρ_e the electric charge density of matter, $\mathbf{u}_\tau$ the velocity of charged matter measured in τ time. This resembles the usual energy balance, except that the energy $\rho_e \mathbf{u}_\tau \cdot \mathbf{E}$ brought by the currents to the field per time unit and the field energy $(E^2 + H^2)/2$ per unit volume are both corrected by a gravitational term corresponding to the potential energy of the associated masses. In other words, an added energy implies not only an added inertial mass but also an added gravitational mass.

To sum up, in these pioneering considerations Einstein investigated the effects of the acceleration of a reference frame on space and time measurement, on electromagnetic processes, and on the energy balance. He thereby introduced two different time variables and used them to derive three promising consequences of general relativity: the gravitational redshift, the gravitational deviation of light rays, and the energy dependence of the gravitational mass. The reasoning was highly ingenious but shaky, for it involved a naive concept of accelerated frame. As we may retrospectively judge, one consequence of this reasoning, the gravitational deviation of light rays, is qualitatively correct; and the two others, the gravitational redshift and the dependence of the gravitational mass on energy content, are exact.

Direct applications of the principle of equivalence (1911)

Four year elapsed before Einstein published again on generalized relativity. He did so for two reasons: he was no longer satisfied with his considerations of 1907, and he now realized that the gravitational deviation of light rays could be observable for the light from stars seen near the surface of the sun (it would then be of the order of a second of arc). His new reasoning was more elementary and rested on direct applications of the equivalence principle, in the strong version of which an accelerated frame simulates uniform gravity in an inertial frame "with respect to all physical processes."[18]

A first consequence of the principle is the equality of the inertial and gravitational masses associated with a given amount of energy. This is immediately seen by noting that

---

[17] Einstein 1907c, pp. 461–462.

[18] Einstein 1911, p. 900. Cf. Pais 1982, pp. 198–200.

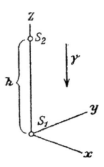

Fig. 9.1.   A source S_2 and an absorber S_1 in the gravitational field γ. From Einstein
        1911, p. 901.

the acceleration of a body carrying this energy should be the same in an inertial frame with
the constant gravity γ and in an accelerated frame with the acceleration $-\gamma$. In addition
to this simple reasoning, Einstein proposes the following thought experiment. The source
S_2 sends radiation of energy E_2 to the absorber S_1 in the direction of the acceleration of
gravity γ (see Figure 8.1) in the inertial frame K. According to the principle of equivalence,
the process should be the same in a gravity-free frame K' with the acceleration $-\gamma$. In the
inertial frame tangent to the K' frame at the time of emission, the radiation reaches S_1 when
S_1 has acquired the velocity $-\gamma h/c$ (in a first approximation). The energy of the radiation
with respect to S_1 therefore is

$$E_1 = E_2(1 + \gamma h/c^2) \qquad (9.10)$$

(using the transformation laws of special relativity). In the frame K, the product γh is the
difference Φ between the gravitational potentials at the locations of S_2 and S_1. Therefore,
the energy of the radiation emitted at S_1 exceeds the energy absorbed at S_2 by the potential
energy $(E_2/c^2)\Phi$ of the mass E_2/c^2. In other words, the energy is conserved in this process
provided that adding energy to a body contributes to its gravitational mass (from which
the potential energy is computed) just as much as it contributes to its inertial mass.[19]

By similar reasoning, monochromatic radiation sent in K' from S_2 with the frequency
v_2 arrives in S_1 with the Doppler-shifted frequency

$$v_1 = v_2(1 + \gamma h/c^2). \qquad (9.11)$$

In the frame K, this means that light from a source placed in a higher gravitational potential
appears to be shifted toward the violet, by the relative amount Φ/c^2 if Φ denotes the excess
of potential. In his reasoning, Einstein makes clear that the frequency of the same spectral
source as judged at the two locations by identically built clocks should be the same. It
is only when the clock used to measure the frequency is at a location different from the
location of the source that the frequency shift occurs.[20]

[19] Einstein 1911, pp. 900–902, 903 (simplest reasoning). On pp. 902–903, Einstein also gives a reasoning based
on a cycle similar to the one used in an early derivation of $E = mc^2$.

[20] Einstein 1911, pp. 903–905.

How could the frequency of the light fail to be conserved during its travel from the source to the observer? In agreement with his two-time reasoning of 1907, Einstein solves this paradox by arguing that v_1 and v_2 are not true frequencies since they are not referred to the "true time" with respect to which the numbers of oscillations of a wave extending between S_2 and S_1 are stationary. It is only with respect to this true time that the laws of physics do not explicitly depend on time. In order to measure this time, clocks must be built so that in a location of potential Φ they run $1 + \Phi/c^2$ times slower than a standard clock brought to this location.[21]

The velocity of light, when measured with a standard clock at any location should always have the same constant value c whatever the gravitational potential. From this consequence of the principle of equivalence, Einstein infers that the light velocity $\hat{c}$ measured with respect to the true time depends on the location $\mathbf{r}$ according to

$$\hat{c}(\mathbf{r}) = c[1 + \Phi(\mathbf{r})/c^2].\tag{9.12}$$

The constancy of the velocity of light, Einstein concludes, no longer holds in the new theory.[22]

Accordingly, the path of light should be deviated when it travels through an intense gravitational field. Einstein computes the amount of the deviation by Huygens's principle. For the deviation by a celestial body of mass M, he finds $2GM/c^2\Delta$, wherein G denotes the gravitational constant and Δ the closest distance of the light ray from the center of the body. This gives 4×10^{-6} in the case of stars appearing close to the surface of the sun. Einstein ends by calling astronomers to check for this effect.[23]

To sum up, in this brief memoir Einstein retrieves his three predictions of 1907: the gravitational redshift of stellar light, the gravitational deviation of light rays, and the relation between gravitational mass and energy content. Although the reasoning is more direct and more solid, it still involves two effects seemingly at variance with modern general relativity: the slowing down of clocks and the increase of the velocity of light in a gravitational potential. Both oddities derive from Einstein's defining time as what would now be the time coordinate in a static metric field, and the velocity of light as what is now called the coordinate velocity.

Einstein still sees the equivalence principle as a step toward a fuller relativity in which all frames of reference would be equally acceptable, with a uniform rule for transforming fields from one frame to another. Just as there is no absolute velocity in special relativity, there is no absolute acceleration in the generalized relativity. Einstein does not give the transformation formulas, although he hints he already has them in the case of constant acceleration.

[21] Einstein 1911, pp. 905–906.

[22] Einstein 1911, p. 906.

[23] Einstein 1911, pp. 906–908.

9.2 The static theory of 1912

Remember that in 1909–1910 Max Abraham amply contributed to the development of relativity theory in Minkowski form. In late 1911, he generalized Minkowski's four-dimensional space–time to accommodate Einstein's gravity-dependent velocity of light, without any recourse to the equivalence principle. This elegant theory, about which more in a moment, challenged Einstein to develop his own field theory of gravitation. In a memoir published in February 1912, he rejected Abraham's theory and proposed an equivalence-based theory in the case of a static gravitational field.[24]

The velocity of light and the static field

Einstein again considers a reference frame K in constant acceleration with respect to the inertial frame Σ (the notation differs from that of 1907). He thereby adopts Born's precise definition of constant acceleration: at every instant the acceleration is constant with respect to the tangent frame to K. The first problem is to find the relation between the coordinates (x, y, z, t) in the accelerated frame and the coordinates (ξ, η, ζ, τ) in the inertial frame. For this purpose, he assumes that the length gauges in K are not affected by its acceleration, and he also assumes the t-time to be defined so that the velocity of light near a given point of K does not depend on the direction of propagation. In passing, he remarks that the first assumption is far from obvious and that it is likely not to hold in the case of a frame rotating around an axis in an inertial frame, because with respect to the latter frame radial rods are not affected by the rotation, while orthoradial rods are subject to the Lorentz contraction. He also notes that the implied rigidity can only be of the kinematic kind.[25]

From these metric assumptions, Einstein first infers that $y = \eta$ and $z = \zeta$ for the coordinates in the directions perpendicular to the acceleration. Then he reasons that for two light-connected, infinitesimally close events one must have $dx^2 + dy^2 + dz^2 - \hat{c}^2 dt^2 = 0$ whenever $d\xi^2 + d\eta^2 + d\zeta^2 - d\tau^2 = 0$. The velocity of light is set to *one* in the inertial frame, and its value $\hat{c}$ in K may depend on the abscissa x. Without further ado, Einstein requires[26]

$$dx^2 - \hat{c}^2(x)dt^2 = d\xi^2 - d\tau^2. \tag{9.13}$$

Einstein then seeks a nonlinear, second-degree transformation for which this relation holds to a sufficient approximation. This transformation exists if and only if

$$\hat{c} = \hat{c}_0 + ax. \tag{9.14}$$

---

[24] Abraham 1911, 1912a; Einstein 1912a. On the latter memoir and its sequel, cf. Pais 1982, pp. 201–206.

[25] Einstein 1912a, pp. 355–357; Born 1909a, 1909b. Einstein had privately discussed Born rigidity and the rotating disk with Born; see p. 224, this volume. Cf. Stachel 1980, 1989a, 2007, who argues that in this context Einstein became aware of the possible lack of direct metric significance of spatial coordinates. In print, however, Einstein was much more concerned with the meaning of the time coordinate.

[26] The two forms must be proportional, and the proportionality coefficient reduces to *one* if the length unit is the same in both frames.

To second order, the transformation is given by

$$\xi = x + a\hat{c}t^2/2, \quad \tau = \hat{c}t \qquad (9.15)$$

if the origins of the two frames coincide and if their relative velocity vanishes at $t = 0$. For small times, the origin of K travels with the constant acceleration $a/\hat{c}_0$ with respect to Σ.[27]

Through the equivalence principle, motions in this accelerated frame are the same as motions in an inertial frame under the gravitational potential $\hat{c} = \hat{c}_0 + ax$. In an arbitrary static field, the function $\hat{c}(\mathbf{r})$ can take any value, and the former transformations must be restricted to domains so small that the linear approximation of $\hat{c}(\mathbf{r})$ can be used. For $\hat{c}$ to play the role of the gravitational potential, its Laplacian $\Delta\hat{c}$ should be proportional to the density ρ of matter. In addition, Einstein requires the field equation to be homogeneous in $\hat{c}$, because the unit for the time t is arbitrary according to the former reasoning (it depends on the location in K of the standard clock with respect to which the other clocks of K are synchronized). As the simplest equation meeting these conditions, he writes

$$\Delta\hat{c} = k\hat{c}\rho, \qquad (9.16)$$

wherein k is 4π times the usual gravitation constant G.[28]

Einstein's next task is to determine the motion of a particle in the static field $\hat{c}$. For this purpose, he considers the motion for a small time t in which the coordinate transformation of Equation (9.15) can locally be used to eliminate the gravitational field. With respect to the transformed coordinates, the motion is rectilinear and uniform. Switching to the (x, y, z, t) coordinates, it is easily seen to satisfy

$$\frac{d}{dt}\left(\frac{\mathbf{v}}{\hat{c}^2}\right) = -\frac{\nabla\hat{c}}{\hat{c}} \quad \text{with } \mathbf{v} = d\mathbf{r}/dt \text{ and } \mathbf{r} = (x, y, z). \qquad (9.17)$$

Although this equation is here established only for small times and for a specific choice of the x axis, Einstein takes it to be valid at any time and for any choice of the function $\hat{c}(\mathbf{r})$, because "the instant $t = 0$ has nothing special compared to other instants."[29]

In modern terms, Einstein's procedure amounts to introducing local geodesic coordinates at any given time t (coordinates for which the metric takes the Minkowski form and the derivatives of the metric coefficients vanish) for the metric $dx^2 - \hat{c}^2 dt^2$ and assuming rectilinear uniform motion with respect to these coordinates in a sufficient approximation.

---

[27] Einstein 1912a, pp. 357–359. I use $\hat{c}$ instead of Einstein's c in order to avoid confusion with the usual constant of nature c. Had Einstein been familiar with the theory of quadratic differential forms or with the theory of Gaussian surfaces, he could have simply required the vanishing of the Gaussian curvature of the form $dx^2 - \hat{c}^2 dt^2$, which is $-\hat{c}''/\hat{c}$. For the exact expression of the desired transformation, he would have found $\xi = a^{-1}(\hat{c}\cosh at - \hat{c}_0)$, $\tau = \hat{c}a^{-1}\sinh at$, in conformity with the Born-rigidity of the accelerated frame (compare with Equation (7.51), p. 223, this volume).

[28] Einstein 1912a, p. 360.

[29] Einstein 1912a, pp. 361–362.

Equivalently, the trajectory in space–time should be a geodesic of the metric manifold. Being unaware of this geometric interpretation, Einstein does not yet realize that his equations of motions derive from the variation

$$\delta \int \sqrt{\hat{c}^2 dt^2 - d\mathbf{r}^2} = \delta \int dt \sqrt{\hat{c}^2 - v^2} = 0. \tag{9.18}$$

Einstein will obtain this result a few months later, presumably by generalizing Planck's Lagrangian formulation of the equation of motion in special relativity. The resulting equation of motion is

$$\frac{d}{dt} \frac{\partial L}{\partial \mathbf{v}} = \frac{\partial L}{\partial \mathbf{r}}, \text{ with } L = \sqrt{\hat{c}^2 - v^2}. \tag{9.19}$$

This gives

$$\frac{d}{dt} \frac{\mathbf{v}}{\hat{c}\sqrt{1 - v^2/\hat{c}^2}} = -\frac{\nabla \hat{c}}{\sqrt{1 - v^2/\hat{c}^2}}, \tag{9.20}$$

which is equivalent to Einstein's Equation (9.17) because both equations imply the constancy of $\hat{c}(1 - v^2/\hat{c}^2)^{-1/2}$ for any given motion.[30]

Einstein exploits this constancy to define the energy of a moving mass m as

$$E = \frac{m\hat{c}}{\sqrt{1 - v^2/\hat{c}^2}} \approx m\hat{c} + mv^2/2\hat{c}. \tag{9.21}$$

(This does not imply any dimensional heterogeneity because time and length have the same dimension in his units). In general, he expects forces and energies to be proportional to $\hat{c}$ and thus to depend on the gravitational potential at the location of the system. He illustrates this point with a compressed spring. A particle of mass m projected by this spring acquires a velocity independent of the location provided this velocity is measured with respect to the local time $l = \hat{c}t$. This is a consequence of the equivalence principle, which Einstein here takes to imply that *the relations between quantities measured at a given location with identically built apparatus should be the same whatever the value of the gravitational potential at this location*. The associated kinetic energy,

$$mv^2/2\hat{c} = m(dx/dt)^2/2\hat{c} = m\hat{c}(dx/dl)^2/2, \tag{9.22}$$

is therefore proportional to $\hat{c}$, and so is the force of the compressed spring since the work of this force is responsible for the velocity of the projectile.[31]

---

[30] Einstein 1912b, p. 458; Planck 1906a. On Planck's formulation, see p. 214, this volume.

[31] Einstein 1912a, pp. 365–367.

Just as in his earlier publications, Einstein regards the time t as more fundamental than the local time. He now calls it the "universal time." Since the rate of physical processes at a given location is independent of location when referred to the local time, all processes run faster in a higher gravitational potential. Again, the premise is a consequence of the principle of equivalence: *gravitation does not affect the measuring contraptions when acceleration does not*. Einstein confirms the resulting clock effect in two simple cases. The first case is the light-clock obtained by counting the successive bounces of light between two parallel mirrors separated by the unit distance. The number of bounces per unit t-time is evidently proportional to $\hat{c}$ and therefore increases with the gravitational potential. The second is the gravitation-clock obtained by counting the revolutions of a mass m around a mass M at a standard distance. Using Equations (9.16) and (9.17), the acceleration d^2r/dt^2 of the mass m should be proportional to $\hat{c}^2$. Kinematically, this acceleration is centripetal and proportional to the square of the angular frequency. This frequency is therefore proportional to $\hat{c}$, as was to be proved.[32]

Electromagnetism and gravitation

In a sequel written in the following month, Einstein investigated the electromagnetic field equations in a static gravitational field. He thereby inaugurated the strategy of covariant field equations, as well as a risky metrology of the transformed coordinates and fields. According to the principle of equivalence, the gravitational field can be locally simulated by acceleration with respect to an inertial frame. In the inertial frame, the Maxwell–Lorentz equations read

$$\nabla' \times \mathbf{H}' = \rho'_e \mathbf{u}' + \frac{\partial \mathbf{E}'}{\partial t'}, \quad \nabla' \times \mathbf{E}' = -\frac{\partial \mathbf{H}'}{\partial t'}, \quad \nabla' \cdot \mathbf{H}' = 0, \quad \nabla' \cdot \mathbf{E}' = \rho'_e. \tag{9.23}$$

The coordinates and fields in the accelerated frame are related to those in the tangent frame by

$$x' = x + a\hat{c}t^2/2, \quad y' = y, \quad z' = z, \quad t' = \hat{c}t \text{ with } \hat{c} = \hat{c}_0 + ax, \tag{9.24}$$

$$\mathbf{H} = \mathbf{H}' - \gamma t' \times \mathbf{E}', \quad \mathbf{E} = \mathbf{E}' + \gamma t' \times \mathbf{H}' \text{ with } \gamma = \hat{c}^{-1}\nabla\hat{c}. \tag{9.25}$$

$$\rho_e = \rho'_e, \quad \mathbf{u} = \hat{c}\mathbf{u}'. \tag{9.26}$$

The field equations in the local accelerated frame at $t = t' = 0$ therefore read

$$\nabla \times \hat{c}\mathbf{H} = \rho_e \mathbf{u} + \frac{\partial \mathbf{E}}{\partial t}, \quad \nabla \times \hat{c}\mathbf{E} = -\frac{\partial \mathbf{H}}{\partial t}, \quad \nabla \cdot \mathbf{H} = 0, \quad \nabla \cdot \mathbf{E} = \rho_e. \tag{9.27}$$

According to the equivalence principle, they are also the field equations in the static gravitational field $\hat{c}(\mathbf{r})$.[33]

[32] Einstein 1912a, pp. 325–327.

[33] Einstein 1912b, pp. 443–446.

In order to find the physical meaning of these equations, Einstein imagines measure-ments done with "pocket" devices, a name suggested by Ehrenfest. For a given kind of measurement, these devices are identically built and they can be transported to the location of the measurement. The aforementioned light-clock is a pocket device for time measure-ment. A spring-scale is a pocket device for the measurement of forces. According to the earlier given argument, the force thus measured is $\hat{c}$ times the true force. Consequently, the electric field measured by the force acting on a unit point charge is $\hat{c}$ times the true field $\mathbf{E}$.[34]

By reasoning à la Poynting, the field equations (9.27) imply the identity

$$\int \rho_e \mathbf{u} \cdot \hat{c}\mathbf{E}\mathrm{d}^3x = -\frac{\mathrm{d}}{\mathrm{d}t}\int \tfrac{1}{2}\hat{c}(E^2 + H^2)\mathrm{d}^3x + \int (\hat{c}\mathbf{E} \times \hat{c}\mathbf{H}) \cdot \mathrm{d}\mathbf{S}, \qquad (9.28)$$

wherein the surface integral is taken over the boundary of the volume integral. The follow-ing interpretation suggests itself: the energy furnished to the field by the currents in a given domain equals the variation of the electromagnetic energy in this domain plus the electro-magnetic energy flux across the boundary of this domain. The quantity $(\hat{c}/2)(E^2 + H^2)$ here plays the role of the electromagnetic energy density. Einstein also shows that the force den-sity $-(1/2)(E^2 + H^2)\nabla\hat{c}$ contributes to the field momentum balance. Both results mean that the quantity $(1/2)(E^2 + H^2)$ plays the role of a gravitational mass, in conformity with the principle of equivalence. As such, this quantity should be added to the density of matter ρ in Equation (9.16) for the gravitational field. At this point Einstein briefly mentions that the variable $\hat{c}$ in the electromagnetic field (9.27) implies the gravitational deflection of light, although he does not repeat the derivation.[35]

Action and reaction

Lastly, Einstein detects and solves a contradiction in the present theory. The equation of motion of a particle in the gravitational field, as given in Equation (9.17), suggests the expression $-\nabla\hat{c}$ for the force acting on a unit mass at rest. Accordingly, the sum of all forces acting on the static mass distribution ρ is $\int -\rho\nabla\hat{c}\mathrm{d}^3x$. According to the principle of equality of action and reaction, this force should vanish (as long as the masses are contained within a finite domain of space). Together with the gravitational field equation (9.16), this condition gives

$$\int \frac{\Delta\hat{c}}{\hat{c}}\nabla\hat{c}\mathrm{d}^3x \equiv \mathbf{0}. \qquad (9.29)$$

Unfortunately, no such identity holds for an arbitrary function $\hat{c}(\mathbf{r})$. Something must be changed in the theory. After considering several possibilities, Einstein decides that the

---

[34] Einstein 1912b, pp. 446–448.

[35] Einstein 1912a, pp. 448–450. Einstein notes that in all rigor including the electromagnetic energy in the source of the gravitational field contradicts its static character. I skip the thermodynamic section of pp. 450–452.

only reasonable option is to modify the field equation (9.16). For this purpose, he uses the identity

$$\frac{2\Delta\sqrt{\hat{c}}}{\sqrt{\hat{c}}} = \frac{\Delta\hat{c}}{\hat{c}} - \frac{1}{2}\left(\frac{\nabla\hat{c}}{\hat{c}}\right)^2 \tag{9.30}$$

thanks to which the modified field equation

$$\frac{\Delta\hat{c}}{\hat{c}} - \frac{1}{2}\left(\frac{\nabla\hat{c}}{\hat{c}}\right)^2 = k\rho \tag{9.31}$$

is compatible with the equality of action and reaction. Indeed, for $u = \sqrt{\hat{c}}$ we have

$$\Delta u = ku\rho/2 \text{ and } \mathbf{f} = -\rho\nabla\hat{c} = -2\rho u\nabla u = -4k^{-1}\Delta u\nabla u. \tag{9.32}$$

By analogy with electrostatics, Einstein writes

$$f_i = -4k^{-1}\partial_i u\partial_k\partial_k u = \partial_j\tau_{ij}, \text{ with } \tau_{ij} = -4k^{-1}(\partial_i u\partial_j u - \tfrac{1}{2}\delta_{ij}\partial_k u\partial_k u). \tag{9.33}$$

Namely, the gravitational force density derives from a stress tensor, and its spatial integral therefore vanishes. The principle of action and reaction is thus satisfied globally. The field equation (9.30) is still homogeneous in $\hat{c}$, and the new term, being of second order in $\nabla\hat{c}$, does not sensibly affect the earlier physical considerations.[36]
    Rewriting the field equation (9.31) as

$$\Delta\hat{c} = k\left[\hat{c}\rho + \frac{1}{2k}\frac{(\nabla\hat{c})^2}{\hat{c}}\right],$$

Einstein recognizes in $\hat{c}\rho$ the gravitational energy density for the mass distribution ρ and suspects that the second term in the square bracket represents the energy of the gravitational field. He confirms this guess by proving that the work of the gravitational forces $-\rho\nabla\hat{c}$ during an infinitesimal displacement of the masses is equal to the variation of the total energy of the masses plus the variation of the total energy of the field. This reasoning increased Einstein's confidence in the new field equation.[37]

[36]Einstein 1912b, pp. 452–457. Cf. Norton 2020, pp. 15–18. On pp. 455–456, Einstein notes he had difficulty accepting the new term because it contradicts the strict equivalence principle (it does not vanish in a uniform field for which $\hat{c}$ is an affine function of the space coordinates); cf. Norton 2020, pp. 19–20. The difficulty is avoided by limiting the equivalence principle to small regions of space for which the affine approximation of $\hat{c}$ is sufficient: cf. Einstein to Besso, March 26, 1912, *ECP* 5, p. 436.
[37]Einstein 1912b, pp. 457–458. Cf. Darrigol 2019b, p. 133.

From the equivalence principle to the metric field

The twin memoirs of 1912 mark a crucial transition from piecemeal applications of the equivalence principle to a metric approach integrating this principle. The first memoir introduces the form $d\mathbf{r}^2 - \hat{c}^2(\mathbf{r})dt^2$ in order to express the isotropy of light propagation in the static field locally interpreted as an inertial acceleration field. Einstein's derivation of the coordinates $(\mathbf{r}', t')$ in a tangent inertial frame, based on equating the previous form to $d\mathbf{r}'^2 - dt'^2$, imitates a then common derivation of the Lorentz transformation through the invariance of the Minkowski interval. Despite his awareness of Minkowski's four-dimensional world, Einstein does not use geometric language or methods in the two memoirs.[38] Also, his derivations do not appeal to the methods of differential geometry.[39]

Yet, there are two reasons to think that by March 1912 Einstein saw the prospects for a metric interpretation of the differential form $\hat{c}^2(\mathbf{r})dt^2 - d\mathbf{r}^2$. The first reason is found in a remark added in the proofs of the second memoir: the motion of a particle in the static field corresponds to an extremum of $\int \sqrt{\hat{c}^2(\mathbf{r})dt^2 - d\mathbf{r}^2}$. Einstein could not possibly have missed the analogy with the determination of geodesics on a surface, and he probably recognized the problem of geodesics on a pseudo-Riemannian four-dimensional manifold with the metric

$$ds = \sqrt{\hat{c}^2(\mathbf{r})dt^2 - d\mathbf{r}^2}. \tag{9.34}$$

Three months earlier, in December 1911, Abraham had tried to integrate Einstein's variable velocity of light into a Minkowski framework. "According to Minkowski's representation," Abraham writes, "we regard x, y, z and $u = i\hat{c}t$ as coordinates in a space of four dimensions." The only difference with Minkowski's theory is that the velocity $\hat{c}$ now depends on the gravitational potential Φ. In modern index notation (with $\alpha = 1, 2, 3, 4$), Abraham's gravitational field equation reads

$$\partial_\alpha \partial_\alpha \Phi = 4\pi G \rho_0, \tag{9.35}$$

wherein G denotes the gravitational constant and ρ_0 the density in the frame tangent to the moving matter. The equation of motion of a particle in this potential reads

$$\ddot{x}_\alpha = -\partial_\alpha \Phi, \tag{9.36}$$

the dot denoting derivation with respect to Minkowski's proper time τ for which

$$\hat{c}d\tau = \sqrt{-dx_\alpha dx_\alpha} \text{ and } \dot{x}_\alpha \dot{x}_\alpha = -\hat{c}^2. \tag{9.37}$$

---

[38] In 1908, Einstein and Laub had given a three-vector reformulation of Minkowski's electrodynamics of moving bodies; see pp. 220, this volume. In texts written after Minkowski's death in 1909, Einstein mentioned the four-dimensional approach favorably; cf. *ECP* 3, pp. 169–170 (1910), 438 (1911): "hoch interessante mathematische Fortbildung."

[39] Yet, at the *ETH* in Zürich Einstein had heard Carl Friedrich Geiser's lectures on infinitesimal geometry, which included Gauss's theory of surfaces: cf. Stachel 2007, pp. 103–104.

Combining $\dot{x}_\alpha \ddot{x}_\alpha = -\hat{c}\dot{\hat{c}}$ with the equation of motion, Abraham gets

$$-\dot{x}_\alpha \, \partial_\alpha \, \Phi = -\hat{c}\dot{\hat{c}} \;\Rightarrow\; \frac{d\Phi}{d\tau} = \frac{1}{2}\frac{d\hat{c}^2}{d\tau} \;\Rightarrow\; \Phi = \frac{\hat{c}^2}{2} - \frac{\hat{c}_0^2}{2}. \qquad (9.38)$$

In a first approximation, this gives $\hat{c} \approx \hat{c}_0(1 + \Phi/\hat{c}_0^2)$, in conformity with the result Einstein has earlier obtained through the equivalence principle. In addition, Abraham finds $m\hat{c}/\sqrt{1 - v^2/\hat{c}^2}$ to be a first integral of the equation of motion, just as in Einstein's subsequent theory.[40]

Einstein was initially impressed by the mathematical elegance of Abraham's theory, although he soon criticized the lack of a proper physical foundation:

> Abraham's theory was created on an empty stomach, namely, through mere consid- erations of mathematical beauty, and it is fully untenable. I cannot at all figure how this intelligent man could let himself be lured into such trifle. At first glance (for 14 days!) I admit I was quite "bluffed" by the beauty and simplicity of his formulas.

As Einstein told Abraham, the characterization of $u = i\hat{c}t$ as the fourth coordinate is not compatible with $\hat{c}$ being the local value of the velocity of light, because it is incompatible with $d\tau = 0$ for $d\mathbf{r}^2 = \hat{c}^2 dt^2$. Abraham published the following corrective:

> Instead of ["we regard x, y, z and $u = i\hat{c}t$ as coordinates in a space of four dimen- sions"], one should read "we regard dx, dy, dz and $du = i\hat{c}dt$ as the components of a displacement ds in four-dimensional space."
> Thus, $ds^2 = dx^2 + dy^2 + dz^2 - \hat{c}^2 dt^2$ is the square of a four-dimensional line-element, wherein the velocity of light $\hat{c}$ is given by $[\hat{c}^2 - \hat{c}_0^2 = 2\Phi]$.

From this statement, Einstein inferred that Abraham believed the Lorentz transformations to apply infinitesimally with the local velocity $\hat{c}$. In the first installment of his static the- ory, he proved this to be mathematically impossible. He recommended the application of the Lorentz group to be restricted to regions of space in which $\hat{c}$ had a uniform value, and he predicted that a larger group and more complicated equations would be needed in the general case. In a subsequent publication, Abraham made clear that he meant to apply the Lorentz group (more precisely, rotations in Minkowski space) to the four-vector $(dx, dy, dz, i\hat{c}dt)$ (whereas Einstein believed Abraham was applying a Lorentz boost of ve- locity $\hat{c}$ to dx, dy, dz, dt). Einstein's chief objection, the incompatibility of Abraham's field equation with the equivalence principle, remained valid. Yet Einstein learned something important in his exchange with Abraham: a Minkowskian interpretation of the variable velocity of light leads to a Riemannian metric.[41]

---

[40] Abraham 1911, p. 678; 1912a, p. 1.

[41] Einstein to Besso, March 26, 1912, *ECP* 5, pp. 436–437; Abraham 1912b; 1912c, p. 433; 1912d, p. 312; Einstein 1912a, pp. 368–369. For a thorough study of Abraham's theory and his polemic with Einstein, cf. Renn 2007a. See also Cattani and De Maria 1989a; Weinstein 2015b, section 2.2.

Altogether, in early 1912 Einstein already had in hand the representation of a gravitational field by a quadratic differential form as well as the now usual method for deriving the equations of motion of a particle from this form. He was aware of a possible geometric interpretation of this form and method. Still, his static theory of 1912 differed from the modern theory in three ways. The first difference is that the quadratic form is of the restricted kind $d\mathbf{r}^2 - \partial^2(\mathbf{r})dt^2$, which Einstein believed to represent exactly a static gravitational field (whereas we now know that the spatial part of a nontrivial static metric is non-Euclidean). The second difference is that Einstein considered only two kinds of coordinates: the coordinates $(\mathbf{r}, t)$ through which the laws of physics in the static gravitational field should be expressed and which Einstein believed to be giving the true space and time relations in some sense (even though the time t differs from the time ∂t given by a standard clock), and the coordinates $(\mathbf{r}', t')$ in a tangent frame in which the gravitational force is eliminated. The third difference concerns the gravitational field equation. With the assumed restriction of the fundamental form, this equation concerns the function $\partial(\mathbf{r})$ only. In order to derive this equation, Einstein relied on three principles: correspondence with Poisson's equation for Newtonian gravitation, homogeneity with respect to ∂ (dictated by kinematics), and equality of action and reaction (impossibility of perpetual motion). He verified its compatibility with two additional principles: energy conservation and the equivalence between energy and gravitational mass. He was convinced that the only field equation compatible with these five principles was Equation (9.31). The fact that so many principles could be satisfied at the same time comforted him in his choice of the restricted form $d\mathbf{r}^2 - \partial^2(\mathbf{r})dt^2$, although he originally doubted that space measurements would remain Euclidean in a gravitational field.

It is usually believed that Einstein's first realization that non-Euclidean geometry may be needed in general relativity came in his remark, at the beginning of the first memoir of 1912, that the tangent frame method of measurement leads to a non-Euclidean perimeter/radius ratio in a rotating frame. In reality, the first way in which non-Euclidean (four-dimensional) geometry truly entered his theory was in his choice of $d\mathbf{r}^2 - \partial^2(\mathbf{r})dt^2$ for what we would now call the metric in a static field. By a somewhat daring generalization from the case of a constantly accelerated frame (for which $\partial(\mathbf{r})$ is an affine function) to an arbitrary static field for which the acceleration varies from place to place, Einstein implicitly introduced a curved four-manifold for which no change of coordinates can globally bring the metric to Minkowski form.

Einstein had great confidence in his static theory of 1912: "I really believe to have found a piece of truth," he then wrote to Ehrenfest. Paradoxically, this theory both initiated and hindered Einstein's quest for a theory of gravitation based on a more general metric. On the positive side, the static theory introduced a few basic formal elements of such a theory as well as a few heuristic principles including correspondence with Newton's theory of gravitation, energy–momentum conservation, and existence of a stress–energy tensor for the gravitational field. On the negative side, it blocked further progress by suggesting that the more general field equation should agree with the equation of 1912 in the static case, and, above all, by conflating coordinate systems with physically predefined reference systems.[42]

---

[42]Einstein to Ehrenfest, March 10, 1912, *ECP* 5.

9.3 The Zürich notebook

Soon after completing his theory of the static gravitational field, Einstein considered its extension to arbitrary fields. This is already apparent in his remark, in the appendix to the second memoir of 1912, that the variational principle (9.18)

$$\delta \int \sqrt{c^2(\mathbf{r})dt^2 - d\mathbf{r}^2} = 0$$

is easily extended to dynamical fields.[43] Although Einstein does not tell us more on this extension, he plausibly reasoned that for a dynamical field more general systems of coordinates and the more general form $g_{\mu\nu}dx^\mu dx^\nu$ would be needed. The resulting principle

$$\delta \int \sqrt{g_{\mu\nu}dx^\mu dx^\nu} = 0 \qquad\qquad (9.39)$$

and the resulting equations are evidently covariant. This induced Einstein to look for a covariant extension of his static field equation. It will soon be clear that he initially did not require covariance with respect to any (mathematically acceptable) change of coordinates. He believed the choice of coordinates had to be restricted in order to preserve some of their physical, metrological meaning. Among the permitted transformations, he included those pertaining to accelerated or rotating frames, because these were needed to implement the equivalence principle. He had already used uniformly accelerated frames in his static theory, and he hoped uniformly rotating frames to be the next simple way to simulate a gravitational field, perhaps with Newton's bucket in mind.[44] In order to retrieve special relativity in a local free-falling frame, he also needed transformations locally turning the form $g_{\mu\nu}dx^\mu dx^\nu$ into the Minkowski form of special relativity.

First tries: Linear covariance

That Einstein reasoned in this or a similar way is confirmed by his first notes on gravitation in a notebook he filled in the years 1912–1913, usually called the Zürich notebook because most of it was written after his move to Zürich in July 1912. There are no dates in this notebook, and it is not easy to time the various steps taken by Einstein. All we know from his correspondence is that in the fall of 1912 he believed he had "the most general equations of gravitation" in hand; that by October at least he had received help from his friend the mathematician Marcel Grossmann and that he had thereby developed a "high regard for mathematics . . ., which [he] heretofore naively considered a luxury in its subtlest parts." At some point, he must have encountered unsuspected difficulties: it was not until March 1913 that he could write to his cousin and future wife that he had "solved the gravitational problem only a few weeks ago after half a year of most strenuous investigation."[45]

[43] Einstein 1912b, p. 458.

[44] See Einstein to Besso, March 26, 1912, *ECP* 5: I am "still far from being able to conceive rotation as rest." From Herglotz 1910, Einstein knew Born-rigid motion to be possible for a *uniformly* rotating frame.

[45] Einstein to Hopf, August 16, 1912 (cited); Einstein to Freundlich, October 27, 1912; Einstein to Sommerfeld, October 29, 1912 (cited); Einstein to Löwenthal, March 23, 1913 (cited). These letters are in *ECP* 5.

In his notebook [p. 39L], Einstein starts from the expression

$$ds^2 = g_{\mu\nu}dx^\mu dx^\nu \tag{9.40}$$

and writes the equations for the transformation of $g_{\mu\nu}$ and $\partial_\mu \equiv \partial/\partial x^\mu$ under a linear change of coordinates. Although the ds notation indicates analogy with the length element on a (hyper)surface, Einstein proceeds unaware of the general theory of Riemannian manifolds and the attached tensor calculus. This may explain why he confines himself to linear changes of coordinates, which are much easier to handle than general changes.[46]

Einstein's strategy [pp. 39L–40L] is to find linearly covariant generalizations of the equations

$$\frac{\partial \hat{c}}{\partial t} = 0, \quad \frac{\Delta \hat{c}}{\hat{c}} - \frac{1}{2}\left(\frac{\nabla \hat{c}}{\hat{c}}\right)^2 = 0, \quad \text{with } g_{00} = \hat{c}^2 \tag{9.41}$$

for the static field. For the first equation he tries the divergence condition $\partial_\mu g_{\mu\nu} = 0$ and fails since this is not covariant by linear transformations.[47] He then [pp. 40L–40R] tries to construct generalizations of the second equation (the field equation proper) by taking the most symmetric possible combinations of $g_{\mu\nu}$ and ∂_μ that are of second order with respect to ∂_μ, but he soon gives up. Something will be left from this first skimpy attempt: for a while Einstein will keep requiring the equations (9.41) for the static case to be special cases of the general equations. In particular, in addition to the second-order field equation for $g_{\mu\nu}$ he will often require a first-order equation.

The stress–energy tensor of a dust

Einstein was well aware of the energy–momentum tensor that Minkowski and Laue had introduced for fields in Minkowski space.[48] At some point of his quest for a gravitational field equation, he introduced energy–momentum considerations to guide his conjectures [pp. 5R, 43L$_B$].

Rewriting the geodetic equation (9.39) as

$$\delta \int L dt = 0 \quad \text{with } L = ds/dt, \tag{9.42}$$

---

[46]The entire contents of the Zürich notebook have been published both in *ECP* 4 with rich annotation by John Norton, and in *GGR* 1 (including a facsimile). The most detailed and accurate interpretation is given in *GGR* 2, pp. 489–714, under the joint efforts of Janssen, Renn, Sauer, Norton, and Stachel. The page numbering is the one used in these sources (Einstein filled the book from both ends, so these numbers are not always a growing function of time). I use slightly modernized notation (with the now usual conventions for indices and summation).

[47]A linearly covariant condition would be $\partial_\mu g^{\mu\nu} = 0$, later used by Einstein. My analysis here differs from *GGR* 2, p. 507.

[48]See pp. 217, 225, this volume.

and interpreting $\partial L/\partial \dot{x}_i$ (with $i = 1, 2, 3$) as the spatial components of the momentum of the particle, the equations of motion read

$$\frac{dp_\mu}{ds} = \frac{1}{2}\partial_\mu g_{\nu\rho}\frac{dx^\nu}{ds}\frac{dx^\rho}{ds} \quad \text{with} \quad p_\mu = g_{\mu\nu}\frac{dx^\nu}{ds}. \tag{9.43}$$

For a quasi-continuous dust of spatial density ρ in the small volume V, Einstein introduces the true density ρ_0 and the volume V_0 in the tangent frame. Combining the invariance of $\sqrt{-g}d^4x$ (with $g = \det g_{\mu\nu}$) and the invariance of mass, he gets

$$\sqrt{-g}V dt = V_0 ds \quad \text{and} \quad \rho V = \rho_0 V_0 \implies \rho = \rho_0\sqrt{-g}dt/ds. \tag{9.44}$$

The four-force acting on the unit volume of the dust is

$$\rho\frac{dp_\mu}{dt} = \rho_0\sqrt{-g}\frac{dp_\mu}{ds} = \frac{1}{2}\sqrt{-g}\partial_\mu g_{\nu\rho}\theta^{\nu\rho} \quad \text{with} \quad \theta^{\nu\rho} = \rho_0\frac{dx^\nu}{ds}\frac{dx^\rho}{ds}. \tag{9.45}$$

The four-momentum flux of the dust is

$$\rho p_\mu\frac{dx^\nu}{dt} = \rho_0\sqrt{-g}g_{\mu\rho}\frac{dx^\rho}{ds}\frac{dx^\nu}{ds} = \sqrt{-g}g_{\mu\rho}\theta^{\nu\rho}. \tag{9.46}$$

According to energy–momentum conservation, the four-divergence of this quantity should be equal to the four-force:

$$\partial_\nu\left(\sqrt{-g}g_{\mu\rho}\theta^{\nu\rho}\right) = \frac{1}{2}\sqrt{-g}\partial_\mu g_{\nu\rho}\theta^{\nu\rho}. \tag{9.47}$$

As Einstein correctly surmises, this equation is equivalent to the equation of motion (9.43) because according to mass conservation we have [pp. 43LB, 19R]

$$\partial_\mu(\rho dx^\mu/dt) = \partial_\mu(\rho_0\sqrt{-g}dx^\mu/ds) = 0. \tag{9.48}$$

Einstein calls $\theta^{\nu\rho}$ the motion tensor, and $g_{\mu\rho}\theta^{\nu\rho}$ the energy–momentum tensor (although he later favored "stress–energy tensor"). He asserts [p. 5R][49] that

$$V_\mu = \partial_\nu(\sqrt{-g}g_{\mu\rho}S^{\nu\rho}) - \frac{1}{2}\sqrt{-g}\partial_\mu g_{\nu\rho}S^{\nu\rho} \tag{9.49}$$

is a covariant vector (*zugeordneter Vektor*) for any symmetric tensor $S^{\mu\nu}$. This is almost true since, as he does not yet know, the quotient $V_\mu/\sqrt{-g}$ is the covariant divergence $D_\nu S^{\mu\nu}$. At any rate, the four-divergence equation (9.47) and its physical derivation will condition Einstein's ulterior considerations in several manners. First, they reinforce his idea that co-variance should play a central role in constructing the gravitational field equation. Second, the interpretation of $\frac{1}{2}\sqrt{-g}\partial_\mu g_{\nu\rho}\theta^{\nu\rho}$ as the four-force acting on a dust of stress–energy $\theta^{\mu\nu}$

[49] Einstein also notes that $V_\mu = 0$ for $S^{\mu\nu} = g^{\mu\nu}$.

gives to the derivatives $\partial_\mu g_{\nu\rho}$ prominence in expressing the strength of the gravitational field (thus generalizing the relation between force, mass, and gravitational potential). Third, these considerations suggest that the gravitational field equation might have the form $G^{\mu\nu} = \kappa T^{\mu\nu}$, in which $G^{\mu\nu}$ is a tensor combination of the derivatives of the metric tensor and $T^{\mu\nu}$ is the stress–energy tensor of matter (including the electromagnetic field). Fourth, they will induce Einstein to demand that the product $\sqrt{-g}\partial_\mu g_{\nu\rho}G^{\nu\rho}$ be expressible as the four-divergence of a (pseudo-)tensor[50] representing the stress–energy of the gravitational field (up to a numerical factor). This last condition is a generalization of Einstein's earlier demand that the sum of all gravitational forces should vanish in the static case. It will henceforth be called *the stress principle*.

Scalar and tensor generalizations of the d'Alembertian

Einstein does not immediately explore the tensor option $G^{\mu\nu} = \kappa T^{\mu\nu}$, presumably because it is much easier to consider the scalar option in which the field equation in the absence of matter is obtained by setting to zero a scalar combination of the metric tensor and its first and second derivatives [pp. 6R–11L].[51] He tries various would-be scalar combinations of g, $g_{\mu\nu}$, $g^{\mu\nu}$, and ∂_μ, sometimes with additional conditions such as $\partial_\mu(\sqrt{-g}g^{\mu\nu}) = 0$ (harmonic coordinates), $\partial_\mu g^{\mu\nu} = 0$, or $g = -1$, thanks to which the candidates become invariant for the transformations compatible with the conditions. Ideally, these conditions should be compatible with the static metric and they should be satisfied by the metric of a flat space–time in a uniformly accelerated or rotating frame—henceforth called "acceleration metric" and "rotation metric," respectively—in order to allow for the frames implied in the equivalence principle. As Einstein will realize sooner or later, they are not.[52]

At some point, Einstein abandons the scalar option and turns to the tensor option in which the field equation takes the form $G^{\mu\nu} = \kappa T^{\mu\nu}$.[53] He hits upon [p. 12L]

$$G^{\mu\nu} = \frac{1}{\sqrt{-g}}\partial_\rho(\sqrt{-g}g^{\rho\sigma}\partial_\sigma g^{\mu\nu}) \qquad (9.50)$$

by analogy with the scalar invariant Beltrami had earlier introduced. This reduces to

$$G^{\mu\nu} = g^{\rho\sigma}\partial_\rho\partial_\sigma g^{\mu\nu} \equiv \Box g^{\mu\nu} \qquad (9.51)$$

[50] I will henceforth drop the "pseudo" in conformity with Einstein's usage (as we will see in Sections 9.6 and 9.7, Einstein was unaware of this distinction until late 1913).

[51] For a more detailed analysis, cf. *GGR 2*, pp. 526–531.

[52] Using Equations (9.16–17), the acceleration metric has the form $ds^2 = (c_0 + \mathbf{a} \cdot \mathbf{r})^2 dt^2 - d\mathbf{r}^2$. Injecting $\xi = x\cos\omega t + y\sin\omega t, \eta = -x\sin\omega t + y\cos\omega t, \zeta = z, \tau = t$ into the Minkowski metric $ds^2 = c^2 d\tau^2 - d\xi^2 - d\eta^2 - d\zeta^2$, we get $ds^2 = (c^2 - \omega^2\mathbf{r}^2)dt^2 - d\mathbf{r}^2 - 2\omega y dx dt + 2\omega x dy dt$ for the rotation metric.

[53] According to *GGR 2*, pp. 600–602, he rather has $G^{\mu\nu} = \kappa(T^{\mu\nu} + t^{\mu\nu})$ in mind. If this were the case, I think he would have considered $\partial_\mu G^{\mu\nu}$ instead of $D_\mu G^{\mu\nu}$ on p. 13R.

for harmonic coordinates. He then [p. 13R] computes $D_\mu G^{\mu\nu}$ under the simplifying conditions $\sqrt{-g} = 1$ and $\partial_\mu g^{\mu\nu} = 0$, probably because he knows that $D_\mu G^{\mu\nu}$ should vanish as a consequence of the equations $D_\mu T^{\mu\nu} = 0$ and $G^{\mu\nu} = \kappa T^{\mu\nu}$. The attempt stops here. This is the end of Einstein's first naive guesses at a conditionally covariant field equation.[54]

Building on the Riemann tensor

Around that time, Grossmann directed Einstein to the absolute differential calculus of Christoffel, Ricci, and Levi-Civita.[55] Einstein thus became aware of the Riemann tensor $R_{\mu\nu\rho\sigma}$, which is the simplest fully covariant combination of second- and first-order derivatives of the metric tensor (up to a constant coefficient). In a new section of his notebook, Einstein forms the Ricci tensor[56] $R_{\mu\nu} = R^\rho_{\ \mu\nu\rho}$ and tries [p. 14L]

$$R_{\mu\nu} = -\kappa T_{\mu\nu} \tag{9.52}$$

for the field equation in the presence of matter with the stress–energy tensor $T_{\mu\nu}$. In continuity with his previous approach, he tries and fails to reduce $R_{\mu\nu}$ to a simple combination of $g_{\mu\nu}$, $g^{\mu\nu}$, g, and ∂_μ, even under the restriction $g = -1$. He deplores that in the first, linear approximation (corresponding to the two $\partial \cdot \Gamma$ terms in the expression of $R^\mu_{\ \nu\rho\sigma}$), the Ricci tensor reduces to

$$R_{\mu\nu}^{(0)} = \tfrac{1}{2}g^{\rho\sigma}(\partial_\rho\partial_\sigma g_{\mu\nu} + \partial_\mu\partial_\nu g_{\rho\sigma} - \partial_\mu\partial_\rho g_{\nu\sigma} - \partial_\mu\partial_\sigma g_{\nu\rho}), \tag{9.53}$$

which contains unwanted terms besides the desired d'Alembertian $\tfrac{1}{2}g^{\rho\sigma}\partial_\rho\partial_\sigma g_{\mu\nu}$. Manipulating the curvature scalar $g^{\mu\nu}R_{\mu\nu}$ under $g = -1$ he tries to extract from it a gravitation tensor that would not have this defect, to no avail [pp. 14R–18R].[57]

At some point [p. 19L], Einstein realizes that the unwanted terms can be eliminated (in the linear approximation) by the harmonic coordinate condition $g^{\mu\nu}\{^\rho_{\mu\nu}\} = 0$ (alias $\partial_\mu(\sqrt{-g}g^{\mu\nu}) = 0$, since $D_\mu g^{\mu\nu} = 0$). In the weak-field approximation for which $g_{\mu\nu}$ differs from the Minkowski metric $\eta_{\mu\nu}$ by the small tensor $h_{\mu\nu}$, and for a dust of stress–energy $\theta^{\mu\nu}$, he arrives at the field equation [p. 19R]

$$\Box h_{\mu\nu} \equiv \eta^{\rho\sigma}\partial_\rho\partial_\sigma h_{\mu\nu} = -\kappa\eta_{\mu\rho}\eta_{\nu\sigma}\theta^{\rho\sigma}, \quad \text{or} \quad \Delta_4 h_{\alpha\beta} = \kappa\theta_{\alpha\beta} \tag{9.54}$$

in the imaginary system of coordinates $(x_\alpha = x^1, x^2, x^3, icx^0)$ for which $\eta_{\mu\nu}$ becomes the unit matrix $\delta_{\alpha\beta}$ and the d'Alembertian becomes the four-dimensional Laplacian $\Delta_4 \equiv \partial_\alpha\partial_\alpha$.[58]

---

[54] On Beltrami's invariant, see p. 256, this volume.

[55] Cf. Stachel 2007, pp. 106–107; Sauer 2013.

[56] Einstein's definition of the Ricci tensor differs from the now common $R_{\mu\nu} = R^\rho_{\ \mu\rho\nu}$ by a change of sign.

[57] Cf. GGR 2, pp. 614–622. On the Einstein–Grossmann collaboration, cf. Sauer 2013; Weinstein 2015b, section 2.4. For the Riemann tensor, see pp. 242, 247, 266, this volume.

[58] Cf. Renn and Sauer 1999, pp. 109–114.

In the same approximation, the energy–momentum equation $D_\mu \theta^{\mu\nu} = 0$ reduces to $\partial_\mu \theta^{\mu\nu} = 0$ because the product of $\theta^{\mu\nu}$ by a Christoffel symbol is of second order; and the harmonic condition $g^{\mu\nu} \{^\rho_{\mu\nu}\} = 0$ reduces to

$$2\partial_\alpha h_{\alpha\beta} = \partial_\beta h_{\gamma\gamma} \text{ or } \partial_\alpha(h_{\alpha\beta} - \tfrac{1}{2}\delta_{\alpha\beta}h_{\gamma\gamma}) = 0 \tag{9.55}$$

in the imaginary coordinate system. Combining the latter equation with $\Delta_4 h_{\alpha\beta} = \kappa\theta_{\alpha\beta}$ and $\partial_\alpha \theta_{\alpha\beta} = 0$, we get

$$0 = \partial_\alpha(\theta_{\alpha\beta} - \tfrac{1}{2}\delta_{\alpha\beta}\theta_{\gamma\gamma}) = -\tfrac{1}{2}\partial_\beta\theta_{\gamma\gamma}, \tag{9.56}$$

which is incompatible with $\theta_{\gamma\gamma} = -\rho_0$. Einstein removes this contradiction by means of the modified field equation [p. 20L]

$$\Delta_4(h_{\alpha\beta} - \tfrac{1}{2}\delta_{\alpha\beta}h_{\gamma\gamma}) = \kappa\theta_{\alpha\beta}. \tag{9.57}$$

He also gives the equivalent form[59]

$$\Delta_4 h_{\alpha\beta} = \kappa(\theta_{\alpha\beta} - \tfrac{1}{2}\delta_{\alpha\beta}\theta_{\gamma\gamma}), \tag{9.58}$$

which is easily obtained by means of the contraction $-\Delta_4 h_{\gamma\gamma} = \kappa\theta_{\gamma\gamma}$ of the former equation.[60]

As we saw, according to Einstein's stress principle a good field equation of the form $G_{\mu\nu} = -\kappa\theta_{\mu\nu}$ should make the force density $\tfrac{1}{2}\sqrt{-g}(\partial_\mu g_{\nu\rho})G^{\nu\rho}$ a four-divergence. Einstein verifies this for the field equation (9.58) in the linear approximation [pp. 19R, 21L]: $\partial_\delta h_{\alpha\beta}\Delta_4(h_{\alpha\beta} - \tfrac{1}{2}\delta_{\alpha\beta}h_{\gamma\gamma})$ is a four-divergence by analogy with $\nabla\varphi\Delta\varphi$ in electrostatics. Alas, for the d'Alembertian term $\tfrac{1}{2}g^{\rho\sigma}\partial_\rho\partial_\sigma g_{\mu\nu}$ of the exact field equation, Einstein fails to bring the product $\sqrt{-g}(\partial_\mu g^{\nu\rho})g^{\sigma\tau}\partial_\sigma\partial_\tau g_{\nu\rho}$ to the form of a four-divergence [p. 21L]. Moreover [p. 21R], his older static field theory turns out to be incompatible with the harmonic condition:[61] the only nonvanishing coefficients of Einstein's static metric are $g_{00} = \hat{c}^2$, $g_{11} = g_{22} = g_{33} = -1$, so that $\partial_\mu(\sqrt{-g}g^{\mu\rho}) = (0, \nabla\hat{c}) \neq 0$.[62]

---

[59] I have corrected an algebraic error leading to the wrong sign in front of $\tfrac{1}{2}$.

[60] As we may retrospectively judge, the new field equation is the linear approximation of the equation $R_{\mu\nu} - \tfrac{1}{2}g_{\mu\nu}R = -\kappa\theta_{\mu\nu}$ on which Einstein based his final theory of gravitation.

[61] Einstein writes only that the static case is "impossible because of the divergence condition." I take this condition to be the harmonic condition. An alternative candidate is the weak-field energy–momentum condition $\partial_\mu\theta^{\mu\nu} = 0$, which leads to the weak-field equation (9.58) $\Delta h_{ij} = -(\kappa/2)\delta_{ij}\theta_{00}$ in the static case, in contradiction with Einstein's static metric of 1912 (for which $h_{ij} = 0$). Still another reading is given in *GGR 2*, p. 643.

[62] At this point [p. 21R], Einstein gives a new argument for his form of the static metric: the spatial coefficients of the metric must be spatially uniform in order that the energy and force of a gravitating particle vary in the same manner with respect to the velocity.

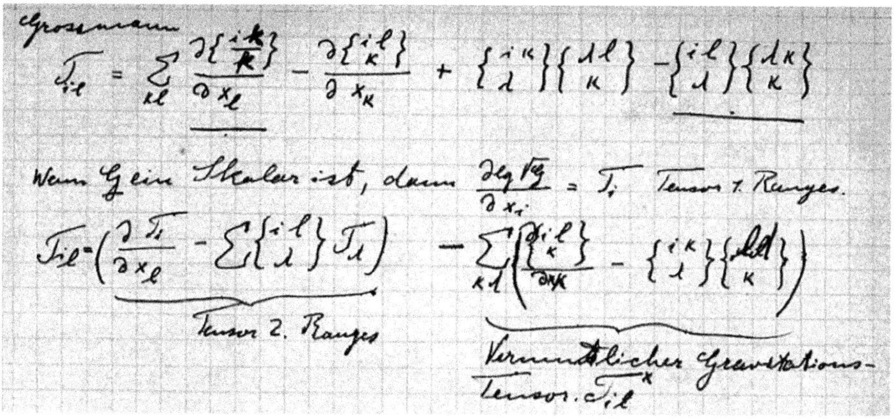

Fig. 9.2. Early manuscript occurrence (1912–13) of the gravitational tensor Einstein will adopt in November 1915. The tensor on the first line is the Ricci tensor, as given by Grossmann to Einstein. On the second line, Einstein builds the unimodular vector $\partial_\mu \ln \sqrt{-g}$ from the determinant g. On the third line, he subtracts the unimodular tensor $D_\nu \partial_\mu \ln \sqrt{-g}$ from the Ricci tensor to get the "presumed gravitation-tensor" (*Vermutlicher Gravitations-Tensor*) called $R^\times_{\mu\nu}$ in this chapter. Courtesy of the Albert Einstein Archives. © The Hebrew University of Jerusalem (also in *GGR* 2, p. 451).

After stumbling over this difficulty, Einstein tries to modify the field equation in such a manner that its second-order terms can be reduced to the d'Alembertian form without the harmonic condition. His strategy is to subtract from the Ricci tensor a unimodular tensor built from g and its derivatives (a unimodular tensor is a tensor with respect to coordinate transformations of Jacobian 1; g is a unimodular scalar). He thus arrives at the tensor [p. 22R] (Figure 9.2)[63]

$$R^\times_{\mu\nu} = R_{\mu\nu} - D_\nu \partial_\mu \ln \sqrt{-g} = -\partial_\rho \{^\rho_{\mu\nu}\} + \{^\sigma_{\mu\rho}\} \{^\rho_{\nu\sigma}\}, \qquad (9.59)$$

whose second-order part reduces to the d'Alembertian $\frac{1}{2} g^{\rho\sigma} \partial_\rho \partial_\sigma g_{\mu\nu}$ under the condition $\partial_\mu g^{\mu\nu} = 0$. Unlike the harmonic condition, this condition is compatible with Einstein's static metric. Yet Einstein does not pursue the consequences of the field equation

$$R^\times_{\mu\nu} = -\kappa T_{\mu\nu}, \qquad (9.60)$$

probably because of his prejudice that the true field equation should involve a simple combination of $g_{\mu\nu}$ and its derivatives.

In the following pages of his notebook, Einstein restricts the coordinate transformations to those for which the symmetric derivative

$$\vartheta_{\mu\nu\rho} = \partial_\rho g_{\mu\nu} + \partial_\mu g_{\nu\rho} + \partial_\nu g_{\rho\mu} \qquad (9.61)$$

[63] Einstein called this tensor $T^\times_{\mu\nu}$. I changed the notation in order to avoid confusion with the energy–momentum tensor.

behaves like a tensor. With respect to these transformations, the simple expression [p. 23L]

$$R_{\mu\nu}^{\times\times} = \tfrac{1}{2}\partial_\rho(g^{\rho\sigma}\partial_\sigma g_{\mu\nu}) + g^{\rho\alpha}g^{\sigma\beta}\partial_\alpha g_{\mu\rho}\partial_\beta g_{\nu\sigma} \tag{9.62}$$

is a tensor. It is easy to see why the new condition simplifies the field equation so much: it allows replacing the Christoffel symbol $\{^\rho_{\mu\nu}\}$ with the much simpler $-g^{\rho\sigma}\partial_\sigma g_{\mu\nu}$. Plausibly, Einstein thought of this condition because he had earlier studied the ϑ-metric fields for which $\vartheta_{\mu\nu\rho} = 0$, with the hope they would include the rotation metric [pp. 42L–42R]. They do not. Nor do the simple alternatives he could imagine to this condition [pp. 43LA, 23L, 25R]. But he still hoped to conciliate the ϑ-covariance with the equivalence principle, by showing that the motion of a particle in a ϑ-metric resembles the motion in a centrifugal force field [pp. 42R, 43LA].[64]

In general, Einstein then regarded his coordinate conditions as restrictions on the choice of physically permitted coordinate systems. This is clear from the fact that he usually tries to determine the transformations satisfying these conditions or the resulting field equation—with little or no success because the relevant differential equations are too complex. For instance, he tries [p. 22L] to determine the transformations for which the condition $\partial_\mu g^{\mu\nu} = 0$ is preserved (he knows that the transformations for which the condition $g^{\mu\nu}\{^\rho_{\mu\nu}\} = 0$ is preserved are those for which the new coordinates are harmonic functions of the old coordinates, hence the name harmonic condition), and he does the same [p. 23R] for the transformations under which $\vartheta_{\mu\nu\rho}$ is a tensor.[65]

For the sake of the equivalence principle, Einstein wanted these transformations to include uniform rotation in Minkowski space.[66] There is no relevant calculation for the two conditions $g^{\mu\nu}\{^\rho_{\mu\nu}\} = 0$ and $\partial_\mu g^{\mu\nu} = 0$ in the section devoted to the Riemann tensor. This may be because Einstein remembered his earlier proof [p. 11L] that the infinitesimal-rotation metric satisfies the first condition (then the second condition should also be satisfied, for it is equivalent to the first for unimodular transformations).[67] Probably for a similar reason (his past considerations of motion in a ϑ-metric), there is almost nothing in the Riemann tensor-based section of the notebook regarding the compatibility of the ϑ-covariance with rotation.[68] In contrast, Einstein is repeatedly concerned with the existence of a stress–energy tensor for a metric field obeying the conditioned field equations. This is the stress principle, so far acting as a test for a physically acceptable field equation.

---

[64] Cf. *GGR* 2, pp. 652–679; Darrigol 2019b, p. 137.

[65] Cf. Norton 2005, pp. 88–90 for the importance of coordinate restrictions in the rejection of the Riemann-based tensor. However, I tend to disagree with his opinion, amplified in Norton 2007, that Einstein initially considered coordinate conditions and turned them into restrictions after being misled by a reification of coordinate systems. The distinction between coordinate condition and coordinate restriction belongs to Renn and Sauer 2007, p. 109. It is emphasized in *GGR* 1, p. 11.

[66] One may wonder why Einstein does not consider the simpler acceleration metric. The reason is that it is incompatible with the harmonic condition (being a subcase of the static metric of 1912); and the unimodular restriction used by Einstein for his other field equations is incompatible with accelerated frames.

[67] Unfortunately, this is not true for finite rotations; cf. Norton 2005, pp. 89–90, and *GGR* 2, pp. 574–577.

[68] The exception is p. 24L; cf. *GGR* 2, pp. 674–679.

The Entwurf strategy

In the case of the last Riemann tensor-based candidate, in which the field operator is $R_{\mu\nu}^{\times\times}$ of Equation (9.62), Einstein fails to re-express the force density $-\frac{1}{2}(\partial_\rho g^{\mu\nu})R_{\mu\nu}^{\times\times}$ as the four-divergence of a stress–energy tensor. This test involves a repeated application of the identity $X\partial_\mu Y = \partial_\mu(XY) - Y\partial_\mu X$ to the product $(\partial_\rho g^{\mu\nu})\partial_\sigma(g^{\sigma\tau}\partial_\tau g_{\mu\nu})$, to which the first term of the field operator leads. Einstein thus generates the identity [p. 24R]

$$(\partial_\rho g^{\mu\nu})[\partial_\sigma(g^{\sigma\tau}\partial_\tau g_{\mu\nu}) - \tfrac{1}{2}\partial_\nu g_{\sigma\tau}\partial_\mu g^{\sigma\tau}] = \partial_\sigma(g^{\sigma\tau}\partial_\rho g^{\mu\nu}\partial_\tau g_{\mu\nu}) - \tfrac{1}{2}\partial_\rho(g^{\sigma\tau}\partial_\tau g_{\mu\nu}\partial_\sigma g^{\mu\nu}), \qquad (9.63)$$

which suggests to him that the true vacuum field equation might be

$$\partial_\sigma(g^{\sigma\tau}\partial_\tau g_{\mu\nu}) - \tfrac{1}{2}\partial_\nu g_{\sigma\tau}\partial_\mu g^{\sigma\tau} = 0. \qquad (9.64)$$

He finds the rotation metric to be a solution of this equation [p. 25R].

These results are incorrect: identity (9.63) does not hold,[69] and the rotation metric is not a solution of Equation (9.64).[70] Despite these errors, Einstein had in hand an efficient strategy to derive a gravitational field equation: start with a simple generalization of the d'Alembertian operator, and introduce additional first-order terms so that the associated four-force density becomes a four-divergence. The remaining pages of the notebook [pp. 26L–29L] contain fragments of a rigorous implementation of this strategy, soon published jointly with Grossmann as "Outline [*Entwurf*] of a generalized theory of relativity and of a theory of gravitation."[71]

To sum up, in his notebook Einstein first tried naive combinations of the metric tensor and its derivatives in order to get conditionally covariant expressions of the gravitational field equation. Conditional covariance here means covariance with respect to transformations compatible with simple, first-order differential conditions on the metric field. At some point, Einstein became aware of the Riemann tensor, and tried to build the field equation from it, using the fully covariant Ricci tensor as well as related, conditionally covariant tensors. None of the resulting candidates passed the four tests of correspondence with the d'Alembertian equation for weak fields, compatibility with the static metric of 1912 (which contains the acceleration metric as a subcase), compatibility with the rotation metric, and existence of a stress–energy pseudo-tensor for the gravitational field. At this point, Einstein realized that this last test could be used to guide the construction of the modified d'Alembertian in the gravitational field equation. This was the final, winning strategy leading to the published *Entwurf*.

All these considerations were of an utterly formalistic nature, at variance with the more intuitive approach of Einstein's earlier memoirs on generalized relativity. The equivalence

---

[69] As noted in *GGR* 2, pp. 683–686, this identity holds to second order in $h_{\mu\nu}$ (the departure of $g_{\mu\nu}$ from the Minkowskian metric) because in this approximation the index-raising and index-lowering tensors needed to convert $g_{\mu\nu}$ into $g^{\mu\nu}$ (and vice versa) can be regarded as constants. It does not seem, however, that Einstein was reasoning in this approximation (Darrigol 2019b, p. 138).

[70] The error is explained in *GGR* 2, pp. 699–702.

[71] Einstein and Grossmann 1913.

principle, being now encapsulated in the geodesic principle, no longer played a direct role in the construction (save for the rotation-metric test). There were no thought experiments, and there was much tensor algebra. Not being accustomed to this more mathematical way of thinking, Einstein encountered a number of difficulties. He lacked familiarity with tensor calculus on a differentiable manifold, and he was relying on conflicting heuristic principles. On the one hand, the full covariance of the equation of motion of particles and of energy–momentum conservation for a dust suggested to him the full covariance of the gravitational field equation. On the other hand, his earlier theory of the static field, the correspondence with the d'Alembertian operator, and the physical interpretability of the coordinates suggested a conditional covariance.

Earlier commentators on the Zürich notebook did not fail to notice that in the second, Riemann tensor-based stage of his research, Einstein wrote down the correct weak-field equations and even the equations from which he would later derive the correct value of the relativistic precession of the Kepler ellipse. He did not pursue their empirical consequences at that time because they did not pass his tests of correspondence and stress principle. Moreover, Einstein was laboring under the prejudice that the field equations should be simple combinations of the metric tensor and its derivatives. The Ricci tensor did not have this kind of simplicity since it was built from the Christoffel symbols, to which Einstein then accorded little significance (he did not even use them in the geodesic equation).

9.4 The *Entwurf* theory of 1913

Fundamental equations

The *Entwurf* memoir of spring 1913 has a "physical part" authored by Einstein, and a "mathematical part" authored by Grossmann. The latter part is mostly a tutorial on the absolute differential calculus. In the physical part, Einstein starts with the generalized d'Alembertian $(1/\sqrt{-g})\partial_\alpha(g^{\alpha\beta}\sqrt{-g}\partial_\beta g^{\mu\nu})$ of Equation (9.50), contracts it with $\sqrt{-g}\partial_\sigma g_{\mu\nu}$, and repeatedly applies the identity $X\partial_\mu Y = \partial_\mu(XY) - Y\partial_\mu X$ to the resulting force density until it becomes a four-divergence. The resulting field identity reads[72]

$$\sqrt{-g}(\partial_\sigma g_{\mu\nu})G^{\mu\nu} = \partial_\nu(\sqrt{-g}g_{\sigma\rho}X^{\nu\rho}), \quad \text{with} \tag{9.65}$$

$$X^{\mu\nu} = g^{\alpha\mu}g^{\beta\nu}\partial_\alpha g_{\rho\sigma}\partial_\beta g^{\rho\sigma} - \tfrac{1}{2}g^{\mu\nu}g^{\alpha\beta}\partial_\alpha g_{\rho\sigma}\partial_\beta g^{\rho\sigma}, \quad G^{\mu\nu} = \Delta^{\mu\nu} + \tfrac{1}{2}X^{\mu\nu}, \tag{9.66}$$

$$\Delta^{\mu\nu} = (1/\sqrt{-g})\partial_\alpha(g^{\alpha\beta}\sqrt{-g}\partial_\beta g^{\mu\nu}) - g^{\alpha\beta}g_{\rho\sigma}\partial_\alpha g^{\mu\rho}\partial_\beta g^{\nu\sigma}. \tag{9.67}$$

The linear tensor $G^{\mu\nu}$ is the new d'Alembertian for which the gravitational field equation reads[73]

$$G^{\mu\nu} = \kappa T^{\mu\nu}. \tag{9.68}$$

---

[72] Einstein (1913, p. 15) claims that this identity is unique. It may be true that it is the only identity that can be obtained by repeated partial integration in the expression of the four-force. However, as John Norton (1984, p. 282) pointed out, this is not the only possible identity of the desired form; it is only the simplest one.

[73] The plus sign is needed on the right-hand side because in the linear approximation $G^{\mu\nu} = \Box g^{\mu\nu}$ and $G_{\mu\nu} = -\Box g_{\mu\nu} = \kappa T_{\mu\nu}$, which gives the correct Newtonian limit.

Einstein interprets the tensor $t^{\mu\nu} = (-1/2\kappa)X^{\mu\nu}$ as the stress–energy tensor of the gravitational field, so that the field equation can be rewritten as

$$\Delta^{\mu\nu} = \kappa(T^{\mu\nu} + t^{\mu\nu}), \qquad (9.69)$$

which means that the energy–momentum of matter and the energy–momentum of the gravitational field both act as sources of the gravitational field. In addition, the equation

$$\tfrac{1}{2}\sqrt{-g}\partial_\mu g_{\nu\rho}T^{\nu\rho} = \partial_\nu(\sqrt{-g}g_{\mu\rho}T^{\nu\rho}) \qquad (9.70)$$

for the four-force acting on matter and the field equation together lead to

$$\partial_\nu[\sqrt{-g}g_{\mu\rho}(T^{\nu\rho} + t^{\nu\rho})] = 0, \qquad (9.71)$$

which is the local expression of the conservation of the energy and momentum of matter and field together.[74]

Einstein insists that the weak-field approximation of the new theory produces the desired d'Alembertian. We also know that he privately believes the rotation metric to be a solution of his field equation. This equation is linearly covariant by construction, and he hopes for a broader covariance in harmony with the equivalence principle. He admits he has not yet been able to specify the relevant class of transformations even though he regards this task as "the most important one." Nor has he solved any concrete problem. The theory truly is nothing but an outline (*Entwurf*)—so much so that one may wonder why Einstein and Grossmann published it. A plausible reason is Einstein's strong belief in a strategy based on seeking a generalized d'Alembertian compatible with the stress principle.[75]

Einstein has arguments against the three alternatives that immediately come to mind. First, one cannot obtain a field equation simply by taking the double covariant derivative of $g^{\mu\nu}$ since the first covariant derivative of this tensor vanishes identically. Second, a scalar theory of gravitation would not comply with the equivalence principle. Third, the Ricci tensor cannot serve to write the field equation because this operator does not reduce to the d'Alembertian in the weak-field approximation. Although Einstein does not give details here, we know from his notebook that he has convinced himself that no coordinate restriction would solve this difficulty.[76]

---

[74] Einstein and Grossmann 1913, pp. 15–17, 37–38.

[75] Einstein and Grossmann 1913, p. 13 (weak field), 18 (citation). For the rotating frames, relevant (inexact) calculations are found on pp. 24R and 25R of the Zürich notebook, and also in the Einstein–Besso MS of 1913, *ECP* 4, pp. 442–445. On the latter, cf. Janssen 1999, 2007. Besso seems to have reached the opposite conclusion by August 1913 (cf. Janssen 2007), and Einstein must have agreed with him for a while since in early 1914 he believed the *Entwurf* theory to be covariant under linear transformations only. In a letter to Lorentz of August 1913 (cited in Janssen 2007, p. 833), Ehrenfest reported that Einstein had already changed his mind five or six times on this issue.

[76] Einstein and Grossmann 1913, pp. 12 ($D_\mu g^{\mu\nu} \equiv 0$), 20–22 (scalar theory), and 11 and 36 (Riemann tensor). On p. 11, Einstein writes: "It must be emphasized that it proves to be impossible to find a [second-order] differential expression $G^{\mu\nu}$ that is a tensor generalization of $\Delta\varphi$" This is compatible with the statement that the Ricci tensor does not agree with the d'Alembertian of the metric field in the weak-field limit. On p. 36, Grossmann writes: "It turns out that [the Ricci tensor] does not reduce to $\Delta\varphi$ in the limit of an infinitely weak static field." This is incorrect if taken literally. Taking the Zürich notebook into account, Grossmann (and Einstein) probably meant that the static weak-field case should be the static subcase of a d'Alembertian, which does not agree with the weak-field limit of the Ricci tensor. For a different interpretation, see Stachel 1989b, p. 67.

From the covariance of the geodesic equation $\delta\int ds = 0$ and of the energy–momentum equation of a dust, Einstein judges it plausible that the exact gravitational equation would be fully covariant. At the same time, he insists that "we lack any clue for a general covariance of the equations of the gravitational equations." He hopes that in a future theory in which the gravitational field equation would include derivatives of order higher than two, general covariance might be reached. But in a theory limited to second-order derivatives, he is convinced that full covariance is impossible.[77]

Measurement

In the first three paragraphs of his contribution to the *Entwurf*, Einstein lays out the interpretive basis of his theory. He first recalls that the equivalence principle, when applied to the static gravitational field, leads to a value of the velocity of light depending on the gravitational field. These considerations have suggested to him an extension of the relativity principle in which a larger class of coordinate systems would be considered and in which the element ds would take the general form $\sqrt{g_{\mu\nu}dx^\mu dx^\nu}$. Owing to its role in determining the motion of particles in the field, this element has to be an absolute invariant and $g_{\mu\nu}$ is a covariant tensor determining the gravitational field. In a section entitled "Meaning of the fundamental tensor for the measurement of time and space," Einstein goes on:[78]

> From the previous considerations, we may already infer that between the space-time coordinates x^μ and the results of measurements through rulers and clocks there cannot be relations as simple as in the old theory of relativity. Regarding time, this feature was already apparent in the case of the static gravitational field. Thus we must raise the question of the physical meaning (principal measurability) of the coordinates x^μ.
>
> We remark that ds is to be regarded as the invariant measure of the distance between two infinitesimally close points of space-time. Therefore, ds must have a physical meaning independent of the selected system of reference. We assume that ds is the "naturally measured" distance of the two points, to be understood in the following manner.

Einstein here introduces a system of coordinates ξ^μ in which ds^2 takes the Minkowski form at a given point, and he assumes that the usual measuring prescriptions of special relativity, based on a rigid frame, rulers, and clocks, can be used locally to determine the value of these coordinates.[79] The metric significance of the coordinates x^μ can then be determined by means of the relation between the differentials dx^μ and the differentials $d\xi^\mu$. This relation being a function of the $g_{\mu\nu}$ coefficients, Einstein comes to the conclusion:

[77] Einstein and Grossmann 1913, pp. 12 (citation), 18.

[78] Einstein and Grossmann 1913, p. 8.

[79] He overlooks the additional requirement that the derivatives of the metric coefficients should vanish at the selected point.

> We see that by given dx^μ, the natural distance belonging to these differences can be specified only if the quantities $g_{\mu\nu}$ that determine the gravitational field are known: the gravitational field influences the measuring bodies and the clocks in a definite manner.

In this citation, we observe the persistence of Einstein's viewpoint, first expressed in his static field theory, that the coordinates in some sense represent the readings of clocks and rulers. These readings being different from the "naturally measured" ds, the relation between ds and dx^μ being dependent on the gravitational field, the rulers and clock are "influenced" by this field.[80]

Mercury's perihelion

In June 1913, Einstein collaborated with Besso to calculate the advance of Mercury's perihelion in the *Entwurf* theory. The calculation involved a center-symmetric, second-order, weak-field solution of the gravitational field equation. Contrary to the static theory of 1912, Einstein and Besso found that the metric had nondiagonal elements in this approximation (although these do not contribute to the equation of motion in the same approximation). Injecting this metric into the equation of motion of the planet, and solving to second order in dx^μ (v being the velocity of the planet), they arrived at the expression

$$\delta\theta = \frac{5\pi}{4}\frac{\alpha}{a(1-e^2)}, \text{ with } \alpha = 2GM/c^2, \tag{9.72}$$

for the relativistic anomaly in the advance of the perihelion, where M is the mass of the sun, G the constant of gravitation, a the semi-major axis of the elliptic orbit, and e its eccentricity. This differs from the expression in the final theory only by having $5\pi/4$ instead of 3π. In the case of Mercury, the formula gives 18″ instead of the 43″ needed to explain the observed anomaly. Einstein does not seem to have taken this failure too seriously, for he did not mention it in later accounts of the *Entwurf* theory. He may have judged that there were too many simultaneous causes of perihelion precession (ordinary perturbations by other planets, relativistic correction, and the rotation of the sun) for it to be reliably computed. This is an interesting case of asymmetry between confirmation and refutation: whereas a negative result in the *Entwurf* context did not suffice to reject the theory, the positive result of November 1915 significantly contributed to the credibility of the final theory.[81]

---

[80] Einstein and Grossmann 1913, p. 9.

[81] *ECP* 4, pp. 344–473, 630–682. Cf. Earman and Janssen 1993. Einstein and Besso originally made a mistake by a factor of hundred in the numerical estimate of the anomaly, but later corrected this error. On earlier attempts to explain the anomaly in the advance of Mercury's perihelion, see pp. 170, 201, 254, this volume. Urbain Le Verrier discovered the anomaly in 1859 (Le Verrier 1859, with a rough estimate of 38″ per year).

9.5 The scalar theory

Nordström's theory

In 1912, the Finnish theorist Gunnar Nordström published a scalar, Minkowskian theory of gravitation in which the motion of a particle does not depend on its mass. The evident Minkowskian generalization of Poisson's equation for the scalar field ψ reads

$$\Box\psi = -\kappa\bar{\rho}, \tag{9.73}$$

wherein $\bar{\rho} = \rho\sqrt{1 - v^2/c^2}$ is the density of matter in the local rest frame. The naive Minkowskian generalization of the Newtonian law of acceleration for a particle of mass m moving in the field ψ reads

$$f_\mu = \frac{\mathrm{d}p_\mu}{\mathrm{d}\tau} = m\partial_\mu\psi \text{ with } p_\mu = m\frac{\mathrm{d}x_\mu}{\mathrm{d}\tau} \text{ and } \mathrm{d}\tau = \sqrt{\eta_{\mu\nu}\mathrm{d}x^\mu\mathrm{d}x^\nu}. \tag{9.74}$$

This law being incompatible with

$$2\frac{\mathrm{d}x^\mu}{\mathrm{d}\tau}\frac{\mathrm{d}p_\mu}{\mathrm{d}\tau} = m\frac{\mathrm{d}}{\mathrm{d}\tau}\left(\frac{\mathrm{d}x^\mu}{\mathrm{d}\tau}\frac{\mathrm{d}x_\mu}{\mathrm{d}\tau}\right) \equiv 0, \tag{9.75}$$

Nordström assumes a variable mass m for which the equation of motion becomes

$$m\frac{\mathrm{d}^2x_\mu}{\mathrm{d}\tau^2} + \frac{\mathrm{d}m}{\mathrm{d}\tau}\frac{\mathrm{d}x_\mu}{\mathrm{d}\tau} = m\partial_\mu\psi. \tag{9.76}$$

Contraction with $\mathrm{d}x^\mu/\mathrm{d}\tau$ then yields $\mathrm{d}m = m\mathrm{d}\psi$ and $m = m_0e^{\psi-\psi_0}$. The equation of motion is still independent of the mass m since it may be rewritten as[82]

$$\frac{\mathrm{d}^2x_\mu}{\mathrm{d}\tau^2} = \partial_\mu\psi - \frac{\mathrm{d}\psi}{\mathrm{d}\tau}\frac{\mathrm{d}x_\mu}{\mathrm{d}\tau}. \tag{9.77}$$

Einstein's objections

In private Einstein told Nordström he had already considered the scalar theory but did not pursue it for it led to a slower fall of rotating bodies, against the spirit of the equivalence principle.[83] In the *Entwurf*, he briefly discussed a variant of Nordström's theory in which the equation of motion of a particle in the gravitational field φ reads

---

[82] Nordström 1912. For a thorough study of this theory and Einstein's contribution to it, cf. Norton 1992a; for a broader discussion, cf. Giulini 2008.

[83] See Nordström 1912, p. 1129 (addendum). Einstein probably obtained this result by (a misleading) analogy with the slower fall of a body with initial horizontal velocity in a scalar theory; see p. 274, this volume.

$$\varphi \frac{\mathrm{d}^2 x_\mu}{\mathrm{d}\tau^2} = \partial_\mu \varphi - \frac{\mathrm{d}\varphi}{\mathrm{d}\tau} \frac{\mathrm{d}x_\mu}{\mathrm{d}\tau} \tag{9.78}$$

(as would result from Nordström's equation for $\psi = \ln \varphi$) and the field equation reads

$$\varphi = -\kappa T_\mu^\mu \tag{9.79}$$

If T_μ^μ denotes the trace of the Minkowskian stress–energy tensor. Einstein derives the equation of motion from the variational principle

$$\delta \int \varphi \mathrm{d}\tau = 0, \tag{9.80}$$

and uses the more general "Laue scalar" T_μ^μ instead of the density of a dust (more on this in a moment). He does not repeat his earlier objection based on a free-falling spinning body, because he must have understood that in such a body the stresses balancing the centrifugal force contribute to the effective mass . Instead, he briefly describes a thought experiment in which cavity radiation is used to construct a perpetual motion. Here is a reconstruction of his argument.[84]

Consider a system made of electromagnetic radiation and of a (quasi-)rigid closed vessel whose internal walls are perfect mirrors. As is well known, the trace of the stress–energy tensor for electromagnetic radiation vanishes. However, the walls of the vessel contribute to the total energy–momentum of the system. According to Laue's theorem, for a closed system in equilibrium the volume integrals $\int T_1^1 \mathrm{d}^3 x$ etc. of the spatial components of the trace vanish. Therefore, the net source of the gravitational field is the total energy $E = \int T_0^0 \mathrm{d}^3 x$. From this result Einstein infers that the gravitational mass of the system is E/c^2.[85]

Einstein then imagines a long rigid cylinder and two rigidly connected pistons that can slide within the cylinder in a frictionless manner. The internal walls of the cylinder and the pistons are perfect mirrors and they contain a fixed amount of electromagnetic radiation. Consider the system made of the radiation and the double piston. Taking the first coordinate axis along the axis of the cylinder, for this system we have

$$\int T_1^1 \mathrm{d}^3 x = 0 , \quad \int T_2^2 \mathrm{d}^3 x = \int T_3^3 \mathrm{d}^3 x = -E/3, \tag{9.81}$$

because according to Maxwell's theory the electromagnetic pressure is one third of the energy density. Therefore, the effective source of the gravitational field is $\int T_\mu^\mu \mathrm{d}^3 x = E/3$, and the associated gravitational mass is $E/3c^2$. Imagine a cycle in which (1) the double piston is raised to the height h in the constant gravity g, (2) the cylinder is replaced by a new wall rigidly attached to the piston, (3) the resulting box is brought to the original elevation, and (4) the new wall is replaced with the sliding cylinder. In this cycle the work

---

[84] Cf. Norton 1993a, pp. 18, 20.

[85] On Laue's theorem, see p. 226, this volume.

$(2E/3c^2)gh$ is produced without compensation. Einstein regards this result as a sufficient reason to reject the scalar theory, although his strongest reason is the lack of relativity with respect to arbitrary frames of reference.[86]

Natural measures

Einstein gave Nordström's theory a new hearing in a review of the gravitation problem for the *Naturforscherversammlung* of September 1913 in Vienna. By then he had worked out the scalar option and changed his mind on its value. By analogy between the variational principles of the scalar and tensor theories, which both involve replacing the Minkowskian form $\eta_{\mu\nu}dx^\mu dx^\nu$ with more complex forms ($\varphi^2 \eta_{\mu\nu}dx^\mu dx^\nu$ and $g_{\mu\nu}dx^\mu dx^\nu$, respectively), he surmised that in the scalar theory just as in the tensor theory the "natural" lengths and times measured by transportable devices differed from the "coordinate-" lengths and times expressed in coordinate differences.[87]

   Then the cycle in the former thought experiment is no longer a cycle if the coordinate section of the cylinder is invariable and if the natural distance between the two pistons is a constant. Under these assumptions, the true volume V of the radiation expands by $2V\delta\alpha$ during step 1, wherein α is the ratio between natural length and coordinate length in the field φ. In order to complete the cycle, an additional step is needed in which the box built in step 2 is contracted back to the volume V. The work needed for the contraction is the product of $2V\delta\alpha$ by the radiation pressure $E/3V$. It compensates for the work $(2E/3c^2)gh$ produced by the rise and fall of the radiation if and only if $\delta\alpha = gh$. In other words, the impossibility of perpetual motion requires $\alpha = \varphi$ (+ constant) for the ratio between natural length and coordinate length.[88]

   Since in Nordström's theory the (coordinate-)velocity of light is a constant, the same ratio holds between natural time and coordinate time. In his communication Einstein asserts that a gravitational clock (made of two bodies orbiting around each other under gravitational pull) gives the natural time for $\alpha = \varphi$.[89] This is seen as follows. The mass in the scalar theory being proportional to the potential φ, the centrifugal force is proportional to φRT^{-2} if R denotes the distance between the two bodies (for a circular orbit) and T the period of the motion. This force is balanced by the gravitational pull, which is proportional to R^{-2}. The balance is unaffected by the gravitational field if and only if $\varphi\alpha^{-1}\alpha^2 = \alpha^2$, or $\alpha = \varphi$.

   The same result is obtained directly by assuming that the form $\varphi^2 \eta_{\mu\nu}dx^\mu dx^\nu$ in the variational principle of the scalar theory represents the square ds^2 of the naturally measured

---

[86] Einstein and Grossmann 1913, pp. 20–22.

[87] Einstein 1913, p. 1252.

[88] Einstein 1913, p. 1253. Einstein does not explain how the absurdity is removed. However, Nordström (1913b, pp. 544–545) details a similar thought experiment in which Einstein imagined a horizontal stressed rod to be lowered, then raised unstressed to its original height, and stressed again. The rod being heavier when stressed than when unstressed, work would be gained in this cycle if it were not for the work associated with the change of length of the rod (this work is done during the descent of the rod if this descent is done at constant natural length). For a lucid discussion of the latter experiment, cf. Norton 1993a, pp. 18–24.

[89] Einstein 1913, p. 1254.

interval between two events of coordinates x^μ and $x^\mu + dx^\mu$, just as in the tensor theory. As we will see in a moment, Einstein developed this view a few months later in collaboration with the young Dutch theorist Adriaan Fokker.[90]

In his Vienna talk, Einstein judges the scalar theory to have nearly as much potential as the *Entwurf* theory. Not only has his previous objection turned out to be invalid, but the scalar theory is compatible with the four basic demands of energy–momentum conservation, universality of free fall, local validity of the Minkowskian structure, and observable laws independent of the value of the gravitational potential (in regions of uniform potential). Its only drawbacks are the lack of relativity with respect to arbitrary reference frames and the incompatibility with Mach's principle.[91]

Einstein had earlier remarked that in his theory the inertia of a particle increases when it is brought near large masses, because the gravitational potential energy of the particle contributes to its inertia. This goes well with Mach's idea that the inertia of a body should be traced to its interaction with all other bodies and not to any mysterious property of space. In the scalar theory, the mass still depends on the gravitational potential but it *diminishes* when the particle is brought near large masses. Although this incompatibility with Mach's principle lowers the probability of the scalar theory in Einstein's eyes, he admits that only experience can decide between the scalar and tensor options, and he urges astronomers to test the gravitational deviation of light during the solar eclipse of 1914.[92]

9.6 Bridled covariance

The fatal rope?

In the part of his Vienna talk devoted to the *Entwurf*, Einstein dwells on the meaning of coordinates. He recalls that all the information needed to determine the motion of particles and to measure lengths and times is contained in the ds^2. To be sure, he still interprets ds/dx^0 as the rate of clocks and he still has the deviation of light hinge on a variable (coordinate-) velocity. But he insists that from the observational point of view and in the spirit of the absolute differential calculus "the coordinates by themselves have no physical meaning" and are mere "auxiliary variables." It is therefore natural to require the general covariance of the fundamental equations of the theory:[93]

> The space-time coordinates thereby degenerate into intrinsically meaningless, freely selectable auxiliary variables. Then the entire problem of gravitation would receive a satisfactory solution if we succeeded in finding generally covariant equations for the quantities that determine the gravitational field.

[90] Einstein and Fokker 1914.

[91] Einstein 1913, p. 1250. See also Einstein to Freundlich, mid-August 1913, in which Einstein judges Nordström's theory "very reasonable."

[92] Einstein 1912c, p. 39; Einstein and Grossmann 1913, p. 6; Einstein 1913, pp. 1254, 1260–1261, 1262 (eclipse). On Einstein and Mach's principle, cf. Renn 2007b. On Mach's views, see pp. 68–70, this volume.

[93] Einstein 1913, pp. 1256–1257 (citations), 1260 (Newtonian approximation).

Einstein goes on to admit that he has not been able to find fully covariant field equations of second differential order. In the *Entwurf*, he had expressed the hope that full covariance would be obtained at a higher differential order. In a letter to Lorentz written on August 14, 1913, he deplored his incapacity to reach general covariance or even to find any nonlinear transformation that left his equations invariant. Two days later, he told Lorentz he had just realized that energy conservation could serve to justify the restriction to linear transformations: "Now that this ugly dark spot [an unjustified restriction of the covariance] has been eliminated, the theory can at last bring me some pleasure."[94]

In September in Vienna, Einstein explained that the equation

$$\partial_v \left[\sqrt{-g}(t_\mu^v + T_\mu^v) \right] = 0 \tag{9.82}$$

for energy conservation was covariant only through linear transformations if t_μ^v had the same transformation properties as T_μ^v (as Einstein then believed it should). In the discussion following his talk, Gustav Mie denied that a wagon's frame was equivalent to the earthbound frame because fluctuations in the wagon's motion (owing to irregularities of the rails etc.) could not be assimilated to a gravitational field. Einstein replied that this lack of equivalence was to be expected in the *Entwurf* theory. In a footnote to the published version of his talk, he mentioned that he had just discovered an argument excluding a generally covariant field equation. Almost certainly, this was the hole argument he would soon publish in reply to other criticism by Mie.[95]

In the latter publication, Einstein reasserts that the "kernel of the equivalence principle," which is the geodesic equation $\delta \int ds = 0$, leads to the demand of general covariance, and he presents the lack of covariance of the *Entwurf* equations as the "fatal rope" with which his colleagues hope to strangle the theory. He distinguishes two ways in which an equation may lack general covariance: (1) there exists a covariant equation from which the given equation can be obtained by specializing the coordinates, or (2) there is no such covariant generalization. In the first case, we have a genuine field equation acting as a restriction on possible fields. In the second case, we have only a restriction on the choice of coordinates. An example of the first case would be Equation (9.60), which is a restriction of $R_{\mu v} = -\kappa T_{\mu v}$ for $\sqrt{-g} = 1$. An example of the second case is $\sqrt{-g} = 1$, because any metric structure (any system of geodesics) can be expressed under this restriction. Einstein firmly believes that the field equation of the *Entwurf* belongs to the first category and that his colleagues' objection therefore reduces to the weaker objection that no sufficient reasons have yet been given for specializing the system of coordinates.[96]

---

[94] Einstein to Lorentz, August 14, 1913, August 16, 1913, *ECP* 5. See also Einstein to Ehrenfest, November 1913, *ECP* 5; Einstein to Mach, December 1913, *ECP* 5: "The reference frame is so-to-say fitted [*angemessen*] to the existing world by means of the energy principle and thus loses its aprioristic nebulous existence ⋯ [The coordinates] are otherwise quite arbitrary."

[95] Einstein 1913, pp. 1258 (linear covariance), 1264 (reply to Mie), 1257n (hole argument). On the argument for linear covariance, cf. Norton 1984, pp. 285–286.

[96] Einstein 1914a, pp. 177–178.

The hole argument

Einstein then gives two such reasons, a "logical" one, and an "empirical" one. We have already encountered the empirical reason, which is the alleged impossibility of expressing energy–momentum conservation in covariant form. The logical reason is what will later be known as the "hole argument." Einstein considers a finite portion of space time in which the stress–energy tensor of matter $T^{\mu\nu}$ vanishes. If the field equation is fully covariant, from a given solution $g_{\mu\nu}(x)$ of this equation we may construct another solution $g'_{\mu\nu}(x)$ by the following procedure (x stands for the quadruplet x_0, x_1, x_2, x_3). Perform a (regular) change of coordinates in which the new coordinates agree with the old ones outside the hole. This leads to the field $g'_{\mu\nu}(x')$, which satisfies the transformed field equation by general covariance. In the transformed equation, we have $T'^{\mu\nu}(x') = T'^{\mu\nu}(x')$ everywhere because the coordinates are unchanged outside the hole and because $T^{\mu\nu}$ vanishes within the hole. Consequently, $g'_{\mu\nu}(x)$ is a solution of the original field equation. As Einstein wants the field $g_{\mu\nu}(x)$ to be uniquely determined by the field equation for a given value of $T^{\mu\nu}(x)$, he feels compelled to specialize the reference frames. The argument is repeated in an appendix to his contemporary republication of the *Entwurf*. Einstein will hold fast to it for two more years.[97]

Bridging the scalar and tensor theories

In early 1914, Einstein and Fokker published a new study of the scalar theory in which the methods of the absolute differential calculus were applied to the form

$$ds^2 = \varphi^2 \eta_{\mu\nu} dx^\mu dx^\nu. \tag{9.83}$$

From this point of view, the scalar theory differs from the tensor theory only by a specialization of the $g_{\mu\nu}$ coefficients. The characteristic assumption is the existence of a system of coordinates for which the (coordinate-)velocity of light is the constant c. For all such systems of coordinates the above-given form of the metric element applies. As Einstein knows, the simplest scalar invariant that can be derived from the metric is the twice-contracted

---

[97] Einstein 1914a, p. 178; 1914b, p. 260. In modern language, Einstein performs an active diffeomorphism instead of a mere change of coordinates. This is made clear in footnote (1) of Einstein 1914a, and also in Einstein 1914c, p. 1067. Cf. Norton 1984, pp. 287–289; Stachel 1989b. The argument almost certainly originated in a private question from Besso to Einstein, as recorded in a memo of August 28, 1913 by Besso; cf. Janssen 2007, pp. 789, 819–830; Renn and Sauer 2007, pp. 237–243. There Besso notes that if the field equation were fully covariant, then the metric field outside a domain containing all matter would not be uniquely determined. This looks like the hole argument, inside out. Besso adds that observable phenomena (e.g., the motion of a material point) may not be affected by this indetermination (not quite the modern reply, because the motion of the points in a given coordinate system is not an intrinsic notion). In a subsequent insert plausibly recording Einstein's reaction, this escape is excluded by arguing that different solutions of the field equation yield different motions in a given reference system. This response betrays the reification of reference systems implicit in the hole argument. Renn and Sauer suggest that Besso's consideration emerged out of concern for the uniqueness of the solutions of the Mercury problem (in which the matter of the sun is indeed contained within a finite domain). I suggest it may be more directly connected to the rotation problems studied by Einstein and Besso: the metric generated by a rotating sun, the metric of a nonrotating sun in a rotating frame, and the possibility of substituting one for the other. The comparison naturally suggests the intermediate case of coordinates agreeing with those of the rest frame inside the sun, and agreeing with those of the rotating frame outside the sun.

Riemann tensor R (the trace of the Ricci tensor). Given the form (9.83) of the metric, a straightforward calculation yields

$$R = -6\varphi^{-3}\Box\varphi. \tag{9.84}$$

For the field equation, Einstein simply takes

$$R = \kappa'T^{\mu}_{\mu}. \tag{9.85}$$

In the dust case, $T^{\mu}_{\mu} = \rho_0$, with $\rho_0 = \bar{\rho}\varphi^{-3}$ since the density ρ_0 is defined with respect to the natural volume and the density $\bar{\rho}$ with respect to the coordinate volume. The field equation is therefore the same as the Einstein–Nordström Equation (9.80), with $\kappa = \kappa'/6$.[98]

These considerations reinforced Einstein's conviction that in any theory compatible with the equivalence principle, the basic field equation could be obtained from a fully covariant field equation by specializing the choice of coordinates:

> Since Nature does not provide us with reference systems to which we could relate things [*Dinge*], we refer the four-dimensional manifold to fully arbitrary coordinates (corresponding to the Gaussian coordinates in the theory of surfaces) and we limit the choice of the reference system only when the treated problem induces us to do so.

In the scalar case, the specialization of coordinates rests on the light principle. In the tensor case, it rests on the energy–momentum principle.[99]

Most important, Einstein now believes that the Riemann tensor could serve to derive the field equation both in the scalar and in the tensor case. As he indicates in a footnote, he no longer holds that a field equation based on this tensor would be incompatible with Poisson's equation in the weak-field static limit. He even hopes to retrieve the *Entwurf* field equation by specializing the coordinate system in a generally covariant, Riemann tensor-based equation, thus providing a derivation independent of the more "physical" requirements such as energy–momentum conservation and correspondence with Newton's theory:[100]

> Finally, the role played by the Riemann-Christoffel tensor in the present investigation [of the scalar theory] makes it plausible that this tensor may also open the way to deriving the Einstein-Grossmann equations in a manner independent of physical assumptions. A proof of the existence or inexistence of a connection of this sort would be a significant theoretical progress.

[98] Einstein and Fokker 1914.

[99] Einstein and Fokker 1914, p. 321.

[100] Einstein and Fokker 1914, p. 328.

9.7 Justified transformations and adapted coordinates

The restricted covariance of the Entwurf theory

A proof of a connection between the Riemann tensor and the *Entwurf* field equation would be conceivable only after determining the transformations that preserve this equation. In the *Entwurf* memoir, Einstein had already declared this question as the most important on his agenda. In particular, he hoped that these transformations would include accelerated and rotating frames. As we saw, in mid-August 1913 he abandoned this hope and he argued that the equation for momentum–energy conservation restricted the covariance to linear transformations. At some point, he realized that this argument was invalid, because it rested on the false assumption that the stress–energy "tensor" t_μ^ν of the gravitational field was a genuine tensor. In March 1914, Einstein wrote to his friend Heinrich Zangger that, contrary to his earlier view, the *Entwurf* theory allowed for "arbitrarily moving reference frames." The remark made the theory so "harmonious" that he "no longer had the slightest doubt about the correctness of the theory."[101]

The reason for this volte-face was Einstein's renewed collaboration with Grossmann. The two friends had recently proved that the "justified" (*berechtigte*) transformations leaving the *Entwurf* field equation invariant were those preserving the equation

$$B_\sigma \equiv \partial_\nu \partial_\alpha (\sqrt{-g} g^{\alpha\beta} g_{\sigma\mu} \partial_\beta g^{\mu\nu}) = 0. \tag{9.86}$$

From the form

$$\partial_\alpha (\sqrt{-g} g^{\alpha\beta} g_{\sigma\mu} \partial_\beta g^{\mu\nu}) = \kappa \sqrt{-g} (t_\sigma^\nu + T_\sigma^\nu), \tag{9.87}$$

of the *Entwurf* field equation (9.69) and from the conservation equation (9.71), it is clear that any justified transformation leaves Equation (9.86) invariant. The reciprocal statement is less evident. Following advice from the Swiss mathematician Paul Bernays, Einstein and Grossmann based their proof (about which more in a moment) on the invariance properties of the field action

$$S_F = \frac{1}{2} \int g^{\alpha\beta} \partial_\alpha g_{\rho\sigma} \partial_\beta g^{\rho\sigma} \sqrt{-g} d^4 x, \tag{9.88}$$

whose variation leads the *Entwurf* field equation through[102]

$$\delta S_F - \kappa \int T_{\mu\nu} \delta g^{\mu\nu} \sqrt{-g} d^4 x = 0. \tag{9.89}$$

Bernays probably reasoned that it was easier to investigate the covariance properties of a scalar than those of a tensor. Having attended Planck's lectures in Berlin and Hilbert's

---

[101] Einstein and Grossmann 1913, p. 18; Einstein to Zangger, March 10, 1914, *ECP* 5. The argument for linear transformations is given in Einstein 1913, p. 1258; 1914a, p. 178; 1914b, p. 260. It is rejected in Einstein and Grossmann 1914, p. 218n.

[102] Einstein has 2κ instead of κ in the following equation.

in Göttingen, he generally understood the power of the principle of least action. He did a great service to Einstein in directing him to the action of the gravitational field. So far Einstein had used the principle of least action only for the motion of particles and had based his field heuristics on energy–stress considerations.[103]

At first glance, the reduction of the covariance of the *Entwurf* field equation to the co-variance of $B_\sigma = 0$ does not seem to be huge progress. In particular, as Einstein admitted to Lorentz, the transformations preserving $B_\sigma = 0$ do not constitute a group, since they depend on the metric $g_{\mu\nu}(x)$ to which they are applied.[104] Einstein accepted this "nonau-tonomous" character of the transformations and did not worry much about the complexity of the condition $B_\sigma = 0$. What mattered to him was that this condition would allow for non-linear transformations, in harmony with the equivalence principle.[105] In private, he told Besso that the justified transformations would include those of the rotation metric.[106]

Formal foundation of the Entwurf theory

In November 1914, Einstein published a bulky memoir entitled "The formal foundation [*formale Grundlage*] of general relativity theory." After a brief introduction of the equiv-alence *principle* and the fundamental metric field comes the "theory of covariance"; then the stress–energy tensor of matter, the vanishing of its covariant divergence, and its expres-sions for a dust and for the electromagnetic field; and next, a new "formal" derivation of the *Entwurf* equations based on the invariance properties of the field action. This deriva-tion capitalizes on Einstein and Grossmann's earlier characterization of the restricted covariance of the *Entwurf* equations.[107]

A little more needs to be said on this characterization. It relies on the fact that the con-dition $B_\sigma = 0$ (Equation (9.86)) is satisfied if and only if the coordinates x^μ are chosen so that $\delta_x S_F = 0$ under any infinitesimal transformation $x^\mu \to x^\mu + \delta x^\mu$. In Einstein and Gross-mann's terms, such coordinates are "adapted" (*angepasst*) to the manifold with the metric $g_{\mu\nu}(x)$. The *Entwurf* field equation turns out to be invariant with respect to a transforma-tion preserving the adaptation of coordinates, because the variation $\delta_g S_F$ of the action for $g_{\mu\nu} \to g_{\mu\nu} + \delta g_{\mu\nu}$ (whose vanishing yields the field equation) is invariant under such transfor-mations. In Einstein's eyes, the adapted coordinates have the additional virtue of solving the hole paradox: the adaptation provides four conditions through which the coordinates in the hole are fully determined when they are known on the boundary of the hole.[108]

---

[103] Einstein and Grossmann 1914, p. 219n (Bernays).

[104] Einstein to Lorentz, August 14, 1913, *ECP* 5, pp. 546–548.

[105] Einstein and Grossmann 1914, p. 8. Einstein there asserts the possibility of nonlinear transformations without giving a proof. He may have reasoned, as he did in Einstein to Lorentz, January 23, 2015, *ECP* 8, p. 80, that $B_\sigma = 0$ was compatible with any choice of the coordinates at the boundary of a finite domain.

[106] Einstein to Besso, March 10, 1914, *ECP* 5. Einstein long hesitated on the validity of this calculation (see note 76), until he firmly decided against it in Einstein to Freundlich, September 30, 1915, *ECP* 8.

[107] Einstein 1914c.

[108] Einstein and Grossmann 1914, pp. 219–224. Remember that for Einstein the existence of two different solu-tions of the field equation with the same coordinates derives from the possibility of expressing the same solution in two different systems of coordinates.

In his *formale Grundlage* of November 1914, Einstein no longer assumes the expression (9.87) of the field action and instead tries to derive this expression by "formal" considerations. For the generic action

$$S_F = \int \Lambda(g_{\mu\nu}, \partial_\rho g_{\mu\nu})\sqrt{-g}\,\mathrm{d}^4x, \qquad (9.90)$$

the variation $x^\mu \to x^\mu + \delta x^\mu$ leads to[109]

$$\delta_x S_F = \int (\delta_x \Lambda)\sqrt{-g}\,\mathrm{d}^4x, \quad \text{with } \delta_x \Lambda = \frac{\partial \Lambda}{\partial g^{\mu\nu}}\delta_x g^{\mu\nu} + \frac{\partial \Lambda}{\partial(\partial_\rho g^{\mu\nu})}\delta_x(\partial_\rho g^{\mu\nu}), \qquad (9.91)$$

$$\delta_x g^{\mu\nu} = g^{\mu\alpha}\partial_\alpha \delta x^\nu + g^{\nu\alpha}\partial_\alpha \delta x^\mu, \quad \delta_x \partial_\rho g^{\mu\nu} = \partial_\rho(g^{\mu\alpha}\partial_\alpha \delta x^\nu + g^{\nu\alpha}\partial_\alpha \delta x^\mu) - \partial_\alpha g^{\mu\nu}\partial_\rho \delta x^\alpha. \qquad (9.92)$$

A simple calculation gives

$$\delta_x \Lambda = X^\nu_\mu \partial_\nu \delta x^\mu + 2g^{\nu\alpha}\frac{\partial \Lambda}{\partial(\partial_\rho g^{\mu\nu})}\partial_\rho \partial_\alpha \delta x^\mu. \qquad (9.93)$$

At this point, Einstein assumes the Lagrangian Λ to be invariant through linear transformations. Then the coefficients X^ν_μ must vanish, and the variation $\delta_x \Lambda$ reduces to the second term of Equation (9.93). A double integration by parts of the resulting expression of $\delta_x S_F$ leads to

$$\delta_x S_F = 2\int B_\mu \delta x^\mu \mathrm{d}^4x, \quad \text{with } B_\mu = \partial_\rho \partial_\alpha \left[g^{\nu\alpha}\frac{\partial \Lambda}{\partial(\partial_\rho g^{\mu\nu})}\right], \qquad (9.94)$$

under proper boundary conditions. Hence, the adapted coordinates are those for which $B_\mu = 0$ (I use bold face instead of Einstein's gothic script for the field density $\mathbf{X} = X\sqrt{-g}$ associated with the field X).[110]

Einstein next proves that $\delta_g S_F$ is invariant under transformations between adapted coordinates. Consequently, the field equation

$$G_{\mu\nu} = \kappa T_{\mu\nu} \quad \text{with } \mathbf{G}_{\mu\nu} = \frac{\partial \Lambda}{\partial g^{\mu\nu}} - \partial_\rho \frac{\partial \Lambda}{\partial(\partial_\rho g^{\mu\nu})}, \qquad (9.95)$$

which derives from $\delta_g S_F = 0$ through Equation (9.89), is covariant under such transformations.[111] In order to restrict the choice of Λ, besides its invariance under linear

[109] The correctness of the following formulas is easily judged from $\int F[X'^\mu(x')]\sqrt{-g'}\mathrm{d}^4x' = \int F[X'^\mu(x)]\sqrt{-g}\mathrm{d}^4x$, $X'^\mu(x) = \partial_\nu x'^\mu X^\nu(x)$, with $x' = x + \delta x$ and $\delta X^\mu(x) = X'^\mu(x) - X^\mu(x)$.

[110] Einstein 1914c, pp. 1069–1071. The similarity with Noether's theorem should be obvious to the modern reader.

[111] As Levi-Civita told Einstein, this seemingly obvious implication does not hold for infinitesimal transformations. Einstein believed he could save his proof by appealing to finite transformations. See the correspondence in *ECP* 8, and the commentary in Cattani and De Maria 1989b; Weinstein 2015b, section 2.8.

transformations Einstein assumes that it is of first differential order, that it is quadratic with respect to $\partial_\rho g^{\mu\nu}$, and that the resulting field equation $G_{\mu\nu} = \kappa T_{\mu\nu}$ (together with $D^\nu T_{\mu\nu} = 0$) should automatically yield the condition $B_\mu = 0$ for adapted coordinates. This last requirement is needed because the ten $g^{\mu\nu}$ coefficients would otherwise be overdetermined by the fourteen equations $G_{\mu\nu} = \kappa T_{\mu\nu}$ and $B_\mu = 0$.[112]

The identity $D^\nu T_{\mu\nu} = 0$ for the stress–energy tensor $T_{\mu\nu}$ of matter leads to $D^\nu G_{\mu\nu} = 0$, which is equivalent to $B_\rho = \partial_\nu S_\rho^\nu$, with

$$S_\mu^\nu = g^{\nu\rho}\frac{\partial \Lambda}{\partial g^{\mu\rho}} + \partial_\rho g^{\nu\sigma}\frac{\partial \Lambda}{\partial(\partial_\rho g^{\mu\sigma})} + \frac{1}{2}\delta_\mu^\nu \Lambda - \frac{1}{2}\partial_\mu g^{\rho\sigma}\frac{\partial \Lambda}{\partial(\partial_\nu g^{\rho\sigma})}. \tag{9.96}$$

Therefore, the condition $B_\mu = 0$ will be automatically satisfied if $S_\mu^\nu \equiv 0$. Einstein uses the latter condition as a constraint on the form of the Lagrangian Λ. Unfortunately, he does not see that the quantities S_μ^ν are identical with the coefficients X_μ^ν in the full expression (9.93) of $\delta_x \Lambda$ (his earlier reasoning on adapted coordinates did not require computing these coefficients). With this knowledge, it is clear that the condition $S_\mu^\nu \equiv 0$ is nothing but the condition for the invariance of Λ with respect to linear transformations. Without this knowledge and with false guesses on the value of S_μ^ν for the various candidate actions, Einstein wrongly concludes that the *Entwurf* action (9.88) is the only one to meet this condition.[113]

Einstein's error is easily understood from a retrospective point of view. His condition for adapted coordinates, $\delta_x S_F = 0$, is equivalent to $D^\nu G_{\mu\nu} = 0$ owing to the identities[114]

$$\delta_x g^{\mu\nu} = D^\nu \delta x^\mu + D^\mu \delta x^\nu \text{ and } \delta_x S_F = \int G_{\mu\nu}\delta_x g^{\mu\nu}\sqrt{-g}d^4x, \tag{9.97}$$

which together imply

$$\delta_x S_F = -2\int (D^\nu G_{\mu\nu})\delta x^\mu \sqrt{-g}d^4x \tag{9.98}$$

by partial integration. Hence the adapted coordinates are just those for which the field equation $G_{\mu\nu} = \kappa T_{\mu\nu}$ is compatible with the equation $D^\nu T_{\mu\nu} = 0$, and this is true for any choice of the Lagrangian density.

Toward the end of his memoir, Einstein offers a few remarks about the possibility that the same coordinate, say x_0, may be time-like in some portion of space–time and space-like in another; about his inability to exclude closed time-like paths in space–time even though they "badly hurt physical intuition," and about the necessity of abandoning Euclidean geometry because of its implicit appeal to direct action at a distance (through the notion of rigid body). He also gives the Newtonian approximation of the metric as

---

[112] Einstein 1914c, pp. 1071–1074.

[113] Einstein 1914c, pp. 1074–1076.

[114] See Landau and Lifshitz 1951, section 94.

$$g_{ij} = \delta_{ij}, \quad g_{i0} = 0, \quad g_{00} = 1 + 2\Phi/c^2. \tag{9.99}$$

Again, he regards ds as the naturally measured time and ds/dx^0 as the rate of a clock.[115]

Since his first introduction of the stress–energy tensor of matter in the Zürich notebook, Einstein sought equations in which the quantity $\partial_\mu g_{\nu\rho}$ occurred in a simple manner, because in Equation (9.70),

$$\partial_\nu(\sqrt{-g}g_{\mu\rho}T^{\nu\rho}) = \tfrac{1}{2}\sqrt{-g}\partial_\mu g_{\nu\rho}T^{\nu\rho},$$

for the stress–energy tensor of matter, the right side represented the four-force density act-ing on matter. Granted that $T^{\nu\rho}$ in the field equation for the potentials $g_{\mu\nu}$ is the counterpart of the mass density in Poisson's equation for the potential Φ, then $\tfrac{1}{2}\sqrt{-g}\partial_\mu g_{\nu\rho}$ should be a generalization of the gravitational field $-\nabla\Phi$. Soon after publishing the *Entwurf*, Einstein favored the simpler form

$$\partial_\nu \mathbf{T}_\mu^\nu = \tfrac{1}{2}g^{\nu\rho}\partial_\mu g_{\rho\sigma}\mathbf{T}_\nu^\sigma \tag{9.100}$$

of Equation (9.70), which gives a prominent role to the quantities

$$\Gamma_{\mu\sigma}^\nu = \tfrac{1}{2}g^{\nu\rho}\partial_\mu g_{\rho\sigma}. \tag{9.101}$$

In his *formale Grundlage*, Einstein introduces these Γ coefficients and calls them the "components of the gravitational field." He rewrites the *Entwurf* field equation as[116]

$$\partial_\alpha(\sqrt{-g}g^{\alpha\beta}\Gamma_{\nu\beta}^\sigma) = -\kappa(\mathbf{t}_\nu^\sigma + \mathbf{T}_\nu^\sigma) \tag{9.102}$$

with[117]

$$\mathbf{t}_\nu^\sigma = -\kappa^{-1}\sqrt{-g}(g^{\sigma\tau}\Gamma_{\mu\nu}^\rho\Gamma_{\rho\tau}^\mu - \tfrac{1}{2}\delta_\nu^\sigma g^{\alpha\beta}\Gamma_{\mu\alpha}^\rho\Gamma_{\rho\beta}^\mu) \tag{9.103}$$

The field action also is a simple function of the Γ components:

$$S_F = -\int g^{\sigma\tau}\Gamma_{\sigma\beta}^\alpha\Gamma_{\tau\alpha}^\beta\sqrt{-g}\mathrm{d}^4x, \tag{9.104}$$

The relative simplicity of these equations and their analogy with the electromagnetic field equations probably contributed to Einstein's faith in the *Entwurf* theory.[118]

---

[115] Einstein 1914c, pp. 1078–1080 (remarks), 1080–1082 (Newtonian approximation).

[116] The kappa in the following equations is half the kappa of the original *Entwurf* equations.

[117] I have added the missing minus sign.

[118] Einstein 1914c, p. 1058. Einstein had earlier introduced the "field strengths" $G_{\mu\sigma}^\nu = \tfrac{1}{2}\sqrt{-g}g^{\nu\rho}\partial_\mu g_{\rho\sigma}$ (*ECP* 4, p. 568). The form (9.102) of the field equations is derived from the original *Entwurf* equation (9.69) by replacing the equated contravariant tensors with the corresponding mixed tensor densities (as suggested by the simplified form of the divergence equation for the total energy–momentum tensor). I do not believe Einstein got it by analogy with the electromagnetic field equation $\partial_\mu F^{\mu\nu} = j^\nu$. Rather, similar principles produced a similar equation.

Remember that right before determining with Grossmann the transformation properties of the *Entwurf* theory, Einstein had speculated that the field equation of this theory might be obtained by specializing the coordinates in an equation based on the Riemann tensor. In early April 1914, Einstein wrote to Ehrenfest: "Grossmann wrote to me that he is now able to derive the gravitational equations from the generally covariant theory. That would be a nice complement to our investigation." Whatever Grossmann did, there is no mention of it in the *formale Grundlage* (November 1914). Plausibly, the concept of adapted coordinates and Einstein's erroneous belief that it could be used to formally determine the field action had again diverted him from the Riemann tensor.[119]

9.8 November 1915

Returning to the Riemann tensor

Einstein had long known that the *Entwurf* theory gave too small a value for the relativistic precession of Mercury's perihelion but he did not seem to worry much about that. In September 1915, he firmly decided that the *Entwurf* field equation was incompatible with the rotation metric, after much wavering on this issue. In a letter to his favorite astronomer Erwin Freundlich, he called this result "a logical contradiction of the quantitative kind." But he did not yet suspect the foundations of his theory. He opined that "a computational error must be hidden somewhere in the edifice" and that the same error might be responsible for the Mercury perihelion failure. He hesitated between a mistake in the coefficients of the field equation and a misapplication of this equation.[120]

The most serious blow came a few weeks later, in early October 1915, when Einstein realized that his condition $S_\mu^\nu \equiv 0$ did not further restrict the choice of the field action. Indeed the condition $B_\mu = 0$ for adapted coordinates follows from the field equation and the vanishing divergence of the energy–stress tensor, for any choice of the Lagrangian. In the letter in which Einstein announced this failure to Lorentz, he concluded that the correspondence with Newton's theory was the only way to justify the *Entwurf* choice for the field action. However, he soon switched to a more formal way to restrict this choice, based on general covariance. In the ensuing communication of November 4, 1915, he explains:[121]

> Thus I returned to the requirement of general covariance for the field equations, from which I had departed with a heavy heart three years ago, during my collaboration with my friend Grossmann. In fact, at that time we came quite close to the forthcoming solution of the problem···· The charm of this theory will not escape anyone who has

---

[119] Einstein to Ehrenfest, early April 2014, *ECP* 8, #2. It seems impossible to derive the *Entwurf* field equation from the Riemann tensor with coordinate conditions (because there are vacuum solutions of this equation for which $R_{\mu\nu} \neq 0$). It could well be that Grossmann got similar but different equations, as Einstein himself did in November 1915.

[120] Einstein to Freundlich, September 20, 1915, *ECP* 8. For two space dimensions, Einstein found $g_{00} = 1 - (3/4)\omega^2(x_1^2 + x_2^2)$ to second order in the angular velocity ω by injecting the first-order rotation metric into the *Entwurf* field equation, whereas for the true rotation metric $g_{00} = 1 - \omega^2(x_1^2 + x_2^2)$.

[121] Einstein to Lorentz, October 12, 1915, *ECP* 8; Einstein 1915a, pp. 778–779.

truly grasped it. It means a true triumph of the methods of the general differential calculus founded by Gauss, Riemann, Christoffel, Ricci-Curbastro, and Levi-Civita.

All the obstacles Einstein had originally imagined against the covariant approach based on the Riemann tensor had successively vanished: since mid-1913 he no longer expected agreement with his static theory of 1912; since early 1914 he no longer required the linear part of the field equation to agree with the d'Alembertian equation (at least not without a proper coordinate condition); and he had just ceased to believe that the field equation should involve a simple combination of the field derivatives $\partial_\rho g_{\mu\nu}$. In the communication of November 4, Einstein indeed denounces the "fatal prejudice" (*ein verhängnisvolles Vorurteil*) that the quantities $\frac{1}{2}g^{\nu\rho}\partial_\mu g_{\rho\sigma}$ are the natural candidates for the components $\Gamma^\nu_{\mu\sigma}$ of the gravitational field as suggested by the form (9.100) of energy–momentum conservation. He now realizes that the absolute differential calculus and the standard form of the geodesic equation instead favor the Christoffel symbols and he takes $\Gamma^\rho_{\mu\nu} = -\{^\rho_{\mu\nu}\}$ for the field components.[122]

The hole argument should not be counted among the obstacles toward general covariance, because Einstein had long ceased to believe that this argument excluded a fully covariant field equation at the most fundamental level. It only meant that the coordinate system had to be adapted to the metric field. It is not clear whether Einstein had yet renounced the hole argument in early November 1915.

For the sake of general covariance, the gravitational field equation should be built from the Ricci tensor. In order to simplify the equation, Einstein subtracts from the Ricci tensor the unimodularly covariant $D_\nu \partial_\mu \ln \sqrt{-g}$, as he has done in the Zürich notebook to get the tensor $R^\times_{\mu\nu}$ of Equation (9.59). He thus obtains the unimodularly covariant equation

$$\partial_\rho \Gamma^\rho_{\mu\nu} + \Gamma^\sigma_{\mu\rho}\Gamma^\rho_{\nu\sigma} = -\kappa T_{\mu\nu} \tag{9.105}$$

which is easily seen to derive from

$$\delta S_F - \kappa \int T_{\mu\nu}\delta g^{\mu\nu}d^4x = 0, \text{ with } S_F = \int \Lambda d^4x \text{ and } \Lambda = g^{\sigma\tau}\Gamma^\alpha_{\sigma\beta}\Gamma^\beta_{\tau\alpha}. \tag{9.106}$$

Einstein next derives the expression

$$t^\nu_\mu = \frac{1}{2\kappa}\left(\delta^\nu_\mu \Lambda - \partial_\mu g^{\rho\sigma}\frac{\partial \Lambda}{\partial(\partial_\nu g^{\rho\sigma})}\right) \tag{9.107}$$

for the stress–energy tensor of the gravitational field such that $\partial_\nu(t^\nu_\mu + T^\nu_\mu) = 0$. For $\Lambda = g^{\sigma\tau}\Gamma^\alpha_{\sigma\beta}\Gamma^\beta_{\tau\alpha}$, this gives[123]

$$t^\nu_\sigma = \kappa^{-1}(\tfrac{1}{2}\delta^\nu_\sigma g^{\alpha\beta}\Gamma^\rho_{\mu\alpha}\Gamma^\mu_{\rho\beta} - g^{\nu\tau}\Gamma^\rho_{\mu\sigma}\Gamma^\mu_{\rho\tau}). \tag{9.108}$$

---

[122] Einstein 1915a, pp. 782–783.

[123] The κ^{-1} factor is missing in Einstein's text.

Equations (9.106–108) are the same as in the earlier *formale Grundlage* except that the expression of $\Gamma^{\rho}_{\mu\nu}$ has changed and the $\sqrt{-g}$ factors have disappeared.[124]

Combining the field equation, its trace, and the equation for energy–momentum conservation, Einstein gets

$$\partial_{\mu}(\partial_{\alpha}\partial_{\beta}g^{\alpha\beta} - g^{\sigma\tau}\Gamma^{\alpha}_{\sigma\beta}\Gamma^{\beta}_{\tau\alpha}) = 0, \tag{9.109}$$

and integrates it to

$$\partial_{\alpha}\partial_{\beta}g^{\alpha\beta} - g^{\sigma\tau}\Gamma^{\alpha}_{\sigma\beta}\Gamma^{\beta}_{\tau\alpha} = 0. \tag{9.110}$$

In a first, linear approximation, this reduces to

$$\partial_{\alpha}\partial_{\beta}g^{\alpha\beta} = 0. \tag{9.111}$$

This condition, Einstein goes on, "does not yet determine the coordinate system since 4 equations are needed for this purpose. Therefore, in the first approximation we may arbitrarily set $\partial_{\beta}g^{\alpha\beta} = 0$." This leads to the d'Alembertian equation

$$\Box g_{\mu\nu} = 2\kappa T_{\mu\nu}, \tag{9.112}$$

of which Poisson's law is the nonrelativistic approximation. Clearly, Einstein no longer regards $\partial_{\beta}g^{\alpha\beta} = 0$ as a universal condition on physically admissible coordinates. Instead he understands that the linear approximation and the correspondence with Newton's theory in themselves require a specification of the coordinate system. In the end, he notes that unimodular transformations include uniform rotation, as he wishes in order to satisfy the equivalence principle. Interestingly, he no longer restricts the transformations to adapted coordinates (even though he still believes adaptation is needed for the sake of energy conservation) and he focuses on the covariance of the initial field equation.[125]

Addendum

In his communication, Einstein remarks that despite the unimodular covariance of the theory, the coordinates cannot be chosen so that $\sqrt{-g} = 1$, because Equations (9.105) and (9.110) together imply

$$\partial_{\alpha}(g^{\alpha\beta}\partial_{\beta}\ln\sqrt{-g}) = -\kappa T^{\mu}_{\mu} \tag{9.113}$$

and because the trace of the stress–energy tensor of matter does not vanish in general (for a dust, the trace is equal to the density of the dust). In an addendum, Einstein speculates

---

[124] Compare with Equations (9.102–104).

[125] Einstein 1915a, p. 786. Einstein overlooked the unimodular covariance of Equation (9.110) and therefore wrongly believed it to imply a partial adaptation of the coordinates; cf. Darrigol 2019b, p. 147.

that all material energy might be reducible to a combination of electromagnetic and gravitational energy, in which case $T_\mu^\mu = 0$ for the nongravitational part of the stress–energy tensor. Then it becomes possible to adopt the fully covariant

$$R_{\mu\nu} = -\kappa T_{\mu\nu} \qquad (9.114)$$

for the gravitational field equation. The equations of the article are retrieved by specializing the coordinates so that $\sqrt{-g} = 1$.[126]

The gravitational deviation of light and the advance of Mercury's perihelion

Einstein announced his new results (without the addendum) to the Berlin Academy on November 4, 1915. Two weeks later, on November 18, he communicated two basic predictions of the theory: a gravitational deviation of light twice as large as in his earlier theory; and an anomaly of about 43″ in the advance of Mercury's perihelion, in stunning agreement with the 45″ ± 5″ measured by astronomers. In the first approximation, Einstein gives

$$g_{ij} = -\delta_{ij} - \alpha x_i x_j / r^3, \quad g_{i0} = g_{0i} = 0, \quad g_{00} = 1 - \alpha/r \text{ with } r^2 = x_1^2 + x_2^2 + x_3^2 \qquad (9.115)$$

for the center-symmetric solution of the field equation in quasi-Cartesian coordinates (Latin indices run from 1 to 3). The coordinate velocity of light,

$$c(r) = \sqrt{(dx^1/dx^0)^2 + (dx^2/dx^0)^2 + (dx^3/dx^0)^2} \quad \text{for } g_{\mu\nu} dx^\mu dx^\nu = 0, \qquad (9.116)$$

depends on location. By Huygens's principle (in analogy with familiar reasoning for the propagation of light in a medium of variable index), this leads to the deviation $2\alpha/\Delta$ of light rays whose closest distance from the center is Δ, instead of the α/Δ given by Einstein's earlier theories. The coefficient α being $2GM/c^2$ for a central mass M, the deviation of rays passing close to the surface of the sun is about 1.7 ″.[127]

In order to compute the motion of a point-mass in a central gravitational field, Einstein uses the Γ coefficients derived from the expression (9.115) of the metric, except for the Γ_{00}^i coefficients, which he gets to the next order of approximation through $\partial_i \Gamma_{00}^i + \Gamma_{0\rho}^\sigma \Gamma_{0\sigma}^\rho = 0$. He then injects these coefficients into the geodesic equation

$$\frac{d^2 x^\mu}{ds^2} - \Gamma_{\nu\rho}^\mu \frac{dx^\nu}{ds} \frac{dx^\rho}{ds} = 0. \qquad (9.117)$$

Using the techniques of integration earlier elaborated with Besso in the *Entwurf* context, he finds

$$\delta\theta = 3\pi \frac{\alpha}{a(1 - e^2)}, \qquad (9.118)$$

which is 12/5 of the *Entwurf* value in Equation (9.72).

[126] Einstein 1915a, pp. 799–801.

[127] Einstein 1915b. Cf. Earman and Janssen 1993.

The final equation

In the former communication, Einstein still defends the electromagnetic–gravitational reduction of matter for which the trace T_μ^μ vanishes and the gravitational field equation can
consistently be $R_{\mu\nu} = -\kappa T_{\mu\nu}$. Under this hypothesis, he comments, "space and time are de-
prived of the last remnant of objective reality [*objektiver Realität*]." However, in a footnote
to the printed version of his text, he mentions he will soon show how to do without this
hypothesis. The new reasoning is found in a third communication to the Berlin Academy
on November 25.[128]

The basic idea is that a slightly modified version of the field equation,

$$R_{\mu\nu} = -\kappa(T_{\mu\nu} - \tfrac{1}{2}g_{\mu\nu}T) \qquad \text{(with } T \equiv T_\mu^\mu = g^{\mu\nu}T_{\mu\nu}) \qquad (9.119)$$

no longer requires $T_\mu^\mu = 0$. Einstein here uses the kind of calculations with which he
related T_μ^μ to $\sqrt{-g}$ in his first November communication. These involve complex, nonco-
variant manipulations in which the stress–energy tensor of the gravitational field still plays
a prominent role. We would now rely on the contracted Bianchi identity

$$D^\mu(R_{\mu\nu} - \tfrac{1}{2}g_{\mu\nu}R) \equiv 0 \qquad (9.120)$$

which is a consequence of the invariance of the field action (through Equation (9.98)). If the
field equation is taken to be $R_{\mu\nu} = -\kappa T_{\mu\nu}$, then the former identity combined with $D^\mu T_{\mu\nu} = 0$
gives $D^\mu(g_{\mu\nu}R) = 0$, which requires $\partial_\mu R = 0$ since $D^\mu g_{\mu\nu} \equiv 0$. Consequently, R_μ^μ and T_μ^μ must
both vanish. This is avoided by adopting Equation (9.119) instead of $R_{\mu\nu} = -\kappa T_{\mu\nu}$.

Einstein concludes:[129]

> With this, the general theory of relativity is finally made a closed logical construct.
> The relativity principle in its most general conception, according to which the space
> and time coordinates become physically meaningless, inevitably leads to a completely
> definite theory of gravitation explaining the motion of Mercury's perihelion.

What happened to the hole argument?

Whereas in the *formale Grundlage* of November 1914 the hole argument played a role in
justifying adapted coordinates, no mention of this argument appears in Einstein's later
writings. The reason cannot be incompatibility with the general covariance of the fundamental equations, because Einstein already believed any physically meaningful equation
to be the expression of a generally covariant equation in a limited subclass of reference
systems. Nor can it be a new awareness that coordinates are mere labels devoid of physical
meaning, because Einstein had expressed this opinion three years earlier. Yet, the lack of
any reference to adapted coordinates in the two last communications of November 1915

---

[128] Einstein 1915b, p. 831; 1915c. As Einstein mentions in his footnote (1915b, p. 831n), the prediction for the
relativistic precession is unaltered by the change in the fundamental equation (because this prediction relies on the
source-free equation only).

[129] Einstein 1915c, p. 847.

suggest that by then Einstein had abandoned the argument, since the two notions were intimately related in his mind. Most evidently, he could not adopt a fully covariant field equation (already in the addendum to the first communication) without admitting a basic error in the argument.[130]

Einstein spelled out this error in a letter to Ehrenfest of December 26, 1915:[131]

> From the fact that the two systems $g_{\mu\nu}(x)$ and $g'_{\mu\nu}(x)$ in the same reference frame satisfy the conditions of the gravitational field, one cannot deduce any contradiction to the univocality of evolution. The seeming force of this argument is lost as soon as one considers 1) that the reference system is nothing real, and 2) that the simultaneous realization of two different g-systems (better: two different gravitational fields) in the same domain of the continuum is impossible by the very nature of the theory. Section 12 [of the *formale Grundlage*, about the hole argument] should be replaced by the following:
>
> The physically real in the world's evolution (in contrast with what depends on the choice of the reference system) consists in coincidences in space and time. For instance, real is the intersection or the non-intersection of two different world-lines. Consequently, the propositions concerning the physically real are unaltered by (univocal) transformations of coordinates. When two $g_{\mu\nu}$ systems (and all variables necessary to the description of the world) are such that one can be deduced from the other by a space-time transformation, then they are completely equivalent. Indeed, they share the same point-like coincidences in space and time, that is, they share all observables. At the same time, these considerations show how natural the demand of general covariance is.

In brief, the functions $g_{\mu\nu}(x)$ and $g'_{\mu\nu}(x)$ describe the same metric manifold in different systems of coordinates, even though the notation suggests differently (the argument x is the same for the two functions). When the metric field $g_{\mu\nu}(x)$ is given, we know only that there exists a labeling of the points of the manifold for which the interval between two point labeled by x^μ and $x^\mu + dx^\mu$ is given by $\sqrt{g_{\mu\nu}dx^\mu dx^\nu}$, but we have no independent knowledge of how the labeling is done. Now, the functions $g_{\mu\nu}(x)$ and $g'_{\mu\nu}(x)$ are constructed by means of a diffeomorphism $x \to x'$ such that $g_{\mu\nu}(x)dx^\mu dx^\nu = g'_{\mu\nu}(x')dx'^\mu dx'^\nu$. The intervals given by the first function with the labeling x of the manifold are therefore equal to the intervals given by the second function with the labeling x' of the manifold.

Einstein did not publicly reject the hole argument in the early years of general relativity (was he ashamed of it?). In his *Grundlage* of 1916, he simply replaced it with the coincidence argument, just as he told Ehrenfest he would do.[132]

---

[130] For Einstein's earlier beliefs regarding general covariance and the meaning of coordinates, see pp. 289, 291, 296, 298–301, 305–307, 309–310, 314, this volume.

[131] Einstein to Ehrenfest, December 26, 1915, *ECP* 8. See also Einstein to Besso, January 3, 1916, *ECP* 8. For a discussion of possible origins of the coincidence argument, cf. Howard and Norton 1993. For a fuller history of this argument, cf. Giovanelli 2021. On Einstein's exchanges with Ehrenfest, cf. Weinstein 2015b, section 2.10.

[132] Einstein 1916a, pp. 776–777. See p. 352, this volume.

Hilbert

In July 1915, Einstein visited Göttingen to deliver the Wolfskehl lectures on his *formale Grundlage* of general relativity. The mathematicians in the audience welcomed a theory that made so much use of the theory of invariants and differential calculus on a Riemannian manifold. Most impressed was David Hilbert, who saw an opportunity to develop his own program of subsuming physics under a unified system of mathematical axioms, following the model of geometry. In continuity with his old interest in electron theory, he wanted a unified theory of gravitation and electromagnetism in which the electrons would appear as singularities. As he knew, in 1912 Gustav Mie had developed a monistic theory of matter and electromagnetic field, based on a nonquadratic, Lorentz-invariant action (the "world function") for the electromagnetic potential. Hilbert believed he would obtain a higher unification by combining Mie's and Einstein's theories. It is not clear when he started to implement this project. On November 13, 1915, he wrote to Einstein about some of his main results: a derivation of the electromagnetic field equation from the gravitational field equation, an appeal to energy conservation to restrict the system of coordinates, and a new derivation of the electromagnetic energy–momentum tensor. On November 19, he sent to the *Göttinger Nachrichten* a first version of this theory, about which he lectured on the following day. This first version is documented through a set of proofs printed on December 6.[133]

Hilbert's first axiom is the existence of a world function (action)

$$S_F = \int \Lambda(g_{\mu\nu}, \partial_\rho g_{\mu\nu}, \partial_\rho \partial_\sigma g_{\mu\nu}, A_\mu, \partial_\mu A_\nu)\sqrt{-g}\mathrm{d}^4 x \qquad (9.121)$$

from which the equations of motion result by varying the integral with respect to the gravitational potentials $g_{\mu\nu}$ and the electromagnetic potentials A_μ. He includes the second-order derivatives of $g_{\mu\nu}$ in the Lagrangian density for he knows they are needed to form a nontrivial invariant action. This takes us to his second axiom, which is the full invariance of the Lagrangian density $g_{\mu\nu}$ with respect to any (smooth) change of the coordinates. As we saw, before November 1915 Einstein did not require the complete covariance of his field equations and he believed he had good physical reasons to restrict the covariance. Nonetheless, he had repeatedly expressed his confidence that the final theory of gravitation would be based on fully covariant equations, to be complemented with some ulterior physical restriction of the system of coordinates. Hilbert, as a mathematician eager to exploit the resources of the theory of invariants, naturally started with an invariant Lagrangian.[134]

Varying the action, Hilbert obtains the field equations

$$\frac{\delta \Lambda}{\delta g^{\mu\nu}} \equiv \frac{\partial \Lambda}{\partial g^{\mu\nu}} - \partial_\rho \frac{\partial \Lambda}{\partial(\partial_\rho g^{\mu\nu})} + \partial_\rho \partial_\sigma \frac{\partial \Lambda}{\partial(\partial_\rho \partial_\sigma g^{\mu\nu})} = 0, \qquad (9.122)$$

---

[133] Mie 1912a, 1912b, 1913; Hilbert to Einstein, November 13, 1915, in *ECP* 8, p. 185; Hilbert [1915]. Cf. Earman and Glymour 1978b; Corry 1999a, 1999b; Sauer 1999, 2005a; Vizgin 2001; Renn and Stachel 2007. Mie's favorite Lagrangian was $-\frac{1}{4}F^{\mu\nu}F_{\mu\nu} + \alpha(A_\mu A^\mu)^3$ in modern notation.

[134] Hilbert [1915], p. 2.

$$\frac{\delta \Lambda}{\delta A_\mu} \equiv \frac{\partial \Lambda}{\partial A_\mu} - \partial_\rho \frac{\partial \Lambda}{\partial (\partial_\rho A_\mu)} = 0. \tag{9.123}$$

He next asserts an important consequence of the invariance of the field Lagrangian Λ: there are four independent linear relations between the resulting field equations (Theorem I). In the case when the Lagrangian depends on the metric field only (Theorem III), these relations read

$$D^\nu \frac{\delta \Lambda}{\delta g^{\mu\nu}} = 0, \tag{9.124}$$

which Hilbert obtains though manipulations of the "polar" invariant

$$J = \frac{\partial \Lambda}{\partial g^{\mu\nu}} h^{\mu\nu} + \frac{\partial \Lambda}{\partial (\partial_\rho g^{\mu\nu})} \partial_\rho h^{\mu\nu} + \frac{\partial \Lambda}{\partial (\partial_\rho \partial_\sigma g^{\mu\nu})} \partial_\rho \partial_\sigma h^{\mu\nu} \tag{9.125}$$

based on the metric tensor $g_{\mu\nu}$ and the arbitrary symmetric tensor $h_{\mu\nu}$. These manipulations resemble the partial integrations Einstein had used while computing the variations $\delta_g S$ and $\delta_x S$ (see pp. 311–312, this volume) in his *Formale Grunlage*, if one takes $h^{\mu\nu} = \delta g^{\mu\nu}$ in the first case and $h^{\mu\nu} = \delta_x g^{\mu\nu} = D^\mu \delta x^\nu + D^\nu \delta x^\nu$ in the second case.[135]

Hilbert further assumes the Lagrangian Λ to be the sum of a gravitational part $\Lambda_g(g_{\mu\nu}, \partial_\rho g_{\mu\nu}, \partial_\rho \partial_\sigma g_{\mu\nu})$ and an electromagnetic part $\Lambda_e(g_{\mu\nu}, A_\mu, \partial_\mu A_\nu)$. The gravitational field equation then takes the form

$$\frac{\delta \Lambda_g}{\delta g^{\mu\nu}} + \frac{\partial \Lambda_e}{\partial g^{\mu\nu}} = 0. \tag{9.126}$$

Exploiting the invariance of the electromagnetic Lagrangian, Hilbert proves that the field derivatives can enter Λ_e only through the covariant combination $F_{\mu\nu} = \partial_\mu A_\nu - \partial_\nu A_\mu$. In addition, he proves that the derivative $\partial \Lambda_e / \partial g^{\mu\nu}$ is the generally covariant extension of the energy–momentum tensor $T_{\mu\nu}$ introduced by Minkowski, Laue, and Mie for the electromagnetic field in Minkowskian context. He thus has a new, important way to derive the energy–momentum tensor of a nongravitational field.[136]

Although the relevant section of the proofs is missing, Hilbert is likely to have taken $\Lambda_g = R$, since he knew from Riemann this was the simplest nontrivial invariant. As far as we can judge from the extant proofs, he did not derive the corresponding field equation. Had he done so, he would have gotten Einstein's final gravitational field equation. He was more concerned with the general structure of his theory than with physical consequences and predictions. In his eyes, his most important result was the derivation of the electromagnetic field equations from the gravitational ones.

---

[135] Hilbert [1915], pp. 2–3 (Theorem I), p. 9 (Theorem III). As Hilbert did not remark, for $\Lambda = R$ Equation (9.124) is the contracted Bianchi identity.

[136] Hilbert [1915], pp. 9–10.

This derivation begins with the relation

$$2\frac{\partial \Lambda_e}{\partial g^{\mu\nu}}g^{\nu\rho} - A_\mu \frac{\partial \Lambda_e}{\partial A_\rho} - F_{\mu\nu}\frac{\partial \Lambda_e}{\partial(\partial_\rho A_\nu)} = 0, \tag{9.127}$$

which results from the invariance of Λ_e with respect to linear transformations. Using $\mathbf{T}_{\mu\nu} = \partial \Lambda_e / \partial g^{\mu\nu}$ and $\Lambda_e = \sqrt{-g}\Lambda_e$, we also have

$$2\mathbf{T}^\rho_\mu = A_\mu \frac{\partial \Lambda_e}{\partial A_\rho} + F_{\mu\nu}\frac{\partial \Lambda_e}{\partial(\partial_\rho A_\nu)} - \delta^\rho_\mu \Lambda_e. \tag{9.128}$$

Taking the ∂_ρ divergence of this relation, using the identities

$$\partial_\mu \Lambda_e = \partial_\mu g^{\nu\rho}\frac{\partial \Lambda_e}{\partial g^{\nu\rho}} + \partial_\mu A_\rho \frac{\partial \Lambda_e}{\partial A_\rho} + \partial_\mu \partial_\rho A_\nu \frac{\partial \Lambda_e}{\partial(\partial_\rho A_\nu)}, \tag{9.129}$$

$$\partial_\mu g^{\nu\rho}\frac{\partial \Lambda_e}{\partial g^{\nu\rho}} = \partial_\mu g^{\nu\rho}\mathbf{T}_{\nu\rho} = -2\partial_\rho \mathbf{T}^\rho_\mu, \tag{9.130}$$

and exploiting the skew-symmetric character of $\partial \Lambda_e / \partial(\partial_\rho A_\nu)$, Hilbert gets

$$(F_{\rho\mu} + A_\mu \partial_\rho)\frac{\delta \Lambda}{\delta A_\rho} = 0. \tag{9.131}$$

Having thus discovered the vanishing of four independent combinations of the Lagrangian derivative, Hilbert takes it to imply the field equation $\delta \Lambda / \delta A_\mu = 0$. In his words: "*This is the whole mathematical expression of . . . the character of electrodynamics as a consequence of gravitation.*"[137]

Having studied Einstein's *formale Grundlage*, Hilbert was aware of Einstein's claim, based on the hole argument, that covariant field equations were unable to fully determine the fields under given boundary conditions. Echoing this claim, Hilbert writes:

> Since our mathematical theorem [Theorem I] teaches us that the Axioms I and II give only ten mutually independent equations for the 14 potentials, and since in addition general covariance does not allow more than ten independent equations for the 14 potentials $g_{\mu\nu}$ and A_μ, then, in order to give to the fundamental equations of physics the deterministic character expressed in Cauchy's theory of differential equations, we must add to the field equations [Equations (9.122–123)] four non-covariant equations.

---

[137] Hilbert [1915], p. 12 (his emphasis). My Equations (9.127–128) correspond to Hilbert's Equations (9.22–23). I have avoided the useless detour Hilbert makes through Equation (9.126) and Equation (9.132). An even simpler derivation is obtained by directly exploiting the invariance of the total Lagrangian: $\delta_x S = \int \delta_x \Lambda \, \mathrm{d}^4 x = 0$ with $\delta_x \Lambda = (\delta \Lambda / \delta g^{\mu\nu})\delta_x g^{\mu\nu} + (\delta \Lambda / \delta A_\mu)\delta_x A_\mu$ and $\delta_x A_\mu = A_\nu \partial_\mu \delta x^\nu - \partial_\nu A_\mu \delta x^\nu$. The first term of $\delta_x \Lambda$ vanishes when the gravitational field equation is satisfied. Partial integration of the second yields Equation (9.131).

Like Einstein in his early discussions of the hole paradox, Hilbert seeks the needed nonco-variant equations in energy considerations. Through J-based manipulations similar to the ones he used for deriving the identity of Equation (9.124), he arrives at the identity

$$\partial_\nu \mathbf{e}_\mu^\nu = \mathbf{e}_\mu \qquad (9.132)$$

$$\text{with } \mathbf{e}_\mu^\nu = \partial_\mu g^{\rho\sigma} \frac{\partial \Lambda}{\partial(\partial_\nu g^{\rho\sigma})} - 2g^{\nu\rho} \frac{\delta\Lambda}{\delta g^{\mu\rho}} \qquad (9.133)$$

$$\text{and } \mathbf{e}_\mu = \partial_\mu g^{\nu\rho} \frac{\partial \Lambda}{\partial g^{\nu\rho}} + \partial_\mu \partial_\nu g^{\rho\sigma} \frac{\partial \Lambda}{\partial(\partial_\nu g^{\rho\sigma})} \equiv \partial_\mu^{(g)} \Lambda. \qquad (9.134)$$

The identity (9.132) is easily obtained as a consequence of $D^\nu(\delta\Lambda/\delta g^{\mu\nu}) = 0$ (I have omitted terms involving $\partial\Lambda/\partial(\partial_\rho\partial_\sigma g^{\mu\nu})$, for they do not occur in the effective Lagrangian in Einstein's theory). Hilbert wants a divergence relation that can pass for energy–momentum conservation. He therefore requires, as his third axiom, the vanishing of $\partial_\mu^{(g)}\Lambda$, which is the ordinary, noncovariant semi-total derivative of the Lagrangian density (semi-total because A_μ and $\partial_\mu A_\nu$ are kept constant during the derivation).[138]

It is interesting to compare the pseudo-tensor $\mathbf{e}_\mu^\nu$ with the energy–momentum pseudo-tensor of the gravitational and electromagnetic fields in Einstein's theory. For the electromagnetic part of $\mathbf{e}_\mu^\nu$, we have

$$\left[\mathbf{e}_\mu^\nu\right]_e = -2\kappa \mathbf{T}_\mu^\nu. \qquad (9.135)$$

For the gravitational part of $\mathbf{e}_\mu^\nu$, we have

$$\left[\mathbf{e}_\mu^\nu\right]_g = -2\kappa \mathbf{t}_\mu^\nu + \mathbf{X}_\mu^\nu, \qquad (9.136)$$

$$\text{with } \mathbf{t}_\mu^\nu = \frac{1}{2\kappa}\left(\delta_\mu^\nu \Lambda_g - \partial_\mu g^{\rho\sigma} \frac{\partial \Lambda_g}{\partial(\partial_\nu g^{\rho\sigma})}\right) \text{ and } \mathbf{X}_\mu^\nu = \delta_\mu^\nu \Lambda_g - 2\mathbf{G}_\mu^\nu. \qquad (9.137)$$

Neglecting the nonessential factor -2κ, these expressions depart from the more familiar ones through the contribution $\mathbf{X}_\mu^\nu$ to the energy–momentum of the gravitational field. The divergence of this contribution vanishes, because Hilbert's condition $\partial_\nu^{(g)} \Lambda = 0$ and the identity $D^\nu G_{\mu\nu} = 0$ together imply

$$\partial_\nu \Lambda_g = -\partial_\nu^{(g)} \Lambda_e = -\kappa(\partial_\mu g^{\nu\rho})\mathbf{T}_{\nu\rho} = -(\partial_\mu g^{\nu\rho})\mathbf{G}_{\nu\rho} = 2\partial_\nu \mathbf{G}_\mu^\nu. \qquad (9.138)$$

Hilbert's divergence equation $\partial_\nu \mathbf{e}_\mu^\nu = 0$ is therefore equivalent to Einstein's $\partial_\nu(\mathbf{T}_\mu^\nu + \mathbf{t}_\mu^\nu) = 0$ under Hilbert's coordinate condition only, despite the different expressions for the gravitational energy–momentum.

---

[138] Hilbert [1915], pp. 3–4 (citation), 4–8.

After reading Einstein's communications of November 1915, Hilbert must have realized the vacuity of his third axiom. He now understood that energy–momentum conservation could be expressed in any system of coordinates under Einstein's definition of the gravitational energy–momentum. In his final publication, he dropped the third axiom and altered his energy considerations to make them "invariant" (expressible in any system of coordinates). He also gave the gravitational field equation,

$$R_{\mu\nu} - \tfrac{1}{2}g_{\mu\nu}R = -\kappa T_{\mu\nu}, \tag{9.139}$$

which is equivalent to Einstein's Equation (9.119), together with the dubious claim that the field operator can be obtained without calculation, through the mere consideration of the invariants that can be built from the metric tensor and its first and second derivatives.[139] His other results remained, with improved derivations. Again, he regarded his derivation of the electromagnetic field equation from the gravitational one as his most important achievement.[140]

Hilbert's efforts and the publicity he gave them annoyed Einstein. To a friend he complained about the "nostrification" (*nostrifizieren*) of his results. In his view, all Hilbert had done was to give an alternative derivation of his equations in conjunction with Mie's improbable theory of matter, amidst physically opaque, pseudo-energetic considerations:

> Hilbert's presentation does not please me. It is unnecessarily specialized with regard to the concept of "matter," it is unnecessarily complicated, and it is not built in a honest manner à la Gauss (by veiling the methods, it gives the illusion of the super-human).

To make it worse, Hilbert's memoir appeared as "communicated on 20 November 1915," without mention of the important differences between the original communication and the published text. This is the origin of the still widely spread opinion that Hilbert obtained the exact gravitational field equation before Einstein communicated it to the Berlin Academy (on November 25). Leo Corry, Jürgen Renn, and John Stachel, also Tilman Sauer, have discredited this claim in careful analyses of the first set of proofs of Hilbert's paper.[141]

This does not mean that Hilbert's contribution to general relativity should be regarded as minor and derivative. Hilbert was first to understand the principal advantage of basing the theory on the principle of least action: covariant field equations can be (almost)

---

[139] The Lagrangian being of second differential order, there is no obvious reason to anticipate a second-order differential equation. Moreover, there is no obvious reason to privilege Einstein's specific combination of $R_{\mu\nu}$ and $g_{\mu\nu}R$ without knowing the Bianchi identity.

[140] Hilbert 1915. The relevant volume of the *Göttinger Nachrichten* appeared on March 21, 1916. Hilbert could send off prints by mid-February; cf. Sauer 1999, p. 565.

[141] Einstein to Zangger, November 26, 1915; Einstein to Ehrenfest, May 24, 1916, *ECP* 8, p. 288; Corry, Renn, and Stachel 1997. The adverse claim, found in Winterberg 2004, that the short missing section of the November proofs must have contained the Einstein equation, is highly improbable: Hilbert was not concerned with the explicit form of the field equations in his memoir; the derivation from the invariant Lagrangian is not easy and would have needed more space; and the hints given in Hilbert's final publication at such a derivation are incorrect. Cf. Corry, Renn, and Stachel [2004]; Sauer 2005a. See also Weinstein 2015b, section 2.9.

uniquely and simply obtained by requiring the invariance of the action. He was first to exploit this invariance in a derivation of the contracted Bianchi identity, and he was first to obtain the energy–momentum tensor of matter (generally speaking) by taking the derivatives of the action with respect to the metric coefficients. He thus unveiled some of the basic structure of general relativity. He would certainly have obtained the exact field equation of general relativity before Einstein if only he had cared to compute the variation of his invariant field action.

In the letter Einstein wrote to Hilbert after seeing an early version of his memoir, he rightly noted that the discovery of covariant field equations did not suffice to found general relativity as a physical theory. He had known such equations for years without understanding they contained Newtonian gravity in a first approximation as well as verifiable departures from this approximation. Yet there is little doubt that Einstein would have come much faster to this understanding if he had from the start based the theory on an invariant action for the gravitational field, and not on a misleading stress principle.[142]

In a letter to Lorentz written on January 17, 1916, Einstein at last judged it desirable to found his new theory on an invariant action of the gravitational field. As he knew, the only simple candidate is the integral

$$S_F = \int R\sqrt{-g}\,\mathrm{d}^4x \tag{9.140}$$

based on the Riemann scalar R. Einstein also indicated that the undesired dependence of the Lagrangian on second-order derivatives could be eliminated by subtracting a (vanishing) surface integral from the original action. The result of this elimination,

$$S_F = \int \Lambda_F \sqrt{-g}\,\mathrm{d}^4x \text{ with } \Lambda_F = g^{\sigma\tau}(\Gamma^\alpha_{\sigma\beta}\Gamma^\beta_{\tau\alpha} - \Gamma^\alpha_{\sigma\tau}\Gamma^\beta_{\alpha\beta}), \tag{9.141}$$

agrees with the Lagrangian $g^{\sigma\tau}\Gamma^\alpha_{\sigma\beta}\Gamma^\beta_{\tau\alpha}$ Einstein had used in November 1915 under the restriction $\sqrt{-g} = 1$. Lorentz, who had already worked on a Hamiltonian formulation of the *Entwurf* theory, soon published a proof that Einstein's new field equation indeed resulted from varying the invariant action.[143]

On October 26, 1916, Einstein discussed the invariant action and its consequences at the Berlin Academy of sciences, with brief references to Hilbert's and Lorentz's relevant works. In this communication, he first gives the reduction to the effective Lagrangian Λ_F, and then replicates the calculation found in his *formale Grundlage* of November 1914 for the variation $\delta_x S_F$ of a similar action during the change of coordinates $x^\mu \to x^\mu + \delta x^\mu$, with the result of Equation (9.94):

$$\delta_x S_F = 2\int B_\sigma \delta x^\sigma \mathrm{d}^4x, \text{ with } B_\sigma = \partial_\rho \partial_\nu \left(g^{\mu\nu}\frac{\partial \Lambda_F}{\partial(\partial_\rho g^{\mu\sigma})}\right).$$

[142] Einstein to Hilbert, November 18, 1915.

[143] Einstein to Lorentz, *ECP* 8, pp. 245–247; Lorentz 1916, pp. 280–281; Einstein [1916b].

Whereas in the *formale Grundlage* the varied action was invariant with respect to linear transformations only, it is now invariant with respect to any smooth transformation of the coordinates. The variation $\delta_x S_F$ therefore vanishes for any δx^σ, and B_μ must vanish identically. Einstein uses this result to prove that the field equation in the presence of matter of Lagrangian density Λ_M,

$$\partial_\rho \left(g^{\mu\nu} \frac{\partial \Lambda_F}{\partial (\partial_\rho g^{\mu\sigma})} \right) = -\kappa (t^\nu_\sigma + T^\nu_\sigma) \tag{9.142}$$

$$\text{with } t^\nu_\mu = \frac{1}{2\kappa} \left(\delta^\nu_\mu \Lambda_F - \partial_\mu g^{\rho\sigma} \frac{\partial \Lambda_F}{\partial (\partial_\nu g^{\rho\sigma})} \right) \text{ and } T_{\mu\nu} = -\frac{\partial \Lambda_M}{\partial g^{\mu\nu}},$$

implies the equation $\partial_\nu (t^\nu_\sigma + T^\nu_\sigma) = 0$ for the conservation of the total energy–momentum. The stress–energy pseudo-tensor t^ν_μ still plays a role in this reasoning, and there is no trace of Hilbert's more covariant reasoning.[144]

Remember that in November 1914, Einstein defined the adapted coordinates by $\delta_x S_F = 0$. This definition becomes impossible when the varied action is invariant. Using fully covariant reasoning à la Hilbert instead of Einstein's recycled calculations, we find

$$\delta_x S_F = -2 \int D^\nu G_{\mu\nu} \delta x^\mu \sqrt{-g} \, d^4 x \text{ with } G_{\mu\nu} = R_{\mu\nu} - \tfrac{1}{2} g_{\mu\nu} R, \tag{9.143}$$

which yields the contracted Bianchi identity $D^\nu G_{\mu\nu} \equiv 0$. Owing to this identity, the field equations $G_{\mu\nu} = -\kappa T_{\mu\nu}$ are not independent. They determine the ten functions $g_{\mu\nu}(x)$ only up to four arbitrary functions. This freedom exactly corresponds to the indetermination exhibited in the hole argument. As Einstein noted with reference to Hilbert, the four relations $D_\mu T^{\mu\nu} = 0$ constrain the material process and they fully determine the differential equations of this process in the simple case in which four equations suffice to this end.[145]

Conclusions

Having reached the end of Einstein's difficult quest for a generalized relativity, we may now reflect on his heuristics principles, his troubles with the Riemann tensor-based options, his various errors on the way to truth, and the continuities and discontinuities in his efforts. Let us first recapitulate the main steps Einstein took on his way to general relativity.

Overview

In 1907, Einstein introduced the equivalence principle according to which the effects of gravitation of intensity **g** in a small portion of space can be simulated by the acceleration −**g** of the reference frame. In a direct application of this principle, as given by Einstein in

---

[144] Einstein 1916d. Cf. Janssen and Renn 2007, pp. 900–911. Einstein (1916d, p. 1111) accused Hilbert of overspecializing the material action, as if his derivation of the contracted Bianchi identity depended on it.

[145] Einstein 1916a, p. 810.

1911, the frequency of monochromatic light emitted from a terrestrial source at the moderate elevation h and received at sea level will appear to be increased by gh/c^2, because in a frame of upward acceleration g, the receiver acquires the velocity $g(h/c)$ toward the source during the (approximate) traveling time h/c and therefore sees a light Doppler-shifted by the amount $g(h/c)/c$. Yet, for a properly defined time in the accelerated frame, the frequency of light should be conserved during its travel from the emitter to the receiver. Einstein eliminates this discrepancy with the former result by assuming that natural clocks (spectral sources) give a "local time" differing from this truer time. The velocity of light being the constant c with respect to the local time, its true value must depend on the gravitational potential $\Phi = -mg \cdot \mathbf{r}$ according to $\hat{c}(\mathbf{r}) = c(1 + \Phi/c^2)$. Generalizing this law to any form of the potential and using analogy with a transparent medium of variable optical index, Einstein predicts that a heavy spherical body should deviate light by the amount $2GM/c^2\Delta$, wherein G denotes the gravitational constant and Δ the closest distance of the light ray from the center of the body.

In another kind of reasoning developed in 1907 and 1911, Einstein directly compares the time and space measured in an inertial frame with the time and space measured in a constantly accelerated frame. Assuming that the acceleration does not distort the frame and that the velocity of light in the accelerated frame depends on location only, he requires

$$dx^2 - \hat{c}^2(x)dt^2 = d\xi^2 - d\tau^2 \tag{9.144}$$

for corresponding differentials of the coordinates in the two frames (Greek letters refer to the inertial frame, Latin letters to the accelerated frame, the x axis is parallel to the acceleration, and the time unit is chosen so that $c = 1$). This is possible if $\hat{c}(x)$ has the affine form

$$\hat{c} = \hat{c}_0 + ax. \tag{9.145}$$

For sufficiently small values of the coordinates, the transformation then is

$$\xi = x + a\hat{c}t^2/2, \quad \tau = \hat{c}t. \tag{9.146}$$

Together with the equivalence principle, this transformation enables Einstein to predict the effects of a uniform gravitation field on physical phenomena.

In early 1912, Einstein generalized this way of reasoning to an arbitrary static field by considering small neighborhoods and small time lapses in which the variation of the gravitational force may be neglected. The interval between two infinitesimally close events now takes the more general form

$$ds^2 = \hat{c}^2(\mathbf{r})dt^2 - d\mathbf{r}^2. \tag{9.147}$$

Einstein first obtains the equations of motion in this field by local transformation to a free-falling frame, although he later realized that they derive from

$$\delta \int ds = 0. \tag{9.148}$$

They may be written as

$$\mathbf{f} = d\mathbf{p}/dt, \quad \text{with} \quad \mathbf{p} = \frac{\mathbf{v}}{\hat{c}\sqrt{1 - v^2/\hat{c}^2}} \quad \text{and} \quad \mathbf{f} = -\frac{\nabla\hat{c}}{\sqrt{1 - v^2/\hat{c}^2}} \tag{9.149}$$

for a unit mass-point. The function $\hat{c}(\mathbf{r})$ thereby plays the role of the gravitational potential.

For the relation between this potential and the density ρ of matter, Einstein generalizes Poisson's equation to

$$\frac{\Delta\hat{c}}{\hat{c}} - \frac{1}{2}\left(\frac{\nabla\hat{c}}{\hat{c}}\right)^2 = k\rho, \tag{9.150}$$

The form of the first-order term is dictated by homogeneity, and the second term is needed so that the force density $-\rho\nabla\hat{c}$ on matter of density ρ derives from the stress system

$$\sigma_{ij} = -4k^{-1}(\partial_i\partial_j\sqrt{\hat{c}} - \tfrac{1}{2}\delta_{ij}\partial_k\sqrt{\hat{c}}\partial_k\sqrt{\hat{c}}), \tag{9.151}$$

and therefore satisfies the equality of action and reaction.

With this theory, Einstein believed he had found the unique static theory that simultaneously satisfies the equivalence principle, Newtonian correspondence, and the stress principle. He published it in early 1912 and then tried to generalize it to an arbitrary gravitational field. His successive trials are found in the Zürich notebook written in 1912–1913.

In Einstein's eyes, higher generality of the gravitational field means higher generality of the permitted reference frames. This is why he looks for a theory in which the motion of particles would be given by the geodesics for the general metric

$$ds^2 = g_{\mu\nu}dx^\mu dx^\nu. \tag{9.152}$$

The geodesic equation being covariant with respect to any change of coordinates, he wants the gravitational field equation to be a covariant generalization of the Poisson equation. In addition, he expects the metric field to satisfy a first-order differential equation that would generalize the equation $\partial_0 g_{00} = 0$ for the static case. That is to say, he is willing to restrict the covariance to physically meaningful coordinates (in the static case, he would exclude coordinates for which the metric coefficients become time-dependent). At the same time, for the sake of the equivalence principle he wants the covariance to be broad enough to include transformations to accelerated or rotating frames.

At some point in his notebook, Einstein introduces the energy–stress tensor

$$\theta^{\nu\rho} = \rho_0 \frac{dx^\nu}{ds} \frac{dx^\rho}{ds} \tag{9.153}$$

for a dust of density $\rho_0\sqrt{-g}$, as the natural counterpart of the mass density in a tensor generalization of Poisson's equation. The particles of the dust satisfy the geodesic equation if and only if

$$\partial_\nu(\sqrt{-g}g_{\mu\rho}\theta^{\nu\rho}) = \tfrac{1}{2}\sqrt{-g}\partial_\mu g_{\nu\rho}\theta^{\nu\rho}. \tag{9.154}$$

Einstein observes the general covariance of this equation—which we would now write as $D_\nu\theta^{\mu\nu} = 0$—and this gives him another reason to seek a covariant gravitational field equation. Interpreting $\tfrac{1}{2}\sqrt{-g}\partial_\mu g_{\nu\rho}\theta^{\nu\rho}$ as the four-force acting on the dust, he requires it to derive from a gravitational stress tensor, as he has already done in his static theory. This is the stress principle.

A tensor generalization of the Poisson equation would read

$$G^{\mu\nu} = \kappa\theta^{\mu\nu}, \tag{9.155}$$

wherein the field operator $G^{\mu\nu}$ is a second-order combination of ∂_μ and $g_{\mu\nu}$. Einstein explores a few naive combinations until he learns from Grossmann about the Riemann tensor, which offers a straightforward way to build a fully covariant field operator. Einstein first tries the Ricci tensor $R_{\mu\nu}$, which is the first contraction of the Riemann tensor. In a weak-field approximation, he expects the gravitational field to propagate at the velocity c and he therefore wants the field equation to agree with the d'Alembertian equation $\Box g_{\mu\nu} = -\kappa\theta_{\mu\nu}$. For this purpose, he introduces the harmonic coordinate condition $g^{\mu\nu}\{^\rho_{\mu\nu}\} = 0$, which reduces the second-order, linear terms of the field operator to the d'Alembertian operator. He gives up after finding this condition to be incompatible with his earlier static metric. He next tries the unimodularly covariant field operator

$$R^\times_{\mu\nu} = R_{\mu\nu} - D_\nu\partial_\mu \ln\sqrt{-g} = -\partial_\rho\{^\rho_{\mu\nu}\} - \{^\sigma_{\mu\rho}\}\{^\rho_{\nu\sigma}\}, \tag{9.156}$$

whose second-order part takes the d'Alembertian form under the condition $\partial_\mu g^{\mu\nu} = 0$. He does not dwell on this option and switches to the much simpler

$$R^{\times\times}_{\mu\nu} = \tfrac{1}{2}\partial_\rho(g^{\rho\sigma}\partial_\sigma g_{\mu\nu}) + g^{\rho\alpha}g^{\sigma\beta}\partial_\alpha g_{\mu\rho}\partial_\beta g_{\nu\sigma}, \tag{9.157}$$

which has the same limited covariance as the expression $\vartheta_{\mu\nu\rho} = \partial_\rho g_{\mu\nu} + \partial_\mu g_{\nu\rho} + \partial_\nu g_{\rho\mu}$.

Einstein hopes his various restrictions of the covariance still allow for rotating frames. He also wants his tentative field operators to comply with the stress principle: the product $\sqrt{-g}\partial_\mu g^{\nu\rho}G_{\nu\rho}$ should be expressible as the four-divergence of a tensor built from the metric field and its first derivatives. This seems implausible to Einstein for $G_{\mu\nu} = R_{\mu\nu}$ and for $G_{\mu\nu} = R^\times_{\mu\nu}$, and he can easily see that it is impossible for $G_{\mu\nu} = R^{\times\times}_{\mu\nu}$. In addition, $R_{\mu\nu}$ and $R^\times_{\mu\nu}$ do not meet his expectation that the gravitational operator should be a simple function of the derivatives $\partial_\mu g^{\nu\rho}$. In his opinion, these quantities naturally represent the components of the gravitational field because their contraction with the field-stress tensor of matter yields the four-force acting on matter.

While testing whether the field operator $R^{\times\times}_{\mu\nu}$ is compatible with the stress principle, Einstein discovers that he can easily build a compatible operator by correcting the d'Alembertian with nonlinear terms generated by partial integration. This leads him to the equations

$$\partial_\alpha(\sqrt{-g}g^{\alpha\beta}\Gamma^\sigma_{\nu\beta}) = -\kappa(\mathbf{t}^\sigma_\nu + \mathbf{T}^\sigma_\nu), \text{ with} \qquad (9.158)$$

$$\Gamma^\nu_{\mu\sigma} = \tfrac{1}{2}g^{\nu\rho}\partial_\mu g_{\rho\sigma} \text{ and } \mathbf{t}^\sigma_\nu = -\kappa^{-1}\sqrt{-g}(g^{\gamma\tau}\Gamma^\rho_{\mu\sigma}\Gamma^\mu_{\rho\tau} - \tfrac{1}{2}\delta^\sigma_\nu g^{\alpha\beta}\Gamma^\rho_{\mu\alpha}\Gamma^\mu_{\rho\beta}). \qquad (9.159)$$

The $\Gamma^\nu_{\mu\sigma}$ are the "field components"; the tensor densities $\mathbf{T}^\sigma_\nu$ and $\mathbf{t}^\sigma_\nu$ represent the stress–energy of matter and gravitational field, respectively, and they satisfy the equation of conservation

$$\partial_\nu(\mathbf{T}^\nu_\mu + \mathbf{t}^\nu_\mu) = 0. \qquad (9.160)$$

Einstein believes these relatively simple equations to be the only ones compatible with the stress principle and d'Alembertian correspondence. This is why he publishes them with Grossmann in early 1913 as an outline (*Entwurf*) of a new theory of gravitation, even before testing their compatibility with the equivalence principle.

The *Entwurf* equations are easily seen not to be generally covariant. This does not bother Einstein for he (erroneously) finds them to be compatible with the rotation metric while collaborating with Besso on the relativistic precession of a planet's perihelion. In the same collaboration, he arrives at the hole argument, following which the metric field within a matter-free hole cannot be uniquely determined by the distribution of matter outside the hole if the field equation is fully covariant: if a given field is a solution of a covariant equation any field obtained by diffeomorphic deformation of this field within the hole is also a solution. Einstein originally believed the hole argument to exclude full covariance at any level of the theory. In light of his and Fokker's geometric interpretation of the Einstein–Nordström scalar theory of gravitation, he changes his mind and he comes to hope that the *Entwurf* field equations could be obtained by specializing the coordinate systems in a Riemann tensor-based field equation.

Resuming his collaboration with Grossmann in early 1914, Einstein succeeds in characterizing the transformations under which the *Entwurf* field equation is covariant. The key, provided by the mathematician Paul Bernays, is to reduce the covariance properties of the field equation to the invariance properties of the field action from which it derives. Einstein and Grossmann find

$$S_F = -\int g^{\sigma\tau}\Gamma^\alpha_{\sigma\beta}\Gamma^\beta_{\tau\alpha}\sqrt{-g}\mathrm{d}^4 x \qquad (9.161)$$

for the relevant field action, introduce "adapted coordinates" such that the action is stationary under an infinitesimal variation of the coordinates, and prove that the field equation is invariant under any transformation between adapted coordinates. The four conditions of adaptation,

$$B_\nu \equiv \partial_\alpha\partial_\sigma(\sqrt{-g}g^{\alpha\beta}\Gamma^\sigma_{\nu\beta}) = 0, \qquad (9.162)$$

block the diffeomorphic freedom in the hole. Einstein believes them to be compatible with some nonlinear transformations, including rotating frames.

By November 1914, Einstein exploits these covariance considerations to design a new "formal" derivation of the *Entwurf* field equation. For this purpose, he starts with an arbitrary, linearly invariant field Lagrangian, and requires that the derived field equation $G_{\mu\nu} = \kappa T_{\mu\nu}$, when combined with the energy–momentum conservation $D^\mu T_{\mu\nu} = 0$, should automatically yield the condition for adapted coordinates (in the *Entwurf* case the conditions (9.162) indeed derive from Equations (9.158) and (9.160)). Otherwise, there would be $10 + 4$ equations for determining the 10 unknowns of the metric field. Einstein errs in developing this requirement, and he falsely concludes that it is compatible with the *Entwurf* action only.

Einstein becomes aware of this error in October 1915. He now understands that the choice of the action is very free and that any linearly covariant Lagrangian leads to a field equation compatible with the stress principle. Taking $\Gamma^\rho_{\mu\nu} = -\{{}^\rho_{\mu\nu}\}$ instead of $\Gamma^\nu_{\mu\sigma} = \frac{1}{2} g^{\nu\rho} \partial_\mu g_{\rho\sigma}$ in Equation (9.160) for the field action, he retrieves the unimodularly covariant field operator $R^\times_{\mu\nu}$ he had briefly considered in the Zürich notebook. He reconsiders this option in November, now understanding that the condition $\partial_\mu g^{\mu\nu} = 0$ can be used in the weak-field limit to get a d'Alembertian field equation and the static Newtonian subcase without compromising the unimodular covariance of the theory. He next derives a center-symmetric solution of the field equation $R^\times_{\mu\nu} = 0$ (with $g = -1$) and solves the geodesic equation for a particle moving in this field. The first approximation yields the Kepler motion, the second a precession compatible with the observed $43''$ anomaly in Mercury's case.

Instead of the unimodularly covariant field operator $R^\times_{\mu\nu}$, Einstein briefly considers the fully covariant $R_{\mu\nu}$, with the coordinate restriction $g = -1$. This choice is incompatible with the nonvanishing trace of the stress–energy tensor of usual matter: in modern terms, the equation $R_{\mu\nu} = -\kappa T_{\mu\nu}$, the contracted Bianchi identity $D^\mu(R_{\mu\nu} - \frac{1}{2} g_{\mu\nu} R) \equiv 0$, and the divergence law $D^\mu T_{\mu\nu} = 0$ together imply the undesired $\partial_\mu T = 0$. Einstein first avoids this difficulty by assuming that all matter is of electromagnetic–gravitational nature, in which case $T_{\mu\nu}$ is purely electromagnetic and has a vanishing trace. In the last week of November, he realizes that a simple variant of the field equation,

$$R_{\mu\nu} = -\kappa(T_{\mu\nu} - \tfrac{1}{2} g_{\mu\nu} T), \quad \text{alias } R_{\mu\nu} - \tfrac{1}{2} g_{\mu\nu} R = -\kappa T_{\mu\nu}, \tag{9.163}$$

is compatible with any value of the trace of the energy–stress tensor. This is his final choice.

As Hilbert soon indicated in his own, fully covariant field theory of matter, this equation simply derives from the invariant field action $\int R\sqrt{-g}\,\mathrm{d}^4x$ built from the Riemann scalar R. Using reasoning formally similar to Einstein's adaptation of coordinates to the noninvariant *Entwurf* Lagrangian, Hilbert also realized that the invariance of the gravitational action led to the identity $D^\mu G_{\mu\nu} \equiv 0$, as is needed for compatibility of the gravitational field equation with energy–momentum conservation. Had Einstein originally based his search for a gravitational equation on an invariant gravitational field Lagrangian, he could have taken this royal road to general relativity much earlier. This would not have annihilated the difficulties he had conciliating the field operator with the correspondence principle, but he would have thus avoided the chief and most persistent obstacle in his three years of struggle: the seeming incompatibility of covariant field operators with the stress principle.

Heuristic principles

As recognized by most commentators, Einstein relied on a few heuristic principles and on mathematical techniques and results he learned from Grossmann, Bernays, and others while he was constructing his theory. Although he was a novice in the theory of Riemannian manifolds and although he had to surmount a spontaneous dislike of abstract formalism, he gradually mastered the mathematics he needed for his purpose. His approach was thoroughly algebraic, with rare hints at geometric interpretation. Where we see geodesics, connections, and curvature, he saw trajectories, field strengths, and a covariant four-tensor, in conformity with the "absolute differential calculus" he inherited from Christoffel, Ricci-Curbastro, and Levi-Civita. A more geometric approach, based on analogy with the theory of surfaces or on Levi-Civita's later concept of affine connection, would plausibly have oriented Einstein's research differently.[146]

In contrast to the mathematical techniques, Einstein's heuristic principles had physical and intuitive meaning. They guided the construction of the theory, either by suggesting constructive elements, or by testing tentative constructions. We may consider five principles: the equivalence principle, general covariance, the stress principle, the correspondence principle, and the principle of least action.[147]

The equivalence of a gravitational field with the inertial field in an accelerated frame plays a capital role in Einstein's theory. It is based on an extrapolation of the empirically known equality of gravitational and inertial mass. Einstein used it in a variety of manners: to directly derive observable phenomena including the gravitational redshift, the gravitational curving of light rays, and the dependence of the gravitational mass on energy content; to determine motion in a uniform gravitational field; to justify tangent inertial frames and to introduce the invariant ds^2 first in the static case and then in the most general case; and to test possible field equations through their compatibility with the rotation metric.

Nowadays, physicists still use the equivalence principle as a springboard to the pseudo-Riemannian manifold of general relativity and to the geodesic principle, but they hurry to say that coordinates are mere labels devoid of physical significance in a generic metric field. In contrast, for Einstein coordinate systems remained tied to reference systems (he indifferently used *Koordinatensystem* and *Bezugsystem*), and he never ceased to believe that accelerated frames were physically meaningful (at least in small domains), in conformity with our common use of the earth-bound reference frame despite its not being in free fall. It is with respect to this frame that the familiar free-falling elevator is falling.[148]

---

[146] According to Lehmkuhl 2014, Einstein never conceived general relativity as a kind of geometrical reduction.

[147] I do not include electromagnetic analogy among Einstein's heuristic principles, although I agree with Janssen and Renn that there are evident formal analogies between Einstein's gravitational equations and the Minkowski formulation of electromagnetic theory. The reason is that I do not believe Einstein obtained his equations through this analogy. Rather, the analogy was produced by shared principles, for instance the stress principle, linear covariance, and least action. The stress principle is usually conflated with broader considerations of energy–momentum conservation. I believe it must be singled out and emphasized.

[148] Cf. Norton 1985 for a lucid discussion of Einstein's version of the equivalence principle and for Einstein's persisting defense of this version. On accelerated frames (local tetrads) in modern general relativity, cf. Synge 1960, pp. 114–118. On the consistency of the equivalence principle with modern general relativity, cf. Muñoz and Jones

Einstein arrived at general covariance through the equivalence principle. First, the principle removes the restriction to inertial frames that is characteristic of special relativity. Second, the application of the principle by means of local tangent frames leads to the invariant ds^2 and to the equation of motion $\delta \int ds = 0$, which is generally covariant. Third, the motion of a dust is described by a generally covariant equation, $D_\mu \theta^{\mu\nu} = 0$, wherein $\theta^{\mu\nu}$ denotes the stress–energy tensor of the dust. A pure mathematician like Hilbert would, at this point, focus on the intrinsic structure of the Riemannian manifold for space and time and require general covariance for all meaningful field equations on the manifold. Einstein did not do so because he kept interpreting general covariance physically through the equivalence principle. In this view, it could happen that a mathematically acceptable coordinate system would not correspond to a physically acceptable reference frame. Among the permitted frames, he considered local frames in which the metric takes the Minkowski form, and also frames that accelerate or rotate in the former frames. Although it is not clear what he meant a frame to be for the global manifold, one may easily imagine the reference "mollusk" he later popularized: a space-filling array of pre-clocks.[149]

At any rate, in his earliest writings on generalized relativity Einstein insisted on the frame dependence of space, time, energy, force, and field measurements. This metrological aspect helps understand why Einstein, in his subsequent quest for a gravitational field equation, could easily imagine a covariance restricted to physically and metrically significant frames. It also explains the implicit reification of coordinate systems behind the hole argument.[150]

In conformity with received field theories, Einstein required the conservation of the total energy and momentum of matter and field. In addition, he required the gravitational force density acting on a dust or on any matter to derive from stresses in the gravitational field, just as electromagnetic forces acting on electrically charged matter derive from Maxwell's stress tensor. This is what I called the stress principle. In the static case, this condition comprehends the vanishing of the total force acting on matter and thereby excludes perpetual motion (by Newton's scholium to his third law). Einstein elevated the four-dimensional generalization of this field-stress assumption to a capital principle of his new theory of gravitation. Today, we would be content with requiring $D_\nu T^{\mu\nu} = 0$ for the energy–stress tensor of matter $T^{\mu\nu}$, and most of us would be suspicious of any attempt at defining energy, momentum, and stress within the gravitational field. In contrast, Einstein wrote $D_\nu T^{\mu\nu} = 0$ under the form (9.100)

$$\partial_\nu \mathbf{T}^\nu_\mu = \tfrac{1}{2} g^{\nu\rho} \partial_\mu g_{\rho\sigma} \mathbf{T}^\sigma_\nu,$$

2010; Darrigol 2015, pp. 186–187. Einstein and Leopold Infeld introduced the free-falling elevator picture in *The Evolution of Physics* (Cambridge: Cambridge University Press, 1938), pp. 226–235.

[149] Einstein 1917a, p. 67.

[150] Although historians (e.g. Janssen 2014) and philosophers have frequently discussed Einstein's thought experiments based on the equivalence principle, they have usually ignored his operational definitions of physical quantities in accelerated frames (optical synchronization and "pocket" measuring contraptions). On the reification of coordinate systems in the hole argument, cf. Norton 2005.

interpreting the left-hand side as the momentum variation of matter, and the right-hand side as the force acting on it. He required this force to derive from the stress–energy tensor t^v_μ for the gravitational field. In the tentative field equation $G_{\mu\nu} = \kappa T_{\mu\nu}$, the field operator $G_{\mu\nu}$ built from $g_{\mu\nu}$ and its derivatives must be such that $g^{v\rho}\partial_\mu g_{\rho\sigma}\mathbf{G}^\sigma_v$ can be expressed as the four-divergence of a symmetric two-tensor.

This stress principle played an enormous role in Einstein's quest for a field equation, first as a test for tentative equations, then as a means to construct this field equation when it was further assumed to yield a d'Alembertian equation in the weak-field limit (the *Entwurf* strategy). Einstein read the product $\frac{1}{2}g^{v\rho}\partial_\mu g_{\rho\sigma}\mathbf{T}^\sigma_v$ as a generalization of the Newtonian product of the gravitational vector field by the mass density matter. For this reason, he long regarded the quantities (9.101)

$$\Gamma^v_{\mu\sigma} = \tfrac{1}{2}g^{v\rho}\partial_\mu g_{\rho\sigma}$$

as the "gravitational field components" of which the gravitational field operator would hopefully be a simple combination. This is the "fatal prejudice" he would condemn in November 1915.

Soon after introducing the invariant element $\mathrm{d}s^2$, Einstein realized that it could serve to construct the action $\int \mathrm{d}s$ for a particle in the $g_{\mu\nu}$ field. From the associated Lagrangian he derived the energy and momentum that guided his construction of the energy–stress tensor of a dust. Yet he did not consider an action for the gravitational field until, in early 1914, the mathematician Paul Bernays advised him to do so for studying the covariance properties of the *Entwurf* theory. It is not clear why Einstein did not earlier think of basing his theory on an expression for the field action. We know, however, that he undervalued or ignored the works that best exemplified the power of least action in field theory: Helmholtz's memoir of 1892 on least action in electrodynamics, Poincaré's memoir of 1906 on the dynamics of the electron, and Mie's memoir of 1912 on an electromagnetic theory of matter. Had he better appreciated these works, he would have seen, as he did with much delay in his *formale Grundlage*, that the principle of least action offered the simplest and most efficient way to derive field equations that automatically admit a stress–energy tensor. Retrospectively, we can see that some of the $G_{\mu\nu}$ field operators Einstein considered in 1913 derived from a field action and were therefore compatible with the existence of an energy–stress tensor, whereas Einstein then believed the contrary.

Lastly, the correspondence principle often guided Einstein's constructions. He relied on four different correspondence conditions. First, the new theory had to be compatible with special relativity in a local free-falling frame. In the metric approach, this is automatically warranted by the fact that in every small neighborhood of the space–time manifold there is a system of coordinates for which the metric takes the Minkowskian form (Einstein did not specify to which approximation, although Riemann had done so long ago in the Euclidean case by means of geodesic coordinates). Second, Einstein required that in the weak-field limit the gravitational field equation should take the d'Alembertian form $G_{\mu\nu}$, in evident analogy with the electromagnetic field equations. The failure of a given $G_{\mu\nu}$ to meet this condition was not eliminatory, because Einstein felt free to add coordinate conditions that eliminated the unwanted terms in the linearized field operator. Third, for a while Einstein required the static solutions of the tentative field equation to agree with his static theory of

1912, in which the metric element has the simple diagonal form $ds^2 = \bar{c}^2(\mathbf{r})dt^2 - d\mathbf{r}^2$. By mid-1913, however, he admitted static solutions with nonvanishing off-diagonal elements in the *Entwurf* theory.[151] More consistently, Einstein required the Newtonian theory of gravitation to hold in the nonrelativistic limit of the static case. Einstein used this correspondence principle either as a construction tool in the *Entwurf* theory, or as a test in the *formale Grundlage* and in the final theory. The second alternative was the better one in his mind, because he did not want the new theory to be based on extraneous knowledge.[152]

Early failure with the Riemann tensor

Today we know there are ways of interpreting Einstein's various heuristic principles so that they become mutually compatible and lead to Einstein's field equations of November 1915. We also know that general covariance and the principle of least action are sufficient for this purpose. Unfortunately, Einstein did not rely on a field action before 1914, and he understood his other principles in ways that made them mutually incompatible. This can be seen by examining the reasons why Einstein early rejected gravitational field equations built from the Riemann tensor.[153]

In the Zürich notebook, Einstein briefly considered the fully covariant equation $R_{\mu\nu} = -\kappa T_{\mu\nu}$ and two variants obtained by subtracting from the Ricci tensor terms invariant through transformations that preserve the determinant g and the symmetric derivative $\vartheta_{\mu\nu\rho} = \partial_\rho g_{\mu\nu} + \partial_\mu g_{\nu\rho} + \partial_\nu g_{\rho\mu}$, respectively. He rejected the fully covariant choice because the harmonic coordinate condition that turned $R_{\mu\nu}$ into $\frac{1}{2}\Box g_{\mu\nu}$ in the linear approximation was incompatible with his static metric of early 1912 and because he could not reshape the product $\sqrt{-g}\partial_\mu g_{\nu\rho}R^{\nu\rho}$ as a four-divergence. The unimodular and the ϑ-based variants did not have the first defect but they still had the second.

For the sake of the equivalence principle, Einstein also wanted the rotation metric to be a solution of his field equation and coordinate conditions. On the basis of earlier inexact considerations, he probably believed this was the case for the coordinate conditions used in the Ricci and unimodularly covariant options. He knew the ϑ-covariant option to be incompatible with the rotation metric, but he still hoped there was sufficient analogy between particle motion in a rotating frame and particle motion in a metric of vanishing $\vartheta_{\mu\nu\rho}$.

Thus we see that Einstein abandoned the Riemann tensor-based options essentially for two reasons: incompatibility with the static metric of early 1912 (for the first option only), and incompatibility with the existence of a gravitational stress–energy tensor. In the

---

[151] *ECP* 4, p. 370. Being of second order, the nondiagonal terms do not contribute to the first correction to the Newtonian motion of planets (whereas they do in the final theory in which they are of first order). They still conflict with the static theory of 1912, because Einstein originally regarded his expression of the static metric as exact (not as an approximation). Stachel 1989b [1980] and Norton 1984, pp. 299–300 (corrected in Norton 2020, pp. 24–25) have judged differently.

[152] Einstein's attitude toward the correspondence principle was similar to Laue's (for the correspondence between relativistic and Newtonian mechanics); see p. 189, this volume.

[153] For a review of the reasons evoked in earlier studies, cf. Weinstein 2018.

Entwurf memoir and in later reminiscences, he suggested that the main reason was incompatibility with the d'Alembertian form in the linear approximation. Before John Norton's study of the Zürich notebook, this statement was commonly interpreted as Einstein's ignorance of coordinate conditions that bring the desired form. In reality, he was fully aware of such conditions, but he believed they failed to solve the d'Alembertian correspondence difficulty because they conflicted with other heuristic requirements: compatibility with the older static metric and, most important, the existence of a gravitational energy–stress tensor.[154]

As noted by Norton, Renn, and Sauer, Einstein thereby understood coordinate conditions as universal restrictions of the class of physically admissible reference frames. Accordingly, the gravitational field is determined by the original field equation *together with* the coordinate condition in any physical situation. In contrast, the coordinate conditions of general relativity as we know it are adapted to specific problems and they should not be included among the fundamental equations of the theory. Had Einstein used modern coordinate conditions instead of coordinate restrictions, he would still have had reasons to reject the Riemann tensor-based field operators: the equation $R_{\mu\nu} = 0$ allows for nondiagonal static solutions (e.g., the Schwarzschild solution) incompatible with his earlier static theory; he would not have seen, without evoking the field action, that the product $\sqrt{-g}\partial_\mu g_{\nu\rho}R^{\nu\rho}$ can be turned into a four-divergence (for $R = 0$); and the field operator is not a simple combination of the expressions $g^{\nu\rho}\partial_\mu g_{\rho\sigma}$ that Einstein then regarded as the natural field components.

Einstein's early objections to a Riemann tensor-based field operator gradually subsided, while new objections emerged. Incompatibility with the static metric of 1912 could no longer be alleged after Einstein and Besso found, in mid-1913, that the center-symmetric solution of the *Entwurf* field equation was nondiagonal. At some point, Einstein might have discovered that none of his coordinate conditions was compatible with the rotation metric, but he had no reason to investigate these conditions in the *Entwurf* context and there is no evidence that he did.

In the late summer of 1913, Einstein discovered the hole argument following which a generally covariant field equation does not fully determine the metric field. Originally, he believed this argument excluded any generally covariant field equation. He changed his mind in January 1914: he now argued that a genuine field equation should be obtainable from a generally covariant equation by specializing the coordinate system. In his subsequent memoir with Fokker on the metric reformulation of the scalar theory, he expressed the hope that the *Entwurf* field equation would derive from the fully covariant $R_{\mu\nu} = -\kappa T_{\mu\nu}$ by such specialization. He was thus willing to resurrect the Riemann tensor, although what truly mattered to him was the effective field equation in the specialized coordinate systems. To his mind this specialization was a necessary precondition for constructing a gravitational stress–energy tensor.

In the same memoir with Fokker, Einstein indicated that he no longer held the non-d'Alembertian character of the linear approximation against a field operator based on the Riemann tensor. Indeed the Einstein–Nordström field equation could be derived from

[154] Cf. Norton 1984.

$R = \kappa' T^{\mu}_{\mu}$ even though it had the desired d'Alembertian form. Of course, Einstein already knew that proper specialization of the coordinate systems could generate the d'Alembertian from the Ricci tensor. He now believed this could be done without contradicting his other heuristic principles. In particular, the specialized field equation could be compatible with the existence of a gravitational stress tensor as the *Entwurf* field equation was, and it could admit the rotation metric as a solution, as Einstein still hoped.

In the end, what blocked Einstein from using a Riemann tensor-based field equation was not any prejudice against the heuristic value of this tensor, but the conviction that it could not be used without coordinate restrictions that would necessarily lead to the *Entwurf* field equation. The restrictions seemed necessary in order to avoid the hole paradox. They had to lead to the *Entwurf* field equation because this equation seemed to be the only one compatible with d'Alembertian correspondence and the stress principle.

In order to escape from the *Entwurf* spell and move toward a truly general relativity, Einstein had to get rid of two prejudices: the incompatibility of the Ricci field operator with the existence of the energy–stress tensor, and the necessity for a universal restriction of admissible coordinate systems.

Errors

The spotting of errors in works of the past is notoriously risky. Yet there is a clear sense in which Einstein committed errors he would have himself recognized if anyone had been able to tell him. These are of variable gravity and subtlety.

Let us begin with the most trivial errors. The Zürich notebook and other manuscript sources contain numerous errors of calculation, sometimes corrected sometimes not. Most of them are inconsequential. A few are more significant. For instance, Einstein repeatedly erred in judging the compatibility of his field equations with the rotation metric, even though the relevant calculations are not especially difficult. Probably, he did not bother to redo a calculation when the result met his expectations. When it did not, he still hoped some minor modification would save the situation. It could also be that, despite the large number of relevant calculations, the compatibility of his theory with rotating frames was not as crucial as most commentators have assumed. For a short while, in the fall of 1913, he was willing to limit the covariance of his theory to linear transformations. The rest of the time he would perhaps have contented himself with nonlinear transformations not necessarily including rotation.[155]

Another trivial error occurred in Einstein's argument that energy–momentum conservation, written as $\partial_{\nu}(t^{\nu}_{\mu} + T^{\nu}_{\mu}) = 0$ was covariant under linear transformations only. There Einstein implicitly assumed the gravitational stress–energy tensor t^{ν}_{μ} to be generally covariant whereas from its expression it is easily seen not to be so. A more consequential and most embarrassing error occurred in the *formale Grundlage* of November 1914, when

---

[155] A nonlinear transformation $x \rightarrow x'$ can be seen as implying the mutual acceleration $\partial^2 x^{\prime i}/\partial^2 x^0$ of local frames, as is noted in Einstein and Grossmann 1914, p. 8: "[The principle of equivalence] is especially convincing when the 'apparent' gravitational field . . . [in an accelerated frame] can be conceived as a 'real' gravitational field, which is the case when accelerative transformations (that is, non-linear transformations) belong to the justified transformations of the theory."

Einstein wrongly asserted that a certain formal condition ($S^\nu_\mu \equiv 0$) was compatible with the *Entwurf* field action only, when it is easily seen to be compatible with any linearly invariant action.

Let us move to a less trivial error. Although he had no rigorous proof of this, Einstein strongly suspected the Riemann tensor-based field operators to be incompatible with the stress principle. This is not true for the Ricci tensor and for the $R^\times_{\mu\nu}$ tensor of Equation (9.59), but it would be very difficult to derive the associated stress–energy tensor without knowing that the associated field equation derives from an invariant action. This is true for the $R^{\times\times}_{\mu\nu}$ tensor of Equation (9.62), but too much covariance is lost on the way. A crucial turning point occurred in Einstein's program when he realized, in the late fall of 1914, that there existed a stress–energy tensor for any field equation deriving from a field action.

The hole argument is Einstein's subtlest error, although it was long judged to be based on the trivial error of regarding the expressions of the same metric field in two different coordinate systems, $g_{\mu\nu}(x)$ and $g'_{\mu\nu}(x')$, as two different fields.[156] In reality, Einstein clearly indicated that he was dealing with two different fields $g_{\mu\nu}(x)$ and $g'_{\mu\nu}(x)$ in the same reference frame, the latter field being the same function of x as $g'_{\mu\nu}(x')$ is a function of x'. Implicitly, Einstein thereby assumed that the reference system was given before expressing the metric field $g_{\mu\nu}(x)$. Analogously, in Euclidean space an orthonormal basis must be given before expressing a field as a function of the Cartesian coordinates. But this is a false analogy.[157]

As we saw, Einstein had reasons to restrict covariance even before he introduced the hole argument: he believed that some physical requirements such as energy–momentum conservation or Newtonian correspondence could impose a general restriction on the co-ordinate system, even if general covariance held at a more fundamental and formal level. Should we call this an error? It was so from a strategic point of view, since it interfered with the heuristic exploitation of general covariance. From a logical point of view, any problem of general relativity can be solved under any given coordinate restriction since the choice of coordinates is fully arbitrary and untied to the system under consideration. All we can say is that for a given problem some restrictions are more convenient than others. Well after Einstein arrived at his final equations, there were attempts to reintroduce coordinate restrictions on a physical basis, for instance Vladimir Fock's in the 1950s with harmonic coordinates.[158]

Continuities and discontinuities

A superficial reading of Einstein's struggles toward a generalized theory of relativity from 1911 to 1915 could leave the impression that he moved through a chaotic succession of failed attempts until he reached the correct solution. In reality, at every stage of his re-search Einstein learned something useful for the later stages, both physically and formally. During his first naive guesses at equations for the metric field, he familiarized himself with

---

[156]Stachel 1980 first remarked on the nontrivial character of the hole argument.

[157]See comments on Eddington's discussion, on pp. 344–346, this volume.

[158]Fok [Fock] 1959.

tensor calculus and a few algebraic properties of the metric tensor, and he already conceived useful coordinate conditions. He soon introduced the energy–stress tensor of matter and the accompanying stress principle, which never ceased to be essential components of his theory. In his aborted attempts at a Riemann tensor-based field equation, he mastered the relevant calculus with Grossmann's help, and he developed the use of coordinate conditions. His failure to conciliate the resulting field equations with the stress principle led him to the *Entwurf* strategy. While working out the relativistic precession of Mercury's perihelion in the *Entwurf* theory, he and Besso developed all the techniques he would need for the similar calculation in his final theory. While consolidating the *Entwurf* theory, he developed the action-based approach that ultimately allowed him to satisfy the stress principle with a Riemann tensor-based field equation.

In November 1915, Einstein announced he was giving up the long-favored *Entwurf* theory to return to general covariance. This statement has induced most commentators to exaggerate the discontinuity of the transition from the *Entwurf* to the final theory. In reality, there was much continuity, which explains why it took Einstein so little time to complete his theory after returning to full covariance. This continuity is best captured by using Einstein's distinction of November 1914 between a "physical" and a "formal" approach. The physical approach is the one he originally used to derive the *Entwurf* field equation by means of the stress principle combined with the correspondence principle: alter the d'Alembertian equation so that it becomes compatible with the existence of a gravitational stress–energy tensor. The formal approach is based on the covariance properties of the field equation, as investigated by means of the field action. Einstein judged this approach superior to the physical approach for it did not rely on the correspondence principle. He originally believed he could re-derive the *Entwurf* field equation in this formal approach. When he discovered a fatal error in this derivation, he saved the formal approach by replacing conditional covariance with general covariance. All he had to do was to change the expression of the field components $\Gamma^\rho_{\mu\nu}$ on which his field action depended. In terms of these quantities, most equations of the theory were unchanged, but the equation for the metric field was altered to become unimodularly covariant. The continuity lies in the formal approach shared by the *formale Grundlage* and the November 1915 theory.[159]

Some of the difficulties that had haunted Einstein's theory suddenly disappeared. The hole argument and the dilemmas of restricted covariance went to Einstein's dustbin; the coordinate restrictions became occasional coordinate conditions; and the stress principle became compatible with a Riemann tensor-based field equation. Yet there also was some continuity in the experienced difficulties. The theory remained an evolving, imperfect construct with persisting roots in earlier approaches. Einstein maintained the correspondence between general covariance and the equivalence principle, thus preserving an unexplained reification of coordinate systems; in his discussions of the gravitational redshift or of the gravitational deflection of light, he continued to rely on coordinate-based quantities as if they were truer than the invariants built from them; and he maintained

---

[159] Janssen and Renn similarly emphasize continuity, although they use different categories; cf. Janssen and Renn 2007, p. 840; also *GGR* 2, pp. 500–501.

the gravitational stress tensor as an essential component of his theory whereas most physicists now regard it as a physically ill-defined formal intermediate in better founded energy–momentum considerations.[160]

No more than any other theory did general relativity emerge fully equipped in the mind of a single genius. Yet, no one would deny that by November 1915 Einstein had the correct general field equations for general relativity (without the cosmological term) and that he knew how to use them to derive the Newtonian approximation and three crucial departures from it.

[160] Cf. Janssen and Renn 2007. On persisting difficulties with the meaning of coordinates, see Chapter 10. On Einstein and the stress–energy (pseudo-)tensor for the gravitational physicists, cf. Cattani and De Maria 1993. A few physicists later defended the physical significance of this tensor; see Trautmann 1962.

10

MESH AND MEASURE IN EARLY GENERAL RELATIVITY

> There is no fundamental mesh-system. In particular problems, and more partic-
> ularly in restricted regions, it may be possible to choose a mesh-system which
> follows more or less closely the lines of absolute structure in the world, and so
> simplify the phenomena which are related to it. But the world-structure is not of a
> kind which can be traced in an exact way by mesh-systems, and in any large region
> the mesh-system drawn must be considered arbitrary. In any case the systems
> used in current physics are arbitrary.[1] (Arthur Eddington, 1920)

A few months after the final breakthroughs of November 1915, Einstein published a synthetic account of his theory, *Die Grundlage der allgemeinen Relativitätstheorie*, meant to
be the foundation of any further research on general relativity. This long memoir largely
served its purpose and now counts among the great landmarks of the history of science. Yet
one cannot say that in 1916 general relativity was a completed, fully understood theory.
A few of the vicissitudes of Einstein's long quest were still perceptible, and now-essential
features of the theory, for instance the role of singularities and topological issues, were discovered much later. Some aspects of the theory, for instance the existence of gravitational
waves, long remained unsettled. Others, for instance the status of energy–momentum conservation, remain controversial. The purpose of this chapter is not to describe the slow
maturation of general relativity from 1916 to this day. It is, in harmony with the relativity
theme of this book, to recount the troubles Einstein's early readers had making sense of
what Arthur Stanley Eddington called the "mesh-system" of general relativity, namely, the
coordinate systems and reference frames. As we will see, the needed clarification occurred
in the first years of the theory, thanks to the efforts of Einstein's cleverest followers.[2]

On the early reception more broadly conceived, it will be sufficient to know that despite
its difficulty and originality, general relativity was a lively research topic even before the
spectacular confirmation of the predicted deflection of light by the sun's mass in 1919.
Einstein himself was quick to develop consequences of his theory: in addition to the three
major predictions of 1915, he considered gravitational waves in 1916 and cosmological

---

[1]Eddington 1920b, p. 150.

[2]Einstein 1916a. There is much valuable literature on the more persistent difficulties of general relativity. On
the Schwarzschild singularity, cf. Eisenstaedt 1982. On topological anomalies, cf. Earman 1999. On gravitational
waves, cf. Kennefick 2007a; Weinstein 2015b, §3.1. On energy–momentum conservation, cf. Cattani and De Maria,
1993; Brading, 2005. For a global overview, cf. Ray 1987.

Relativity Principles and Theories from Galileo to Einstein. Olivier Darrigol, Oxford University Press.
© Olivier Darrigol (2022). DOI: 10.1093/oso/9780192849533.003.0010

models in 1917. He also responded, privately and publicly, to the contributions of a few sympathetic mathematicians, astronomers, and physicists in the years 1916–1918.[3]

In Göttingen, Hilbert and Klein developed invariant-theoretic and geometric consequences of the theory, and the astronomer Karl Schwarzschild derived the exact solution of the center-symmetric problem. In Berlin, another astronomer, Erwin Freundlich, authored a popular book on general relativity and agreed to collaborate with Einstein on a verification of the gravitational redshift. From the ETH in Zürich, Hermann Weyl brought his philosophical and mathematical acumen to the service of the new theory, and authored the first major treatise on it. In Vienna, under Friedrich Hasenöhrl's lead, Friedrich Kottler, Hans Thirring, Josef Lense, Ludwig Flamm, and Erwin Schrödinger all worked on the new theory, triggering exchanges with Einstein on fundamental issues of equivalence and energy–momentum. In Italy, Levi-Civita injected geometric intuition into Einstein's theory with his concept of parallel transport, and he clarified the static solutions of Einstein's field equations. In the Netherlands, Lorentz, his student Fokker, and his younger Leiden colleague Ehrenfest, who had all participated in the genesis of general relativity, remained active contributors after 1915. Lorentz reformulated the theory through invariant variational procedures and through intrinsic geometry in the space–time manifold. His student Johannes Droste independently discovered the Schwarzschild solution and pioneered the n-body problem. His colleague the astronomer Willem de Sitter offered a penetrating exposition of the theory as well as his own cosmological solution in 1917. He communicated Einstein's and his own papers to his British colleague Arthur Eddington, who soon became "the fountainhead of relativity in England."[4]

In 1919, Eddington led the British solar-eclipse expeditions that confirmed the gravitational deviation of light rays and instantaneously made Einstein a popular hero. This publicity won Einstein new allies and new enemies. Here we will be concerned only with the physicists or mathematicians who seriously engaged with the contents of his theory, before or soon after this crowning event. We will focus on the temporary difficulties they had in understanding how he derived physical predictions from impalpable formalism. These difficulties mostly concerned the meaning of coordinates and their relation to concrete measurement. They affected the early attempts to test Einstein's predictions, and, as we will see in a discussion of Einstein's *Grundlage*, they are intimately related to Einstein's takes on the equivalence principle, Mach's principle, and chrono-geometrical measurement, which have been amply and competently discussed by other scholars.[5]

---

[3] Einstein 1916c, 1917b; *ECP* 8–9. My focus is on the scientific reception of general relativity. On the popular and philosophical receptions of relativity theory, cf. Hentschel 1990; Ryckman 2005.

[4] Hilbert 1915b, 1917; de Sitter 1916a, 1916b; Droste 1916a, 1916b; Flamm 1916; Kottler 1916; Lorentz 1916; Schwarzschild 1916; Freundlich 1916; Droste and Lorentz 1917; Levi-Civita 1917a, 1917b, 1918; Weyl 1917, 1918; Klein 1918a, 1918b; Lense 1918; Lense and Thirring 1918; Thirring 1918; Schrödinger 1918a, 1918b; Dirac 1977, p. 115 (fountainhead). On the early reception of general relativity in Germany, Vienna, and the Netherlands, cf. Goenner 2016; Havas 1999; Kox 1992. On Einstein and Freundlich, cf. Hentschel 1994.

[5] On the eclipse expeditions, cf. Kennefick 2007b, 2009. On the history of tests of Einstein predictions in general, cf. Earman and Glymour 1980a, 1980b; Hentschel 1994, 1998; Stanley 2003; Crelinsten 2006. On general covariance and the equivalence principle, cf. Norton 1992b, 1993b; Stachel 1993; Howard 1999. On Mach's principle, cf. Barbour and Pfister 1995; Janssen, 2004. On the measurement problem, cf. Brown 2005; Ryckman 2005; Giovanelli 2014.

As Eddington remarked, the difficulties with the meaning of coordinates in general relativity have much to do with a much simpler problem: the characterization of the intrinsic geometry of an ordinary curved surface. This is why Section 10.1 of this chapter is a concise explanation of the meaning of coordinates and metric coefficients in the Gaussian theory of surfaces already discussed in Chapter 8. Section 10.2 is a point-by-point discussion of Einstein's *Grundlage* of the spring of 1916, which was the principal source of the confusions that affected the comprehension of Einstein's three basic new predictions regarding the gravitational redshift, the deflection of light rays, and Mercury's perihelion. These predictions and early commentary or developments by other theorists are discussed separately in Sections 10.4 and 10.5 of this chapter.[6]

As we saw in the previous chapter, Einstein built his theory in a mostly nongeometrical manner, by means of the absolute differential calculus of Christoffel and Ricci-Curbastro. Most of the difficulties encountered in the interpretation of coordinates have to do with the lack of geometrical intelligibility in this conception. It is therefore no wonder that the two men who most significantly contributed to the clarification of Einstein's theory, Weyl and Eddington, also were the promoters of a geometrical understanding of this theory. In 1917 Weyl recast the theory within his *Nahegeometrie*, based on the concept of connection between group properties at two infinitesimally close points of the manifold. His *Raum ·Zeit ·Materie* of 1918 was both a treatise on the foundations of infinitesimal geometry and a systematic exposition of general relativity worthy of Einstein's admiration: "I am reading the proofs of your book ... with true enthusiasm. It is like a master's symphony. Every little word relates to the whole, and the layout of the work is grandiose." Weyl's philosophical turn of mind helped him clarify the basic interpretive issues of the theory.[7]

Eddington shared this quality and some of the mathematical brilliance. In addition, his competence as an astronomer helped him clarify the relation of the theory to concrete observation. His endeavor to explain general relativity to a broad audience brought him to dissolve the difficulties of the concept of metric manifold through simple geometric illustrations. His beautifully written books on general relativity, the philosophical *Space, Time and Gravitation* of 1920 and the more technical *The Mathematical Theory of Relativity* of 1923, long were the best English sources for learning general relativity, and they remain highly recommended reading to this day. Other important sources of interpretive lucidity were Laue's articles and books on relativity theory, which combined technical prowess and philosophical profundity; the young Wolfgang Pauli's encyclopedia article of 1921, which concisely and competently synthesized anterior contributions to the theory; and Guido Beck's encyclopedia article of 1929, which gave special attention to the meaning of coordinates in observational predictions.[8]

The materials in this chapter offer a rich illustration of the ways in which the concrete application of a new theory, no matter how much its inventor strove to define the

---

[6]Einstein 1916a.

[7]Weyl 1917, 1918; Einstein to Weyl, March 8, 1918, *ECP* 8. On Weyl's background and motivation, cf. Ryckman 2005, chap. 5.

[8]Eddington 1920b, 1923; Laue 1921; Pauli 1921; Beck 1929. On Eddington's background and motivation, cf. Laguens 2018, introduction.

foundations, still needs much conceptual analysis in order to yield unambiguous, incontrovertible results. Einstein's account of the new predictions of general relativity was at best incomplete, and at least confusing. The critical interpretive work of some of his followers, especially Eddington's and Weyl's, was no mere "mopping up" à la Thomas Kuhn. It shed a new light on the fundamental issues of covariance, well beyond what Einstein had succeeded in conveying, and it clarified the presuppositions of any concrete test of Einstein's predictions.[9]

10.1 A Gaussian preliminary

In order to understand the meaning of coordinates and the related issue of covariance, it will help to have in mind the Gaussian theory of surfaces in which very similar issues occur in a more intuitive way.

For a two-dimensional surface immersed in ordinary Euclidean space, a possible description of a (small enough portion of a) surface is obtained by associating each point $\mathbf{r}$ of the surface with two coordinates x_1 and x_2, for instance the Cartesian coordinates of the projection of this point on a given plane. The surface has metric properties induced by the Euclidean structure of the space in which it is immersed. For instance, the distance between two neighboring points $\mathbf{r}$ and $\mathbf{r} + \mathbf{dr}$ is given by

$$ds^2 = g_{ij}dx_i dx_j, \qquad \text{with } g_{ij} = \frac{\partial \mathbf{r}}{\partial x_i} \cdot \frac{\partial \mathbf{r}}{\partial x_j} \tag{10.1}$$

(the indices i and j take the values 1 and 2, summation over repeated indices is understood, and the centered dot is the ordinary scalar product). Geodesics can be defined as the line of shortest length between two different points, and the distance between two points as the length of the geodesic line that connect them.[10]

For a being living on the surface, the only accessible metric properties are those based on the concept of distance between two points. Some properties of the immersed surface are not conserved by bending without stretching, and they are therefore lost to such a being. The intrinsic metric structure is completely defined by the metric ds^2. It contains the geodesics defined as the lines for which the integral of the element ds is stationary, and the Gaussian curvature defined through the excess of the sum of the angles of a small geodesic triangle over a flat angle. It is also possible to define intrinsic coordinates, for instance by taking the distances u_1 and u_2 of the running point M from two reference points A and B (on a sufficiently small portion of the surface). In this case, the metric coefficients g_{ij} cannot be chosen arbitrarily because they must satisfy the conditions

$$u_1 = \text{Min} \int_A^M \sqrt{g_{ij}du_i du_j}, \quad u_2 = \text{Min} \int_B^M \sqrt{g_{ij}du_i du_j}. \tag{10.2}$$

---

[9] Kuhn 1962, chap. 3. Modern texts on general relativity are still often unclear on how the coordinates in the Schwarzschild solution relate to concrete observations, and the offered derivations of the gravitational redshift are often defective (even sometimes duplicating Einstein's pseudo-proof of 1916). On the latter point, cf. Scott 2015.

[10] See pp. 239–240, this volume.

To avoid these awkward constraints, it is more convenient to choose the system of coordinates arbitrarily (any invertible differentiable pair of functions of the intrinsic variables u_1 and u_2 will do) and to determine the metric meaning of the coordinates in an a posteriori manner by inverting the relations

$$u_1 = \text{Min} \int_A^{M(\bar{x})} \sqrt{g_{ij} dx_i dx_j}, \quad u_2 = \text{Min} \int_B^{M(\bar{x})} \sqrt{g_{ij} dx_i dx_j} \qquad (10.3)$$

between $u = (u_1, u_2)$ and $\bar{x} = (\bar{x}_1, \bar{x}_2)$ for the given choice of the functions $g_{ij}(x_1, x_2)$ in the x coordinates.

In the latter approach, any function $g_{ij}(x)$ defines a portion of a Gaussian surface, or of a metric manifold of dimension 2. There is a price to be paid for this freedom: two different choices of the function $g_{ij}(x)$ may define the same intrinsic metric structure on the surface. Indeed for any diffeomorphism $x' = h(x)$ leaving the two reference points A and B unchanged, the functions

$$g'_{ij}(x) = g_{kl}(x') \frac{\partial x'_k}{\partial x_i} \frac{\partial x'_l}{\partial x_j} \qquad (10.4)$$

yield the same metric structure as the functions $g_{ij}(x)$. To see this, consider two arbitrary points M and N of coordinates $x = (x_1, x_2)$ and $y = (y_1, y_2)$. As a consequence of Equations (10.3) and (10.4), the intrinsic coordinates $u = (\text{AM}, \text{BM})$ and $v = (\text{AN}, \text{BN})$, and the distance MN satisfy the relations

$$u'(x) = u[h(x)], \quad v'(y) = v[h(y)], \quad \text{and} \quad \text{MN}'(x, y) = \text{MN}[h(x), h(y)], \qquad (10.5)$$

wherein the prime is used for distances calculated from g'_{ij} instead of g_{ij}. Consequently, the distance MN is the same function of the distances AM, BM, AN, and BN in both cases, and the intrinsic geometry is the same.

Another way to see the equivalence of g_{ij} and g'_{ij} is to observe that the second choice differs from the former by a mere relabeling of the points of the surface. There is a subtlety in this statement. In order to imply well-defined metric properties, the functions $g_{ij}(x)$ must be well-defined functions and therefore a given value of the coordinates x must correspond to a well-defined point of the surface. This seems to suggest that two different sets of $g_{ij}(x)$ functions necessarily lead to two different metrics: they give different values for the distances between the points of coordinates x and $x + dx$. The fallacy of this reasoning comes from the fact that a given value x of the coordinates in general does not correspond to the same point of the manifold when the metric coefficients are altered diffeomorphically. The choice of the functions $g_{ij}(x)$ not only determines the metric properties, it also (to some extent) determines to which points of the manifold the various values of the coordinates correspond. When we write $g_{ij}(x)$, the variable x runs through well-defined points of the manifold, but we do not know *which ones* until the geodetic consequences of the associated ds^2 are explored.

To summarize, when we say that a manifold has the metric structure induced by the metric field $g_{ij}(x)$ we really mean the following: *There exists at least one way of labeling the points of the manifold such that the distance between any two neighboring points labeled by x and x + dx is given by* $\mathrm{d}s^2 = g_{ij}\mathrm{d}x_i\mathrm{d}x_j$. This definition makes it obvious that $g_{ij}(x)$ and the diffeomorphic variant $g'_{ij}(x)$ lead to the same metric structure, because for the labeling x such that $\mathrm{d}s^2 = g'_{ij}(x)\mathrm{d}x_i\mathrm{d}x_j$, the labeling $x' = h(x)$ yields $\mathrm{d}s^2 = g_{ij}(x')\mathrm{d}x'_i\mathrm{d}x'_j$.

The take-home message is that the coordinates in the usual theory of surfaces and metric manifolds have no metric significance a priori. Since the only intrinsic way to locate points on a manifold is through the metric properties, this implies in particular that the coordinate values do not a priori tell us where we are on the manifold. They only tell us that we are at a certain point. Consequently, two diffeomorphically equivalent choices of the metric coefficients lead to the same intrinsic metric structure.[11]

In order to illustrate these remarks, consider the metric

$$\mathrm{d}s^2 = \mathrm{d}x_1{}^2 + x_1{}^2\mathrm{d}x_2{}^2. \tag{10.6}$$

A first important point is that this expression determines the metric structure even if we do not have any a priori knowledge of the meaning of the coordinates. This is why, following Eddington, I have used the neutral names x_1 and x_2 instead of the loaded names r and θ. In order to determine the structure, we may note that the lines of constant x_2 are geodesics (this immediately follows from the minimum condition). These geodesics all intersect at a single point for which $x_1 = 0$ and x_2 is arbitrary, because the metric formula (10.6) implies a vanishing distance between the points (x_1, x_2) and $(x_1, x_2 + \mathrm{d}x_2)$ when x_1 reaches zero. On any such geodesic, the distance of a point from the origin is simply given by the value of x_1. The lines of constant x_1 intersect the former geodesics at a right angle because owing to the metric formula the vectors $(\mathrm{d}x_1, 0)$ and $(0, \mathrm{d}x_2)$ are perpendicular. On one of these lines, the distance between two points is x_1 times the variation of x_2. This implies that the mesh of lines of constant coordinate value can be drawn on a flat piece of paper without any distortion. In this way we realize that the metric formula belongs to a flat surface. Of course, my learned reader expected this result from the start since in Equation (10.6) he or she recognized the expression of the Euclidean metric in polar coordinates. My point is that no such knowledge is a priori necessary: the coordinates x_1 and x_2 acquire their metric, polar-coordinate meaning as a consequence of the metric formula. It thus becomes clear that the metric $\mathrm{d}s^2 = \mathrm{d}x_1{}^2 + \mathrm{d}x_2{}^2$ and the metric $\mathrm{d}s^2 = \mathrm{d}x_1{}^2 + x_1{}^2\mathrm{d}x_2{}^2$ differ only by the relabeling $(x_1, x_2) \rightarrow (x_1 \cos x_2, x_1 \sin x_2)$.[12]

Now suppose that beings on the surface do some thermal physics in addition to mere geometry. Specifically, we will assume that they are able to measure temperature distributions and that they are looking for the equilibrium distribution of the temperature T for a given distribution of heat sources of density ρ. They may write down any relation between the two distributions (fields) in a given system of coordinates and then obtain the counterpart in any system of coordinates by mere substitution. The result is a so-called covariant

---

[11] Lack of understanding of this point led Einstein to the hole argument of 1913, discussed on pp. 307, 318–319, this volume.

[12] Cf. Eddington 1920b, pp. 79–80.

law. In this way of reasoning, it is clear that general covariance per se does not restrict the choice of laws. Yet it would seem that covariance excludes equations that do not enjoy tensor homogeneity. How do we solve this contradiction?

Consider, for instance, the equation $\partial_1 T = \rho$. Even though it does not look homogeneous, it can be rewritten under the covariant form $a^i \partial_i T = \rho$, where a^i is a vector field that has the constant components $(1, 0)$ in a given system of coordinates (the density is here defined with respect to the invariant volume element $d^3 x \sqrt{\det[g_{ij}]}$, so that it is a scalar). A more challenging example is the equation $g^{ij} \partial_i \partial_j T = \rho$. This choice seems excluded because $\partial_i \partial_j T$ is not a tensor. Yet it can be rewritten under the covariant form $g^{ij} D_i^{(h)} D_j^{(h)} T = \rho$, wherein $D_i^{(h)}$ is the covariant derivative formally associated with a pseudo-metric field h_{ij} that has vanishing derivatives in the original system of coordinates. Thus we see that the general covariance of any differential field equation can always be achieved by means of additional background fields. Nevertheless, the laws are severely restricted by tensor homogeneity *if no such fields are allowed*.

In our model, we may require that on a flat surface the laws of physics respect the rotational and translational symmetry of a plane. If, in addition, we require the laws to be local laws involving only the derivatives of the various fields at a particular point, the simplest nonabsurd law of temperature equilibrium will be given by $\Delta T = \rho$ (in proper units) in Cartesian coordinates on the flat surface. This may be rewritten in arbitrary coordinates as

$$g^{-1/2} \partial_i (g^{1/2} g^{ij} \partial_j T) = \rho, \quad \text{with} \quad g = \det[g_{ij}]. \tag{10.7}$$

In a curved space, the easiest generalization of this law is obtained by assuming that the former equation remains valid for a nonflat choice of the g_{ij} coefficients. This is not as daring a move as it might seem. As Riemann noted, Gaussian surfaces and Riemannian manifolds enjoy the remarkable property of remaining very nearly flat in the vicinity of any given point.[13] This local flatness should not be confused with the trivial remark that there is a tangent plane at any point of the surface. It means much more: the square of the diagonal of a rectangular triangle differs from the sum of the squares of the two other sides by a term proportional to the square of its area (and to the curvature). Formally, this property results from the existence of a choice of coordinates for which the derivatives of the metric coefficients vanish at any given point. Under this circumstance, it seems natural to assume that the laws of physics on a Riemannian manifold share the local Euclidean symmetry of this manifold. As long as their expression does not involve derivatives of the g_{ij} coefficients higher than the first, this requirement fully determines their expression. This corresponds to a modern form of the strong equivalence principle in general relativity.

To go on with our model, let us now assume that the metric depends on the temperature distribution. For instance, imagine that the surface is a thin, flexible, but mechanically inextensible metal sheet. Excess heating at one point will cause a local dilation of the metal and this will create a bulge of non-zero Gaussian curvature on the surface. We require the relation between the metric field g_{ij} and the temperature field to be covariant and to involve

---

[13]See pp. 244–245, 252, this volume.

these fields and their derivatives only. The simple option $g_{ij} = \alpha \partial_i T \partial_j T$ is excluded because it would lead to $g_{ij} = 0$ for uniform temperature. The next simple option, relying on the invariance of the Gaussian curvature C, posits

$$C = \alpha g^{-1/2} \partial_i (g^{1/2} g^{ij} \partial_j T). \tag{10.8}$$

It fits the physical intuition that the bulge at a point of the surface should depend on an excess of the temperature at this point over the average temperature in the surroundings (which is a way to define a Laplacian).[14] Equations (10.7) and (10.8) jointly imply $C = \alpha \rho$: the curvature is proportional to the density of heat production. Evidently, there are infinitely many choices of the metric coefficients that satisfy this equation for a given heat production, but they all lead to the same intrinsic geometry since the value of the Gaussian curvature at every point completely determines this geometry.

To sum up, the intrinsic geometry of a surface is given by the metric form $ds^2 = g_{ij} dx_i dx_j$ in which the g_{ij} coefficients are defined up to a diffeomorphism that can be interpreted as a relabeling of the points of the surface. The coordinates x_i lack any a priori metrical meaning, although the metric form indirectly gives them such meaning. The laws of physics for beings living on the surface can depend only on the intrinsic geometry, not on any specific labeling of the points of the surface. Therefore, their formal expression must be covariant. Although this property is compatible with any imaginable expression of the laws in a given system of coordinates, it severely restricts the form of the laws if they are further assumed to be differential equations that involve only a small number of fields of low tensor rank. Covariant laws can determine the metric coefficients only up to a diffeomorphism. This formal underdetermination has no physical consequence since it does not affect the intrinsic geometry. These simple characteristics of the intrinsic physico-geometry of surfaces all have counterparts in general relativity.

10.2 Einstein's *Grundlage* of 1916

Variable rods and clocks

Both in Newtonian mechanics and in special relativity, Einstein tells us, there is a kinematics based on invariable rigid rods and clocks. The length of the rods and the period of the clocks depend neither on location nor on time as long as they are at rest with respect to a permitted reference frame. Einstein warns us that "general relativity cannot maintain this simple physical interpretation of space and time."[15]

The statement is misleading, for it suggests that invariable clocks and rods make no sense in general relativity. On the contrary, a basic assumption of general relativity is that there are clocks whose period, measured in proper time, is an absolute constant, and small invariable rods can then be defined by chrono-optical control of rigidity. What is truly incompatible with general relativity is the existence of extended rigid bodies that could

---

[14] It still cannot be regarded as a realistic model of the bulging, because such a model would require a zone of negative curvature to connect the bulge and its flat surroundings.

[15] Einstein 1916a, p. 770.

serve as global reference frames. Locally, that is, in a sufficiently small neighborhood of a given event, it is still possible to define a rigid reference frame. Einstein assumes so much later in his memoir, since he regards special relativity as valid in local free-falling frames.[16]

Mach's argument

Einstein goes on to discuss Mach's famous objection to Newtonian mechanics: to some observable modifications of a system, this theory gives "a merely fictitious cause" such as the rotation of the system with respect to a "Galilean space." For instance, a spherical mass of fluid assumes an oblate shape under rotation in empty space. A proper application of the principle of causality, Einstein tells us, should appeal to an "experimentally observable cause," for instance distant masses. All spaces, Galilean or not, should be equally acceptable as reference frames, since the choice of a reference frame cannot be regarded as a physical cause of anything:[17]

> Of all thinkable spaces R_1, R_2, etc. that move with respect to each other in an arbitrary manner, none can be a priori favored without reanimating the given epistemological paradox [Mach's]. *The laws of physics must be so constituted that they hold with respect to arbitrarily moving systems of reference.*

The argument is debatable. As Eddington noted in 1920, the exclusion of "space" or any immaterial entity as a physical cause is a materialist prejudice that should have little force for a post-Maxwellian physicist:[18]

> It is probable that here we part company from many of the continental relativists, who give prominent place to a principle known as the law of causality—that only those things are to be regarded as being in causal connection which are capable of being actually observed. This seems to be interpreted as placing matter on a plane above geodesic structure in regard to the formulation of physical laws, though it is not easy to see in what sense a distribution of matter can be regarded as more observable than the field of influence in surrounding space which makes us aware of its existence. The principle itself is debateable; that which is observable to us is determined by the accident of our own structure, and the law of causality seems to impose our own limitations on the free interplay of entities in the world outside us. In this book the tradition of Faraday and Maxwell still rules our outlook; and for us matter and electricity are but incidental points of complexity, the activity of nature being primarily in the so-called empty spaces between.

Indeed, in Newtonian mechanics it is possible to regard space (better, the affine structure of Galilean space–time) as a physical cause, as Mach himself recognized.[19] In the context

---

[16] For a clear expression of these points, cf. Synge 1960.

[17] Einstein 1916a, pp. 771–772. This point and other fundamental issues of Einstein's *Grundlage* are discussed in Janssen 2004. It does not occur in Einstein's earlier arguments for a generalized relativity.

[18] Eddington 1920b, p. 156.

[19] Mach 1883, pp. 215–216 for space as a subtle medium; see p. 73, this volume.

of general relativity, the geodesic structure (as Eddington puts it) plays the same role. The true difficulty in Newtonian mechanics is that the presumed cause of inertial behavior, Galilean space, is not simply or directly related to any observable circumstance. It is only by trial and error that we know how to choose a reference frame in which Newton's laws have consequences in agreement with astronomical observations. This state of affairs is not a sufficient argument for placing all conceivable reference frames on the same footing. Einstein's argument is at best a heuristic subterfuge, since the implied "spaces" or frames are concrete, unlimited rigid frames that are incompatible with general relativity.

The equivalence principle

Next comes Einstein's favorite argument in favor of a generalized principle of relativity, the one based on the principle of equivalence. He begins with a Galilean reference system (*Bezugsystem*) K, in which free particles move rectilinearly and uniformly "at least in the four-dimensional domain under consideration" and introduces a second reference system K' (*Koordinatensystem* or *Bezugsystem*) in uniform acceleration with respect to K. An observer bound to K' has no means to decide whether his reference frame is accelerated because, as a consequence of the universality of free fall, he could just as well interpret the acceleration of free particles as the effect of a uniform gravitational field. Hence K and K' are "equally justified reference systems for the description of the processes."[20]

A first remark: the equivalence between inertial and gravitational fields seems limited to small space–time domains, since a gravitational field is generally nonuniform. Einstein's "at least in the four-dimensional domain under consideration" probably implies a restriction of this kind. It remains true, however, that there is no way of distinguishing between the inertial and gravitational components of an acceleration field. As Eddington remarked, if the whole universe as we know it were a free-falling system in a larger, mostly invisible universe, then in the Newtonian view we would regard the Galilean frame of this larger universe as the legitimate K frame and we would ascribe to the known universe a global acceleration under the effect of the gravitational pull from the larger universe. So there is truly no way to decide whether the current astronomical frame is accelerated or not, and there is no absolute distinction between the gravitational and inertial causes of observed motions.[21]

According to the preceding argument, there is no reason to favor the K frame over the K' frame if both frames are applied to the entire known universe. In contrast, if the reference systems are as small as an elevator cage, then it should be natural to privilege the K frame for which the motion of free particles is rectilinear and uniform within the cage. So Einstein's argument implies nearly the opposite of what Einstein means to prove. The local indistinguishability of inertial and gravitational fields suggests that gravitation is only inertia in disguise. If that is true, then the definition of inertial reference frames is a merely

---

[20] Einstein 1916a, pp. 772–773.

[21] Eddington 1920b, p. 68. The remark is very similar to one made by Newton with regard to the determination of absolute space; see p. 18, this volume.

local affair and it makes no sense to speak about global, rigid reference systems. It is not the relativity of such systems that the equivalence principle suggests, it is their impossibility.

Einstein's persisting reliance on accelerated frames in 1916 (and later) derives from his early exploitation of the equivalence principle by means of such frames. At the end of his plea for an extended relativity principle, Einstein indeed asserts that "a gravitation field can be 'synthesized' through a mere change of the coordinate system" and that, as a consequence, "the principle of the constancy of the velocity of light in a vacuum must undergo a modification." The idea is simple: if the propagation of light is rectilinear in the inertial frame K, it must be curved in the frame K' in which a uniform gravitational field has been synthesized by acceleration. These assertions make no sense in general relativity (as we now understand it) since the intrinsic physics should be independent of the choice of coordinates. A variable velocity of light occurs only when the velocity is measured with respect to coordinates devoid of direct metric significance. The only valuable residue of this old way of thinking is the idea that the laws of gravitation-free physics very nearly apply in a free-falling frame of small extension in space–time.

Rotating frames

In the next section of his memoir, Einstein readily admits that in his previous arguments he has relied on a concept of rigid extended frame that is no longer sustainable in general relativity. His proof for this impossibility is based on the consideration of a frame K' uniformly rotating with respect to an inertial frame K. As he had already noted in 1912, a circle of K' centered on the axis of rotation and the diameter of this circle have a ratio larger than π when measured with a tiny unit rod at rest in K', because this rod, as seen from K, is contracted when it lies on the circumference and unchanged when it lies on the diameter. Therefore, the geometry is non-Euclidean in K' and the coordinates in this frame can no longer be defined by orthogonal projection on three orthogonal axes. Similarly, the period of a clock of K' as measured from K depends on the distance from the axis, and an observer on the axis (thus belonging to both frames) judges that the rate of this clock depends on its location (since he assumes that successive light pulses from this clock take a constant time to reach him).[22]

Einstein takes this observation to imply that it is generally impossible to rely on space and time coordinates that can be individually measured by standard rods and clocks. The deduction is incorrect: the non-Euclidean or non-Minkowskian nature of space and time measurements in a rotating frame does not imply the non-Minkowskian nature of space–time. A change of frame or a change of coordinates does not affect the intrinsic geometry of space–time.

[22]Einstein 1916a, pp. 773–775. Cf. Stachel 1989a, and pp. 224, 280, this volume. Einstein later combined the rotating-disk clock effect and the equivalence principle to derive the gravitational redshift in a nonuniform field: Einstein 1917a, pp. 54–55; also Eddington 1918, pp. 57–58.

General covariance

From the alleged impossibility of choosing the coordinates so that the interval between two events acquires a simple (pseudo-Euclidean) form, Einstein generalizes to the impossibility of choosing the coordinates so that the laws of physics globally take an especially simple form. He is left with a nearly total freedom in the choice of coordinates: "The only remaining possibility is to regard all conceivable coordinate systems as equally justified in principle for the description of nature." This freedom leads him to the requirement of general covariance:

> *The general laws of nature must be expressed by equations that hold in every system of coordinate, that is, by equations that are covariant with respect to arbitrary changes of coordinates.*

This demand, Einstein tells us, "takes away the last remnant of physical objectness from space and time." Its natural character derives from the possibility of reducing any space–time determination to the coincidence of point-like events such as the intersection of the world-lines of two material particles. Coordinates are nothing but a convenient way to express such coincidences.[23]

There is much to be said about this train of reasoning. First, Einstein's exclusion of (pseudo-)Cartesian coordinates does not imply the broader exclusion of any coordinates that have direct metric significance. This can be seen by analogy with the theory of curved surfaces: as we saw, we can define a point of a surface by its distance from two reference points (as long as the points remain in a sufficiently small domain). But there are infinitely many ways to choose the reference points, and it is mathematically more convenient to begin with uninterpreted coordinates for which the choice of the metric coefficients is completely free (save for regularity conditions). Similarly, in general relativity it is possible to (locally) map point-like events by coordinates that have immediate metric significance. For instance, we may rely on the proper times at which two free-falling observers send and receive light signals that connect to the event. John Synge introduced such radar coordinates in 1921, and they have proved very useful in discussing the physical content of general relativity.[24]

Einstein's next point remains unaffected: there is no system of coordinates in which the laws of physics acquire a special simplicity in general relativity. In the surface analogy, this corresponds to the fact that on a curved surface there is no choice of coordinates for which the g_{ij} coefficients have the same value at every point. The same is true for the metric tensor $g_{\mu\nu}$ of general relativity. Moreover, the value of this tensor field depends on the contingent distribution of matter, so that one cannot hope to simplify its form by an adequate choice of coordinates in a manner independent of this distribution. Simplifying choices of coordinates are only possible for specific solutions of the field equations corresponding to specific distributions of matter.

---

[23] Einstein 1916a, p. 776 (his emphasis). This is the point-coincidence argument that first appeared in Einstein's correspondence in response to the failure of the hole argument; see p. 319, this volume.

[24] Synge 1921.

Next comes Einstein's demand for general covariance. Einstein misleadingly ties this demand to the lack of privileged coordinates. The existence of systems of coordinates in which the fundamental equations of physics generically take an especially simple form does not imply that other systems of coordinates are impossible. On the contrary, the equations of physics can always be subjected to widely free changes of coordinates, and every law of physics can be expressed in general covariant form. In the surface analogy, the existence of simplifying Cartesian coordinates on a plane surface does not preclude recourse to curvilinear coordinates. In physics, the reduction of every physical phenomenon to sets of point-like coincidences, to which Einstein appeals in order to justify general covariance, does not depend on the kind of phenomenon or the kind of theory considered. It should equally apply to Newtonian gravitation and to general relativity. Planck's former doctoral student Erich Kretschmann made these points soon after the publication of Einstein's memoir. In his reply, Einstein readily accepted Kretschmann's criticism, with an important inflection:

> Although it is true that every empirical law can be brought to generally covariant form, the principle [of general covariance] still has a significant heuristic power that has already been strikingly confirmed for the gravitation problem. This power rests on the following: from two theoretical systems compatible with experience we must favor the one that is the simplest and the most transparent from the viewpoint of the absolute differential calculus.

More than general covariance seems to be here at stake: Einstein requires the possibility of treating space–time as a metric manifold on which there is an absolute differential calculus. In other words, the physical properties of space–time should be reduced to the intrinsic geometry of a four-dimensional manifold. Or else: the laws of physics should depend only on the intrinsic metric properties of the space–time manifold and not on the coordinate system with which the events are arbitrarily labeled.[25]

Does this stronger demand truly exclude theories other than general relativity? The discussion of intrinsic surface-based physics in the first section of this article suggests a negative answer. History indeed brought contradiction. As was first shown by Elie Cartan in 1923 and as is now well known, Newton's theory of gravitation can be recast as the theory of geodesics on an adequate space–time manifold, and the resulting form is neither complicated nor awkward. In the context of general relativity, Einstein's simplicity requirement does not prevent us from replacing the covariant equation $R_{\mu\nu} = 0$ for the gravitational field in a region devoid of matter with the stronger condition $R_{\mu\nu\rho\sigma} = 0$ for the curvature tensor. This would lead us to a flat space–time in which the metric is uniform for a proper class of coordinates. In order to exclude the Newton–Cartan space–time or the flat space–time, other reasons need to be invoked. The Newton–Cartan option is excluded by Einstein's demand that the metric should be locally Minkowskian, and the flat option is excluded by

---

[25] Kretschmann 1917; Einstein 1918, p. 242. Cf. Norton 1992b, 1993b; Rynasiewicz 1999; Weinstein 2015b, §2.12. On Kretschmann, cf. Gebhardt 2016,

his demand that the geodesics of space–time should represent the path of free-falling bodies. The moral is that in order to restrict the choice of theories, general covariance must be used in conjunction with other structural requirements and principles.[26]

The immateriality of space–time

We now come to Einstein's intriguing statement that general covariance robs space and time of "their last remnant of physical objectness [*Gegenständlichkeit*]," which has been abundantly commented on by philosophers since Ernst Cassirer's relativity book of 1921. Contrary to this claim, general covariance cannot be the source of any change in our concepts of space and time since it is compatible with any theory of phenomena in space and time. The source of Einstein's confusion is his conflating coordinate systems and reference systems. General covariance, he tells us, implies the equivalence of reference systems that move with respect to each other in any manner: "It is clear that a physics that satisfies this postulate also satisfies the general principle of relativity. For among *all* the changes of coordinates, there are those which correspond to any relative motion of the (three-dimensional) systems of coordinates." And it is the absence of a privileged class of reference systems that ruins space–time as an independent physical object. The problem with this line of thought is that, again, extended rigid reference frames are incompatible with general relativity. A more acceptable interpretation of Einstein's statement would be that there is no intrinsic space–time structure that is given independently of the distribution of matter. But is this truly the case in general relativity? In a footnote, Einstein tells us that he will not discuss the restrictions of univocality and continuity for transformations of coordinates. This decision conceals the assumption of a continuous, differentiable manifold of events. This basic structure is required, whatever the distribution of matter.[27]

In addition to this minimal structure, there is a metric structure that depends on the contingent distribution of matter in the universe. This is to be contrasted with the cases of Galilean relativity and special relativity for which the metric structure is an a priori given background. In sum, Newton's mechanics, special relativity, and general relativity rely on three different space–time structures that correspond, in Kleinian spirit, to three different groups of transformations: the Galilean group, the Lorentz–Poincaré group, and the groups of diffeomorphisms of a four-dimensional manifold. If the "objectness" (*Gegenständlichkeit*) of space–time is read as the existence of a permanent metric structure, then this quality is lost in general relativity. This is a modernized version of Cassirer's interpretation of Einstein's statement.[28]

Altogether we see that Einstein's heuristic arguments in favor of general relativity suffer from a tension between the metric and label conceptions of coordinates. In his long quest for a relativistic theory of gravitation, he began with a metric conception and gradually

---

[26] Cf. Havas 1964. For an in-depth historico-critical discussion of the meaning of general covariance, cf. Norton 1992b, 1993b. For a modern philosophical assessment, cf. Brown 2005, §9.2.

[27] Einstein 1916a, p. 776n. To translate *Gegenständlichkeit*, I prefer the neologism "objectness" to the equally misleading "objectivity" and "materiality." In a similar statement in November 1915 (see p. 318, this volume) Einstein used *objective Realität* instead of *physikalische Gegenständlichkeit*.

[28] Cassirer 1921, pp. 81–86, 106. Cf. Ryckman 2005, §2.5; Darrigol 2018b, §1.3.

moved toward the mere label conception in analogy with the coordinates of Riemannian manifolds. This conception is necessary if one wants to treat the $g_{\mu\nu}$ coefficient as independent variables. Nevertheless, Einstein could not quite get rid of the idea that coordinates somehow refer to concrete reference systems. This is why his heuristics of 1916 still involved rigid reference frames whose extension is not clearly defined, at least as a prelude to a demonstration of their absurdity. He was clearer on these points in his popular presentation of relativity theory in 1917. There he emphasized that extended rigid bodies were a mere fiction in general relativity, and he introduced the "reference mollusk," namely a body of arbitrary, variable shape with a dense array of attached pre-clocks that yield nearly the same time when they are in the same small neighborhood. The Gaussian coordinates are four numbers giving the position of a clock in the mollusk and its pre-time. Any such mollusk is acceptable and so is any (sufficiently regular) coordinate system.[29]

The formal apparatus

In the next section of his *Grundlage*, Einstein comes to the systematic construction of the theory. He begins with the existence of local coordinate systems for which the metric is Minkowski's, in agreement with the assumption that in small enough free-falling reference frames and for short enough lapses of time the laws of special relativity still hold. This implies in particular that in such frames (small) rigid rods and clocks can be used to measure the interval between two events. In an arbitrary coordinate system the metric takes the form

$$ds^2 = g_{\mu\nu}dx^\mu dx^\nu, \tag{10.9}$$

with the signature $(+,-,-,-)$. In general there is no coordinate system for which the $g_{\mu\nu}$ coefficients are the same everywhere. At any given point of space–time, however, there are coordinates such that $g_{\mu\nu}$ takes the Minkowskian form at this point and for which the derivatives of these coefficients also vanish. As was mentioned, this important property means that (pseudo-)Riemannian manifolds can be regarded as locally flat. In the context of general relativity, this means that special relativity applies in a small enough neighborhood, in agreement with the way Einstein justifies the form of the metric.[30]

Einstein goes on with the remark that in the flat case of special relativity, free particles follow straight lines. The natural generalization of this statement to the curved space–time is that free particles follow the geodesics of the manifold. Although Einstein does not quite say so, this geodesic principle results from the fact that special relativity is locally valid together with the fact that geodesics can be defined as locally straight lines (lines along which the tangent vector remains parallel to itself).[31]

---

[29] Einstein 1917a, p. 67. On Einstein's struggle with the meaning of coordinates, cf. Stachel 1993.

[30] Einstein 1916a, p. 778. Einstein is rather vague about local flatness.

[31] Einstein 1916a, p. 779.

A few sections of the memoir are devoted to the needed apparatus of absolute differential calculus on a manifold. Einstein then gives the equation of geodesics

$$\frac{d^2x^\mu}{ds^2} + \{^{\ \mu}_{\nu\rho}\}\frac{dx^\nu}{ds}\frac{dx^\rho}{ds} = 0, \tag{10.10}$$

and the simplified form

$$\partial_\rho\Gamma^\rho_{\mu\nu} + \Gamma^\sigma_{\mu\rho}\Gamma^\rho_{\nu\sigma} = -\kappa T_{\mu\nu} \ \text{ with } \ \Gamma^\rho_{\mu\nu} = -\{^{\ \rho}_{\mu\nu}\} \tag{10.11}$$

of the gravitational field equation for $g = -1$. Einstein also gives the generating action

$$S_F = \int \Lambda d^4x \ \text{ with } \Lambda = g^{\sigma\tau}\Gamma^\alpha_{\sigma\beta}\Gamma^\beta_{\tau\alpha}, \tag{10.12}$$

and the associated stress–energy pseudo-tensor

$$t^\nu_\sigma = \kappa^{-1}(\tfrac{1}{2}\delta^\nu_\sigma g^{\alpha\beta}\Gamma^\rho_{\mu\alpha}\Gamma^\mu_{\rho\beta} - g^{\nu\tau}\Gamma^\rho_{\mu\sigma}\Gamma^\mu_{\rho\tau}). \tag{10.13}$$

He mentions the possibility of a fully covariant formulation, without giving the corresponding field equation[32]

$$R_{\mu\nu} - \tfrac{1}{2}g_{\mu\nu}R = -\kappa T_{\mu\nu}. \tag{10.14}$$

The Newtonian approximation

In general relativity even more than in other theories, the true difficulties begin with the physical applications of the fundamental equations. Einstein's first concern is to retrieve Newtonian gravitation in a first approximation. For this purpose, he assumes that "for a proper choice of coordinates" the metric tensor $g_{\mu\nu}$ differs little from its Minkowskian value over the domain of space–time under consideration. He further assumes that for the particles moving in this field the three quantities dx^i/ds ($i = 1,\ 2,\ 3$) are negligible compared with dx^0/ds, as would be the case if the space–time were Minkowskian and if the velocity of the particle in this space–time were negligible compared with the velocity of light. This assumption sounds legitimate as long as the true metric does not much differ from the Minkowski metric.[33]

---

[32] Einstein 1916a, pp. 779–808. See p. 315, this volume.

[33] Einstein 1916a, pp. 816–818.

In these approximations and for a static field in which $\partial_0 g_{\mu\nu} = 0$, Einstein contends that the geodesic equation reduces to

$$\frac{d^2 x^i}{c^2 dt^2} = -\Gamma^i_{00} = \frac{1}{2}\partial_i g_{00}, \quad \text{wherein } ct = x^0. \tag{10.15}$$

This is not quite correct: at the same order of approximation there could be nonnegligible Γ^i_{j0} coefficients corresponding to Coriolis forces in a rotating system of coordinates.[34] It remains true that there is a system of coordinates for which Einstein's simplification is valid. In the same approximation, the field equations are dominated by the T_{00} component of the energy–momentum of matter, which is to be interpreted as the mass density ρ multiplied by c^2. The resulting equation reads

$$\Delta g_{00} = \kappa \rho c^2, \tag{10.16}$$

which has the same form as the Poisson equation for the Newtonian gravitation potential. The equation of motion of a particle can therefore be rewritten as

$$\frac{d^2 \mathbf{x}}{dt^2} = -\nabla\Phi, \quad \text{with } \Phi(\mathbf{x}) = -\frac{\kappa c^4}{8\pi}\int \frac{\rho d^3 x'}{|\mathbf{x} - \mathbf{x}'|}, \tag{10.17}$$

in conformity with the Newtonian equations of gravitation for the value $G = \kappa c^4/8\pi$ of the gravitational constant.[35]

Space–time geodesy

In order to be physically meaningful, this reasoning should be accompanied by a proof that to a sufficient approximation the coordinates x^μ can be regarded as the time and space coordinates measured in Minkowski space–time. Einstein's next section, entitled "Behavior of measuring rods and clocks in a static gravitational field," is plausibly meant to fill this gap. There he gives the following central-symmetric solution of the gravitational field equation generated by a central mass M in the needed approximation:[36]

$$g_{ij} = -\delta_{ij} - \alpha x_i x_j/r^3, \quad g_{i0} = g_{0i} = 0, \quad g_{00} = 1 - \alpha/r, \tag{10.18}$$

with $r^2 = x_1{}^2 + x_2{}^2 + x_3{}^2$, and $\alpha = \kappa M c^2/4\pi$. Einstein presumably arrived at this expression by acting as if the spatial coordinates x_i were coordinates in a fictitious Euclidean space and by conflating isotropy in this fictitious space with isotropy in the true metric space

---

[34] Silberstein 1922, pp. 48–49.

[35] Einstein has different powers of c in some of these equations because he takes $t = x^0$ instead of $t = x^0/c$. The gravitational constant G is defined so that the Newtonian attraction between the point mass M and the point mass m at the distance r from the former is GMm/r^2.

[36] Einstein has $\alpha = \kappa M/8\pi$ (his equation 70a) instead of $\alpha = \kappa Mc^2/4\pi$. The omission of the c^2 factor comes from his special choice of units. The 8 instead of 4 is a slip. I will later explain (notes 40 and 61) why this slip does not affect his final conclusions.

associated with the metric $-g_{ij}$. Unlike modern theorists, he did not have Killing vectors and Lie derivatives to deal with the global symmetries of a metric manifold. He could not even rely on polar coordinates because the condition $g = -1$ forbids them.

Later expositors of general relativity dropped the condition $g = -1$, expressed the central symmetry through the polar coordinates form

$$ds^2 = \alpha(r)dt^2 + \beta(r)dr^2 + \gamma(r)(d\theta^2 + \sin^2\theta d\phi^2), \qquad (10.19)$$

and determined the α, β, γ coefficients through the homogeneous subcase of the field equation (10.11). This procedure, just as Einstein's, has the defect of presuming a partial metric interpretation of the coordinates before the metric is given. It could induce Einstein's readers to believe that he inconsistently mixed the usual metric interpretation of the polar coordinates with the non-Euclidean geometry derived from the ds^2. So did for instance the French mathematician Jean Marie Le Roux in the *Comptes rendus* for June 1921, even though Eddington had properly elucidated this point three years earlier in his own derivation of the Schwarzschild solution. Here is Eddington's remark:[37]

> We shall now find a solution of [Einstein's] equations corresponding to the field of a particle at rest at the origin of space-co-ordinates. We choose polar co-ordinates $[x_0 = t, x_1 = r, x_2 = \theta, x_3 = \phi]$. In making this statement we are departing somewhat from the standpoint of general relativity. Strictly speaking, we can only define a system of co-ordinates by the form of the corresponding expression for ds^2; that is, by the gravitation-potentials. So that to specify the co-ordinates that are used involves solving the problem. Further, we have at present no knowledge of a particle of matter, except that it must be a point where the equations $[R_{\mu\nu} = 0]$, which hold at points outside matter, break down; we can only distinguish a particle from other possible singularities, such as doublets, by the symmetry of the resulting field. Thus the logical course it to find a solution, and afterwards discuss what distribution of matter and what system of co-ordinates it represents. We shall, however, find it more profitable to accept the guidance of our current approximate ideas in order to arrive at the required solution inductively.

After giving the solution (10.18) of his field equations, Einstein obtains its full chrono-geometric interpretation in the following manner. He first considers spatial intervals for which $dx^0 = 0$. A radial unit rod is obtained by setting $x_2 = x_3 = 0$, $dx_2 = dx_3 = 0$, and

$$-1 = ds^2 = g_{11}dx_1{}^2 = -(1 + \alpha/r)dx_1{}^2, \quad \text{or} \quad dx_1 \approx 1 - \alpha/2r. \qquad (10.20)$$

An orthoradial rod is obtained by setting $x_2 = x_3 = 0$, $dx_1 = dx_3 = 0$, and

$$-1 = ds^2 = g_{22}dx_2{}^2 = -dx_2{}^2, \quad \text{or} \quad dx_2 = 1. \qquad (10.21)$$

---

[37] Le Roux 1921; Eddington 1918, p. 43. In 1923, Le Roux demolished Einstein's relativity in *L'ouest-éclair* (April 2, front-page): "Les conséquences de la théorie d'Einstein sont même tellement singulières qu'il est impossible de lui attribuer une valeur scientifique quelconque. On y découvre des erreurs grossières et flagrantes démontrant qu'Einstein ne possède pas une culture mathématique suffisante pour apprécier exactement la signification des calculs, ni pour interpréter et discuter les résultats. Ce n'est, je le répète, qu'une grossière contrefaçon de la science."

In Einstein's words, the radial unit rod "appears to be shortened by the given amount with respect to the system of coordinates in the presence of the gravitation field," and the "gravitational field has no effect" on the orthoradial unit rod. Einstein concludes:

> Euclidean geometry does not hold in a gravitational field even in first approximation, provided that the same rod concretizes the same length independently of its location and of its orientation. . . . However, the expected deviations [from Euclidean geometry] are too small to affect terrestrial geodesy.

Plausibly, Einstein also meant that the deviations from Euclidean geometry did not affect astronomical observations in the Newtonian approximation, thus completing the argument of his former section.[38]

There is some oddity in this way of discussing the non-Euclidean properties of space: On the one hand, it is based on the sound idea that true metric relations are defined by the metric element ds^2; on the other hand, it seems to retain a concept of length with respect to a system of coordinates, as if the coordinates still had an independent physical existence. Perhaps Einstein was reasoning by (false) analogy with his derivation of the non-Euclidean character of space measurements in a rotating frame.

There are two ways of reconciling Einstein's utterances with the lack of a priori metric significance of the coordinates. First, we can imagine that the radial segment $r \leq |x_1| \leq r + dr$, $x_2 = x_3 = 0$ and the circle $x_1^2 + x_2^2 = r^2$, $x_3 = 0$ are covered with small adjacent unit rods. Independent of any preconceived metric meaning of the coordinates, the ratio of the numbers of rods on the circle and on the radial segment will differ from $2\pi r/dr$ by the amount $1 - a/2r$, which is an intrinsic proof of the non-Euclidean character of space.

Second, following Eddington's suggestion, we may observe that the metric (10.14) can be obtained by starting from a Euclidean space in which the radial rods are contracted by the factor $1 - a/2r$ under the effect of the gravitational field. As Eddington puts it:

> We must correct the measured length . . . in the radial direction, multiplying it by $[1 - a/2r]$ in order to obtain a length dr which will fit into Euclidean space. Or we may say that our measuring rod contracts when placed radially; transverse measures require no correction.

In this view, space is originally Euclidean but it appears to be non-Euclidean when the measuring rods are affected by a gravitational field. This is of course contrary to Einstein's mature view, in which gravitation-free measurement is a mere fiction. As Eddington explains, the latter view is preferable because the other view is highly degenerate:

> There is more than one way of correcting the measures to fit Euclidean space, so that we are not really justified in making precise statements as to the behaviour of our clocks and measuring rods. It is better not to discuss their defect, and to accept the measures and examine the properties of the corresponding non-Euclidean space and time.

---

[38] Einstein 1916a, p. 820.

From a formal point of view, it is nonetheless possible to characterize the non-Euclidean features of space by invoking a perverted geodesy.[39]

10.3 The gravitational redshift

Einstein's reasoning

In the next section of the *Grundlage*, Einstein extends his perverted-measurement approach to the measurement of time in the same approximation. Namely, he defines the interval between two beats of a unit-clock through

$$1 = ds^2 = g_{00}dx_0{}^2 = (1 - \alpha/r)dx_0{}^2, \quad \text{or} \quad dx^0 \approx 1 + \alpha/2r = 1 - \Phi/c^2. \tag{10.22}$$

From the latter equation, he abruptly concludes:[40]

> The clock runs slower when it is set in the vicinity of masses. Consequently, the spectral lines in the light from the surface of a big star should appear to be shifted toward the red end of the spectrum.

This is a misleading statement: the clocks *do not run slower*; rather, the rate of a given clock as judged from another distant clock depends on the gravitational potentials in which the two clocks are immersed. In analogy with the case of spatial geometry, Einstein seems to be starting from a fictitious Minkowski space–time and to be treating the departure from Minkowskian geometry as an effect of the gravitational field on the rate of clocks. Or he may still have in mind his first derivation (1907) of the clock effect, based on the distinction between different times: the local time given by a standard clock brought to the given location, and the true time with respect to which the propagation of light is defined. In the present reasoning, ds corresponds to the elapsed local time and x_0 to the true time.[41]

Whatever was concealed behind Einstein's quick reasoning, he could not doubt the result for he had known since 1911 that he could get it through a direct application of the equivalence principle. The reasoning otherwise looks problematic and confusing, because it is literally incompatible with general covariance. The expression $g_{00} = 1 - \alpha/r$ of the coefficient of $(dx^0)^2$ in the metric formula indeed varies with a change of coordinates. According to Einstein's reasoning, the period of a unit-clock is the value $(g_{00})^{-1/2}$ of the interval dx^0 for which $ds^2 = -1$, so that the retardation depends on the choice of coordinates. This is why several of Einstein's early readers, including Jean Chazy, Allvar Gullstrand, and

---

[39] Eddington 1918, pp. 27–28. That Einstein relied on perverted geodesy is confirmed in Einstein (1917a, p. 63), where he compares the non-Euclidean geometry of general relativity with the apparent geometry of a tabletop of uneven temperature when gauged with dilatable rods.

[40] Einstein 1916a, p. 820. In Einstein's notation, $dx_4 = 1 + (\kappa/8\pi)\int \rho d\tau/r$ (his equation 72), which agrees with our $dx^0 = 1 - \Phi/c^2$. This means that Einstein here uses the correct value of α.

[41] For a criticism of Einstein's reasoning, cf., e.g., Earman and Glymour 1980a, p. 182: "To the modern eye, Einstein's derivation is no derivation at all." On the derivation of 1907, see pp. 275–276, this volume. For a critical history of Einstein's derivations of the redshift, cf. Bacelar Valente 2018.

Paul Painlevé, declared that his theory led to an indeterminate value of the gravitational redshift.[42]

Weyl's cosmic time

In order to save Einstein's derivation, some physical interpretation of the favored "time coordinate" x^0 is needed. As Levi-Civita remarked in his "Statica Einsteiniana" of 1917, Einstein's weak-field metric and the Schwarzschild metric belong to a broader class of "static" solutions of Einstein's field equations for which the metric element takes the form

$$ds^2 = g_{00}(dx^0)^2 - \gamma_{ij}dx^i dx^j, \tag{10.23}$$

while the metric coefficients g_{00} and γ_{ij} do not depend on the "time coordinate" x^0, the g_{00} coefficient is positive, and the quadratic form $\gamma_{ij}dx^i dx^j$ is positive definite. Clearly, the intervals for which $dx^1 = dx^2 = dx^3 = 0$ are time-like and those for which $dx^0 = 0$ are space-like. This explains why Levi-Civita calls x^0 a time and why he calls this kind of solution static, even though proper time intervals differ from x^0 time intervals by the space-dependent factor $\sqrt{g_{00}}$.[43]

As Weyl remarked the following year, in order to preserve the static form of the metric, a change of coordinates can imply only a change of unit or origin ($x^0 \to ax^0 + b$) for the time coordinate (it may of course also imply any time-independent change of the spatial coordinates). This means that in a static gravitational field the time coordinate plays a genuinely special role. Weyl called it the "cosmic time," in a probable allusion to Edmund Husserl's distinction between "phenomenological time" and "cosmic time." In Husserl's *Ideen* of 1913, the phenomenological time is associated with the flux of consciousness, while the cosmic time is the time measured by external physical means. At first glance, it would seem that Weyl permuted the two notions, since the measured time of general relativity is the proper time ds. He truly did not, because in his view the metric ds^2 pertains to a simplified form of individual consciousness while the cosmic time belongs to the entire static universe. In a later discussion of the global properties of space–time, Weyl explicitly referred to the cosmic/phenomenological time distinction. The context was the logical possibility of closed time-like world-lines allowing someone to meet his old self in the past:[44]

> That would imply a radical doppelgängerness, as E. T. A. Hoffmann has conceived. In reality the needed high variability of the $g_{\mu\nu}$ does not occur in the world domain in which we live. Yet it is rather interesting to analyze these possibilities in regard to the philosophical problem of the distinction between cosmic and phenomenological time. Provided something paradoxical here occurs, there is never any actual contradiction with immediate given facts in our life.

[42] Chazy 1921; Gullstrand 1921, p. 14; Painlevé 1921a, p. 680; 1921b, p. 887.

[43] Levi-Civita 1917b.

[44] Weyl 1918, p. 192; Weyl 1917, p. 132; Weyl 1919, p. 236 for the citation; Husserl 1913, §81. On Weyl's Husserlian motivations, cf. Ryckman 2005, chap. 5.

Weyl's concept of cosmic time directly informed his discussion of the gravitational red-shift. In the three first editions of *Raum · Zeit · Materie* he asserted that "the light waves emitted from an atom, when measured in *cosmic* time, naturally have the same frequency everywhere." In addition, he assumed that the period of oscillation of an atom was an absolute constant when measured through the proper time $ds = \sqrt{g_{00}}\,dt$. If the atom is located on the surface of a star, the interval dt of cosmic time is preserved during the travel of the light to a terrestrial observer. For this observer, cosmic time and proper time are very nearly the same because $g_{00} \approx 1$. Therefore, the observed frequency of a spectral line from a stellar source is $\sqrt{g_{00}}$ times the frequency of the same line from a terrestrial source.[45]

One remaining weakness of this reasoning is the lack of proof that the frequency of a light wave is a constant when measured in cosmic time. In the fourth edition of his treatise (1921), Weyl explained: "This is so because in a *static* metric field the Maxwell equations have a solution for which the time dependence is given by the expression $e^{i\nu t}$ with an arbitrary *constant* frequency ν." The proof, which Weyl does not give, is easily supplied by injecting the static metric (10.20) into the electromagnetic action and observing that the resulting field equation has time-independent coefficients and contains only the square of the time derivative ∂_0 besides the spatial derivatives.[46]

Eddington's lights

The first review of general relativity in English was the one given by the Dutch astronomer Willem de Sitter in 1916, soon after the publication of Einstein's *Grundlage*. De Sitter's derivation of the gravitational redshift follows Einstein's very closely. He regards the interval ds of an atomic oscillation as a "constant" and then uses the relation $dt/ds = 1/\sqrt{g_{00}}$ to conclude that "the measure of time is different at different places in the gravitational field." He thus follows Einstein in regarding dt as the measured time corresponding to an oscillation of the source. Eddington's first derivation of the redshift in 1918 is similar. He first states that an atom is "a natural clock which ought to give an invariant measure of an interval δs" and then uses the resulting equality $\sqrt{g_{00}}\,dt = \sqrt{g'_{00}}\,dt'$ between the "periods of two similar atoms vibrating at different parts of the field" to derive the redshift.[47]

No more than Einstein do de Sitter and Eddington explain why measured periods or frequencies should be associated with the time coordinate t. Fundamentally, they compare the periods of two oscillations occurring in different locations without stating the basis of the comparison. In 1919, Larmor and Cunningham both exploited this logical gap to argue that Einstein's theory actually implied a vanishing spectral shift, in agreement with the contemporary failure of astronomers to confirm Einstein's prediction. In Cunningham's argument, the comparison of the distant oscillations is done by imagining an observer who carries the solar atom to the earth. Larmor more realistically considers that the rays of light play the role of "messengers across space" and he deplores the neglect of this role:

[45] Weyl 1918, p. 197; 1919, p. 212.

[46] Weyl 1921, p. 223. For details, cf. Darrigol 2015, p. 173.

[47] de Sitter 1916a, p. 719; Eddington 1918, pp. 56–57. Cf. Earman and Glymour 1980a, p. 183.

The difficulty was to recognise how a theory which professes to supersede an aether with its definite space and time, by concepts purely relativist, could manage to effect direct comparison, at a distance without tracing transmission across the intervening space, of the radiations of a molecule at the sun and those of a molecule of the same substance at the earth.

A physics lecturer at Liverpool, James Rice, remedied this defect in the following month by asserting that the interval between two flashes of light sent from a given location was conserved when the flashes were received in a different location. By extension, the interval ds of an atomic oscillation occurring on the sun should be conserved by the light conveying this oscillation to a terrestrial observer, and there should not be any shift of the observed spectral lines when compared with the lines of a terrestrial source.[48]

This communication prompted an insightful reply by Eddington in the February issue of *Nature*. The astronomer denied that the interval ds was preserved by the light propagating from the distant source, and instead asserted:

> The rule deduced from Einstein's theory for comparing the passage of two light-pulses at the points A and A' respectively is not $ds = ds'$ but $dt = dt'$, *provided the co-ordinates used are such that the velocity of light does not change with t.*

Indeed, the italicized condition, which is evidently met for the coordinates used in the Schwarzschild solution, implies that successive light pulses take the same t-time to travel from A to A'. Accordingly, the period $dt = (g_{00})^{-1/2}ds$ associated with the interval ds of a solar spectral source is also the period of the light from this source when it reaches a terrestrial observer. It therefore exceeds by $(g_{00}/g'_{00})^{-1/2}$ the period dt' of the light from a similar terrestrial source. As Eddington notes, the value of this ratio is intrinsic since a ratio of times at a given location is the same as the ratio of the corresponding proper times.[49]

Eddington went on to refute Larmor's suggestion that the time independence of the co-ordinate velocity of light was an absolute requirement without which any comparison of time intervals at a distance became illusory. He recalled that there was an infinity of equivalent solutions for the central-symmetric problem obtained by a mere change of coordinates, and commented:

> Whether we use [Schwarzschild's expression for the center-symmetric metric field] or any other expression, we have to find out from the expression itself the meaning of the co-ordinates introduced. In the limiting case [of a vanishing central mass], [Schwarzschild's] expression agrees with the formula for polar co-ordinates and time in a Euclidean world; hence it is usual to call r the distance from the sun and t the time. But there can be no exact identification of variables in a non-Euclidean world with quantities the definition of which presupposes a Euclidean world; and the only exact definition of r and t is that they are mathematical intermediary quantities which satisfy [the Schwarzschild formula]. The variable t is in no sense an absolute time; it is

---

[48] Cunningham 1919; Larmor 1919; 1920, p. 530; Rice 1920.

[49] Eddington 1920a, p. 598. Cf. Earman and Glymour 1980a, pp. 185–186. Eddington's statement about the intrinsic character of the redshift should not be confused with a wrong statement of covariance for $(g_{00}/g'_{00})^{-1/2}$.

specifically associated with the sun, which in [the Schwarzschild formula] is regarded
as the only mass in the universe worth considering.

As we will see in a moment, many of Einstein's readers ignored Eddington's warning
against a pre-metric interpretation of coordinates. His last remark against absolute time
may have been intended for Weyl, who seemed to ascribe a special physical significance to
the coordinate $t = x^0$ in the static system by calling it "cosmic time."[50]

The discussion in *Nature* informed the derivation of the redshift that Eddington gave in
his treatises of 1920 and 1923. There he repeated his earlier argument that the intervals ds
and ds' for the oscillations of similar spectral sources occurring at two different locations
were equal and the associated periods dt and dt' were therefore unequal. He now added:

> There is one important point to consider. The spectroscopic examination must take
> place in the terrestrial laboratory; and we have to test the period of the solar atom
> by the period of the waves emanating from it when they reach the earth. Will they
> carry the period to us unchanged? Clearly they must. The first and second pulse have
> to travel the same distance (r), and they travel with the same velocity (dr/dt); for the
> velocity of light in the mesh-system used is $1 - a/r$ [for the Schwarzschild solution],
> and though this velocity depends on r, it does not depend on t. Hence the difference
> dt at one end of the waves is the same as that at the other end.

Eddington again warned against a metric interpretation of the t coordinate:

> The quantity dt is merely an auxiliary quantity introduced through the equation [of
> the Schwarzschild metric] which defines it. The fact that it is carried to us unchanged
> by light-waves is not of any physical interest, since dt was *defined* in such a way that this
> must happen. The absolute quantity ds, the interval of the vibration, is not carried to
> us unchanged, but becomes gradually modified as the waves take their course through
> the non-Euclidean space-time. It is in transmission through the solar system that the
> absolute difference is introduced into the waves, which the experiment hopes to detect.

Altogether, Eddington offered the first fully satisfactory derivation of the gravitational redshift and clearly identified the obstacles in correctly interpreting the physical consequences
of a metric formula.[51]

Laue's calculations

A third important contribution to the theory of the gravitational redshift came in
September 1920 from Max von Laue during a session on general relativity at the *Naturforscherversammlung* in Bad Nauheim. Laue gave the eikonal approximation of Maxwell's

---

[50] Eddington 1920a, p. 599.

[51] Eddington 1920b, pp. 128–129; 1923, p. 92. My assessment of Eddington's contribution somewhat differs from
Earman and Glymour's. In my opinion, the inconsistency and incompleteness they perceive in his approach is an
effect of a presentist perspective. In particular, there is nothing wrong in what they call the "backwards derivation"
(ds = ds' ⇒ dt ≠ dt') if the intervals dt and dt' are understood in Eddington's, Laue's, and Weyl's manner as the
periods, constant through space, of the radiations emitted by an atom on the sun and an atom on earth, respectively.

equations in general relativity, and used it to prove that the frequency of a wave (with respect to the cosmic time) was the same for the receptor and the emitter in a static gravitational field. While this approach does not have the simplicity of Eddington's considerations, it has the merit of showing the consistency of the electromagnetic field equations with the constancy of the velocity of light in static coordinates.[52]

Laue also gave his own derivation of the gravitational redshift, taking into account the constancy of frequency. Laue defines ds_{sun} and ds_{earth} as the measures of the same time interval dt with two identical clocks on the sun and on the earth, respectively. From $ds = \sqrt{g_{00}}dt$ he derives $ds_{sun}/ds_{earth} \approx 1 - M/R$, wherein M is the mass of the sun and R its radius (for $G = c = 1$). He then notes that "the frequencies, reckoned with respect to t, behave like the proper times." Indeed the proper time ds corresponding to the time dt is the number of oscillations performed by the clock during dt, which is given by vdt if v denotes the frequency reckoned with respect to t. There follows $v_{sun}/v_{earth} = 1 - M/R$, in conformity with Einstein's prediction as long as the implied frequencies conserve their value when conveyed by light through space. The argument is correct, and it is equivalent to Eddington's, although the conventions and modes of reasoning differ.[53]

During the discussion in Bad Nauheim, the mathematician Georg Hamel noted that Laue's derivation of the gravitational redshift required the oscillating atom to behave like an ideal clock for measuring the interval ds. Was there a rigorous proof of this behavior? To this question Einstein replied:

> It is a logical weakness of relativity theory in its present state that it must introduce the measuring rods and clocks separately instead of constructing them as solutions of differential equations. However, regarding the validity of the consequences affecting the empirical basis of the theory, the consequences that concern the behavior of rigid bodies and clocks are the most secure ones. Since the emitting atoms must be regarded as "clocks" in the sense of the theory, the redshift belongs to the most secure results of the theory.

Einstein's conviction rested on his basic assumption that the laws of gravitation-free physics hold in a local free-falling frame. In particular, this implies that reputedly good clocks such as oscillating atoms give the proper time when they are free falling. In addition, Einstein implicitly assumed that good clocks still gave the correct proper time when their world-line no longer was a geodesic.[54]

Eddington made these points most clearly. After recalling that (according to Kretschmann and to Weyl)[55] the metric $g_{\mu\nu}$ is completely determined by the motion of light rays and free-falling particles independently of any consideration of clocks and rods, he wrote:

---

[52] Laue 1920a.

[53] Laue 1920a, pp. 661–662; 1921, pp. 188–190. Earman and Glymour 1980a, pp. 188–189 interpret Laue's ds as the interval of an oscillation of the clock, which leads to contradictions.

[54] Laue 1920a; Hamel's and Einstein's remarks are reported ibid., p. 662. See also Laue 1921, pp. 189–191. On the Bad Nauheim event, cf. Giovanelli 2014, pp. 30–31. On Laue's eikonal approximation, see p. 372, this volume.

[55] See Kretschmann 1917, pp. 585–590; Weyl 1921, pp. 206–207, 285–286.

To proceed from this to determine exactly what is measured by a scale and a clock, it would at first seem necessary to have a detailed theory of the mechanisms involved in a scale and clock. But there is a short-cut which seems legitimate. This short-cut is in fact the Principle of Equivalence. Whatever the mechanism of the clock, whether it is a good clock or a bad clock, the intervals it is beating must be something absolute; the clock cannot know what mesh-system the observer is using, and therefore its absolute rate cannot be altered by position or motion which is relative merely to a mesh-system. Thus wherever it is placed, and however it moves, provided it is not constrained by impacts or electrical forces, it must always beat equal intervals as we have previously assumed. Thus a clock may fairly be used to measure intervals, even when the interval is defined in the new manner; any other result seems to postulate that it pays heed to some particular mesh-system.

To this day, the assumption that good concrete clocks measure the proper time is a basic component of general relativity on which every concrete space–time geodesy necessarily relies. It is for instance the basis of Synge's chronogeometrical interpretation of general relativity theory.[56]

Soon after the Bad Nauheim meeting, Laue answered Einstein's challenge of theoretically "constructing" clocks that yield the proper time. He first noted that a light-bouncing clock (*Lichtuhr*) in which light is reflected back and forth between two rigidly connected mirrors should yield the proper time since the velocity of light is a constant by assumption. This suggestion, which Laue says he got from Einstein in Bad Nauheim, is ill-conceived because it assumes without proof the possibility of constructing rigid bodies. More to the point, Laue gives a general-relativistic theory of elasticity and uses it to construct a spring-clock that yields the proper time (I have not checked the validity of his reasoning). Lastly, he offers a proof that Bohr's relation $\Delta E = h\nu$ between the energy variation ΔE of an atom and the frequency ν of the emitted radiation yields an invariant proper time if the energy of the atom behaves in agreement with the transformation properties of the energy–momentum tensor in general relativity. Laue thereby assumes the relations $E = m\sqrt{g_{00}}$ and $\Delta E = h\nu$ between energy, mass, and frequency in the static gravitational field. These choices are justified by their being the time component of the covariant relations $p_\mu = mg_{\mu\nu}\mathrm{d}x^\nu/\mathrm{d}s$ and $\Delta p_\mu = \hbar k_\mu$, with $\mathrm{d}s = \sqrt{g_{00}}\mathrm{d}x^0$ along the world-line of an atom at rest, and $k_0 x^0 = 2\pi\nu x^0$ for the emitted wave. The frequency of the spectral line emitted during a quantum jump is therefore given by $\nu = \bar{\nu}\sqrt{g_{00}}$, wherein $\bar{\nu} = (c^2/h)\Delta m$ is the frequency of the line when the gravitation field vanishes at the location of the atom (neglecting the recoil of the atom). This agrees with Einstein's prediction.[57]

Einstein, Pauli, and Beck

Although Einstein had heard Laue in Bad Nauheim, he kept teaching the gravitational redshift as an effect on the *t*-time measured with respect to the static coordinate system.

[56]Eddington 1920b, p. 131; Synge 1960. The difficulties in testing the redshift prediction fueled speculation on a possible violation of this assumption; see, e.g., Silberstein 1922, pp. 104–105.

[57]Laue 1920b. In Darrigol 2015, p. 175, I used $E = p^0$ instead of Laue's more adequate $E = p_0$.

Weyl's, Eddington's, and Laue's complementary arguments could not matter much to him, for he had known since 1907 that the t-time was the one with respect to which "the laws of physics [including those for the propagation of light] do not vary in time but may depend on location." In his influential encyclopedia article of 1921, Pauli gave the following explanation of the shift:

> In a static gravitational field, we can always choose the time coordinate so that the $g_{\mu\nu}$ do not depend on it. Then the number of waves of a light beam between two points P_1 and P_2 must also be independent of time, and therefore the frequency of the light beam must be the same in P_1 and P_2 . . . when measured in the given time scale.

This is quite similar to Eddington's argument. In a footnote, Pauli referred to Laue's formal demonstration of the constancy of the frequency. Pauli also explained, in Eddington's manner, that a strong version of the equivalence principle justified the agreement between the ds^2 used in determining geodetic motion and the ds^2 derived from clock-based surveys.[58]

In his own encyclopedia article, published in 1929, Guido Beck provided a detailed proof that the electromagnetic field equations in a static gravitational field had solutions of constant frequency. He also offered an alternative derivation of the redshift based on the now accepted concept of lightquantum. An atom of mass m in its excited state emits a lightquantum of energy $h\nu$ at the distance r from a large mass M, and the light quantum travels a long way from this mass. From a Newtonian point of view, energy conservation reads:

$$mc^2 - GMm/r = (mc^2 - h\nu) - GM(m - h\nu/c^2)/r + h\nu_\infty, \tag{10.24}$$

where mc^2 is the internal energy of the atom in its original excited state, $-GMm/r$ its gravitational energy in the same state, $mc^2 - h\nu$ its internal energy after emission of the quantum $h\nu$, $-GM(m - h\nu/c^2)/r$ the corresponding gravitational energy, and $h\nu_\infty$ the energy of the lightquantum in its final remote location. This balance requires

$$\nu_\infty = \nu(1 - GM/c^2r), \tag{10.25}$$

in conformity with Einstein's redshift formula. The basic ingredients of this reasoning are the equivalence between mass and energy and the identity of gravitational mass and inertial mass. A more direct but less persuasive reasoning would directly ascribe the gravitational mass $h\nu/c^2$ to the lightquantum. In this view the gravitational redshift results from the conservation of the sum of the kinetic and potential energies of the light quantum considered first at the time of emission and second in a remote location:[59]

$$h\nu - GM(h\nu/c^2)/r = h\nu_\infty. \tag{10.26}$$

---

[58] Einstein 1923, p. 100; Einstein 1907c, p. 458; Pauli 1921, p. 714 (citation), p. 708 (geodesics and clocks).

[59] Beck 1929, pp. 366–367, p. 375 (constant frequency). The frequency of the lightquantum should not be confused with the frequency of the associated wave, which is a constant in the static-GR context.

Beck's reasoning contributed to the robustness of the redshift prediction. Indeed the grav-itational redshift can be derived in several distinct, sometimes even incompatible manners with the same result. Each reasoning has its own subtlety. The most fundamental deriva-tion, the one based on Einstein's full theory of gravitation, requires special attention to the meaning of the implied coordinates. The Dutch physicist Hendrik Casimir witnessed the following incident, which occurred in 1932 while Pauli was lecturing on general relativity in Zürich:

> I remember that once when he was speaking about the so-called red shift . . . he obtained an expression with the wrong sign, which means a shift towards the vio-let instead of towards the red. He then began to walk up and down in front of the blackboard, mumbling to himself, wiping out a plus sign and replacing it by a minus sign, changing it back into a plus sign, and so on. This went on for quite some time until he finally turned to the audience again and said: "I hope that all of you have now clearly seen that it is indeed a red shift."

Even to Pauli, deriving the gravitational redshift could be a nerve-racking experience.[60]

10.4 The gravitational deflection of light

Einstein's derivation

In Einstein's *Grundlage* of 1916, the derivation of the path of light rays in a static gravita-tional field immediately follows the derivation of the redshift. It goes as follows. For a light ray in the isotropic metric of Equation (10.18), the condition $ds^2 = 0$ gives a propagation velocity $v = \sqrt{(dx^1/dx^0)^2 + (dx^2/dx^0)^2 + (dx^3/dx^0)^2}$ that depends on the location and on the direction of propagation. Consequently, wavelets emanating from a plane wave front at time t have different velocities along this front and their envelope at time $t + \tau$ makes an angle with the original wave. By Huygens's construction, this envelope represents the posi-tion of the wave plane at time $t + \tau$. Therefore, with respect to the x^i coordinates the rays are curved. According to Einstein, the curvature is approximately given by the velocity gradi-ent in the direction perpendicular to the direction of propagation. For a light ray roughly propagating in the direction of the x_2 axis in the (x_1, x_2) plane at the minimal distance a from the center of the gravitational field, the metric (10.18) yields

$$v = \sqrt{-\frac{g_{00}}{g_{22}}} = 1 - \frac{\alpha}{2r}\left(1 + \frac{x_2{}^2}{r^2}\right), \tag{10.27}$$

and the total deflection is[61]

[60] Casimir 1983, pp. 137–138.

[61] Einstein 1916a, pp. 336–337. Cf. Earman and Glymour 1980b, pp. 53–55. For the application of Huygens's principle, cf. Einstein 1911, pp. 906–908 (there the propagation is isotropic, the curvature formula is correct, but the net deviation is half the true value). There is an inconsequential sign error in Einstein's version of Equation (10.27). Owing to the slip in his expression of α (see note 36), Einstein gives $\kappa M/4\pi a$ instead of the correct $\kappa M/2\pi a$ for the total deviation (in my notation). However, he must have used the correct value of α in his numerical calculation.

$$\Delta\theta \approx \int_{-\infty}^{+\infty} (\partial v / \partial x_1)_{x_1=a} dx_2 = 2\alpha/a. \tag{10.28}$$

Although the result is undoubtedly correct, Einstein's account is flawed. The propagation being anisotropic, it is not true that the curvature is given by the transverse gradient of the velocity. Einstein's expression for the total deviation is nevertheless correct, because the value he gives to the curvature of rays truly corresponds to the rotation of the wave planes. As the difference between these two quantities vanishes in the asymptotic regions for which $v = 1$, the total bending of a ray equals the total rotation of the associated wave normal. In sum, Einstein's only error was to confuse ray and wave normal. His calculation becomes correct if what he calls "ray curvature" is replaced by "rotation of the wave normal."[62]

Fermat's principle

Another defect of Einstein's reasoning is that it is based on the unproven validity of Huygens's or Fermat's principles for the propagation of light with respect to the x^i coordinates. Again, Einstein seems to have relied on the intuition that the effect of gravitation can be seen as a modification of the behavior of rods and clocks in a fictitious Minkowski space–time. He may also have unconsciously conserved features of his earlier theory of gravitation in terms of a variable velocity of light. At any rate, the validity of Huygens's and Fermat's principles in general relativity is far from obvious. In 1919, Larmor opined that Fermat's notion of least time lost its meaning because of the "heterogeneity of time" in Einstein's theory. He probably did not know that in 1917 Levi-Civita and Weyl had both given a satisfactory proof of the validity of Fermat's principle in general relativity.[63]

For a static metric as defined by Levi-Civita, we have

$$ds^2 = V^2 dt^2 - d\sigma^2, \quad \text{with } V = c\sqrt{g_{00}} \text{ and } d\sigma^2 = \gamma_{ij} dx^i dx^j, \tag{10.29}$$

wherein the coefficients g_{00} and γ_{ij} depend on the spatial coordinates x^i only. Levi-Civita obtains the trajectory of light rays by considering the limit of the motion of material particles when the velocity is very close to the velocity of light. In general, the motion of a particle corresponds to the geodesics of the metric manifold. It is therefore obtained by requiring the integral $\int ds$ to be stationary, or, equivalently and more conveniently for the present discussion, by requiring the stationary character of the integral

$$I = \int g_{\mu\nu} \frac{dx^\mu}{d\tau} \frac{dx^\nu}{d\tau} d\tau, \tag{10.30}$$

---

[62] For details, cf. Darrigol 2015, pp. 176–177. De Sitter and Eddington may have been aware of this difficulty. This would explain why in their calculation of the deflection of light they relied on the shifted radial variable $\bar{r} = r - \alpha/2$ for which the propagation of light becomes (approximately) isotropic; see de Sitter 1916a, pp. 717–719 and Eddington 1918, pp. 53–54. In his Princeton lectures, Einstein (1921, pp. 102–103) still relied on the same curvature formula, but he now used coordinates for which the propagation of light is isotropic.

[63] Larmor 1919.

wherein τ is an arbitrary parameter. For the static metric, this integral reads

$$I = \int \left[\left(V\frac{dt}{d\tau}\right)^2 - \left(\frac{d\sigma}{d\tau}\right)^2\right] d\tau = 0. \tag{10.31}$$

Separating the spatial and temporal variations of this integral, and integrating the temporal variation by parts we get

$$\delta I = \left[2V^2\frac{dt}{d\tau}\delta t\right]_{\tau=\tau_1}^{\tau=\tau_2} - \int \frac{d}{d\tau}\left(2V^2\frac{dt}{d\tau}\right)\delta t\, d\tau + \delta J, \tag{10.32}$$

with

$$\delta J = \int \left[\left(\frac{dt}{d\tau}\right)^2 \delta(V^2) - \delta\left(\frac{d\sigma}{d\tau}\right)^2\right] d\tau \tag{10.33}$$

for the spatial variation. The vanishing of the temporal variation for fixed extremities of the path requires the constancy of $V^2 dt/d\tau$ on the particle's path. The scale of the parameter being arbitrary, we may take

$$V^2 dt/d\tau = 1. \tag{10.34}$$

With this constraint, the spatial variation can be rewritten as the variation of the integral

$$J = -\int \left[\left(\frac{d\sigma}{d\tau}\right)^2 + \frac{1}{V^2}\right] d\tau, \tag{10.35}$$

Consequently, the condition $\delta J = 0$ has the same form as the equation of motion of a mechanical system for which the kinetic energy is $\gamma_{ij}\dot{x}_i\dot{x}_j$ and the potential energy is $-1/V^2$. Calling E the energy constant for a motion of this system, its trajectory between two fixed points is given by Maupertuis's principle (in Jacobi's form):[64]

$$\delta \int \sqrt{E + 1/V^2}\, d\sigma = 0. \tag{10.36}$$

For a particle moving with a velocity infinitely close to the velocity of light, the condition $ds = 0$ and the constraint (10.34) yield $E = 0$. The trajectories therefore satisfy

$$\delta \int \frac{d\sigma}{V} = 0, \tag{10.37}$$

which is Fermat's principle of least time.[65]

---

[64] Cf. Landau and Lifshitz 1960, §44.

[65] Levi-Civita 1917b, pp. 469–470. A direct proof that null geodesics obey Fermat's principle is found in Levi-Civita 1918.

Weyl's reasoning is subtle, and its condensed formulation is an obstacle to the reader. Here is an explanation. Weyl assumes, without any comment, the general principle that "the world line of a light signal is a singular geodesic line" for which the condition $\delta I = 0$ (for fixed extremities of the world-line) and the condition $ds = 0$ (all along the world-line) are both verified.[66] Now consider a given geodesic and an infinitely close world-line that has the same spatial extremities but different temporal extremities t_1 and t_2. According to Equation (10.32), the integral I on the latter world-line exceeds its value on the geodesic by

$$\delta I = \left[2V^2 \frac{dt}{d\tau} \delta t \right]_{\tau=\tau_1}^{\tau=\tau_2} = 2[\delta t]_{\tau=\tau_1}^{\tau=\tau_2}. \tag{10.38}$$

Suppose, in addition, that the two compared lines are null lines along which $ds = 0$. Then the integral I vanishes on both of them, and the elapsed time is the integral of $d\sigma/V$. For such lines the previous relation yields the expression (10.37) of Fermat's principle.[67]

Null geodesics

Neither Levi-Civita nor Weyl care to explain why light rays should follow null geodesics. In 1920, Eddington tells us that "by the principle of equivalence . . . the track of a ray of light . . . agrees with that of a material particle moving with the speed of light." Namely, this agreement is known in special relativity, and by the principle of equivalence special relativity holds in a local free-falling frame. From a mathematical point of view, this implies the existence of a parameter τ for which the relations $d^2x^\mu/d\tau^2 = 0$ and $\eta_{\mu\nu}(dx^\mu/d\tau)(dx^\nu/d\tau) = 0$ hold when the coordinates are chosen so that the metric satisfies $g_{\mu\nu} = \eta_{\mu\nu}$ and $\partial_\rho g_{\mu\nu} = 0$. In arbitrary coordinates, this statement is easily seen to imply the null-geodesic character of light rays and fast free-falling particles.[68]

Granted that light rays follow null geodesics, they also obey Fermat's principle of least time and their trajectory can be derived from this principle. This is not the most convenient procedure, however. In 1916 Flamm obtained the trajectory of light rays directly as a limiting case of the equation of ordinary geodesics. In the plane of the motion and in Eddington's notation, the expression of the Schwarzschild metric is

$$ds^2 = f dt^2 - f^{-1} dr^2 - r^2 d\theta^2, \quad \text{with } f = 1 - 2M/r \quad (G = 1, \; c = 1) \tag{10.39}$$

The resulting equations of geodesics,

$$\frac{d^2\theta}{ds^2} + \frac{2}{r}\frac{dr}{ds}\frac{d\theta}{ds} = 0, \quad \frac{d^2t}{ds^2} + \frac{df}{f dr}\frac{dr}{ds}\frac{dt}{ds} = 0, \tag{10.40}$$

[66] Weyl (1917, p. 127). That light rays follow null geodesics is assumed in Flamm 1916, p. 452, in Hilbert 1917, p. 71 (with credit to Einstein), and in Levi-Civita 1918.

[67] Weyl 1917, pp. 127–128; Weyl 1919, pp. 209–210; Pauli 1921, pp. 716–717. Although Weyl's later accounts and Pauli's account are a little more detailed than in Weyl 1917, they do not make the conditions of the relevant variations entirely clear.

[68] Eddington 1920b, p. 207.

have the first integrals

$$r^2\frac{\mathrm{d}\theta}{\mathrm{d}s} = B, \quad \frac{\mathrm{d}t}{\mathrm{d}s} = \frac{1}{f},$$ (10.41)

which can be obtained directly by exploiting the symmetries of the problem as Schwarzschild did in 1916. Eliminating t and s, and differentiating the resulting relation between $u = 1/r$ and $\mathrm{d}u/\mathrm{d}\theta$, Schwarzschild and Eddington arrive at

$$\frac{\mathrm{d}^2 u}{\mathrm{d}\theta^2} + u = \frac{M}{B^2} + 3Mu^2,$$ (10.42)

which is a modification of Binet's equation for the Kepler motion. For null geodesics, Flamm and Eddington take the limit $B = \infty$, which corresponds to particles moving at a velocity very close to the velocity of light. It is then easy to find an approximate integral of the resulting equation and thus to retrieve Einstein's prediction for the net deflection of light by the central mass. This is how the deflection is derived in most of today's treatises.[69]

The assumption that light rays follow null geodesics is the common basis of all such calculations. Unlike Eddington, Laue did not regard the equivalence principle as a sufficient proof of this fact. In 1921, he established the behavior of light rays through the eikonal approximation of Einstein's covariant generalization of Maxwell's equations. In this approximation, the electromagnetic field can be regarded as the product of a slowly varying amplitude and a rapidly oscillating phase factor $e^{i\psi}$. As could be expected from general covariance only, the phase ψ satisfies the approximate equation

$$g^{\mu\nu}\partial_\mu\psi\partial_\nu\psi = 0.$$ (10.43)

The world-lines of light rays are the orthogonal trajectories of the surfaces of constant ψ. In other words, the four-vector $g^{\mu\nu}\partial_\nu\psi$ must be constantly tangent to the world-line of the ray. Covariant derivation of the previous equation gives[70]

$$D_\rho(g^{\mu\nu}\partial_\mu\psi\partial_\nu\psi) = 2g^{\mu\nu}\partial_\mu\psi D_\rho\partial_\nu\psi = 2g^{\mu\nu}\partial_\mu\psi D_\nu\partial_\rho\psi = 0.$$ (10.44)

The last equality means that the covariant derivative of the vector $\partial_\mu\psi$ in the direction of the same vector vanishes. Therefore, the tangent to the light ray is parallel-transported along the light ray, which is the general definition of a geodesic. Equation (10.43) further requires this geodesic to be of the null kind. This ends the proof that the light rays follow null geodesics.[71]

---

[69] Flamm 1916, p. 452; Schwarzschild 1916, p. 195; Eddington 1920b, pp. 206–207. See also Pauli 1921, p. 730.

[70] Use is made of the identities $D_\rho g^{\mu\nu} = 0$ and $D_\rho\partial_\nu\psi = \partial_\rho\partial_\nu\psi - \Gamma^\tau_{\rho\nu}\partial_\tau\psi$.

[71] Laue 1921.

Earth-based goniometry

A last difficulty concerns the interpretation of the computed deviation of the light path in the static field of a large mass. Einstein reasons as if the deviation, when referred to the coordinates in which he expresses his metric, were a deviation in the ordinary geometrical sense. This is not necessarily true since in the approximation for which the deviation occurs the spatial part of the metric is non-Euclidean. It may be noticed, however, that the chosen coordinates retain their ordinary metric significance everywhere except in the immediate vicinity of the deflecting star. Consequently, the angle coordinate θ genuinely measures the direction of a ray when it arrives on earth. For rays emanating from a given star, this direction correctly indicates the position of the star in the sky unless they happen to pass near a big mass. In this exceptional case, the star will be observed at some angle from its true position, which is of course unaffected by the interposition of a distant massive body on its line of sight.

One component of this reasoning is the idea that any astronomical test of Einstein's theory is based on the observation of angles between rays reaching a point-like observer. This idea appears in Weyl's memoir of 1917 in the section entitled "Relation with the observations." In Husserlian guise, Weyl writes:

> The "objective" world which physics strives to extract from the immediately perceived reality [*Wirklichkeit*] can only be grasped through mathematical concepts. However, in order to identify the meaning of this mathematical system of concepts in regard to reality, we must somehow try to describe its relationship with the immediately given. This is a problem of the theory of knowledge, which of course cannot be solved by physical concepts alone but requires constant appeal to what is intuitively experienced in consciousness. . . . Here I would like to describe this state of affairs somewhat more precisely in the case of a very simplified relation between subject and object.

Weyl thereby assumes the world to consist of mass-points, stars, and rays that follow singular geodesics. He goes on in Husserlian terms: "The comprehending consciousness, the 'monad', we simplify to a 'point-eye'." This point-eye follows a time-like world-line at each point of which a tangent time-like direction and an orthogonal three-space can be defined. The rays from two stars make a definite angle in this monadic three-space. This angle defines the difference of the direction of the stars, first directly and intuitively in visual perception, then more accurately in indirect, instrument-aided perception. "At any rate," Weyl concludes, "this simple scheme in principle suffices for the description of the art and manner in which the observation of stars can be used to test Einstein's theory."[72]

In *Raum · Zeit · Materie* Weyl developed this point for the solar-deviation test, with full awareness of the merely "graphic" character of space coordinates:[73]

> If we use for a static gravitation field any spatial coordinates x_1, x_2, x_3, for the pur-
> pose of graphic representation we may rely on a Euclidean image-space [*Bildraum*]
> by associating the image point of Cartesian coordinates x_1, x_2, x_3 with the point of

[72] Weyl 1917, pp. 126–127.
[73] Weyl 1918, pp. 196–197.

coordinates x_1, x_2, x_3. If you mark the location of two stars at rest S_1 and S_2 and an observer at rest B in this image space, the angle under which the two stars appear to be observed is not the angle of the straight connecting lines BS_1 and BS_2. We must connect B to S_1 and S_2 through the curved lines of least time and a further auxiliary construction must be used to transform the angle that these lines make in B in Euclidean measure into the angle that they make in the measure given by the metric $d\sigma^2$.[74] The angles determined in this manner are those corresponding to the intuitively perceived separation of the stars; they are the angles that are read on the graduated circle of the instrument of observation.

Similarly, Eddington made clear that the coordinates used in the light-deflection problem lacked direct instrumental meaning:

> After we have traced the course of the light ray in the coordinates chosen, we have to connect the results with experimental measures, using the corresponding formula for ds^2. This final connection of mathematical and experimental results is, however, comparatively simple, because it relates to measuring operations performed in a terrestrial observatory where the difference of $[f = 1 - 2M/r]$ from unity is negligible.

Another of Eddington's insights was that astronomical position measurements implicitly involved the assumption that concrete clocks and rods measure the interval ds, against the well-spread view that the gravitational redshift is the only of Einstein's three tests that involves concrete clocks or rods:[75]

> Without some geometrical interpretation of s our conclusions as to the courses of planets and light-waves cannot be connected with the astronomical measurements which verify them. The track of a light-wave in terms of the coordinates r, θ, t cannot be tested directly; the coordinates afford only a temporary resting-place; and the measurement of the displacement of the star-image on the photographic plate involves a reconversion from the coordinates to s, which here appears in its significance as the interval in clock-scale geometry.

Weyl's and Eddington's caution in distinguishing coordinates from measured angles and distances was rarely emulated in later accounts of the gravitational deflection of light. The Polish-American physicist Ludwik Silberstein was one of the very few authors who warned against a too literal understanding of the coordinate velocity dx^i/dx^0 in Einstein's reasoning. Einstein kept reproducing his naive reasoning of 1916. Others reasoned in terms of null geodesics and ignored the subtleties in the interpretation of the relevant coordinates. The same can be said of many modern treatises.[76]

---

[74] This step is not necessary if the coordinates are chosen so that $d\sigma^2 = dx_1{}^2 + dx_2{}^2 + dx_3{}^2$ in the vicinity of the observer.

[75] Eddington 1920b, pp. 108n, 127. Rigid rods can be constructed by clock-based optical control, exploiting the constancy of the velocity of light.

[76] Silberstein 1922, p. 25.

10.5 The advance of Mercury's perihelion

Einstein's and Schwarzschild's derivations

In the *Grundlage* of 1916, Einstein gives only the final formula for the general-relativistic contribution to the advance of Mercury's perihelion. For the calculations, he refers to his relevant communication of November 1915 and to Karl Schwarzschild's subsequent contribution. In the November article, Einstein begins with the approximate center-symmetric metric of Equation (10.18), he computes the corresponding Γ coefficients, and he injects them into Equation (10.10) for the geodesics. He then obtains the first integrals

$$r^2 \frac{d\theta}{ds} = B, \quad \frac{dt}{ds} \approx 1 + \frac{\alpha}{r} \quad \text{and} \quad \frac{1}{2}\left(\frac{dr}{ds}\right)^2 + \frac{1}{2}r^2\left(\frac{d\theta}{ds}\right)^2 - \frac{\alpha}{2r} = A. \tag{10.45}$$

Eliminating s, θ, and t and introducing the variable $u = 1/r$, he gets

$$\left(\frac{du}{d\theta}\right)^2 + u^2 = \frac{\alpha}{B^2}u + \alpha u^3 + \frac{2A}{B^2}, \tag{10.46}$$

which is the first integral of the modified Binet equation (10.42). The small αu^3 term induces a precession of the major axis, at the rate

$$\Delta\theta \approx 3\pi \frac{\alpha}{a(1 - e^2)} \tag{10.47}$$

per Kepler period, if a denotes the semi-major axis and e the eccentricity of the Kepler ellipse.[77]

Schwarzschild improved on Einstein's calculation in two manners: by starting with the exact central-symmetric solution (10.39) of Einstein's field equations,[78] and by directly deriving the first integrals of the motion from the symmetries of the metric. Pauli used the same strategy in his encyclopedia article. From then on, this derivation has been popular owing to its avoiding the tedious calculation of the gamma coefficients in the equation of geodesics.[79]

The astronomer's coordinates?

Any well-trained mathematical physicist could follow Einstein's or Schwarzschild's calculation of the advance of the perihelion. Here too the real difficulties reside in the choice and interpretation of the coordinates. Again, Einstein implicitly assumed that the coordinates he used matched the usual coordinates of astronomers in the approximation needed to interpret his predictions. In his encyclopedia article of 1929, Guido Beck explained why Einstein's coordinates met this implicit condition: these coordinates, or the associated polar coordinates, are coordinates for which the intrinsic symmetries

---

[77] Einstein 1915a. Cf. Earman and Janssen 1993. See p. 317, this volume.

[78] This is true up to a shift of the r variable, for which the singularity occurs at $r = 0$; cf. Eisenstaedt 1982, §2.

[79] Schwarzschild 1916; Pauli 1921, pp. 730–731.

of the problem are respected. This implies, in particular that the light from a planet travels rectilinearly to the earth in this system of coordinates, even though its coordinate velocity varies along the path. In addition, the coordinates are such that the metric coefficients depart very little from their flat Minkowski value. This condition is needed to preserve the usual geometric interpretation of the elements of the Kepler ellipse.[80]

As Droste already remarked in 1916 while solving the center-symmetric problem, one may re-express the Droste–Schwarzschild solution in a different system of coordinates, and this may be advantageous from a formal point of view; for instance, it is possible to choose the coordinates so that the spatial part of the metric becomes isotropic. On this freedom Droste commented:

> To which [metric] formula should we give our preference?. . . This is a matter of personal inclination. However, me must remember that the *coordinate r* does not represent the *measured interval*. The choice of a coordinate is free (as long as the values it can take allow us to reach every point), although one choice may prove to be more judicious than another.

The remark is echoed in Eddington's frequent remarks about the lack of a priori metric significance of the coordinates, some of which have been cited earlier. Yet it did not register in the mind of some of Einstein's critics.[81]

Painlevé's criticism

In October 1921, the French mathematician Paul Painlevé argued that in the derivation of the Schwarzschild metric Einstein implicitly assumed a Newtonian, metric significance of the coordinates since he directly expressed space–time symmetries (central symmetry and time independence) in terms of coordinate transformations. He then noted that the Schwarzschild formula was not the only solution of Einstein's field equations that complied with these symmetries. For instance, the metric formula could be

$$ds^2 = (1 - a/r)dt^2 + 2drdt\sqrt{a/r} - (dr^2 + r^2d\theta^2 + r^2\sin^2\theta d\phi^2). \qquad (10.48)$$

The point is more obvious than Painlevé makes it look, because this metric formula derives from the Schwarzschild formula through a mere change of coordinates (namely, $t = t' + \tau(r)$ with $d\tau/dr = \sqrt{a/r}(1 - a/r)^{-1}$). Imitating Einstein's argument about contracted rods, Painlevé notes that for $dt = 0$ the coordinate length of a rod such that $ds^2 = -1$ is independent of its orientation. Einstein's alleged metric distortion thus turns out to depend

---

[80] Beck 1929, p. 323.

[81] Droste 1916a, p. 20, cited in Eisenstaedt 1982, p. 168. As Eisenstaedt remarks, Schwarzschild originally used a radial coordinate different from the one in the now standard Droste–Schwarzschild formula.

on the choice of coordinates. Painlevé concludes: "It is pure fantasy to pretend to deduce this sort of consequence from the ds^2."[82]

In a November sequel to this critical note, Painlevé gives the more general symmetric solution of Einstein's equation (in units for which $G = 1$ and $c = 1$)

$$ds^2 = \left(1 - \frac{2M}{\rho(r)}\right)(dt - \chi(r)dr)^2 - \rho^2(r)(d\theta^2 + \sin^2\theta d\phi^2) - \left(1 - \frac{2M}{\rho(r)}\right)^{-1}\left(\frac{d\rho}{dr}\right)^2 (dr)^2.$$

(10.49)

Again, this solution results from a mere substitution of coordinates in the Schwarzschild solution: $t = t' - \int \chi(r)dr$ and $r = \rho(r')$. Painlevé requires $\chi = 0$ for the sake of reversibility and $d\rho/dr > 0$ and $\rho(0) = 0$ for metric consistency. Under these conditions as in the Schwarzschild case, the geodesics are nearly periodic functions of period T and the coordinate r oscillates between the minimum value r_1 and the maximum value r_2. Painlevé derives

$$\frac{M}{1 + 6M/\bar{a}} = 4\pi^2 \frac{\bar{a}^3}{T^2} \text{ with } \bar{a} = \frac{1}{2}[\rho(r_1) + \rho(r_2)]$$

(10.50)

for the counterpart of Kepler's third law, and

$$\Delta\phi = 3\pi M\left(\frac{1}{\rho(r_1)} + \frac{1}{\rho(r_2)}\right),$$

(10.51)

for the advance of the perihelion. These formulas are correct and they are easily derived from the corresponding formulas in the Schwarzschild case by replacing r with $\rho(r)$ (exploiting the covariance of the equation of geodesics).[83]

When $\rho(r) = r$, the motion is quasi-elliptical and the constant $\bar{a}$ is half the large axis of the ellipse. For this reason, Painlevé judges that agreement with astronomical observations requires the function $\rho(r)$ to differ very little from r. He comments: "Einstein assumes that $\rho(r) \equiv r$. This identification is not a consequence of the theory of relativity but it is *approximately* required by a first confrontation with astronomic observations." He goes on to show that the predictions for the advance of Mercury's perihelion and for the solar deflection of the light from stars remain the same as Einstein's if the deviation of $\rho(r)$ from r is of the order of M/r, although such a deviation could lead to a significant violation of Kepler's third law.[84]

The fallacy of these remarks derives from Painlevé's assumption that in the new theory the variable r still represents the distance of the planet from the sun. In reality, this distance is principally determined by spatial geodesy based on the approximate metric $d\sigma^2 = (d\rho/dr)^2 dr^2 + \rho^2(r)[d\theta^2 + \sin^2\theta d\phi^2]$, as long as all length measurements are performed far from the Schwarzschild singularity. It is therefore equal to $\rho(r)$, and the constant $\bar{a}$ approximately represents the semi-major axis of the Kepler ellipse, whatever the choice of the

[82] Painlevé 1921a, p. 680. On Painlevé and general relativity, cf. Eisenstaedt 1982, pp. 174–175 and the thorough study in Fric 2013.

[83] Painlevé 1921b.

[84] Painlevé 1921b, p. 884.

function $\rho(r)$. Painlevé's assertion that different choices of this function lead to different violations of Kepler's third law is wrong for the same reason.

In December 1921, Painlevé received a letter from Einstein including the following clarification:

> When in the ds^2 of the central symmetric static solution r is replaced by any function of r, this does not give a *new* solution, because the quantity r does not have physical meaning in itself. Only the quantity ds or, better, the net of all ds in the four-dimensional manifold does so. We must always keep in mind that the coordinates do not have physical meaning in themselves, namely: they do not represent the results of a measurement. Physical meaning can be given only to results reached by elimination of the coordinates. In addition, the metric interpretation of ds is not "pure fantasy": it is the core of the whole theory.

This letter did not suffice. When Einstein visited Paris in the spring of 1922, Painlevé publicly rehearsed his criticism. The astronomer Charles Nordmann describes the ensuing debate in these words:

> There followed an extremely brilliant and animated discussion, so animated that now and again everyone was speaking at the same time. . . . It truly was beautiful battling and enjoyable sport. . . . While this rattle of colliding arguments, verbal or written, all quick and stunning, was filling the room with turmoil and the blackboard with elegant integrals, . . . Einstein was sitting and silently smiling in the middle of the storm. Then, suddenly, raising his hand as a schoolboy who seeks a favor from the teacher: "May I also say a little thing?", he asked softly. Everyone laughed; Einstein spoke in the suddenly recovered silence; and in a few minutes everything became clear.

The main issue, Nordmann tells us, was the interpretation of the quantities occurring in Einstein's theory, especially the coordinate r. Einstein explained that absolute time and space determinations were no longer possible in his theory, that clocks and measuring rods were affected by the gravitational field. In a central static field such as the field of the sun one had to imagine a contiguous alignment of rods between the center of the field and the observed point, and it was the number of these rods, not the value of the coordinate r, that generally gave the physical distance between these two points. Thus, "there are only rods and clocks, no observer is left, everything subjective is eliminated." We may use different systems of coordinates to write down the equations of the problem, but the results of concrete geodesy do not depend on the selected system. They are absolute.[85]

The semi-Einsteinian theory

Einstein's reply to Painlevé might have been more convincing if it had employed a more realistic, light-based geodesy instead of filling space with ideal rods and clocks. At any rate,

---

[85] Einstein to Painlevé, December 7, 1921 (*ECP* 12, pp. 368–369), partly cited in Eisenstaedt 1982, p. 175; Nordmann 1922, pp. 156–159 (discussion held on April 7, 1922). Some of the details of Nordmann's rendering of Einstein's arguments seem dubious. On Einstein's Paris visit, cf. Biézunsky 1991 and *ECP* 13, pp. xlvii–l, 299, apps. A–B.

Einstein clearly made the essential point that physical observations depend only on the intrinsic geometry of the space–time manifold and not on a specific system of coordinates. A few months later, Painlevé published a third note in which he admitted that the theory discussed in his previous notes departed from Einstein's true theory. He now recognized that in Einstein's theory distance measurements were given by the spatial part of the metric and not by the r coordinate. However, he still defended the possibility of the "semi-Einsteinian theory" in which the isotropic and static ds^2 is used to determine the trajectory of planets but in which Euclidean geometry remains valid in astronomic measurements. The lack of gravitational redshift in this theory was not a problem, since this effect remained unverified at that time. The theory predicted the same advance of Mercury's perihelion and the same gravitational deflection of light as Einstein's theory for a wide range of choices of the arbitrary function $\rho(r)$, and it could more easily accommodate violations of Kepler's third law.[86]

Painlevé's semi-Einsteinian approach and his earlier misunderstanding of the meaning of coordinates in Einstein's theory largely depended on a geometric analogy between Newtonian and Einsteinian gravitation.[87] As was earlier mentioned and as Weyl and Levi-Civita first noted (see Equation (10.35)), the equations of geodesics for a static metric in Einstein's theory can be regarded as the equation of motion of a classical mechanical system for a certain Lagrangian. The elimination of time in Jacobi's manner then leads to the version

$$\delta \int \sqrt{E + 1/g_{00}(r)}\sqrt{-g_{ij}(r)dx^i dx^j} = 0 \qquad (10.52)$$

of Maupertuis's principle. This is to be compared with the Newtonian version of the same principle:

$$\delta \int \sqrt{E - U(r)}\sqrt{dx_1^2 + dx_2^2 + dx_3^2} = 0. \qquad (10.53)$$

In both cases, the trajectories are the geodesics of a spatial manifold, with the respective metrics

$$\gamma_{ij} = -g_{ij}(E + 1/g_{00}) \quad \text{and} \quad \gamma_{ij} = (E - U)\delta_{ij}. \qquad (10.54)$$

This remark, or the simpler remark that the radial equation of motion in the two theories only differ by the contribution $-MB^2/r^3$ to the effective potential, suggested that Einstein's and Newton's theory were not so different after all. It invited conservative physicists to reinterpret Einstein's ds^2 in a persistently Euclidean context. Painlevé was not alone in thinking on those lines. Several commentators on Einstein's theory noted that the two best

[86] Painlevé 1922.

[87] Cf. Fric 2013, chap. 2.

verified predictions of his theory, the value of Mercury's perihelion and the deflection of starlight by the sun, did not require Einstein's radical geometric interpretation of the ds^2.[88]

Privileged coordinates

Whatever the value of the semi-Einsteinian option, it remains true that in Einstein's treatment of the advance of the perihelion, the coordinates approximately retain their ordinary astronomical significance. Einstein did not care to explain this point, and in the introduction to the *Grundlage* he rather emphasized the freedom in the choice of coordinates. Yet he clearly favored a specific system of coordinates in which the results of calculations could be read directly as astronomical predictions. In general, the large freedom in the choice of coordinates does not exclude the existence of coordinate systems in which the coordinates have a fairly direct physical meaning, even though direct pseudo-Euclidean interpretation is not an option. Several lucid commentators on Einstein's theory, including Kretschmann, Gustav Mie, and Cornelius Lanczos, tried to determine such systems. After reading Mie, the physics professor at the Museum d'histoire naturelle Jean Becquerel cricitized Painlevé for introducing coordinates that did not have physical significance, unlike the coordinates originally used by Einstein:

> Mr. Painlevé has used other coordinates and has naturally found another exact expression of ds^2. However, whereas the *mathematician* considers all systems of coordinates as equally good, the *physicist* cannot do the same for he must interpret the results and must therefore introduce the quantities that can be measured with his instruments: the choice of coordinates is then necessarily subjected to certain conditions. The formula of Mr. Painlevé cannot be interpreted physically because it contains a term in $drdt$ that is incompatible with the symmetry in "time."

The exclusion of such terms still leaves much freedom in the choice of spatial coordinates. Becquerel explained why Einstein's coordinates or their polar variant was still to be preferred: "We could express the line element ds with other coordinates, but the coordinates we have used, after Schwarzschild, are *the closest to the Euclidean polar coordinates.* Practically, r and t are 'the distance' and 'the time.'"[89]

These remarks appeared in Becquerel's lectures on relativity theory, which he published in 1922. The following year, he deepened his reflections on the physical meaning of the Schwarzschild solution and he again denounced Painlevé's error:

> All this discussion [Painlevé 1922b] rests on a misunderstanding, more exactly on the fact that the question is taken in reverse because the meaning of coordinates is *imposed a priori* and assumed to be the same in all formulas and because the ideas that dominate the discussion are impregnated with Euclidean conceptions.

---

[88] On the semi-Einsteinian option, cf. Hamel 1921; Chazy 1928–1930, Vol. 2, pp. 33–35. The theory of Whitehead 1922 has a fixed background Minkowski metric, but unlike Painlevé's theory it does not require any formal exploitation of Einstein's metric; cf. Bain 1998.

[89] Kretschmann 1917; Mie 1920; Lanczos 1922, 1923; Becquerel 1922, pp. 224n, 229. Cf. Eisenstaedt 1982, pp. 172–173, 176–177.

Becquerel then echoed Eddington's and Einstein's comments on the priority of the metric over the coordinates:

> As we a priori do not know the geometrical structure of the universe as determined by the presence of a material center, we cannot posit a priori the geometrical meaning of the coordinates and look for the corresponding expression of the ds^2. But we can take the question by the opposite end, namely: a formula like Schwarzschild's being established in agreement with the law of gravitation, one can afterwards *try* to find what the coordinates mean geometrically or physically and one can succeed in this task if the coordinates happen to be entities measurable with our instruments (at least theoretically). In one word, we should not start with the coordinates in order to establish the ds^2; on the contrary, we should start with the ds^2 in order to investigate the meaning of the coordinates.

Becquerel went on to explain two ways of providing the Droste–Schwarzschild coordinates with a posteriori meaning: geometrically by imitating Mie, and physically by discussing the behavior of ideal clocks and rods. According to Mie, the Schwarzschild coordinates are the coordinates obtained by orthographic projection of a center-symmetric four-dimensional submanifold in a pseudo-Euclidean ten-dimensional manifold, the projection being done perpendicularly to the asymptotic flat space–time far away from the center.[90]

On the physical side, Becquerel showed that distant clocks could be consistently synchronized by exchange of light signals, with the convention that the bouncing of a light signal from a first clock on a second clock was simultaneous with the marking of the average of the emission and reception times by the first clock. Accordingly, two distant events are simultaneous if and only if they occur at the same value of the cosmic time t; and the proper time difference between two events at a given point of space differs from the proper time difference between two synchronous events at a different point of space. As for distance measurements, they are solely determined by the condition that an ideal unit rod covers any spatial interval such that d$s^2 = -1$. Consequently, the space associated with the Schwarzschild solution is slightly non-Euclidean; for instance, the increase in the measured circumference of a centered circle is not 2π times the measured increase of its radius. Becquerel warned against Einstein's way of expressing this property as a consequence of the contraction of rods in the gravitation fields:

> This interpretation is not too correct, because no length standard allows us to directly compare the length of a radially oriented rod with the length of a transversally oriented rod: we should consider the rod as representing the unit length in any case and conclude from the measurements that the metric of space is not Euclidean.

With the latter prescription, the variations of the radial variable r turn out to be approximately proportional to the variations of the Euclidean distance from the center (far enough from the singularity). If Painlevé's solution (10.49) is used instead of Schwarzschild's, it is the variable $\rho(r)$ that has this meaning: "Therefore, [Painlevé's] formula does not essentially

---

[90] Becquerel 1923, pp. 5–6; Mie 1920.

differ from Schwarzschild's: it is only a disguised form of this formula . . . and it does not tell us anything more *at least insofar as the function* [$\rho(r)$] *is left undetermined.*"[91]

During one of the discussions in Einstein's presence in the spring of 1922, Becquerel praised Mie's attempt to justify Schwarzschild's coordinates by immersion of the space–time manifold in a ten-dimensional Minkowskian space. Einstein's reaction was most concise: "One can always select a preferred representation, if one believes that this representation is more convenient for the work to be done; but this does not have any objective meaning."[92]

Gullstrand's criticism

Painlevé was not alone in misinterpreting the multiplicity of static isotropic solutions of Einstein's field equations. The priority in this sort of misreading goes to the Swedish Nobel-prize winning ophthalmologist Allvar Gullstrand. A little before Painlevé, Gullstrand questioned Einstein's ability to derive a definite value for the relativistic advance of the perihelion, and wrote long negative reports on Einstein and general relativity for the Nobel committee. As a result, the 1921 prize was postponed to 1922, and Einstein got it for his theory of the photoelectric effect, not for relativity theory.[93]

In the spring of 1921, Gullstrand privately labored to show that Einstein's field equations allowed for an infinite variety of static isotropic solutions of the form

$$ds^2 = L(r)dt^2 + 2M(r)dtdr + N(r)dr^2 + O(r)r^2(d\theta^2 + \sin^2\theta d\phi^2) \tag{10.55}$$

in polar coordinates. Under the constraint $g = -1$ on the corresponding metric coefficients in Cartesian coordinates, he found

$$O = -(1 + \beta/r^3)^{2/3}, \quad L = 1 - \alpha/r\sqrt{-O}, \text{ and } O^2(LN - M^2) + 1 = 0. \tag{10.56}$$

The case $\beta = 0$ yields the Schwarzschild metric formula if $M = 0$, and the Painlevé spatially Euclidean metric (10.48) for $M = \sqrt{\alpha/r}$. Further heavy calculations gave him the equation of geodesics for his generic metric. As the modified Binet equation now contained a β-dependent term, he surmised that Einstein's theory led to an arbitrary value for the advance of the perihelion.[94]

Before Gullstrand publicly communicated this disastrous conclusion, his theoretical-physicist friend Carl Oseen showed him that the allegedly new solutions of the field equations could be obtained by a mere change of variables in the Schwarzschild solution.

---

[91] Becquerel 1923, pp. 18, 22. Becquerel's definition of simultaneity resembles Einstein's in 1907; see pp. 275–276, this volume.

[92] Einstein, in "Séance du 6 avril 1922 : la théorie de la relativité," *Bulletin de la Société Française de Philosophie*, 17: 91–113, on p. 99.

[93] On the Nobel prize context, cf. Friedman 2001, pp. 132–133 for 1921, and pp. 134–135 for 1922. On Gullstrand's criticism and the ensuing controversy, cf. Eisenstaedt 1982, pp. 177–179; Earman and Janssen 1993, pp. 159–160.

[94] The story is told in Gullstrand 1921.

Gullstrand's metric is not as general as Painlevé's metric of Equation (10.49), only because Gullstrand imposes the unnecessary restriction $g = -1$ from Einstein's *Grundlage*. Oseen's remark and the covariance of the equation of geodesics together imply that the inverse of the radial variable $r' = r\sqrt{-O}$ must satisfy the same modified Binet equation as the inverse radial variable in the Schwarzschild problem. The advance of the perihelion is therefore given by $3\pi M(r_1'^{-1} + r_2'^{-1})$, which differs from the Schwarzschild value $3\pi M(r_1^{-1} + r_2^{-1})$ by a term proportional to β (r_1 and r_2 being the minimal and maximal values of r in the unperturbed elliptic motion). Gullstrand concluded:[95]

> Since β can be freely chosen, *one can predict an arbitrary advance of the perihelion in Einstein's theory if this advance is computed in the usual manner.* Whether Einstein's gravitation theory agrees with Le Verrier's discovery therefore remains unknown until the influence of the curvature of the world on the relevant astronomic observations and on the theory of perturbations is taken into account and until the resulting advance of the perihelion comes out independent of β. As indeed the transformation affects only the radial coordinate then the true advance of the perihelion must be independent of β for arbitrarily large values of this constant.

The subtleties of this conclusion escaped most of Gullstrand's readers, historians included. By "in the usual manner" Gullstrand means that the r variable is directly interpreted as the observed distance of the planet from the sun, as is implicitly done in Einstein's and Schwarzschild's derivations. However, Gullstrand is perfectly aware that r is not the true distance of the planet from the sun and he anticipates that, after taking spatial curvature into account, the true advance of the perihelion does not depend on β. Oseen's communication, which immediately follows Gullstrand's in the same journal, is devoted to a proof of this independence.[96]

After remarking, in Weyl's manner, that any astronomical observation boils down to a terrestrial goniometry of the light rays from celestial bodies, Oseen shows that the equation of these rays near the earth is the same in local Minkowskian coordinates, whatever the choice of the Gullstrand metric (this argument is insufficient, as it does not explain how the terrestrial goniometry is related to intrinsic geometric properties of the planetary motion). Oseen's reader is not helped by his oddly phrased conclusion: "The meaning of the fact found by Prof. Gullstrand seems to me that this fact teaches us in a penetrating manner that verifications of general relativity do not concern the objective statements of the theory; they concern the observations."

By "objective statement" Oseen seems to mean a statement regarding a specific system of coordinates, which is precisely what Einstein would call "subjective." At any rate, he shares with Einstein the opinion that reduction to observable quantities is essential in interpreting general relativity, and he clearly understands that there should be no observational difference between Einstein's and Gullstrand's solutions of the field equations:

[95] Gullstrand 1921, pp. 13–14.

[96] Oseen 1921.

"When applied to the *observations*, Gullstrand's more general solution must lead to exactly the same results as Schwarzschild's special solution."[97]

In January 1922, Kretschmann publicly criticized Gullstrand for confusing the radial coordinate r with the measured distance from the sun. This distance, being intrinsically determined by the radius of curvature of the sphere of all points that share the same value r of the radial coordinate, must be approximately equal to a coordinate r' for which the spatial metric is very nearly Euclidean (far from the Schwarzschild singularity). Consequently, Gullstrand's way of calculating the advance of the perihelion ("in the usual manner") gives the correct value only if the function $-O(r)$ differs very little from unity and if $M = 0$. Kretschmann could have added that Gullstrand's metric always gives the correct advance if the distance from the sun is measured by r'. But he did not. His main point was that Gullstrand was wrong in believing that the curvature of the world was responsible for the compensation of the β dependence of the advance of the perihelion.[98]

In his reply, Gullstrand reasserted this opinion. Curvature being an intrinsic property of the space–time manifold, one may wonder how he could do so without falling into complete absurdity. One reason is that he was relying on the spatial curvature, which is not intrinsic if the metric involves the off-diagonal $2Mdtdr$ term. Unlike Kretschmann, Gullstrand admitted $M \neq 0$, in which case the spatial curvature can widely differ from its small Schwarzschild value and affect measurements. Moreover, in the case $M = 0$ Gullstrand defined curvature in an extrinsic manner, through the contraction of longitudinal and radial rods. In this idiosyncratic view, the discrepancy between the value of r and measured radial distances is always a matter of curvature. For a proper choice of M (the Painlevé–Gullstrand metric), the spatial metric becomes Euclidean and the radial coordinate directly gives the measured value of distances from the sun. In any other case, that is no longer true. In the Schwarzschild case, the discrepancy is a small quantity of the order a/r. This discrepancy being of the same order as the Einsteinian contribution to the advance of Mercury's perihelion, Gullstrand judged that it could affect the (ten times larger) contribution from perturbations by other planets. He concluded his reply with a call for a general-relativistic treatment of the latter contribution.[99]

Another aspect of Gullstrand's reply, the first in the order of exposition, was his defense of the calculation of the advance of the perihelion "in the usual manner." Gullstrand believed that astronomers, since they interpreted their observations in the context of Newtonian space and time, did not measure the true value of the perihelion. What they measured was the value calculated "in the usual manner"; namely, the value obtained by doing as if the coordinate r was the true value of the distance of the planet from the sun. So in Gullstrand's opinion, it was only after proper corrections of the effect of "curvature" that astronomers should obtain a β-independent result.

As Kretschmann observed in his final contribution to this debate, this view presupposes a factual content of the coordinates independent of the choice of the metric form, whereas in reality any physical interpretation of the coordinates must be derived from the

---

[97]Oseen 1921, pp. 5–6.

[98]Kretschmann 1922.

[99]Gullstrand 1922.

metric form. When this is done properly, it turns out that what the astronomers measure *in their current practice* is the coordinate r' in a first approximation, not the coordinate r. This is so, Kretschmann explains, because it is only with respect to the Schwarzschild variables (or slightly different variables, and up to a change of units) that the paths of light rays are very nearly straight. The Schwarzschild and the Gullstrand forms of the metric are perfectly equivalent since they represent the same intrinsic metric relations in different coordinate systems: "The choice of the coordinates in which we express the ds^2 does not make any factual difference. Therefore, whether we work with [Schwarzschild's] system or with [Gullstrand's] system is factually irrelevant and it is only a matter of computational convenience." Kretschmann nonetheless agreed with Gullstrand that the perturbation of Mercury's motion by other planets needed to be treated in the context of general relativity and that the resulting contribution to the advance of the perihelion was likely to differ from the Newtonian value. In general, he believed that "astronomical calculations and observations needed to be more thoroughly examined under the viewpoint of general relativity."[100]

Gullstrand closed the debate with the following concise reply:

> As Mr. Kretschmann in his last point agrees as much with me as I judge it to be factually necessary, I find it superfluous to prolong the discussion. In fact, the worlds used by Mr. Kretschmann, "identical or at least nearly identical," or "in very good approximation," suffice to characterize the difference between our viewpoints.

From this statement, it is not clear whether Gullstrand now accepted Kretschmann's view that the astronomers' present estimate of the advance of Mercury's perihelion referred to the r' coordinate, in which case the prediction of Einstein's theory did not depend on the choice of the coefficients of the isotropic static metric. At any rate, he (and Kretschmann) remained skeptical about the alleged agreement between the theoretical and observed values of the perihelion shift.[101]

Gullstrand and Kretschmann were not alone in criticizing the practice of simply adding the general-relativistic one-body perihelion shift to the Newtonian shift caused by the perturbation of the other planets. So too did, for instance, the Italian anti-relativist Cesare Burali-Forti in the same year 1923. In 1916–1917, de Sitter, Droste, and Lorentz had given approximate equations for the n-body problem in general relativity. In principle, their equations permitted a fully relativistic treatment of Mercury's perihelion problem. Astronomers nonetheless preferred the semi-relativistic approach in which the force to which each body is subjected in the Newtonian approach is modified in accordance with the effective potential representation of motion in the Schwarzschild metric. In 1925 the mathematician Aurel Wintner claimed that in this context Lagrange's method of the variation of constants led to the additional perihelion shift $3\pi\alpha/a$, which differs from Einstein's prediction by the factor $1 - e^2$. In his treatise of 1930 the French astronomer Jean Chazy reached the opposite conclusion: both the de Sitter n-body equations and the semi-relativistic approach agree with Einstein's estimate. This is a difficult problem, involving

---

[100] Kretschmann 1923, pp. 2, 4.

[101] Gullstrand 1923, p. 4.

complex calculations and subtleties in the choice of the proper axes to which the precession is referred. An error in Droste's and de Sitter's original calculations long contaminated the problem, and later attempts by Levi-Civita and others to include self-field effects turned out to be erroneous. Although these errors did not affect the perihelion shift, a rigorous treatment was still lacking. Also, departure from the usual physical assumptions, for instance an oblateness of the sun, could affect the result. For these reasons, as late as 1960 Synge still refused to commit himself on the confirmation of Einstein's perihelion shift prediction.[102]

Conclusions

Einstein meant his *Grundlage* of 1916 to serve as the foundation of subsequent research on general relativity. Yet some features of his exposition made it ill-fitted for this service. Here is a short list: lack of a clear distinction between heuristic and deductive arguments, conceptual obscurities or contradictions, gaps and errors in some deductions, and an opaque nongeometrical approach to tensor calculus. Although these textual flaws did not necessarily reflect misunderstandings on Einstein's part and although they did not prevent him from obtaining essentially correct results, they confused his least receptive readers and they challenged the most perspicacious ones. The most basic difficulty concerned the meaning of coordinates and reference frames. It affected Einstein's three main empirical predictions, and it was resolved mostly in critical discussions of these predictions.

Einstein's first prediction was the redshift of the spectral lines from stars. The derivation given in the *Grundlage* is problematic, because it relies on the coordinate-dependent notion that clocks slow down in an intense gravitational field, a notion at odds with the rest of his theory. Weyl, Eddington, and Laue explained that the gravitational redshift depended on the conservation of frequency (when measured with respect to the time coordinate for which the metric coefficients are time-independent) during the propagation of light from the star to the earth. Eddington and Laue proved this conservation. They also understood that the equivalence principle justified the assumption that an atomic oscillation corresponded to a well-defined value of the proper time ds. In addition, Laue gave proofs that spring-clocks and Bohr-atom clocks measured the proper time. Weyl and Eddington clarified the meaning of the time coordinate in the static solutions of Einstein's field equations, and they highlighted the empirical significance of the proper time. None of these important points could be found in Einstein's *Grundlage*.

Einstein's derivation of the gravitational deflection of light was equally problematic. It tacitly assumed a partial persistence of the old metric interpretation of the coordinates; it was based on the unproven validity of Huygens's principle for propagation referred to a specific system of coordinates; and it contained an error in the expression of the local curvature of light rays (this error fortunately did not affect the final result). Weyl and Levi-Civita proved the validity of Fermat's equivalent principle for null geodesics, and Laue further

---

[102] Burali-Forti 1923; de Sitter 1916b; Droste 1916b; Droste and Lorentz 1917 (with a correct, long-ignored result); Wintner 1925; Chazy 1928–1930, chaps. 3, 13; Synge 1960, p. 296. On the history of the *n*-body problem in general relativity, cf. Damour and Schäfer 1992. On Burali-Forti, cf. Earman and Janssen 1993, pp. 161–162.

proved that light followed null geodesics in the eikonal approximation. After a contribution by Flamm, the usual route to the gravitational deflection of light has been through the limiting case of geodesic motion for which the velocity of the particle reaches the velocity of light. This procedure, which Eddington justified through the equivalence principle, avoids all the difficulties of Einstein's approach except one: it remains to be shown that the usual astronomic measurements yield the deflection expressed in the special system of coordinates of the Schwarzschild solution. Weyl and Eddington provided the missing argument.

Einstein's derivation of the anomaly in the advance of Mercury's perihelion similarly lacks any discussion of the relation between the favored coordinates and astronomic observations. Here as in the light-deflection case, Einstein must have intuitively understood that the favored coordinates depart from the usual Newtonian coordinates in such a way that the measured parameters retain their usual meaning. Beck clarified this point in his encyclopedia article of 1929. Several years earlier, in 1921, Painlevé and Gullstrand had fallen into the error of confusing the coordinates of any given central-symmetric solution of Einstein's field equations with the astronomically measured coordinates, even when the former coordinates significantly differed from those used by Einstein and Schwarzschild. This was perhaps the most glaring manifestation of the difficulty that Einstein's readers (and the pre-1915 Einstein) had in assimilating the basic fact that the coordinates in general relativity have no a priori physical meaning independent of the expression of the metric coefficients.

The moral of this chapter is not that Einstein, in 1916, did not quite know what he was doing, that he was lucky to get correct results, and that his most penetrating followers understood and founded the theory much better than he did. The true point is that the *exposition* he gave in the *Grundlage* was flawed enough to confuse his readers and to require significant conceptual reworking. We will never know how well Einstein understood the interpretive difficulties in private. We only know that he did not publicly solve all of them in 1916. Plausibly, most of the contradictions or gaps in his exposition did not reflect fundamental confusion in his mind.[103]

Consider, for example, the apparent contradiction between his asserting the lack of metric significance of the coordinates on the one hand and his implicit reliance on metrically significant coordinates in his derivation of testable effects on the other hand. We know that by late 1915 he perfectly understood that Gaussian or Riemannian geometry could be described intrinsically through the properties of geodesics and that coordinates did not have any direct metric significance.[104] Having read Poincaré and others on the foundations of geometry, he also knew that the metric properties of a curved space could be described extrinsically as resulting from measurement with variable gauges in a fictitious flat space.[105] In the *Grundlage* he was in effect joggling with these two incompatible descriptions. We do

---

[103] As John Norton told me, there is at least one case, however, in which Einstein got it wrong: the alleged Machian increase of the inertia of a body when masses are piled up in its vicinity (Einstein 1921, pp. 110–113). Einstein derived this effect from the form of the equation of motion of the body in a given system of coordinates and did not realize that it had no intrinsic meaning. Cf. Guth 1970, pp. 190–193.

[104] See, e. g., Howard and Norton 1993.

[105] See also the remark in note 39 above.

not know whether he consciously did so. He certainly did not warn the reader about possible confusions in this regard, and his derivations of testable effects left something to be desired. His intuition properly filled the gaps and yielded correct predictions. But the readers of the *Grundlage* could not read his mind, and they legitimately required full, internally consistent, and conceptually clear derivations.

11

EPILOGUE

In ten chapters, we have traveled through three centuries of physics in search of relativity principles and theories. The scenery varied considerably from Descartes's geometric world to Einstein's general relativity, with vistas on mechanics, gravitation, optics, and electrodynamics. The main purpose of this epilogue is not to reflect on the historical motors of these variations: speculative thought, experiments, mathematics, local cultures, institutionalization, etc. This will be done succinctly in Section 11.1 as a prelude to a more philosophical undertaking, which is to examine the continuities, interconnections, and analogies in the long history of relativity. The running thread will be the reference of motion: how is it defined, for what purpose, by what means, with what restrictions? Sections 11.2–6 condense and comment on the contents of the previous chapters with this question in mind. Section 11.2, on mechanical relativity, addresses the early modern concepts of motion, the tension between Newton's absolute space and Galilean relativity, and the emergence of the relativity principle as a structural, constructive principle in the context of mechanics. Sections 11.3 and 11.4 discuss two other kinds of relativity occurring in optics and electrodynamics, with partial connections to mechanical relativity. Sections 11.5 and 11.6 revisit the two major theories built on relativity principles, special relativity and general relativity, and their connections with the earlier relativities. In particular, we will ponder whether general relativity truly resulted from an extension of the relativity principle.

11.1 Actors and stages

It is impossible to describe in a few words the cultural and institutional circumstances around the developments recounted in this book, as we would have to cover too many different places and periods. What we can do is review the disciplinary profiles of the main actors and frame these reflections in contemporary physics, mathematics, and philosophy, with proper attention to geographical differences. This should suffice to situate our story in the broader history of science, which abundantly documents the evolution of disciplines, the major developments in each discipline, and the eventual effects of national or regional styles.[1]

Relativity and the question of the reference of motion concern some of our most basic experiences of the world. They involve the concepts of space and time, their physical realization, and their mathematical representation. They have been discussed in a great

---

[1] References to broader histories were given in the previous chapters. They will not be repeated here.

Relativity Principles and Theories from Galileo to Einstein. Olivier Darrigol, Oxford University Press.
© Olivier Darrigol (2022). DOI: 10.1093/oso/9780192849533.003.0011

variety of fields including physics, philosophy, mathematics, astronomy, technology, and even theology (in the early modern period). The deepest investigators often had interests and competence in multiple fields, not only in the early modern period in which natural philosophy was broadly defined to include any meditation on nature, but also in modern times when polymaths became a rarity. In the long list of genially versatile figures encountered in this book, we find Galileo, Descartes, Huygens, Newton, Leibniz, Euler, d'Alembert, Maxwell, Riemann, Mach, Poincaré, and Einstein. Also, the central field of interest of the scrutinizers of space, time, and motion was not necessarily physics. It was mathematics for Riemann, Poincaré, and Minkowski; it was philosophy for Leibniz and Kant.

The involvement of mathematicians and philosophers might suggest that experience and experiments played little or no role in reflections about the reference of motion. This was indeed the case for Descartes, Euler, d'Alembert, Leibniz, and Kant, who shared the ideal of a purely rational mechanics or dynamics, and also, paradoxically, for the empiricist Mach as he *a priori* excluded nonmaterial causes in his criticism of inertia. Other authors, in line with Galileo and Newton, regarded the principle of inertia and the associated kind of relativity as the result of a broad induction from observation and experience. That said, after Newton, experimental observations no longer drove the discussions of the reference of motion in mechanics. The true motor was criticism or consolidation of the foundations of Newtonian mechanics: for philosophical reasons in the case of Leibniz and Kant; for methodological reasons in the case of Euler, d'Alembert, Laplace, Maxwell, W. and J. Thomson, Streintz, and Mach; and for pedagogical reasons for less-known textbook writers such as Bélanger or Violle. As is well known, the systematic ordering of knowledge for better application and transmission often induces the scrutiny of foundations. In the case of mechanics, such systematization was necessary to increase the scope of Newtonian mechanics in the eighteenth century (connected systems and deformable bodies), to develop purely mechanical worldviews in the seventeenth and nineteenth centuries, or to understand the relationship between mechanics and the new doctrine of energy conservation in the second half of the nineteenth century.

Leaving the realm of mechanics and entering the domains of optics and electrodynamics, we see that observation and experiments took on a much more important role in the discussion of relativity issues. In optics, we saw that stellar aberration, Arago's experiment of 1810, Fizeau's experiment of 1849, the Michelson–Morley experiment of 1887, and numerous other ether-drift experiments played an important role first as an optical vindication of corpuscular optics, then as a constraint in constructing the relationship between ether and matter. The observations almost always had astronomical import as they concerned the relative motions of celestial bodies, and they were often done by astronomers (Bradley, Arago, Kinklerfues, Ketteler, Airy, Higgins...). They challenged the optical ether theorists, Fresnel and Stokes in the first place, for they pulled the theory in two opposite directions: whereas stellar aberration and the Fizeau experiment suggested a stationary ether, the negative ether-drift experiments suggested a fully dragged ether. For a long time this tension remained unresolved, and the optics of moving bodies was only a minor subfield of wave optics. In the last third of the nineteenth century, the growing interest of astronomers and the rapid progress of optical instrumentation turned this lingering tension into an urgent problem, epitomized in the conflict between the two major experiments of Michelson and Morley in 1886 and 1887.

The relativity considerations in electrodynamics similarly depended on experimental findings: first the relativity of electromagnetic induction (also the Rowland effect for moving charges, the Röntgen effect for moving dielectrics, and the magnetic and electric deflection of cathode rays), then the basic facts of the optics of moving bodies when they entered electrodynamics through the electromagnetic theory of light. The major constructors of electrodynamic theories (Neumann, Weber, Maxwell, Helmholtz, Heaviside, Hertz, Lorentz, and Cohn) all had to decide the extent to which the phenomena depended on relative motion only, and, starting with Maxwell, they had to decide whether or not the electromagnetic ether followed the motion of matter. Whereas electrodynamic phenomena seemed to require a complete drag, the optics of moving bodies increasingly favored the stationary ether.

Thus we see that in optics and electrodynamics, relativity considerations largely depended on observations and experiments, but also on the basic theoretical picture. This picture changed radically over the course of time, from light corpuscles to ethereal waves in optics, from action at a distance to field-mediated action in electrodynamics. The consideration of moving bodies was only one reason among others to evolve the relationship between ether and matter. Boussinesq's and Lorentz's ultimate picture of this relationship through the motion of atomistic entities in a stationary ether responded to difficulties encountered in crystal optics, optical dispersion, magneto-optics, electrolysis, and electric discharge in rarefied gases. Whereas in mechanics relativity considerations had long ceased to depend on experiments and concerned a well-established theory, in optics and electrodynamics they constantly referred to evolving experimental knowledge and they accompanied the ongoing construction of theories.

As for the geographical distribution of relativity considerations, they occurred wherever the foundations of mechanics were a concern, and wherever optics and electrodynamics were being developed. In the case of nineteenth-century mechanics, the motivations depended on the country: the British wanted to consolidate the basis of their matter-and-motion program, the Germans and the Austrians (the British too) wanted to clarify the relationship between mechanics and the newer theories of thermodynamics and electrodynamics, and the French were often driven by pedagogical ambitions. The quasi-philosophical criticism of foundations was strongest in Germany and Austria in the last third of the century, as the continental physicists were facing traumatic changes in theoretical conceptions (especially in electrodynamics) and as theoretical physics prospered in new institutional settings. This critical turn was not specific to physics; it also happened in mathematics and in other natural sciences as a consequence of the creation of new fundaments and new domains.

In electrodynamics and optics, the most successful construct of the late nineteenth century was Lorentz's theory, a Dutch synthesis of French, German, and British elements. The outcome incited criticism regarding the meaning of the ether and the relativity of electrodynamic phenomena, in France by Poincaré, in Germany by Cohn, Bucherer, and Einstein, and in Britain by Larmor. Cultural predispositions do not suffice to explain the nature and content of these reflections. The two most influential critics, Poincaré and Einstein, were singular figures who transcended the milieu in which they were educated. On the one hand, they were addressing issues raised by their elders and contemporaries; on the other hand, they were exceptionally predisposed to question the most entrenched concepts of fundamental physics. They took a God's eye view in which the relativity issues in the distinct fields of mechanics, optics, electrodynamics, and gravitation appeared to be deeply interrelated.

As no one is indispensable for any fundamental breakthrough in physics, we can safely assume that special relativity would have someday emerged from Lorentz's theory even if Poincaré and Einstein had never lived. And general relativity would probably have later emerged as a special tensor-field theory of gravitation. However, the specific forms Poincaré and Einstein gave the special theory and the way Einstein conceived of general relativity intimately depended on their unique intellectual makeup. The relativity theories as we know them were neither the sudden inventions of solitary geniuses nor the inevitable outcomes of a historical determinism. They resulted from the collective maturation of the still incomplete and imperfect theories that Lorentz, Poincaré, Einstein, and a few others built out of numerous resources in a long tradition of concern with the appropriate reference of motion in physical theories.

11.2 Mechanical relativity

Although this section and the corresponding chapters of this book are mostly about the question of the reference of motion in mechanics from the seventeenth century to the nineteenth century, it is instructive to begin with ancient Greek concepts of motion, as a backdrop for the early modern concepts.

Two Greek concepts of motion

For Aristotle, motion is change of a state in a broad sense. Motion in the modern sense is a special kind of motion, called locomotion (φορά) and defined as change of place. The place of an object is the inner boundary of surrounding matter. An object in a void would have no place, because there is nothing in a void to differentiate between different locations. Place is always defined because there is no void. The world is a finite, spherical plenum, with a center located at the center of the earth, and a periphery defined by the celestial sphere. Objects have a natural place, depending on their degree of lightness or heaviness. In the sublunary world, there are four elements of increasing lightness: earth, water, air, and fire. In the natural state of the world, the elements would be arranged in four concentric spherical, homogeneous layers. When an object is not in its natural place, it tends to move to its natural space owing to some sort of buoyancy: earth in air or in water tends to move downward, whereas air in water tends to move upward. Heavier bodies (containing a higher proportion of earth) fall faster than lighter bodies because they are further from being in their natural place and because they are relatively less affected by resistance of the air or water in which they fall. If the bodies were falling in a void, Aristotle tells us, they would all fall together because there would be no reason for one body to move faster than the other. But there cannot be any void.[2]

---

[2] Aristotle, *Physics*, Books 4 (216a for fall in a void), 7, 8. Cf. Lloyd 1970, chap. 8; Machamer 1978.

Falling is the first kind of natural motion. The second kind is the eternal motion of the sun, the planets, and the stars. According to the received mathematical astronomy, these motions are composed of uniform circular motions induced by the primary motion of the stellar sphere. The celestial bodies and the interspace cannot be made of the sublunary elements, because their natural motion would then be radial. They are made of a fifth element Aristotle calls aether. The uniform circular motions must be caused by a "prime mover" acting on the celestial sphere, since there is nothing in the symmetric buildup of the world to otherwise justify this motion. In the sublunary world, there is also "violent motion," that is, artificially induced motion. This is why not all bodies move vertically downward. For Aristotle, any nonnatural motion requires a continuously acting force (the reverse is not true: a small force may not be sufficient to move a heavy body). The velocity of the body is proportional to the force and inversely proportional to its weight. Aristotle is well aware that a projectile, when thrown horizontally, keeps moving horizontally for some time. He even argues that in a void any initial motion would persist indefinitely since there would be no more reason for the horizontal velocity to decrease than to increase. But there would also be no reason for the body to move in one direction instead of any other direction. This absurdity is one more reason to reject the idea of a void. The persisting motion must be caused by the surrounding medium. For instance, the air expelled at the front of a bullet may exert a weaker pressure than the replacing air at the rear of the bullet. Aristotle did not commit himself to this fanciful view, and some of his medieval heirs replaced it with *impetus*, an acquired inclination to move.[3]

Aristotle meant his theory of locomotion to avoid the absurdities of the older atomist concept of motion. For atomist philosophers, primitive matter can only move into a void, since two objects cannot be in the same place. All changes should be reduced to the motion of atoms in a void. The atoms move incessantly and exchange motion when they collide. The underlying definition of motion is unclear. One option would be to refer motion to the position of a macroscopic body (e.g., in atomist optics the atoms of simulacra move with respect to their emitter); another would be to treat the void as an imagined infinite container in which the atoms have a well-defined location at any time. Aristotle rejected this second option as an absurd idealization contradicting the very definition of a void as the absence of matter, and he could not conceive of the first option because in his mind the concept of distance between two objects presupposed matter in the intervening space.[4]

In the fragmentary way in which it has reached us, Greek atomism is too sketchy to offer a well-defined theory of motion. Aristotle's theory is far more developed, it is internally coherent, and it has some factual basis in common observation, although it would fail by Galileo's standards. At any rate, these two theories inspired the two major concepts of motion of seventeenth-century mechanical philosophy. In the first concept, the motion of a body is defined with respect to contiguous matter. In the second concept, motion is defined with respect to space regarded as a physical entity. The first concept is fully relativist, since it involves only motion of matter with respect to matter. It is purely local, since it involves

---

[3] Aristotle, *Physics*, Books 4 (215a for horizontal motion in a void), 7 (Part 5 for the relation between velocity, force, and weight), 8; *On the heavens*. On impetus, cf. Drake 1975; Wallace 1981.

[4] Aristotle, *Physics*, Book 4. On the atomists, cf. Lloyd 1970, chap. 4.

only the relation of contiguous bodies in a continuous plenum. The second involves an immaterial, absolute reference of motion. It nevertheless allows for a limited form of phenomenal relativity: the laws of mechanics remain unchanged in a reference frame moving uniformly and rectilinearly in absolute space. It is essentially nonlocal, since the reference is an indefinitely extended rigid frame.

Inertia and relativity in early modern mechanics

The first concept of motion plays a central role in Descartes's system, as Descartes identifies matter with spatial extension, and traces every change in nature to a rearrangement of the particles of matter thus defined. He thereby preserves three characteristics of Aristotle's theory of motion: the proscription of the void, the purely relative character of motion, and the local character of motion. The similarity ends there, *pace* the Cartesian Jesuits, who portrayed Descartes as a disciple of Aristotle. Although Aristotle regarded locomotion as a necessary prelude to any kind of change, unlike Descartes he believed that motion, as a change of state, was generally irreducible to locomotion. Descartes's vortex-based cosmology has nothing to do with Aristotle's doctrine of natural motion in the ether, and his laws of motion flatly contradict Aristotle's: he assumes a peculiar form of relative inertia, he stipulates the general conservation of motion, and he does not distinguish between natural and violent motion.[5]

Galileo's departure from Aristotle is even more marked, for he defends a heliocentric cosmology and he adopts the atomist concepts of matter and void. His theory of motion is quite at odds with Descartes's: it is based on a different concept of inertia in which bodies naturally move uniformly in circles centered on the celestial bodies or in straight lines. Galileo assumes a general tendency of masses to come together instead of the Aristotelian tendency for bodies to find their natural place. Unlike other neo-atomists, he does not have a definite concept of space, and he always refers motion to concrete rigid bodies: the earth, the sun, any planet, or a boat. Since this reference varies, the question of the dependence of the laws of motion on the choice of the reference naturally arises. Should motion with respect to the earth be affected by the motion of the earth around the sun? In reply, Galileo argues that a uniform motion commonly impressed on a system's bodies does not affect the regularities of motion within this system. This fact of relativity is intimately related to inertia, since rectilinear inertia is obviously unaffected by a common uniform rectilinear motion. In order to be quite coherent in this nexus, Galileo should exclude circular inertia. He does not because, like most of the neo-atomists, he needs this kind of inertia to account for the circular motion of celestial bodies.[6]

Newton avoids this inconsistency by retaining rectilinear inertia only and by tracing the circular or elliptic motion of planets to the combined effect of inertia and attraction by the sun. The result agrees with Kepler's laws if the motion obeys the three laws Newton has attained in a study of collisions, and if the attraction obeys Hooke's inverse square law. As the law of rectilinear inertia cannot be true with respect to every celestial body,

---

[5] See Section 1.3.

[6] See Section 1.1.

Newton refers it to absolute space. This space is the neo-atomist container of all things, it is an imagined rigid frame, and it is not realized by any existing material system. It is chosen so that the three laws of motion, when applied to celestial bodies, reproduce the observed motions. Unfortunately, this criterion is not sufficient to uniquely determine the reference space. Newton indeed shows that his laws of motion imply Galilean relativity (Corollary 5) and the concomitant impossibility of empirically distinguishing between absolute space and a space moving uniformly and rectilinearly in absolute space. Moreover, Newton proves that a common acceleration impressed on a system of bodies does not affect relative motion under given forces within this system (Corollary 6), so that one cannot even distinguish between absolute space and an accelerated space: the entire visible universe could be accelerated toward a huge invisible distant mass, without any observable effect of this acceleration. In order to escape this indeterminacy of absolute space, Newton assumes that the center of mass of the observable universe (supposed to be complete) should be at rest in absolute space.[7]

As a moderate Cartesian with Galilean sympathies, Huygens does not accept any immaterial space. He wants every motion to be referred to material bodies, at least in principle. In his mechanics he assumes rectilinear inertia, the composition of motions, and Galilean invariance. Unlike Newton and more like Galileo, he refers motion to concretely defined frames (typically a boat and the earth) and he sees a complete equivalence between all inertial frames in this concrete sense. This perfect symmetry may explain why he pioneers the use of the relativity principle in the derivation of mechanical laws. What was a fact to Galileo and a theorem to Newton truly is a constructive principle for Huygens. In particular, he uses this principle in ingenious derivations of the laws of elastic collision and the laws of free fall. This practice raises the question of the determination of the concrete reference of motion. Huygens knows from Galileo that the earth and a smoothly navigating boat are reasonably good approximations. In stricter reasoning, a more principled definition is needed. From Huygens's late manuscripts, we know he found it in the notion of free particles at mutual rest.[8]

Although Huygens does not tolerate the idea of a void any more than Descartes and although he considers that any body is subjected to the action of the corpuscles of the medium in which it is immersed, he admits the possibility that the net action of the medium on a particle of matter (a small concrete body) may vanish when this particle is far enough from any observable body. This defines a free particle. He further admits that it is possible to choose a set of free particles so that their mutual distances are invariable. The law of inertia states that in the frame defined by a set of this kind the motion of any other free particle will be rectilinear and uniform. Evidently, the law implies that a frame moving uniformly and rectilinearly in this frame also is an inertial frame, in conformity with the principle of relativity. Huygens is thus able to define the inertial class of referential frames operationally, without any abstract concept of space.

[7] See Section 1.4.

[8] See Sections 1.5 and 3.1.

From Newton's absolute space and time to Galilean kinematics

In retrospect, Newton's conception seems artificial because in his theory he introduces an asymmetry that has no empirical counterpart: the selection of absolute space among all inertial spaces conflicts with the perfect empirical equivalence of all inertial spaces. Yet this conception has one advantage over Huygens's: it identifies a cause for the otherwise mysterious behavior of free particles. Why is it that we can arrange a set of free particles to be perpetually at mutual rest? Why is it that in an arbitrary set of free particles, the mutual distances of these particles vary linearly in time? While Huygens simply accepts this preestablished harmony of motion, Newton requires an explanation. As he does not want local inertial behavior to be the effect of a remote cause (the starry sky or an invisible material body), he introduces a space that can serve as the needed causal entity. The late-medieval concept of space as an immaterial but quasi-substantial container of all things does the job, except that it conflicts with Galilean relativity.

Newton admits absolute space and absolute time in parallel, as homogeneous containers of all events. The most fervent defenders of this notion, Euler and d'Alembert, will argue that the law of inertia then is a mere consequence of the principle of sufficient reason applied to absolute space and time. A body at rest in absolute space remains at rest because it has no more reason to move in one direction than in another, etc. This reasoning echoes Aristotle's argument for rejecting true inertia, except that the void that Aristotle judged impossible is replaced with absolute space. Somehow, Newton is able to reconcile the homogeneity of space with the distinguishability of its parts. Aristotle is not (nor will be Leibniz). In the century following Euler's *Mechanica*, most authors of mechanical texts accepted Newton's absolute space and time uncritically and did not adjudicate on the physical reality of these concepts.[9]

Later, in the growingly empiricist mood of the nineteenth century, a few authors including W. and J. Thomson, Tait, Streintz, and Lange rejected or at least avoided absolute space in favor of a class of operationally defined inertial frames, thus bringing the relativity principle to the foreground. They were unconsciously walking in Huygens's shoes. In their view, concrete "reference frames" (the name comes from J. Thomson), not an imagined "space," should be the proper foundation of spatiality in mechanics. Their chief purpose was to clarify the conditions of application of Newton's law, not to identify the cause of inertial behavior. One exception is Streintz, who compared the class of inertial frames in space and time to the class of Cartesian coordinate axes in space, and thus alluded to an intrinsic space–time structure behind the inertial class. This is close to the modern view of a causally effective space–time structure defined through the Galilean group. Mach's stricter empiricism tolerated none of the contemporary answers to the question of inertia. Mach wanted an empirically ascertainable cause of inertial behavior, and he wanted the laws of mechanics to imply only the motion of matter with respect to matter. He believed he could explain inertia in this material–relational manner, as an effect of motion with respect to the remote masses of the universe. As this theory existed only in rough sketch

---

[9] See Section 2.1.

and as it hypothesized a strange action at a distance, it failed to gain the favor of Mach's contemporaries. It did appeal to one of his younger readers.[10]

In philosophy, Leibniz rejected Newton's absolute space and time in the name of the relational character of any spatial and temporal coordination. He defined space as the order of coexistence of things, and time as their order of succession. He accepted the conservation of *impetus* in horizontal motion, and more generally the conservation of unimpeded motion in harmony with the conservation of *vis viva*. He granted that a well-defined *vis viva* presupposed absolute motion, and thus wavered between complete relativity of motion (in the geometrical context) and complete determination of motion (in the dynamical context), but he did not consider the intermediate Galilean option.[11]

Whereas Leibniz ignored Newton's definition of force and the accompanying laws, Kant deliberately adjusted his philosophy to Newton's mechanics. Having defined space and time as necessary forms of intuition, Kant was free to consider the choice of a specific space as merely conventional, and regarded any space moving with respect to the former space as equally legitimate from a kinematical point of view. Then he selected the inertial subclass of spaces from a mechanical point of view, and the inertial space for which the center of the universe is at rest from a phenomenological point of view. In his own philosophical way, he anticipated the late nineteenth-century reliance on an equivalence class of inertial frames, except that his spaces belonged to the transcendental apparatus of the human mind, whereas frames à la Thomson were just physical objects.[12]

As we just saw in Leibniz's and Kant's case, the rejection of Newton's absolute space went together with the rejection of his absolute time. For the late-nineteenth-century critics of inertia, time had to be defined through the law of inertia, and more exactly through the empirical core of this law. Following an old suggestion by d'Alembert, time can indeed be defined by the distance traveled by a particle in inertial motion. Better, for Thomson and Tait time is defined as proportional to the distance between two free particles. The law of inertia warrants proportionality for any other choice of the two particles. As several authors remarked, this definition is only one possible convention among others, and any other monotonous function of the inertial time could also be called time. However, the inertial time is the one for which the laws of mechanics have the simplest form. Poincaré emphasized this point in his conventionalist take on the laws of mechanics. Most originally, he remarked that not only the measure of time flow but also the simultaneity of two events was a matter of convention. For distant events, the only way we can assert simultaneity is by the exchange of light signals, under the convention that these signals travel at the constant velocity c. Similarly, for Poincaré the law of inertia is only a convention, through which we may define a uniform time flow.

To sum up, nineteenth-century attempts to elucidate Newtonian inertia ultimately brought new kinematical concepts in place of Newton's absolute space and time. These concepts were based on the operational selection of a privileged class of reference frame.

---

[10] See Section 3.2.

[11] Cf. Bernstein 1981.

[12] See Section 3.1.

By assuming the perfect equivalence of all inertial frames, they made the relativity principle an essential part of the new kinematics, whereas Newton only had a relativity theorem in odd company with absolute space.

Bélanger's principle of relative motions

There is another way in which the principle of relativity made it to the front stage of mechanics, independently of any criticism of Newton's space and time. Huygens opened this route in his theory of free fall, when he derived Galileo's laws by applying the same velocity increment in a succession of inertial frames. This kind of reasoning became popular in nineteenth-century France after Laplace adapted it to the original, discrete form of Newton's second law (equating impulse to momentum increment). In 1847, Bélanger derived the continuous version of the second law by means of the stronger relativity principle according to which a commonly impressed rectilinear motion, uniform or not, does not affect the relative motion in a system of bodies under given forces (Newton's Corollary 6).

This principle and the attached derivation can be found in the text the young Einstein used to prepare for the ETH exams. The important point is that the constructive power of relativity principles was recognized within mechanics. These principles were thereby treated as the inductive generalization of the absence of observed effects of a commonly impressed rectilinear motion, on Galileo's boat, on the earth, or on the earth–moon system (which is commonly accelerated by the sun).

Before 1847, the principles of relativity did not have a name. By calling his principle *Principe des mouvements relatifs*, Bélanger initiated the chain of variants leading to Poincaré's *Principe de relativité*. The importance of this gesture should not be undervalued: a principle, by definition, has a higher status than a fact, law, or theorem. Clearly, Bélanger used "principle" to mark a position atop the constructive hierarchy. This marking remained.[13]

11.3 Optical relativity

In optics, the motion of the parts of optical systems would have no effect if the velocity of light were infinite. This is why the optics of moving bodies and the history of the optical principle of relativity should start with the first measurements of the velocity of light. Rømer's pioneering considerations of 1676 were based on a kind of Doppler effect: the apparent frequency of a periodic phenomenon, the eclipses of one of Jupiter's satellites, depends on whether the earth is moving toward or away from Jupiter. Bradley's more reliable measurement of 1728 depended on another effect of the earth's motion, the aberration affecting the direction of observation of a star.

Corpuscles versus waves

Clairaut, Euler, and a few British astronomers understood that in Newton's then dominant theory of light the aberration angle depended only on the relative velocity of the source and the observer. In modern terms, this is a consequence of the Galilean relativity

---

[13] See Chapter 2.

of the interaction between material bodies and light corpuscles. More generally, a common (approximately uniform) motion of the parts of a system should not affect optical phenomena within this system. Although in Newton's mechanics this result was only a theorem, in optics it also worked as a principle for interconnecting or excluding phenomena. Clairaut and Euler used it to derive aberration for a moving observer from the easier case of aberration from a moving source; Wilson and Robison used it to deny any change of the aberration angle when filling a telescope with water.[14]

In the wave theory of light, the motion of the parts of an optical system through the ether should naturally have an effect on the propagation of light. In particular, the directional aberration or the Doppler frequency shift should depend on whether the source or the observer moves through the medium. Strangely, Euler (for aberration in 1739) and Doppler (for the Doppler effect in 1842) found that the difference between the two cases was only of second order with respect to the emitter/light or receptor/light velocity ratio. As Euler noted in the case of aberration, the relativity exactly occurring in the corpuscular theory still holds in the wave theory to a sufficient approximation.[15]

Would this approximate relativity extend to other optical phenomena? An experiment performed by Arago in 1810 in the corpuscular context suggested a positive answer to this question in the case of refraction. In the rest of the nineteenth century, experimental evidence accumulated for the absence of first-order effects of the motion of the earth through the ether on the phenomena of refraction, reflection, interference, and diffraction. Contrary results did not resist replication. This relativity was most surprising in the case of refraction, because one intuitively expected the motion of a prism through the ether to affect the bending of light.[16]

Fresnel versus Stokes

Historically, there were two ways to address this observed relativity of refraction. As Fresnel proposed in 1818, one could assume the ether to be *partially* dragged by moving transparent matter, to an extent adjusted to produce the desired relativity: the ether is strictly stationary in a vacuum, nearly stationary in a light medium like the air, and dragged in the proportion $1-1/n^2$ in a dense transparent medium of index n. Alternatively, as Stokes proposed in 1845, one could assume that the ether remained at relative rest near the surface of the earth, owing to the plausible impermeability of the huge mass of the earth to the ether. Fresnel's conception looked ad hoc, but it had the advantage of simply explaining stellar aberration as an effect of the motion of the earth through the stationary ether. Conversely, Stokes's conception directly predicted the *exact* relativity of terrestrial optics, but it required a further assumption, the irrotational character of the motion of the ether, in order to explain stellar aberration.[17]

For a long time, the French and the Germans favored Fresnel's theory, while the British favored Stokes's theory. Fizeau's direct confirmation of the ether drag in his running-water experiment of 1951 did not suffice to shake the British bias. In 1886, Michelson and

---

[14] See Sections 4.1 and 4.2.

[15] See Section 4.3.

[16] See Sections 4.4 and 4.5.

[17] See Section 4.4.

Morley's clear-cut confirmation of Fizeau's result changed the game. By that time, Boussinesq, Potier, and Michelson had shown that Fresnel's stationary ether, when properly combined with the molecular picture of matter, led to a partial drag of light waves by transparent media, very nearly in the proportion assumed by Fresnel. So by that time the absence of the motion of effects of the earth's motion was a well-established empirical result, and it had a consensual explanation in an extended Fresnel theory, all of that to first order in the earth/light velocity ratio.[18]

This relativity should not be confused with the Galilean relativity of Newtonian mechanics, despite the then nearly universal belief that processes in the ether obeyed the laws of mechanics. In nineteenth-century wave optics, unlike in the older corpuscular optics, the relativity principle of mechanics does not necessarily translate into the relativity of optical phenomena: a commonly impressed motion on the material parts of an optical system does not necessarily concern the ether (it does so only in Stokes's theory). The mechanical ether picture (in Fresnel's version) suggested effects of the ether wind, and the observed lack of effect came as a surprising fact, oddly explained by first-order compensation mechanisms.

This compensation did not occur at second or higher order. In particular, the ether wind implied that the relative velocity of light on earth should depend on the direction of propagation, at second order for round trips. In 1887, Michelson and Morley's failure to detect an interferential consequence of this induced anisotropy brought puzzlement. On the one hand, their earlier confirmation of the Fresnel–Fizeau drag seemed to endorse Fresnel's stationary ether; on the other hand, their new result brought more relativity than expected in Fresnel's theory.[19]

So far the optics of moving bodies had been discussed in a purely optical context. Hertz's contemporary confirmation of Maxwell's theory changed the game. Light being now identified with an electromagnetic wave, and a kind of relativity being associated with electrodynamic theories, the difficulties of the optics of moving bodies became entangled with similar difficulties in the electrodynamics of moving bodies.

11.4 Electrodynamic relativity

The electrodynamics of moving bodies began mostly with Faraday's discovery of electromagnetic induction in 1831. Faraday traced the electromagnetic force in a portion of a conductor to its cutting the lines of force of a magnetic field. This law conveniently summarized all of his observations and it implied that the induced current depended only on the relative motion between the conductor and the source of the magnetic field, at least in the case of closed currents.

Neumann, Weber, and a complete relativity

From then on, the relativity of electromagnetic induction was regarded as a solidly established constraint on any electrodynamic theory. Neumann remarked that in the contrary

---

[18] See Section 4.5.

[19] See Section 4.5.

case the motion of the earth would suffice to induce a current in a coil placed near a magnet in relative rest. He plausibly reasoned that, by analogy with mechanics, the motion of the earth could not affect electrodynamic phenomena. His electrodynamic theory of 1846 and Helmholtz's later variant of 1870 were designed to respect this electrodynamic relativity principle, for they made all electrodynamic phenomena depend on the variations of a quantity, the electrodynamic potential, that depended only on the relative configuration of two objects. Weber's contemporary theory assumed a similar relativity at the micro-level for the force between two particles of electricity. Weber required this force to depend only on the relative distance of the two particles and on its two first time derivatives, in a natural extension of Newton's requirement that elementary forces should depend on relative distance only.[20]

As Neumann and Weber themselves noted, the resulting relativity of electrodynamic interactions is broader than Galilean relativity: it implies the lack of effect of any common rigid motion of the system whereas Galilean relativity restricts the common motion to uniform translation. In particular, an ongoing rotation of the system does not affect electromagnetic induction: if a cylinder magnet and a wire touching the pole of the magnet at one end and the meridian of the magnet at the other end rotate together around the axis of the magnet, there should be no current induced in the circuit thus formed, as several authors remarked while discussing the "unipolar induction" occurring when the wire rotates with respect to the magnet.[21]

Maxwell's mobile ether–matter

In contrast, in Maxwell's electromagnetic theory or in any theory in which the electrodynamic interactions depend on the states of the medium in which material bodies are immersed, one should expect effects of a common motion of these bodies if the medium does not partake in this motion. The situation of these theories vs. continental theories is entirely similar to that of wave optics vs. corpuscular optics. On one side you have a Newtonian theory in which a kind of relativity is fully integrated; on the other side you have a medium-based theory in which effects of absolute motion are naturally expected. Again, in the medium-based approach theorists needed to face the observed relativity of large classes of phenomena.

In Maxwell's theory and in Hertz's fuller electrodynamics of moving bodies, the relativity of electromagnetic induction is obtained by neglecting retardation and by assuming that within conductors and magnets the ether moves together with the material substance. The lack of effects of the earth's motion on terrestrial optics is accounted for by assuming that the ether is fully dragged by the earth and its atmosphere, in accordance with Stokes's theory. However, effects of a commonly impressed motion should generally be expected. For instance, an electric charge should appear at the poles of a rotating magnet. As Maxwell noted, the force between two identical electric charges should diminish when they move

---

[20] See Section 5.2.

[21] Of course, the mechanical motion of magnets and current carriers under electrodynamic forces does not obey this broader relativity since Newton's laws do not.

together in a direction perpendicular to the joining line. At first glance, retardation should make the induction between two rigidly connected coils depend on their orientation with respect to the velocity of the earth. Des Coudres tried and failed to detect the latter effect. The other imagined effects were two small to be measurable.[22]

Lorentz's immobile ether

Our comparison between optical and electrodynamic relativity continues with Lorentz's intervention in the last quarter of the nineteenth century. Lorentz, like Fresnel, assumed the ether to be strictly stationary in vacuum, and, like Boussinesq, Potier, and Michelson, he assumed the atomistic constituents of matter to be moving freely in this stationary ether. The motivation was similar. In older optics and in Maxwell's electrodynamics, there was only one ether–matter medium with no specific structure for the embedded matter and this homogeneity made it difficult or impossible to explain the phenomena that we now know to depend on the molecular structure of matter: dispersion, magneto-optics, electrolysis, and electric discharge in rarefied gases. Boussinesq and Lorentz both found that the fullest and simplest separation of ether and matter best fitted the phenomena. While Boussinesq treated the ether as an elastic body in mechanical interaction with the immersed molecules, Lorentz applied Maxwell's equations and the Lorentz force formula to the ether in interaction with ions or electrons.[23]

By averaging his microscopic equations over macroscopic elements of volume, Lorentz could cover all known electromagnetic, magnetic, and optical phenomena. The atomistic constituents and processes he had assumed for theoretical reasons found direct applications in the new contemporary microphysics of electrons, X-rays, and atomic spectra. In the optical context, Lorentz found that his theory, like Boussinesq's, implied the Fresnel–Fizeau drag and the general absence of first-order effects of the motion of the earth on terrestrial optics.

By 1892 Lorentz had an elegant proof of this first-order relativity of optical phenomena by means of a local time shift that formally brought the Maxwell–Lorentz equations for a system moving in the ether to those of a system at rest in the ether. He also found that a dilation of the coordinate along the velocity of the earth brought the equations of an electrostatic system on earth to the simple form they have when the earth does not move. Extending this correspondence to the cohesive forces of matter, he deduced the contraction that explained the null result of the Michelson–Morley experiment of 1887. At the close of the century, he generalized his formal transformations to any order, and obtained the general invariance of optical phenomena under the additional assumption that all forces, including cohesive forces and the inertial force of electrons, transformed like the Lorentz force. However, he still expected electrodynamic effects of the motion of the earth through the ether.

Lorentz's theory essentially departed from Maxwell's theory by giving atomistic pictures of the basic concepts of charge, polarization, and current in terms of the accumulation,

---

[22] See Sections 5.3, 5.4, and 5.5.

[23] See Section 5.6.

displacement, or circulation of electrons. The Maxwell–Hertz equations for the electrody-namics of moving bodies essentially differ from the Maxwell–Lorentz equations because the former concern macroscopic fields and the latter microscopic fields, and also because they enjoy a different symmetry. As noted by Hertz, the Maxwell–Hertz equations are in-variant under any change of reference frame, whatever the motion of the new frame in the old frame. As Lorentz nearly proved in 1899 (in the optical, dipolar approximation for the source terms), the Maxwell–Lorentz equations are invariant under what Poincaré later called the Lorentz transformations.

Lorentz used his transformations to discuss the effect of a uniform translation of the sys-tem through the ether. They differ from a Galilean transformation by an uninterpreted time shift and by a dilation of the spatial coordinates. The Lorentz transformations, as under-stood by Lorentz, do not directly imply the relativity of phenomena, because for Lorentz the transformed coordinates and states are merely formal and because the Lorentz force is not invariant. This symmetry must be completed with the assumption that all microscopic forces, including inertial forces, transform like the Lorentz force, and special reasoning is needed to show that the Lorentz time shift and the Lorentz contraction have no observable effect in the desired approximation.

Lorentz's theory has the evident advantage over the Maxwell–Hertz theory of be-ing compatible with a much wider range of observed phenomena. The simplicity of its fundamental concepts and equations is appealing, and Lorentz's skill in bridging these fundaments with macroscopic observations was admirable. These qualities sufficed to at-tract the attention of Lorentz's colleagues, not to convince all of them. At the close of the century, there still was a strong resistance to theories based on atomistic assumptions. In addition, Lorentz's ether had contradictory properties: while its internal motion satisfied the general principles of mechanics, it could not undergo the bulk motion expected for any mechanical system. This immobility seemed to exclude momentum carrying, in which case Maxwell's radiation pressure implied a violation of the principle of reaction. Lorentz saw these difficulties, but unlike his critical readers he did not take them seriously. He did not mind hurting mechanical intuition as long as the formal foundations were consistent, com-plete, and true enough to warrant definite and well-verified predictions. The end justified the means.

In retrospect, the most glaring defect of Lorentz's theory is the disparity between the assumption of a stationary ether and the relativity retrieved at the end of long, difficult, indirect reasoning. A keyword of the theory is compensation. Naturally expected effects of the ether wind disappear owing to various compensatory mechanisms, as if nature had conspired to hide the obvious. Yet, at the close of the century no one except Poincaré com-plained about this state of affairs. There are plausible reasons for this silence. With Fresnel's theory, physicists had been accustomed to deriving approximately symmetric results from an asymmetric picture. In the early wave theories of aberration and in the theory of the Doppler effect, the aberration angle and the frequency shift depended only on the relative motion of the source and the observer in a first approximation, despite the obvious asym-metry between the case of a moving source and the case of a moving observer. By deriving approximate relativity for systems moving in the ether, Lorentz was just extending this state of affairs. Before Poincaré's and Einstein's interventions in 1905, he never doubted that the ether wind remained principally observable. He did not worry that the relativity approximation was getting better and better: first-order, then second-order, then any order under the dipolar approximation.

Consider the case of waves propagating in an elastic, incompressible, isotropic medium between a moving source and a receptor. Here we have the same approximate relativity as in the optical case. The wave equation is a d'Alembertian equation just as in Lorentz's theory, and this equation is invariant through Lorentz transformations in which the speed of elastic waves replaces the speed of light. This symmetry can be exploited to derive first-order relativity for the Doppler effect. Voigt did so in 1887 in the elastic-solid theory of light. This did not induce him to imagine that some additional mechanism would produce exact relativity in accord with the Lorentz-group symmetry of the wave equation. Late nineteenth-century ether theorists had no reason to admit in optics a relativity that did not exist in acoustics.[24]

11.5 Special relativity

Poincaré's principle of relativity

Poincaré was alone, before the turn of the century, in judging that the relativity of optical and electrodynamic phenomena should be exact despite the ether wind. As a mathematician confronted with the embarrassing diversity of past and present ether theories, he decided that the ether was a merely imaginary construct, not to be reified as ordinary matter and perhaps bound to disappear someday from physics. Consequently, the motion of the earth through the ether could not have any observable effect, and the ether could not carry any momentum. The repeated failure to detect effects of the ether wind and the lack of empirical evidence for Maxwell's radiation pressure comforted him in this view. Moreover, Poincaré believed that general principles of inductive origin, rather than detailed mechanical pictures, should gauge and guide the construction of physical theories. Four principles of mechanical import, the principle of reaction, the principle of relativity, the energy principle, and the principle of least action, played a crucial role in his criticism of *fin de siècle* electrodynamics. In 1895 he was first to bet on a strict validity of the two first principles for matter alone, and he started to criticize Lorentz's theory for not complying with them. Being aware of the constructive role of the relativity principle in the history of mechanics, in 1900 he borrowed the name *principe du mouvement relatif* from the French mechanical textbook tradition. He insisted that this principle and the principle of reaction were intimately related, because in mechanics the energy principle and the relativity principle together imply the principle of reaction.[25]

In the same year, Poincaré understood that the missing momentum in Lorentz's theory could be formally represented as a momentum distributed through the ether. He thus prepared for the existence of what he wanted to deny. Most important, he interpreted Lorentz's transformed coordinates and field states as those measured by observers ignoring their motion through the ether. In particular, he showed that Lorentz's transformed time was the

---

[24] Voigt 1887. See p. 144, note 46, this volume.

[25] See Sections 6.1 and 6.2.

time measured by moving observers who synchronize their clocks by exchanging light signals while ignoring their motion in the ether. Thanks to these insights, the relativity of electromagnetic phenomena became the direct consequence of the symmetry Lorentz had identified in his field equations. In 1904, Lorentz gave the full Lorentz transformations, with errors in the source terms, and he designed a contractible electron model satisfying his earlier demand that the inertial force of an electron should transform like the Lorentz force. He was thus able to reconcile the relativity of optical phenomena at any order with the recent endeavor to construct all inertia electromagnetically. This electromagnetic worldview, pioneered by Wien in 1899 and implemented by Abraham through a rigid spherical electron model of purely electromagnetic momentum (given by Poincaré's formula), was then flying high in the skies of mathematical physics.[26]

In early 1905, Poincaré saw that he could reach the exact invariance of the Maxwell–Lorentz equations through the Lorentz transformations by correcting Lorentz's expression of the transformed source terms. Interpreting these transformations as a means to transform a system at rest in the ether into a system moving uniformly through the ether, he directly interpreted the Lorentz-group symmetry as the formal realization of the relativity postulate. He required all forces, including cohesive and gravitational forces, to respect this symmetry. He now accommodated the electromagnetic momentum he had earlier rejected, for he understood that its existence after all did not contradict the relativity principle. Based on the principle of least action, he discussed various electron models and showed that Lorentz's contractible electron, properly stabilized through a cohesive pressure, was the only one compatible with the relativity principle. He also proposed relativistic generalizations of Newton's law of gravitation.[27]

Poincaré's *dynamique de l'électron*, as he called his theory, had much in common with special relativity as we know it: the Lorentz-group symmetry as a general symmetry for all interactions, the direct connection between this symmetry and the relativity principle, the use of invariants in theory construction, and the earliest interpretation of the Lorentz transformation in terms of measurements performed by moving observers. Yet this theory is not quite our special relativity, for Poincaré, who had earlier declared the ether moribund, now gave it a new life as the momentum carrier of all matter and fields and as a frame in which the received concepts of space and time still made sense. He regarded the space and time relations in other frames as merely apparent, and he refused to follow Einstein's and Minkowski's redefinitions of time and space–time, even though he of course understood that the relativity principle excluded any empirical determination of the ether frame. In his eyes, the ether frame was only a convention, though one useful to preserve ancestral habits and intuitions.

Einstein's new kinematics

The young Einstein originally entertained views similar to Lorentz's, and imagined several experiments to test effects of the motion of the earth through the stationary ether. At some

---

[26] See Sections 6.2 and 6.3.

[27] See Section 6.4.

point, around 1902, he decided that such tests were in vain and that the relativity principle held exactly. Even if he found inspiration in reading Poincaré, his motivation differed from Poincaré's. Unlike most ether theorists and in the spirit of Hertz's minimalism, he disliked the fact that even for the simplest phenomena, the theoretical representation contained asymmetries without empirical counterpart. The example he picked in 1905 was the current induced in a coil through a magnet, which in Lorentz's theory receives two quite different explanations according to whether the magnet or the coil moves in the ether. In his eyes, adopting the relativity principle implied denying such asymmetries and eliminating the culprit, the ether.[28]

In the spring of 1905, after vainly trying to satisfy the relativity principle in an emission theory for which the velocity of light depends on its emitter, Einstein became aware that Lorentz's local time was the time measured by moving observers who assume the constancy of the velocity of light despite their motion in the ether. This point in itself does not differ from Poincaré's insight of 1900, of which Einstein might have been aware. The important difference is that Einstein, unlike Poincaré, decided that the times thus defined in various inertial frames should all be on the same footing. Poincaré's contrary stance—if Einstein knew it—could only irk Einstein's epistemological consciousness, for it harbored the empirically meaningless distinction between apparent and true time. Einstein then raised the constancy of the velocity of light in a given frame to the status of principle, and derived the Lorentz coordinate transformations from the conjunction of this principle with the relativity principle. He went on in ways familiar to modern students of relativity: derivation of simple consequences of the "new kinematics"; proof of covariance of the Maxwell–Lorentz equation with physical consequences, and derivation of the relativistic dynamics of a particle from Lorentz covariance and Newtonian correspondence. His first argument for the mass–energy equivalence appeared a few weeks later, based on the comparison of a radiation process in two different inertial frames.[29]

Minkowski and Laue: A new world and a new dynamics

One may fruitfully compare the difference between Poincaré's and Einstein's versions of relativity theory with Newton's and Huygens's takes on mechanical inertia. For Newton, inertia can only be understood through the assumption of absolute space, even though there is no empirical way to distinguish between this space and any other inertial space. For Huygens, there is no absolute space, inertial frames are selected operationally, and they are strictly equivalent to each other. Einstein's elimination of the ether resembles Huygens's and others' elimination of absolute space. Minkowski's introduction of an absolute space–time structure through the Lorentz group resembles Streintz's tendency to regard the space and time coordinates in different inertial frames as belonging to the same point in a Galilean space–time. Minkowski's view (unlike Streintz's) was not only a matter of conceptual clarification. it went along with a focus on Lorentz-covariant objects and with the associated tensor calculus that proved to be so efficient in constructing relativistic theories.

[28]See Section 7.1.

[29]See Section 7.3.

Minkowski expressed his grand vision in 1908, at a time when Planck and his disciples were just about the only ones to take Einstein's theory seriously. The formal beauty of the new tensor calculus and the dizzying depth of the four-dimensional world concept did more than anything else to convert electron theorists to relativity theory. There was, however, a lingering dissatisfaction with what may be called the ontological emptiness of Einstein's relativity theory. Einstein prided himself with being able, in his principles-based approach, to derive important physical laws without entering into the detailed constitution of the implied entities. In particular, he could derive the equation of motion of the electron or of any particle without any constructive model, as the only Lorentz-invariant equation that agrees with Newton's law of acceleration in the limit of small velocities. As Laue remarked, this conception implies a lack of autonomy of relativistic dynamics with respect to the received Newtonian dynamics. As electron theorists insisted and as Einstein conceded, the theory remained an empty shell as long as one did not exhibit a construct obeying the relativistic particle dynamics. Poincaré had already given such a construct in 1905. Born tried a Born-rigid electron in 1909, with little success. Einstein briefly and privately tried a singularity in the electromagnetic field, although he generally doubted the possibility of understanding the electron without entering into quantum-theoretical considerations. More fruitfully, Laue developed a relativistic continuum dynamics that could broadly cover the dynamics of any extended object, based on a generalization of the energy–momentum tensor Minkowski had introduced for electromagnetic fields. Laue then proved that any extended object in internal equilibrium globally obeyed Einstein's particle dynamics, with a mass given by the total energy.[30]

Thus we see that the relativity principle had variable functions in the relativity theories of Poincaré, Einstein, Minkowski, and Laue. Although these authors all agreed that the principle should play a central role in the construction of future theories, they conceived of this role differently. Poincaré regarded the principle as a formal constraint on a constructive ether-based dynamics, and not as a challenge to the received concepts of space and time. Einstein rejected the ether constructs, redefined time operationally in conformity with the relativity principle, and used this principle in combination with a correspondence principle in order to avoid detailed modeling. Minkowski used a similar correspondence strategy, but replaced Einstein's multiplicity of frame-dependent quantities with intrinsic objects in a four-dimensional world. Laue adopted Minkowski's world, and brought in continuum dynamics in place of the correspondence strategy.

11.6 General relativity

In one important respect, the new relativistic mechanics did not differ from Newtonian mechanics: both theories required the selection of inertial frames, in conformity with the principle of inertia and the principle of relativity. The mystery Poincaré had seen in this selection remained entire in 1905. For most of the early relativists, there was no reason to worry about something easily admitted by their Newtonian predecessors. The exception was Einstein, when he devised the equivalence principle in 1907.

---

[30] See Section 7.5.

The equivalence principle

According to Galileo, all bodies fall at the same speed irrespective of their weight as long as the resistance of the air can be ignored. Galileo verified this universality of falling speed on pendula. Through a famous thought experiment with two connected falling bodies, this universality also resulted from his implicit assumption that the falling speed depends only on the weight (a generalization of the Aristotelian proportionality between weight and velocity) together with the additivity of weights. In his *Principia*, Newton separated the inertial mass in his second law from the gravitational mass in his gravitation law, thus opening up the possibility that different bodies might fall at different speed. However, his concept of matter as made of atoms of uniform density implied the proportionality between these two masses. For empirical confirmation of this proportionality, he adduced Galileo's pendulum experiments and the fact that the motion of Jupiter's satellites appeared to be unaffected by the attraction of the sun. He exploited this fact through his theorem of accelerative relativity (Corollary 6). In concrete and modern terms, this theorem implies that when Jupiter and its satellites move under the action of the sun, their relative motion is the same as if the sun did not exist. This is the free-falling elevator *avant la lettre*. As was mentioned, in his student years Einstein had seen Newton's theorem turned into a principle for deriving Newton's second law.[31]

By the nineteenth century, Newton's concept of a primitive matter of uniform density was completely obsolete. It seemed natural to distinguish inertial mass from gravitational mass as much as we distinguish inertial mass from electric charge, and the observed equality of the two kinds of mass became a mystery. At the close of the century, Wien justified this equality by reducing the two masses to the same distribution of electromagnetic energy. Being out of sympathy with the electromagnetic worldview, Einstein could not rely on such reasoning. Instead he strengthened Bélanger's principle of accelerative relativity and turned it into the principle that the physical phenomena observed in an inertial frame in a uniform gravitational field **g** should be the same as the phenomena observed in a reference frame moving with the constant acceleration –**g** in the absence of external gravitation. When the observed phenomena are mechanical interactions obeying Newton's laws of motion, this equivalence principle is just a rephrasing of Bélanger's accelerative relativity. Einstein's decisive originality was to extend this relativity to any kind of phenomenon. This move reminds us of his turning the constancy of the velocity of light, a mere consequence of propagation through a stationary ether in Lorentz's theory, into a principle concerning any physical phenomenon independently of any sustaining picture.[32]

The analogy with special relativity goes on when Einstein remarks that the gravitational field, like the electromagnetic field, is a frame-dependent concept. In both cases, the frame-dependence of fields is what allows the laws of physics to retain the same expression in different frames. In the first case, the frames are restricted to be inertial; in the second case, accelerated frames become equally legitimate. Einstein thus uses the equivalence principle as a springboard to a generalized relativity principle. Long ago Newton had noticed that no experience could distinguish between an inertial frame and a rectilinearly accelerated

[31] See Sections 1.1 and 1.4.

[32] See Section 9.1.

frame (to put it in modern terms) because one could always imagine a distant massive body responsible for the common acceleration of free bodies in the latter frame. Newton judged the appeal to an invisible body "precarious" and he rather required the center of mass of the observable universe to be acceleration free. In contrast, Einstein applied the equivalence principle locally, in any region of space small enough for the gravitational field to be approximately uniform, and without considering the distribution of masses outside this region. Although this local character of the equivalence principle is only implicit in Einstein's considerations of 1907, it should be regarded as an essential feature, preparing for his later relying on tangent frames to apply the principle in nonuniform fields.

In special relativity, a simple way to address gravitation is to modify Newton's force law so as to make it Lorentz covariant, as Poincaré and Minkowski did. Another simple way is to make the gravitational field a Lorentz scalar satisfying a d'Alembertian equation, as Nordström did in 1912. Einstein probably tried the latter option before 1907, but he did not retain it because he then believed it to be incompatible with the universality of free fall: it seemed to make the speed of fall of a body depend on its internal kinetic energy. This is a fortunate error, because it allowed Einstein to take the equivalence principle as a fresh starting point. Through ingenious thought-experiments in accelerated frames, he found that the observed frequency and the propagation velocity of the light from a distant source both depended on the local value of the gravitational potential. The first effect is the gravitational redshift; the second implies the deflection of light rays by a large mass.

The metric manifold

In 1912, Einstein obtained a theory of static gravitational fields by applying the equivalence principle locally in a generalized Minkowski space for which the interval between two neighboring events takes the form $d\mathbf{r}^2 - \widehat{c}^{\,2}(\mathbf{r})dt^2$, where $\widehat{c}(\mathbf{r})$ represents a variable velocity of light. In a geometrical language that Einstein did not yet use, the paths of free-falling particles are the geodesics of the associated metric manifold, so that $\widehat{c}(\mathbf{r})$ plays the role of the gravitational potential. Einstein then got the gravitational field equation by requiring asymptotic correspondence with Poisson's equation and compatibility with the vanishing of the sum of gravitational forces (equality of action and reaction).[33]

In the following years, Einstein tried to extend this theory to any gravitational field by means of the more general form $g_{\mu\nu}dx^\mu dx^\nu$ for the interval between two elements. He regarded this form and the concomitant freedom in the choice of coordinates as a consequence of the local freedom to use accelerated frames, freedom implied in the equivalence principle. In order to represent matter in a continuous manner, he built a stress–energy (or energy–momentum) tensor à la Laue for a dust of independent particles. If the particles of the dust move along the geodesics of the metric manifold, then the covariant four-divergence of the energy–momentum tensor of the dust vanishes. Being aware of the covariance of the geodesic and stress–energy relations, Einstein wanted the gravitational field equation to be as covariant as possible. He also required this equation to be asymptotically compatible with the d'Alembertian equation (correspondence principle), with the

[33] See Section 9.2.

static theory of 1912, and with the existence of a stress–energy (pseudo-) tensor for the gravitational field (stress principle). Inspired by analogy with the electromagnetic field, this last requirement played an increasingly important role in Einstein's quest.[34]

After a few naive attempts, Einstein became aware of Ricci's "absolute differential calculus," an invariant-theoretical approach to tensor calculus on Riemannian manifolds. In this framework, the obvious way to obtain a covariant gravitational field equation is to equate the first contraction of the Riemann curvature tensor, the so-called Ricci tensor, to the stress–energy tensor of matter. Although Einstein was willing to consider less covariant modifications of the Ricci tensor, he failed to meet his other heuristic requirements, especially the stress principle.

The Entwurf theory

This is why in 1913 Einstein gave up on the Riemann tensor and instead combined the stress principle and the correspondence principle to produce a field equation of very limited covariance, the so-called *Entwurf* equation. He later found reasons for this limitation in the noncovariance of energy–momentum conservation and in the so-called hole argument; namely, a generally covariant field equation determines the field under given boundary conditions only up to four arbitrary functions of the coordinates, in seeming contradiction with a Cauchy kind of determinism. In November 1914, Einstein thought he could obtain the *Entwurf* equation without the correspondence principle, by making it derive from a reasonably simple field action and by requiring the compatibility of the field equation with energy–momentum conservation.[35]

The final theory

As Einstein realized almost a year later, this compatibility automatically results from the transformation properties of the action, whatever the choice of the action (this is a special case of Noether's theorem). In the meantime, he had learned that for any choice of the field action, there existed an energy–stress tensor for the gravitational field, and he easily saw that a variant of the action for the *Entwurf* theory led to one of the Ricci tensor-based options he had briefly considered and rejected three years earlier. He now realized that this option was after all compatible with the stress principle, and that all his older objections were unjustified. In November 1915, he published the resulting unimodularly covariant equation, showed that a center-symmetric solution of this equation led to the desired relativistic correction to the advance of Mercury's perihelion, gave the now correct expression of the gravitational deflection of light, and lastly wrote the final field equation for which the unimodular restriction is no longer needed.[36]

To summarize, the equivalence principle, understood as a way to locally eliminate the gravitational field in a free-falling frame, and the assumed validity of special relativity in

---

[34] See Section 9.3.

[35] See Sections 9.4, 9.6, and 9.7.

[36] See Section 9.8.

such local frames, led Einstein to the basic metric manifold of general relativity, to the motion of free particles along the geodesics of this manifold, and to representing the source of the gravitational field equation by a divergenceless energy–momentum tensor. The perfect covariance of all these notions formalized the ultimate generalization of the relativity principle as a total freedom in the choice of reference frames in every neighborhood of a continuous manifold. Einstein first looked for a generally covariant field equation, but renounced this when he judged the Ricci tensor-based equations incompatible with his stress principle. For three years he lived with the very limited covariance of the *Entwurf* theory, hoping that the leftover covariance still allowed for the accelerated (rotating) frames of the equivalence principle. After vainly trying to consolidate this theory through reasoning based on an action for the gravitational field, he discovered that the Ricci-based option was after all compatible with the stress principle and quickly removed all remaining difficulties to return to general covariance and general relativity.

Three principles conditioned Einstein's constructive enterprise: the equivalence principle, the stress principle, and the principle of least action. Unfortunately, Einstein did not rely on a field action until Paul Bernays advised him to do so in 1914. He would otherwise have seen that the equivalence principle and the stress principle are easily reconciled by deriving the gravitational field equation from an invariant action. This is indeed the way Hilbert took in the fall of 1915, with important insights into the relation between invariance and conservation laws. Had Einstein thus obtained the correct field equation at an earlier stage of his research, he would still have had to understand the not-so-obvious Newtonian limit of the theory and to solve the fairly complicated equations governing crucial departures from this limit. But the tedious detour through the *Entwurf* theory would have been avoided, and relativity might have sooner become general.[37]

The meaning of coordinates

As was mentioned, Einstein's tolerance for the very limited covariance of the *Entwurf* field equation partly resulted from the hole argument, in which he mistakenly regarded two diffeomorphically equivalent choices of the metric coefficients as physically different. As he explained to friends in late 1915, the error derived from his unconsciously associating the coordinates with a predefined concrete reference, independently of the metric. When Einstein wrapped up his theory in the *Grundlage* of 1916, he insisted that the coordinates were nothing but a means to decide about the coincidence of concrete point-like events. Yet, his derivations of physical consequences of the theory still seemed to rely on a metric interpretation of the coordinates. In particular, in his derivation of the gravitational redshift he perplexed his readers by making the differential time interval $dx^0 = g_{00}^{-1/2}ds$ the measure of the period of a clock immersed in a gravitational field. Probably, he was unconsciously or at least silently relying on his old definition of the time x^0 through optical synchronization in 1907. As Eddington explained, what makes this coordinate time special in a static metric is that the frequency of electromagnetic waves, when measured with respect to it, is conserved as the wave travels through the gravitational field. In the end,

---

[37] See Section 9.8.

what we truly measure is an intrinsic quantity: the value of the proper time ds associated on earth with the proper period of a spectral source on a star's surface. Although the co-ordinates have no a priori physical significance beyond the tracking of coincidences, for a specific choice of the metric form they acquire indirect meaning since their increments are related to the interval ds in a well-defined manner.[38]

Einstein's derivation of the gravitational deflection of light rays and his derivation of the relativistic precession of planets suffer from a similar defect: in interpreting the results of calculation, he implicitly admits that the coordinates of his center-symmetric solution of the field equation preserve their usual astronomical meaning, as if space had remained flat. As Eddington and Beck later explained, this is the case in a sufficient approximation for the specific center-symmetric solution used by Einstein, and not for the alternative center-symmetric solutions Painlevé and Gullstrand used to discredit Einstein's prediction. As Einstein explained to his French critics, all we can measure is relations between intrinsic quantities defined through the values of the invariant ds^2. A coordinate-dependent pre-diction acquires physical meaning only after, for a given solution of the field equations, the coordinate values are related to measurable quantities computed through the metric formula.[39]

Relativities of motion

Behind the practical difficulty in obtaining correct physical predictions in reasoning em-ploying coordinates, there is a fundamental difficulty in defining place and motion in general relativity. For Aristotle and Descartes, place and motion are defined with respect to the immediate surroundings of an object; for Galileo and for many practical thinkers, with respect to an important rigid body (the earth, the sun, a boat, etc.); for Newton and all his successors before general relativity, place is defined with respect to an imaginary rigid frame (a permeable solid or a system of axes) of indefinite extension. If, as the inventors of non-Euclidean geometry imagined, space were a curved manifold preserving its curva-ture for all times, the latter definition of place would still be possible. This is no longer an option in general relativity, for at least three reasons: there is generally no choice of coordi-nates though which the metric is globally Minkowski's, there generally cannot be a uniform choice of the time axis through the manifold, and the spatial geometry generally varies in time. The idea of a global, permanent, rigid reference frame thoroughly collapses. All we can imagine is Einstein's reference "mollusk," namely, a curvilinear three-dimensional lat-tice of pre-clocks serving to record an event through the lattice coordinates and the time of the closest pre-clock. There still are intrinsic ways to locate an event, for instance through Synge's radar coordinates, in analogy with the use of central geodesic coordinates to situ-ate a point on a curved manifold. But there is no permanent structure associated with this determination.[40]

[38] See Sections 10.2 and 10.3.

[39] See Sections 10.4 and 10.5.

[40] Synge 1921.

Owing to this collapse of the received concept of reference frame, it is sometimes argued that general relativity should not be called a theory of relativity: motion is not relative to anything well defined in general relativity. Einstein did not see it that way. In his view, we first have Lorentz's stationary ether theory, in which there is a privileged ether frame. Then we have special relativity, in which there still is a privileged class of reference frames and relativity is confined to changes of frames within the inertial class. General relativity abolishes this privilege in two steps. First, it legitimates accelerated frames and rotating frames through the equivalence principle. Second, it confines the rigid frames to small space–time neighborhoods and denies the existence of extended rigid frames. This second step does not eliminate the need for a reference: we still need coordinates to define the manifold of events, and these coordinates may still be attached to a concrete reference system (*Bezugsystem*), for instance the reference mollusk. This view implies a general relativity principle according to which all imaginable references of motion are fully equivalent. The formal counterpart of this physical principle is the demand for general covariance of the fundamental equations of the theory, just as the formal counterpart of the restricted relativity principle is the demand for Lorentz covariance.

Although no one would question the heuristic value of the equivalence principle and of general covariance, much of Einstein's take on the relativity principle now seems obsolete. In today's general relativity, the equivalence principle is expressed through the geodesic principle and through the condition that the equations for the nongravitational fields should be simple covariant generalizations of their Minkowski version. Recourse to accelerated frames to derive gravitational time dilation is still popular, but only in elementary or propaedeutic accounts. The common wisdom is that rigid frames should better be avoided in general relativity, even in local reasoning. Provided rigid frames can be properly defined for a limited local purpose (they can), a change of frame cannot in itself turn a flat space–time into the curved space–time of general relativity, since curvature is an intrinsic property. Moreover, there is a sense in which inertial frames still play a privileged role in general relativity: they are the local free-falling frames to which the Minkowski structure applies. Then why did Einstein retain his original view of the equivalence principle to the end of his life? Part of the answer may be psychological: he could not simply forget "the happiest thought in [his] life." Also, accelerated frames are realities of common life, since an earth-based laboratory is an accelerated frame with respect to the free-falling inertial frames. Reasoning based on them may now seem unnecessary, but it can be shown to be compatible with modern general relativity.[41]

Another questionable feature of Einstein's relativity principle is the analogy between the constructive roles of the special and general versions of this principle. Although the reality of these roles is undeniable, their similarity is questionable. In special relativity, the selection of a special class of frames matters at least as much as the equivalence of the frames within this class. In general relativity, there is no such selection (beyond differentiability). Whereas the special principle is associated with a universal and immutable metric structure, the general principle forbids any permanent metric structure beyond the structure of differential manifold. This is the so-called background independence. Whereas Lorentz

---

[41] On this last point, cf. Muñoz and Jones 2010.

covariance strongly structures the theories that conform to it, the requirement for general covariance is nearly vacuous: as Kretschmann noted, any field theory can be put into covariant form. Fortunately, Einstein's general principle of relativity contains more than general covariance: it forbids the existence of coordinate systems in which the fundamental equations of the theory take an especially simple form for all the solutions. Background independence, not general covariance per se, is the key to heuristic success: general covariance must be obtained with a limited number of mutually coupled fields, without any background-defining field.[42]

Mach's principle

Historically, for instance in Descartes's philosophy, in Mach's criticism of mechanics, or in the optics and electrodynamics of moving bodies, the laws of physics were sometimes required to depend only on the motion of matter with respect to matter. This strong kind of relativity includes Mach's principle, according to which inertial behavior should be traced to the distribution of matter in the universe and not to any mysterious space structure. Newtonian mechanics and special relativity contradict this version of the relativity principle, for they admit a background space–time structure: absolute space and time for Newton, a Galilean space–time for Streintz, and the Minkowski structure for special relativity. In general relativity, the geodesic motion of particles depends on the metric manifold, the structure of which in turn depends on the distribution of matter. The metric manifold of general relativity plays a role similar to Newton's absolute space and time or to Minkowski space–time in causing inertial motion. The main difference is that the causally effective structure is not fixed; it has its own dynamics partially driven by the distribution of matter.

In 1916, Einstein excluded a causal role of immaterial entities in an argument for general relativity: the cause of the oblate form of a liquid mass in empty space cannot be its rotation with respect to space because space is not a material entity; the true cause should be distant masses and all reference spaces should therefore be equivalent. In 1920, he expressed a different view when Lorentz invited him to speak on the ether in relativity theory. He now argued that distant masses could not be the direct cause of inertial behavior since this would imply a kind of action at a distance. In order to satisfy Mach's principle, one had to imagine a quasi-material entity, the ether, as the direct cause of inertial behavior—an option considered by Mach himself. Surely, Einstein went on, the ether could no longer be the mechanical construct once imagined by Maxwell. Most of its mechanical attributes had to be stripped off. Lorentz had taken a first important step by excluding bulk motion of the ether. As Einstein put it humorously, immobility was the only mechanical property Lorentz had left to the ether. Special relativity robbed the ether of this last attribute, but general relativity resurrected the ether by conveying physical properties to space (space–time):

> According to the general theory of relativity, space is endowed with physical qualities. In this sense, there exists an ether. According to the general theory of relativity, a space without ether is unthinkable; for in such space there not only would be no propagation of light, but also no possibility of existence for measuring-rods and

---

[42]See Section 10.2.

clocks, nor therefore any space-time intervals in the physical sense. But this ether cannot be thought of as endowed with the characteristic property of ponderable media, as consisting of parts that can be tracked through time. The idea of motion cannot be applied to it.

Einstein still believed that distant masses should be the *indirect* cause of inertial behavior. This is why in his early relativistic cosmology he wanted the world to be closed and finite. Eddington held a different view: in his mind matter and electricity were only "incidental points of complexity" in the more primitive "geodesic structure" just as for Faraday and Maxwell there were only field-based constructs. Then there was no reason to require the geodesic structure or Einstein's ether to be completely determined by the distribution of matter.[43]

From Descartes to Einstein

We began with Descartes's fully relativist concept of motion, there being only motion of matter with respect to matter in his world. We then saw how Galileo related his concept of inertial motion to the observed absence of effects of a uniform motion of the reference body. Newton traced rectilinear inertia to absolute space and time, and yet accepted Galilean relativity as a consequence of his laws of motion. In optics and electrodynamics, the immobile ether of Fresnel and Lorentz came to reify Newton's absolute space, with expected infractions to the Galilean relativity of physical phenomena, and yet all attempts to detect these infractions failed. Poincaré made this failure a principle, and used it to construct a new relativistic dynamics. Somewhat like the few critics of Newton's mechanics who replaced absolute space with an operationally defined class of inertial frames, Einstein and Minkowski replaced Lorentz's ether with a Lorentz-invariant inertial structure, thus avoiding an unnecessary asymmetry in the theoretical representation of phenomena. As Newton already understood, Newton's laws of motion apply not only in inertial frames, but also in frames freely falling in the field of a remote massive body. Like Bélanger, Einstein turned this theorem into a principle. Moreover, he extended it to any phenomenon occurring in a portion of space–time in which the gravitational field is approximately uniform. He thus made rectilinear inertia a local, gravity-dependent property in a space–time manifold, and turned gravitation into inertial motion in a curved space–time. Galilean relativity survived in the selection of local inertial frames, general relativity made its entrance for the global coordinate systems. Space–time became a fully physical entity, not only equipped with the causal powers of Newton's space and time, but also having its own dynamics, like the curved space imagined by Riemann and Clifford. In the end, the reference of motion was neither matter (as Descartes would have it) nor a rigid imaginary frame (as Newton would have it); it was a physical being of intermediate kind, Einstein's neo-ether or Eddington's geodesic structure.

[43] Einstein 1920, p. 15; Eddington 1920b, p. 156. Einstein's talk was related to the Leiden appointment arranged by Lorentz; cf. van Dongen 2012. On the argument of 1916, see Section 10.2. Einstein first hinted at a new ether concept in a letter to Lorentz of June 17, 1916. For the history of this concept, cf. Kostro 1992

ABBREVIATIONS

ACP	*Annales de chimie et de physique*
AHES	*Archive for the history of exact sciences*
AJP	*American journal of physics*
AMAF	*Arkiv för matematik, atronomi och fysik*
AN	*Archives néerlandaises des sciences exactes et naturelles*
AP	*Annalen der Physik*
BAR	British Association for the Advancement of Science, *annual report*
BB	Königlich Preußische Akademie der Wissenschaften zu Berlin, mathematisch-physikalische Klasse, *Sitzungsberichte*
BJHS	*The British journal for the history of science*
CR	Académie des sciences, *Comptes-rendus hebdomadaires des séances*
ECP	*The collected papers of Albert Einstein* (Princeton: Princeton University Press)
EE	*L'éclairage électrique*
FER	Michael Faraday, *Experimental researches in electricity*. 3 vols. (1839, 1844, 1855). London: Taylor
GGR	*The genesis of general relativity*, 4 vols. (Dordrecht: Springer, 2007), ed. by Jürgen Renn (general ed.), Michel Janssen, John Norton, Tilman Sauer, Matthias Schemmel, and John Stachel
GN	Königliche Gesellschaft der Wissenschaften und der Georg August Universität zu Göttingen, *Nachrichten*
HEP	Oliver Heaviside, *Electrical papers*. 2 vols. (1892). London: Chelsea
HSPS	*Historical studies in the physical (and biological) sciences*
HWA	Hermann von Helmholtz, *Wissenschaftliche Ahhandlungen*. 3 vols. (1882, 1883, 1895). Leipzig: Barth
JP	*Journal de physique*
JRAM	*Journal für die reine und angewandte Mathematik*
LCP	Hendrik Antoon Lorentz, *Collected papers*. 9 vols. The Hague: Nijhoff (1934–1939)
LMPP	Joseph Larmor, *Mathematical and physical papers*. 2 vols. (1929). Cambridge: Cambridge University Press
MAS	Académie Royale des Sciences, *Mémoires*
PM	*The philosophical magazine*
PRA	Royal Academy of Amsterdam, *Proceedings*
PRS	Royal Society of London, *Proceedings*
PT	Royal Society of London, *Philosophical transactions*
PZ	*Physikalische Zeitschrift*
RAL	Atti della Reale Accademia (Nazionale) dei Lincei (Classe di scienze fisiche, matematiche e naturali), *Rendiconti*
RCMP	*Rendiconti del Circolo Matematico di Palermo*
RGSPA	*Revue générale des sciences pures et appliquées*
RHS	*Revues d'histoire des sciences*
SEP	*The Stanford encyclopedia of philosophy*, ed. by Edward Zalta (https://plato.stanford.edu)
SHPMP	*Studies in history and philosophy of modern physics*
SHPS	*Studies in history and philosophy of science*
VDNA	Gesellschaft Deutscher Naturforscher und Ärzte, *Verhandlungen*
VKA	Koninklijke Akademie van Wetenschappen, Amsterdam, *Verslagen*

REFERENCES

Abardia, Judit, Agustí Reventós, and Carlos Rodríguez
2012. What did Gauss read in the Appendix? *Historia mathematica*, 39: 292–323.

Abraham, Max
1902a. Dynamik des Elektrons. *GN* (1902): 20–41.
1902b. Prinzipien der Dynamik des Elektrons. *PZ*, 4: 57–62.
1903. Prinzipien der Dynamik des Elektrons. *AP*, 10: 105–179.
1904. Die Grundhypothesen der Elektronentheorie. *PZ*, 5: 576–579.
1905. *Theorie der Elektrizität*, Vol. 2: *Elektromagnetische Theorie der Strahlung*. Leipzig: Teubner.
1908. *Theorie der Elektrizität*, 2nd ed. Vol. 2: *Elektromagnetische Theorie der Strahlung*. Leipzig: Teubner.
1909a. Zur Elektrodynamik bewegter Körper. *RCMP*, 28: 1–28.
1909b. Zur elektromagnetischen Mechanik. *PZ*, 10: 737–741.
1910a. Die Bewegungsgleichungen eines Massenteilchens in der Relativtheorie. *PZ*, 11: 527–531.
1910b. Sull'elettrodinamica di Minkowski. *RCMP*, 30: 33–46.
1911. Sulla teoria della gravitazione. *RAL*, 20: 678–682.
1912a. Zur Theorie der Gravitation. *PZ*, 23: 1–4.
1912b. Berichtigung. *PZ*, 23: 176.
1912c. Sulla conservazione dell' energia e della materia nel campo gravitazionale. *RAL*, 21: 432–437.
1912d. Die Erhaltung der Energie und der Materie im Schwerkraftfeld. *PZ*, 13: 311–314.

Acloque, Paul
1991. L'aberration stellaire : un mirage qui a destitué l'éther. *Cahiers d'histoire et de philosophie des sciences*, 36: 1–258.

Adam, Charles, and Paul Tannery
1897. *Œuvres de Descartes*, Vol. 1. Paris: Cerf.

Airy, George Biddell
1871. On a supposed alteration in the amount of astronomical aberration of light, produced by the passage of the light through a considerable thickness of refracting medium. *PRS*, 20: 35–39.

Alexander, Henry Gavin (ed.)
1956. *The Leibniz–Clarke correspondence*. New York: Philosophical Library.

Ampère, André Marie
1820a. Analyse des mémoires lus par M. Ampère à l'Académie des sciences, dans les séances du 18 et 25 septembre, des 9 et 30 octobre 1820. *Annales générales des sciences physiques*, 6: 238–257.
1820b. Sur l'action mutuelle entre deux courans électriques, entre un courant électrique et un aimant ou le globe terrestre, et entre deux aimans. *ACP*, 15: 59–76, 170–208.
1820c. Note sur un mémoire lu à l'Académie Royale des Sciences, dans la séance du 4 décembre 1820. *JP*, 91: 226–230.
1820d. Note sur deux mémoires lus le 26 décembre 1820 et le deuxième les 8 et 15 janvier 1821. *JP*, 92: 160–165.

Ångström, Anders Jonas
1864. Neue Bestimmung der Länge der Lichtwellen, nebst einer Methode, auf optischem Wege die fortschreitende Bewegung des Sonnensystems zu bestimmen. *AP*, 123: 489–505.

Antonello, Elio
2014. Water-filled telescopes. arXiv:1401.5585v1

Appell, Paul
1893. *Traité de mécanique rationnelle*. Paris: Gauthier-Villars.
1902. *Traité de mécanique rationnelle*, 2nd ed. Paris: Gauthier-Villars.

Arabatzis, Theodore
2006. *Representing electrons: A biographical approach to theoretical entities*. Chicago: University of Chicago Press.

Arago, François
1838. Sur un système d'expériences à l'aide duquel la théorie de l'émission et celle des ondes seront soumises à des épreuves décisives. *CR*, 7: 954–965.
1839. Sur un système d'expériences à l'aide duquel la théorie de l'émission et celle des ondes seront soumises à des épreuves décisives. *ACP*, 71: 49–65.
1853. [read on December 10, 1810] Mémoire sur la vitesse de la lumière. *CR*, 36: 38–49.

Arthur, Richard
2007. Beeckman, Descartes and the force of motion. *Journal of the history of philosophy*, 45: 1–28.

Babinet, Jacques
1839. Sur l'aberration de la lumière. *CR*, 9: 774.
1862. De l'influence du mouvement de la Terre dans les phénomènes optiques. *CR*, 55: 561–564.

Bacelar Valente, Mario
2018. Einstein's redshift derivations: Its history from 1907 to 1921. *Circumscribere*, 22: 1–16.

Bain, Jonathan
1998. Whitehead's theory of gravity. *SHPMP*, 29: 547–574.

Barbour, Julian
2001. *The discovery of dynamics: A study from a Machian point of view of the discovery and the structure of dynamical theories*. Oxford: Oxford University Press.

Barbour, Julian, and Herbert Pfister (eds.)
1995. *Mach's principle: From Newton's bucket to quantum gravity*. Boston: Birkhäuser.

Battimelli, Giovanni
1981. The electromagnetic mass of the electron: A case study of a non-crucial experiment. *Fundamenta scientiae*, 2: 137–150.

Beck, Guido
1929. Allgemeine Relativitätstheorie. In Hans Geiger and Karl Scheel (eds.), *Handbuch der Physik* (Berlin: Springer), Vol. 4, pp. 298–407.

Becker, Barbara
2011. *Unravelling starlight: William and Margaret Huggins and the rise of the new astronomy*. Cambridge: Cambridge University Press.

Becquerel, Jean
1922. *Le principe de la relativité et la théorie de la gravitation*. Paris: Gauthier-Villars.
1923. *Gravitation einsteinienne : champ de gravitation d'une sphère matérielle et signification physique de la formule de Schwarzschild*. Paris: Hermann.

Beeckman, Isaac
1939–1953. *Journal tenu par Isaac Beeckman de 1604 à 1634, publié avec une introduction et des notes*, ed. by Cornelis de Waard. 4 vols. La Haye: Nijhoff.

Beichler, James
1988. Ether/or: Hyperspace models of the ether in America. In Goldberg and Stuewer 1988, pp. 206–233.

Bélanger, Jean-Baptiste
1847. *Cours de mécanique*. Paris: Carilian Goeury.
1848. *Lehrbuch der Mechanik und ihrer Anwendungen auf das Ingenieurwesen*. Ludwigsburg: Neubert.

Beltrami, Eugenio
1868a. Saggio di interpretazione della geometria noneuclidea. *Giornale di matematiche*, 6: 284–312.
1868b. Teoria fondamentale degli spazi di curvatura costante. *Annali di matematica pura ed applicata*, 2: 232–255.
1869. Sulla teoria generale dei parametri differenziali. *Memorie dell' Academia Scientifica dell'Istituto di Bologna*, 8: 581–590.

Ben-Menahem, Yemima
2001. Convention: Poincaré and some of his critics. *The British journal for the philosophy of science*, 52: 471–513.

Bernoulli, Daniel
1726. Examen principiorum mechanicae, et demonstrationes geometricae de compositione et resolutione virum. *Commentarii Academiae scientiarum imperialis Petropolitanae*, 1: 126–142.

Bernstein, Howard
1981. Passivity and inertia in Leibniz's dynamics. *Studia Leibnitiana*, 13: 97–113.

Bertoloni Meli, Domenico
1993. The emergence of reference frames and the transformation of mechanics in the Enlightenment. *HSPS*, 23: 301–335.
2006. *Thinking with objects: The transformation of mechanics in the seventeenth century*. Baltimore: Johns Hopkins University Press.

Biezunski, Michel
1981. *La diffusion de la théorie de la relativité en France*. PhD dissertation, Université de Paris VII.
1991. *Einstein à Paris: Le temps n'est plus . . .* Saint-Denis: Presses Universitaires de Vincennes.

Blackmore, John
1972. *Ernst Mach: His work, life, and influence*. Berkeley: University of California Press.

Blair, Robert
[1786]. A proposal for ascertaining by experiments whether the velocity of light be affected by the motion of the body from which it is emitted or reflected; and for applying instruments for deciding the question to several optical and astronomical enquiries. Royal Society of London, Manuscript L&P, VIII.182.

Blondel, Christine
1982. *Ampère et la création de l'électrodynamique, 1820–1827*. Paris: Bibliothèque Nationale.

Blondlot, René
1901. Exposé des principes de la mécanique. *Bibliothèque du Congrès international de philosophie* (Paris: Armand Colin), Vol. 3, pp. 445–455.

Bobis, Laurence, and James Lequeux
2008. Cassini, Rømer, and the velocity of light. *Journal of astronomical history and heritage*, 11: 97–105.

Boltzmann, Ludwig
1897. *Vorlesungen über die Principe der Mechanik*, Vol. 1. Leipzig: Barth.
1904. *Vorlesungen über die Principe der Mechanik*, Vol. 2. Leipzig: Barth.

Bolyai, János Farkas
1832. Appendix: Scientiam spatii absolute veram exhibens. In Wolfgang Bolyai, *Tentamen juventutem studiosam in elementa matheosis purae* (Maros Vásárhelyini: Kali), Vol. 1, appendix (26 pages).

Bonnet, Ossian
1858. *Leçons de mécanique élémentaire*. Paris: Mallet-Bachelier.

Bork, Alfred
1964. The fourth dimension in nineteenth-century physics. *Isis*, 55: 326–338.

Born, Max
1909a. Die Theorie des starren Elektrons in der Kinematik des Relativitätsprinzips. *AP*, 30: 1–56.
1909b. Über die Dynamik des Elektrons in der Kinematik des Relativitätsprinzips. *PZ*, 10: 814–817.
1910a. Über die Definition des starren Körpers in der Kinematik des Relativitätsprinzips. *PZ*, 11: 233–234.
1910b. Zur Kinematik des starren Körpers im System des Relativitätsprinzips. *GN* (1910), 161–179.

Bošković, Ruđer Josip
1748a. *Dissertationis de lumine pars prima*. Rome: Antonius de Rubeis.
1785. De effectu aberrationis luminis respectu objectorum terrestrium. In *Opera pertinentia ad opticam et astronomiam maxima ex parte nova et omnia hucusque inedita in quinque tomos distributa* (Bassano: Prostant Venetiis apud Remondini), Vol. 2, pp. 297–314.

Bottazini, Umberto
1999. *Ricci and Levi-Civita: From differential invariants to general relativity*. In Gray 1999, pp. 241–259.

Bouasse, Henri
1895. *Introduction à l'étude des théories de la mécanique*. Paris: Carré.

Boussinesq, Joseph
1868. [read August 5 1867] Théorie nouvelle des ondes lumineuses. *Journal de mathématiques pures et appliquées*, 13: 313–339, 425–438.
1873. Exposé synthétique des principes d'une nouvelle théorie des ondes lumineuses. *ACP*, 30: 539–565.

Brace, DeWitt Bristol
1904. On double refraction in matter moving through the aether. *PM*, 7: 317–329.
1905. A repetition of Fizeau's experiment on the change produced by the earth's motion in the rotation of a refracted ray. *PM*, 10: 591–599.

Bracco, Christian
2017. *Quand Albert devient Einstein*. Paris: CNRS éditions.

Bracco, Christain, and Jean-Pierre Provost
2009. De l'électromagnétisme à la mécanique: le rôle de l'action dans le mémoire de Poincaré de 1905. *RHS*, 62: 457–493.
2013. Les points de vue de Poincaré sur la «mécanique nouvelle» et leurs rapports à l'enseignement et à sa pratique scientifique. *RHS*, 66: 137–165.

Brading, Katherine
2005. A note on general relativity, energy conservation, and Noether's theorems. In Kox and Eisensteadt 2005, pp. 125–135.

Bradley, James
1728. A letter from the Reverend Mr. James Bradley Savilian Professor of Astronomy at Oxford, and F. R. S. to Dr. Edmond Halley Astronom. Reg. &c. giving an account of a new discovered motion of the fix'd stars. *PT*, 35: 637–661.

Brown, Harvey
2005. *Physical relativity: Space-time structure from a dynamical perspective*. Oxford: Oxford University Press.

Brush, Stephen
1999. Why was relativity accepted? *Physics in perspective*, 1: 184–214.

Bucherer, Alfred
1903. Über den Einfluss der Erdbewegung auf die Intensität des Lichtes. *AP*, 11: 270–283.
1904. *Mathematische Einführung in die Elektronentheorie*. Leipzig: Teubner.
1905. Das deformierte Elektron und die Theorie des Elektromagnetismus. *PZ*, 6: 833–834.
1906. Ein Versuch, den Elektromagnetismus auf Grund der Relativbewegung darzustellen. *PZ*, 7: 553–557.
1907. On a new principle of relativity in electromagnetism. *PM*, 13: 413–420.
1908a. On the principle of relativity and on the electromagnetic mass of the electron. A reply to Mr. Cunningham. *PM*, 15: 316–318.
1908b. Messungen an Bequerelstrahlen. Die experimentelle Bestätigung der Lorentz-Einsteinschen Theorie. *PZ*, 9: 755–762.

Buchwald, Jed
1985. *From Maxwell to microphysics: Aspects of electromagnetic theory in the last quarter of the nineteenth century*. Chicago: University of Chicago Press.
1988. The Michelson experiment in the light of electromagnetic theory before 1900. In Goldberg and Stuewer 1988, pp. 55–70.
1994. *The creation of scientific effects: Heinrich Hertz and electric waves*. Chicago: University of Chicago Press.

Buchwald, Jed, and Robert Fox (eds.)
2013. *The Oxford handbook in the history of physics*. Oxford: Oxford University Press.

Budde, Emil
1890–1891 *Allgemeine Mechanik der Punkte und starren Systeme: ein Lehrbuch für Hochschulen*. 2 vols. Berlin: Reimer.

Bühler, Walter Kaufmann
1981. *Gauss: A biographical study*. New York: Springer.

Buijs Ballot, Christoph Hendrik Diederik
1845. Akustische Versuche auf der Niederländischen Eisenbahn nebst gelegentlichen Bemerkungen zur Theorie des Hrn. Prof. Doppler. *AP*, 66: 321–351.

Burali-Forti, Cesare
1923. Flessione dei raggi luminosi stellari e spostamento secolare del perielio di Mercurio. Accademia delle scienze di Torino, classe di scienze fisiche matematiche e naturali, *Atti*, 58: 149–151.

Burau, Werner
1981. Christoffel und die Invariantentheorie. In Butzer and Fehér 1981, pp. 518–525.

Burstyn, Harold
1965. The deflecting force of the earth's rotation from Galileo to Newton. *Annals of science*, 21: 47–80.

Butzer, Paul Leo
1981. An outline of the life and work of E. B. Christoffel (1829–1900). *Historia mathematica*, 8: 243–276.

Butzer, Paul Leo, and Franziska Fehér (eds.)
1981. *E. B. Christoffel: The influence of his work on mathematics and the physical sciences*. Basel: Birkhäuser.

Cahan, David
2000. The young Einstein's physics education: H. F. Weber, Hermann von Helmholtz, and the Zurich Polytechnic Physics Institute. In Howard and Stachel 2000, pp. 43–82.

Cajori, Florian
1926. Origins of fourth dimension concepts. *The American mathematical monthly*, 33: 397–406.
1962. *Sir Isaac Newton's Mathematical Principles of Natural Philosophy and his System of the World*. 2 vols. Berkeley: University of California Press.

Calinon, Auguste
1885. Étude critique sur la mécanique. *Bulletin de la société des sciences de Nancy*, 7: 87–180.
1896. *Étude sur les diverses grandeurs en mathématique*. Paris: Gauthier-Villars.

Campbell, Norman Robert
1910. The aether. *PM*, 19: 181–191.
1911. The common sense of relativity. *PM*, 21: 502–517.
1913. *Modern electrical theory*, 2nd ed. Cambridge: Cambridge University Press.

Cantor, Geoffrey
1983. *Optics after Newton: Theories of light in Britain and Ireland, 1704–1840*. Manchester: Manchester University Press.

Capecchi, Danilo
2014. *The problem of the motion of bodies: A historical view of the development of classical mechanics*. Heidelberg: Springer.

Carazza, Bruno, and Helge Kragh
1990. Augusto Righi's magnetic rays: A failed research program in early 20th-century physics. *HSPS*, 21: 1–28.

Carrier, Martin
1994. Geometric facts and geometric theory: Helmholtz and 20th-century philosophy of physical geometry. In Lorenz Krüger (ed.), *Universalgenie Helmholtz: Rückblick nach 100 Jahren* (Berlin: Akademie Verlag), pp. 276–291.

Casimir, Hendrik
1983. *Haphazard reality: Half a century of science*. New York: Harper & Row.

Cassini, Alejandro, and Marcelo Leonardo Levinas
2019. Einstein's reinterpretation of the Fizeau experiment: How it turned out to be crucial for special relativity. *SHPMP*, 65: 55–72.

Cassini, Giovanni Domenico
1693. Les hypotheses et les tables des satellites de Jupiter, réformées par de nouvelles observations. In *Recueil d'observations faites en plusieurs voyages par ordre de Sa Majesté, pour perfectionner l'astronomie et la geographie* (Paris: Imprimerie Royale).

Cassini, Jacques
1738. Des variations que l'on observe dans la situation et dans le mouvement de diverses étoiles fixes. *MAS* (1738), 331–346.

Cassirer, Ernst
1921. *Zur Einstein'schen Relativitätstheorie. Erkenntnistheoretische Betrachtungen.* Berlin: Bruno Cassirer.

Cattani, Carlo, and Michelangelo De Maria
1989a. Max Abraham and the reception of relativity in Italy: His 1912 and 1914 controversies with Einstein. In Howard and Stachel 1989, pp. 160–174.
1989b. The 1915 epistolary controversy between Einstein and Tullio Levi-Civita. In Howard and Stachel 1989, pp. 175–200.
1993. Conservation laws and gravitational waves in general relativity. In Earman, Janssen, and Norton 1993, pp. 63–87.

Cauchy, Augustin-Louis
1839. Note sur l'égalité des réfractions de deux rayons lumineux qui émanent de deux étoiles situées dans deux portions opposées de l'écliptique. *CR*, 8: 327–329.

Challis, James
1845a. On the aberration of light. *BAR*, 1845: 9.
1845b. A theoretical explanation of the aberration of light. *PM*, 27: 321–327.
1846a. On the aberration of light, in reply to Mr. Stokes. *PM*, 28: 90–93.
1846b. On the principles to be applied in explaining the aberration of light. *PM*, 28: 176–177.
1846c. On the aberration of light. *PM*, 28: 335–336.
1846d. On the aberration of light, in reply to Mr. Stokes. *PM*, 28: 393–394.
1848. On the course of a ray of light from a celestial body to the earth's surface, according to the hypothesis of undulations. *PM*, 32: 168–170.

Chatzis, Konstantinos
1994. Mécanique rationnelle et mécanique des machines. In Bruno Belhoste, Amy Dahan-Dalmenico, and Antoine Picon (eds.), *La formation polytechnicienne, 1794–1994* (Paris: Dunod), pp. 95–108.
1995. Un aperçu de la discussion sur les principes de la mécanique rationnelle en France à la fin du siècle dernier. *Revue d'histoire des mathématiques*, 1: 235–270.

Chazy, Jean
1921. Sur les fonctions arbitraires figurant dans le ds^2 de la gravitation einsteinienne. *CR*, 173: 905–907.
1928–1930. *La théorie de la relativité et la mécanique céleste.* 2 vols. Paris: Gauthier-Villars.

Christoffel, Elwin Bruno
1868. Allgemeine Theorie der geodätischen Dreiecke. *Abhandlungen der Königlich Preußischen Akademie der Wissenschaften zu Berlin, mathematische Abhandlungen* (1868), 119–176.
1869. Über die Transformation der homogenen Differentialausdrücke zweiten Grades. *JRAM*, 70: 46–70.

Clairaut, Alexis
1737. [read December 11, 1737, pub. 1740] De l'aberration apparente des étoiles, causée par le mouvement progressif de la lumiere. *MAS* (1737), 205–227.

1739. [read July 24 1739, pub. 1741] Sur les explications cartésienne et newtonienne de la réfraction de la lumiere. *MAS* (1739), 259–275.

Clarke, Samuel (ed.)
1717. *A collection of papers, which passed between the late learned Mr. Leibnitz, and Dr. Clarke, in the years 1715 and 1716*. London: Knapton.

Clifford, William Kingdon
1876. On the space-theory of matter [abstract of a communication given on February 21, 1870]. *Proceedings of the Cambridge Philosophical Society*, 2: 157–158.
1885. *The common sense of the exact sciences* [posthumous ed. by Karl Pearson]. London: Kegan Paul, Trench, & Co.

Cohen, Hendrik
2015. *The rise of modern science explained: A comparative history*. Cambridge: Cambridge University Press.

Cohen, Robert, and Raymond Seeger (eds.)
1970. *Ernst Mach, physicist and philosopher*. Dordrecht: Reidel.

Cohn, Emil
1900a. *Das elektromagnetische Feld. Vorlesungen über die Maxwell'sche Theorie*. Leipzig: Hirzel.
1900b. Über die Gleichungen der Elektrodynamik für bewegte Körper. In *Recueil de travaux offerts par les auteurs à H. A. Lorentz à l'occasion du 25ème anniversaire de son doctorat le 11 décembre 1900. AN*, 5: 5: 516–523.
1902. Über die Gleichungen des elektromagnetischen Feldes für bewegte Körper. *AP*, 7: 29–56.
1904. Zur Elektrodynamik bewegter Systeme. *BB* (1904), 1294–1303, 1404–1416.

Coriolis, Gaspard
1829. *Du calcul de l'effet des machines*. Paris: Carilian Goeury.
1844. *Traité de la mécanique des corps solides et du calcul de l'effet des machines*. Paris: Fain et Thunot.

Cornu, Alfred
1875. *Cours de Physique, première division, 1874–1875*. Cours autographié. Paris: Ecole Polytechnique.

Corry, Leo
1998. The influence of David Hilbert and Hermann Minkowski on Einstein's views over the interrelation between physics and mathematics. *Endeavour*, 22: 95–97.
1999a. From Mie's electromagnetic theory of matter to Hilbert's unified foundations of physics. *SHPMP*, 30: 159–183.
1999b. David Hilbert between mechanical and electromagnetic reductionism (1910–1915). *AHES*, 53: 489–527.
2010. Hermann Minkowski, relativity and the axiomatic approach to physics. In Vesselin Petkov (ed.), *Minkowski spacetime: A hundred years later* (Dordrecht: Springer), pp. 3–41.

Corry, Leo, Jürgen Renn, and John Stachel
1997. Belated decision in Hilbert–Einstein priority dispute. *Science*, 278: 1270–1273.
[2004]. Response to F. Winterberg. Withdrawn from *Zeitschrift für Naturforschung*. Available (March 2021) at https://web.archive.org/web/20050313161944/http://www.mpiwg-berlin.mpg.de/texts/Winterberg-Antwort.html

Costabel, Pierre
1960. *Leibniz et la dynamique*. Paris: Hermann.

Crelinsten, Jeffrey
2006. *Einstein's jury: The race to test relativity*. Princeton: Princeton University Press.

Crowe, Michael
1967. *A history of vector analysis: The evolution of the idea of a vectorial system*. Notre Dame: University of
    Notre Dame Press.
1990. Duhem and history and philosophy of mathematics. *Synthese*, 83: 431–447.

Cunningham, Ebenezer
1907. On the electromagnetic mass of a moving electron. *PM*, 14: 538–547.
1908. Principle of relativity and electromagnetic mass of the electron. *PM*, 16: 423–428.
1910. The principle of relativity in electrodynamics and an extension thereof. *Proceedings of the London
    Mathematical Society*, 8: 77–98.
1912. The application of the mathematical theory of relativity to the electron theory of matter. *Proceedings
    of the London Mathematical Society*, 10: 116–127.
1914. *The principle of relativity*. Cambridge: Cambridge University Press.
1919. Einstein's relativity theory of gravitation. III. The crucial phenomena. *Nature*, 104: 394–395.

Cushing, James
1981. Electromagnetic mass, relativity, and the Kaufmann experiments. *AJP*, 49: 1133–1149.

Cuvaj, Camillo
1968. Henri Poincaré's contributions to relativity and the Poincaré stresses. *AJP*, 36: 1102–1113.
1970. *A history of relativity. The role of Henri Poincaré and Paul Langevin*. PhD dissertation, Yeshiva
    University.

Daguin, Pierre Adolphe
1861. *Traité élémentaire de physique théorique et expérimentale*, 2nd ed. Toulouse: Privat.

d'Alembert, Jean le Rond
1743. *Traité de dynamique, dans lequel les loix de l'équilibre et du mouvement des corps sont réduites au plus
    petit nombre possible, et démontrées d'une manière nouvelle, et où l'on donne un principe général pour
    trouver le mouvement de plusieurs corps qui agissent les uns sur les autres, d'une manière quelconque*.
    Paris: David.
1751. Aberration. In Denis Diderot and Jean le Rond d'Alembert (eds.), *Encyclopédie ou dictionnaire
    raisonné des sciences, des arts et des métiers* (Paris: Briasson, David, Le Breton, Durand), Vol. 1,
    pp. 23–25.
1754. Dimension. In Denis Diderot and Jean le Rond d'Alembert (eds.), *Encyclopédie, ou dictionnaire
    raisonné des sciences, des arts et des métiers* (Paris: Briasson, David, Le Breton, Durand), Vol. 4,
    pp. 1009–1010.
1758. *Traité de dynamique*, 2nd ed. Paris: David.

Damerow, Peter, Gideon Freudenthal, Peter McLaughlin, and Jürgen Renn
1992. Concept and inference: Descartes and Beeckman on the fall of bodies. In P. Damerow, G. Freudenthal,
    P. McLaughlin, and J. Renn, *Exploring the limits of preclassical mechanics* (New York: Springer), pp.
    8–67.

Damour, Thibault, and Gerhard Schäfer
1992. Levi-Civita and the general-relativistic problem of motion. In Eisenstaedt and Kox 1992, pp. 393–399.

Damour, Thibault
2005. *Si Einstein m'était conté*. Paris: Le cherche midi.
2017. Poincaré, the dynamics of the electron, and relativity. *Comptes rendus physique*, 18: 551–562.

Darrigol, Olivier

1993a. The electrodynamic revolution in Germany as documented by early German expositions of "Maxwell's theory." *AHES*, 45: 189–280.

1993b. The electrodynamics of moving bodies from Faraday to Hertz. *Centaurus*, 36: 245–260.

1994. The electron theories of Larmor and Lorentz: A comparative study. *HSPS*, 24: 265–336.

1995a. Henri Poincaré's criticism of *fin de siècle* electrodynamics. *SHPMP*, 26: 1–44.

1995b. Emil Cohn's electrodynamics of moving bodies. *AJP*, 63: 908–915.

1996. The electrodynamic origins of relativity theory. *HSPS*, 26: 241–312.

1998. Aux confins de l'électrodynamique maxwellienne : ions et électrons vers 1897. *RHS*, 51: 5–34.

2000a. *Electrodynamics from Ampère to Einstein*. Oxford: Oxford University Press.

2000b. Poincaré, Einstein et l'inertie de l'énergie. *Comptes rendus physique*, 1: 143–153.

2001. God, waterwheels, and molecules: Saint-Venant's anticipation of energy conservation. *HSPS*, 31: 285–353.

2004. The mystery of the Einstein-Poincaré connection. *Isis*, 95: 614–626.

2007. Diversité et harmonie de la physique mathématique dans les préfaces de Henri Poincaré. In Jean-Claude Pont et al. (eds.), *Pour comprendre le XIXe: histoire et philosophie des sciences à la fin du siècle* (Florence: Olschi), pp. 221–240.

2008. The modular structure of physical theories. *Synthese*, 162: 195–223.

2012a. *A history of optics from Greek antiquity to the nineteenth century*. Oxford: Oxford University Press.

2012b. Poincaré's light. In Bertrand Duplantier and Vincent Rivasseau (eds.), *Poincaré, 1912–2012. Séminaire Poincaré 2012* (Basel: Birkhäuser, 2015), pp. 1–50.

2012c. Electrodynamics in the physics of Walther Ritz. In Jean Claude Pont (ed.), *Le destin douloureux de Walther Ritz (1878–1909), physicien théoricien de génie* (Sion: Vallesia), pp. 207–240.

2014a. *Physics and necessity: Rationalist pursuits from the Cartesian past to the quantum present*. Oxford: Oxford University Press.

2014b. The mystery of Riemann's curvature. *Historia mathematica*, 42: 47–83.

2015. Mesh and measure in early general relativity. *SHPMP*, 52: 163–187.

2018a. The unnamed structuralism of four nineteenth-century philosopher-physicists. In Maria de Paz and João Principe (eds.), *Évora studies in the philosophy and history of science*, Vol. 2: *From ontology to structure* (Casal de Cambra: Caleidoscópio), https://dspace.uevora.pt/rdpc/handle/10174/23129

2018b. Constitutive principles versus comprehensibility conditions in post-Kantian physics. *Synthese*, 197: 4571–4616.

2019a. Stokes's optics. In Mark McCartney, Andrew Whitaker, and Alistair Wood, *George Gabriel Stokes: Life, science, and faith* (Oxford: Oxford University Press), pp. 63–114.

2019b. Frames and stresses in Einstein's quest for a generalized theory of relativity. *SHPMP*, 68: 126–157.

Davis, John

1986. The influence of astronomy on the character of physics in mid-nineteenth century France. *HSPS*, 16: 59–82.

de Andrade Martins, Roberto

2012. O éter e a óptica dos corpos em movimento: A teoria de Fresnel e as tentativas de detecção do movimento da Terra, antes dos experimentos de Michelson e Morley (1818–1880). *Caderno Brasileiro de ensino de física*, 29: 52–80.

Dedekind, Richard

[c. 1867]. Analytische Untersuchungen zu Bernhard Riemann's Abhandlungen über die Hypothesen, welche der Geometrie zu Grunde liegen. In Sinaceur 1990, pp. 237–293.

1876. Bernhard Riemann's Lebenslauf. In Riemann 1876, pp. 507–526.

Dedekind, Richard, and Heinrich Weber

1876. Anmerkungen. In Riemann 1876, pp. 384–399.

Dell'Aglio, Luca
1996. On the genesis of the concept of covariant differentiation. *Revue d'histoire des mathématiques*, 2: 215–264.

Deguin, Nicolas
1853. *Cours élémentaire de physique*, 8th ed. Paris: Belin.

Delaunay, Charles
1856. *Traité de mécanique rationnelle*. Paris: Masson.

De Risi, Vincenzo
2015. Introduction. In V. de Risi (ed.), *Mathematizing space: The objects of geometry from antiquity to the early modern age* (Cham: Birkhäuser), pp. 1–13.

Descartes, René
[1633]. Manuscript published posthumously as *Le monde de Mr. Descartes ou le Traité de la Lumiere* (Paris, 1664).
1644. *Principia philosophiae*. Amsterdam: Ludovicus Elzevirius.

Des Coudres, Theodor
1889. Über das Verhalten des Lichtäthers bei der Bewegung der Erde. *AP*, 38: 71–79.

de Sitter, Willem
1916a. On Einstein's theory of gravitation, and its astronomical consequences. Royal Astronomical Society, *Monthly notices*, 76: 699–728.
1916b. On Einstein's theory of gravitation, and its astronomical consequences. Second paper. Royal Astronomical Society, *Monthly notices*, 77: 155–184.

Dirac, Paul Adrien Maurice
1977. Recollections of an exciting era. In Charles Weiner (ed.), *History of twentieth century physics* (New York: Academic Press), pp. 109–146.

DiSalle, Robert
1993. Helmholtz's philosophy of mathematics. In David Cahan (ed.), *Hermann von Helmholtz and the foundations of nineteenth-century science* (Berkeley: University of California Press), pp. 498–521.
2006. *Understanding space-time: The philosophical development of physics from Newton to Einstein*. Cambridge: Cambridge University Press.

Doppler, Christian
1842. *Über das farbige Licht der Doppelsterne und einiger anderer Gestirne des Himmels: Versuch einer das Bradley'sche Aberrations-Theorem als integrirenden Theil in sich schliessenden allgemeineren Theorie*. Prague: Borrosch & André.
1844. *Über eine bei jeder Rotation des Fortpflanzungsmittels eintretende eigenthümliche Ablenkung der Licht- und Schallstrahlen, zunächst angewandt auf mehre theils schon bekannte, theils neue Probleme der praktischen Astronomie, ein weiterer Beitrag zur allgemeinen Wellenlehre*. Prague: Borrosch & André.
1845. *Über die bisherigen Erklärungs-Versuche des Aberrations-Phänomens*. Prague: Haase.

Doran, Connemara
2018. Poincaré's mathematical creations in search of the "true relations of things." In Navarro 2018, pp. 45–66.

Drake, Stilman
1975. Impetus theory reappraised. *Journal of the history of ideas*, 36: 27–46.
1978. *Galileo at work*. Chicago: University of Chicago Press.

Drake, Stillman, and Israel Edward Drabkin
1969. *Mechanics in sixteenth century Italy: Selections from Tartaglia, Benedetti, Guido Ubaldo, & Galileo.* Madison: The University of Wisconsin Press.

Drake, Stillman, and Charles Donald O'Malley (eds.)
1960. *The controversy on the comets of 1618.* Philadelphia: University of Pennsylvania Press.

Droste, Johannes
1916a. *Het zwaartekrachtsveld van een of meer lichamen volgens de theorie van Einstein.* Leiden: Brill.
1916b. The field of *n* moving centers in Einstein's theory of gravitation. *PRA*, 19: 447–455.

Droste, Johannes, and Hendrik Lorentz
1917. De beweging van een stelsel lichamen onder den invloed van hunne onderlinge aantrekking, behandeld volgens de theorie van Einstein. *VKA*, 26: 392–403, 649–660. English in *LCP*, 5, pp. 330–355.

Drude, Paul
1894. *Physik des Aethers auf elektromagnetischer Grundlage.* Stuttgart: Enke.
1895. *Die Theorie in der Physik: Antrittsvorlesung gehalten am 5. December 1894 and der Universität Leipzig.* Leipzig: Hirzel.
1900a. *Lehrbuch der Optik.* Leipzig: Hirzel.
1900b. Zur Ionentheorie der Metalle. *PZ*, 1: 161–165.
1900c. Zur Elektronentheorie der Metalle. *AP*, 1: 566–613; *AP*, 3: 369–402.

Duchesneau, François
1994. *La dynamique de Leibniz.* Paris: Vrin.

Duhamel, Jean-Marie Constant
1845–1846. *Cours de mécanique de l'Ecole Polytechnique.* 2 vols. Paris: Bachelier.
1870. *Des méthodes dans les sciences de raisonnement*, Vol. 4: *Application des méthodes générales à la science des forces.* Paris: Gauthier-Villars.

Dühring, Eugen Karl
1873. *Kritische Geschichte der allgemeinen Prinzipien der Mechanik.* Berlin: Grieben.

Earman, John
1989. *World enough and space-time: Absolute versus relational theories of space and time.* Cambridge: MIT Press.
1999. The Penrose-Hawking singularity theorems: History and implications. In Goenner et al. 1999, pp. 235–267.

Earman, John, and Clark Glymour
1978a. Lost in the tensors: Einstein's struggles with covariance principles 1912–1916. *SHPS*, 9: 251–278.
1978b. Einstein and Hilbert: Two months in the history of general relativity. *AHES*, 19: 291–308.
1980a. The gravitational red shift as a test of general relativity: History and analysis. *SHPS*, 11: 175–213.
1980b. Relativity and eclipses: The British eclipse expeditions of 1919 and their predecessors. *HSPS*, 11: 49–85.

Earman, John, and Michel Janssen
1993. Einstein's explanation of the motion of Mercury's perihelion. In Earman, Janssen, and Norton 1993, pp. 129–172.
Earman, John, Michel Janssen, and John Norton (eds.)
1993. Einstein studies, Vol. 5: *The attraction of gravitation: New studies in the history of general relativity.* Boston: Birkhäuser.

Eckert, Michael, Helmut Schubert, and Gisela Torkar
1992. The roots of solid state physics before quantum mechanics. In Lilian Hoddeson *et al.* (eds.), *Out of the crystal maze* (New York: Oxford University Press), pp. 3–87.

Eden, Alec
1992. *The search for Christian Doppler*. Vienna: Springer.

Eddington, Arthur Stanley
1918. *Report on the relativity theory of gravitation*. London: Fleetway Press.
1920a. [Reply to Rice 1920]. *Nature*, 104: 598–599.
1920b. *Space, time and gravitation*. Cambridge: Cambridge University Press.
1923. *The mathematical theory of relativity*. Cambridge: Cambridge University Press.

Ehlers, Jürgen
1973. Survey of general relativity theory. In Werner Israel (ed.), *Relativity, astrophysics and cosmology* (Dordrecht: Reidel), pp. 1–125.
1981. Christoffel's work on the equivalence problem for Riemannian spaces and its importance for modern field theories of physics. In Butzer and Fehér 1981, pp. 526–542.

Ehrenfest, Paul
1907. Die Translation deformierbarer Elektronen und der Flächensatz. *AP*, 23: 204–205.
1909. Gleichförmige Rotation starrer Körper und Relativitätstheorie. *PZ*, 10: 918.
1910. Zu Herrn v. Ignatowskys Behandlung der Bornschen Starrheitsdefinition. *PZ*, 11: 1127–1129.

Einstein, Albert
[1895]. Über die Untersuchung des Aetherzustandes im magnetischen Felde. In *ECP* 1, pp. 6–9.
1905a. Zur Elektrodynamik bewegter Körper. *AP*, 17: 891–921.
1905b. Ist die Trägheit eines Körpers von seinem Energieinhalt abhängig? *AP*, 18: 639–641.
1906. Das Prinzip der Erhaltung des Schwerpunktsbewegung und die Trägheit der Energie. *AP*, 20: 627–633.
1907a. Bemerkungen zu der Notiz von Hm. Paul Ehrenfest: "Die Translation deformierbaren Elektronen und der Flächensatz." *AP*, 23: 206–208.
1907b. Über die vom Relativitätsprinzip geforderte Trägheit der Energie. *AP*, 23: 371–384.
1907c. Über das Relativitätsprinzip und die aus demselben gezogenen Folgerungen. *Jahrbuch der Radioaktivität und der Elektronik*, 4: 411–462.
1909. Zum gegenwärtigen Stand des Strahlungsproblems. *PZ*, 10: 185–193.
1911. Über den Einfluss der Schwerkraft auf die Ausbreitung des Lichtes. *AP*, 35: 898–908.
1912a. Lichtgeschwindigkeit und Statik des Gravitationsfeldes. *AP*, 38: 355–369.
1912b. Zur Theorie des statischen Gravitationsfeldes. *AP*, 38: 443–458.
1912c. Gibt es eine Gravitationswirkung, die der elektrodynamischen Induktionswirkung analog ist? *Vierteljahrsschrift für gerichtliche Medizin und öffentliches Sanitätswesen*, 44: 37–40.
1913. Zum gegenwärtigen Stande des Gravitationsproblems. *PZ*, 14: 1249–1266.
1914a. Prinzipielles zur verallgemeinerten Relativitätstheorie und Gravitationstheorie. *PZ*, 15: 176–180.
1914b. Bemerkungen [to a republication of the *Entwurf*]. *Zeitschrift für Mathematik und Physik*, 62: 260–261.
1914c. Die formale Grundlage der allgemeinen Relativitätstheorie. *BB* (1914), 1030–1085.
1915a. Zur allgemeinen Relativitätstheorie. *BB* (1915), 778–786, 799–801 (*Nachtrag*).
1915b. Erklärung der Perihelbewegung des Merkur aus der allgemeinen Relativitätstheorie. *BB* (1915), 831–839.
1915c. Die Feldgleichungen der Gravitation. *BB* (1915), 844–847.
1916a. Die Grundlage der allgemeinen Relativitätstheorie. *AP*, 49: 769–822.
[1916b]. Anhang. Darstellung der Theorie ausgehend von einem Variationsprinzip [unpublished appendix to Einstein 1916a]. In *ECP*, 6, pp. 340–346.
1916c. Näherungsweise Integration der Feldgleichungen der Gravitation. *BB* (1916), 688–696.
1916d. Hamiltonsches Prinzip und allgemeine Relativitätstheorie. *BB* (1916), 1111–1116.

1917a. *Über die spezielle und die allgemeine Relativitätstheorie. (Gemeinverständlich).* Braunschweig: Vieweg.

1917b. Kosmologische Betrachtungen zur allgemeinen Relativitätstheorie. *BB* (1917), 142–152.

1918. Prinzipielles zur allgemeinen Relativitätstheorie. *AP*, 55: 241–244.

[1920]. Grundgedanken und Methoden der Relativitätstheorie in ihrer Entwicklung dargestellt. In *ECP*, 7, pp. 245–281.

1920. *Äther und Relativitätstheorie* [Leiden lecture of May 5, 1920]. Berlin: Springer.

[1921]. The development and present position of the theory of relativity [London, King's College lecture of June 13, 1921]. In *ECP*, 7, pp. 431–433.

[1922]. How I created the theory of relativity [English transl. of Jun Ishiwara's notes of Einstein's lecture at Kyoto University on December 14, 1922]. In *ECP*, 13, pp. 629–639.

1923. *The meaning of relativity: Four lectures delivered at Princeton University, May 1921.* Princeton: Princeton University Press.

[1933]. Einiges über die Entstehung der allgemeinen Relativitätstheorie. In Carl Seelig (ed.), *Mein Weltbild* (Zürich: Europa Verlag), pp. 176–181.

1949. Autobiographisches. In Schilpp 1949, pp. 1–94.

Einstein, Albert, and Adriaan Fokker
1914. Die Nordströmsche Gravitationstheorie vom Standpunkt des absoluten Differentialkalküls. *AP*, 44: 321–328.

Einstein, Albert, and Marcel Grossmann
1913. *Entwurf einer verallgemeinerten Relativitätstheorie und einer Theorie der Gravitation.* Leipzig: Teubner.

1914. Kovarianzeigenschaften der Feldgleichungen der auf die verallgemeinerte Relativitätstheorie gegründeten Gravitationstheorie. *Zeitschrift für Mathematik und Physik*, 63: 215–225.

Einstein, Albert, and Jakob Laub
1908a. Elektromagnetische Grundgleichungen für bewegte Körper. *AP*, 26: 532–540.

1908b. Die im elektromagnetischen Felde auf ruhende Körper ausgeübten ponderomotorischen Kräfte. *AP*, 26: 541–550.

Eisenstaedt, Jean
1982. Histoire et singularités de la solution de Schwarzschild (1915–1923). *AHES*, 27: 157–198.

1991. De l'influence de la gravitation sur la propagation de la lumière en théorie newtonienne. L'archéologie des trous noirs. *AHES*, 42: 315–386.

1996. L'optique balistique newtonienne à l'épreuve des satellites de Jupiter. *AHES*, 50: 117–156.

2005a. *Avant Einstein : Relativité, lumière, gravitation.* Paris: Le Seuil.

2005b. Light and relativity, a previously unknown eighteenth-century manuscript by Robert Blair. *Annals of science*, 62: 347–376.

Eisenstaedt, Jean, and Anne Kox (eds.)
1992. *Einstein studies*, Vol. 3: *Studies in the history of general relativity.* Boston: Birkhäuser.

Euler, Leonhard
1736. *Mechanica, sive motus scientia analytice exposita.* 2 vols. Petersburg: Typographia Academiae Scientiarum.

1739. Explicatio phaenomenarum quae a motu successivo lucis oriuntur. *Commentarii Academiae scienciarum Petropolitanae*, 11 (pub. 1750), 150–193.

1746a. Nova theoria lucis and colorum. In L. Euler, *Opuscula varii argumenti* (Berlin: Spener), pp. 169–244.

1746b. Mémoire sur l'effet de la propagation successive de la lumiere dans l'apparition tant des planètes que des comètes. *Mémoires de l'Académie de Berlin*, 2: 141–181.

1748. Reflexions sur l'espace et le tems. *Histoire de l'Académie royale des sciences et belles lettres de Prusse*, 4: 324–333.

1767. Recherches sur la courbure des surfaces. *Mémoires de l'Académie de Berlin*, 16: 119–143.

Faraday, Michael
1821. On some new electro-magnetical motions, and on the theory of magnetism. *The quarterly journal of science and the arts*. Also in *FER*, 2, pp. 127–147.
1831. Experimental researches in electricity. Series I. *PT*. Also in *FER*, 1, pp. 1–41.
1832. Experimental researches in electricity. Series II. *PT*. Also in *FER*, 1, pp. 42–75.

Farwell, Ruth, and Christopher Knee
1990a. The missing link: Riemann's "Commentatio," differential geometry and tensor analysis. *Historia mathematica*, 17: 223–255.
1990b. The end of the absolute: A nineteenth-century contribution to general relativity. *SHPS*, 21: 91–121.

Faye, Hervé
1874. *Cours d'astronomie, 1ᵉ division, 1873–1874*. Cours autographié. Paris: Ecole Polytechnique.

Fermi, Enrico
1922. Über einen Widerspruch zwischen der elektrodynamischen und der relativistischen Theorie der elektromagnetischen Masse. *PZ*, 23: 340–344.

Fernflores, Francisco
2018. *Einstein's mass-energy equation*. 2 vols. New York: Momentum Press.

Finocchiaro, Maurice
1989. *The Galileo affair: A documentary history*. Berkeley: University of California Press.

Firode, Alain
2001. *La dynamique de d'Alembert*. Paris: Vrin.

Fisher, John
2010. Conjectures and reputations: The composition and reception of James Bradley's paper on the aberration of light with some reference to a third unpublished version. *BJHS*, 43: 19–48.

Fisher, Saul
2014. Pierre Gassendi. *SEP*, Spring 2014 edition. https://plato.stanford.edu/archives/spr2014/entries/gassendi/

FitzGerald, George Francis
1880. On the electromagnetic theory of the reflection and refraction of light. *PT*, 171: 691–711.
1882. On the electromagnetic effects due to the motion of the earth. *Transactions of the Royal Dublin Society*, 1: 319–324.
1889. The ether and the earth's atmosphere. *Science*, 13: 390.

Fizeau, Hippolyte
1848. Acoustique et optique. Société Philomatique, procès-verbaux des séances de 1848, 81–83.
1849. Sur une expérience relative à la propagation de la lumière. *CR*, 29: 90–92.
1851. Sur les hypothèses relatives à l'éther lumineux, et sur une expérience qui paraît démontrer que le mouvement des corps change la vitesse avec laquelle la lumière se propage dans leur intérieur. *CR*, 33: 349–355.
1852. Mouvement de la terre autour du soleil. *Cosmos*, 1: 690–692.
1859a. Sur les hypothèses relatives à l'éther lumineux, et sur une expérience qui paraît démontrer que le mouvement des corps change la vitesse avec laquelle la lumière se propage dans leur intérieur. *ACP*, 57: 385–404.

1859b. Sur une méthode propre à rechercher si l'azimut de polarisation du rayon réfracté est influencé par le mouvement du corps réfringent. Essai de cette méthode. *CR*, 49: 717–723.

1860. Sur une méthode propre à rechercher si l'azimut de polarisation du rayon réfracté est influencé par le mouvement du corps réfringent. Essai de cette méthode. *ACP*, 58: 129–163.

1870. Des effets du mouvement sur le ton des vibrations sonores et sur la longueur d'onde des rayons de lumière. *ACP*, 19: 211–221.

Fizeau, Hippolyte, and Louis Breguet

1850a. Note sur l'expérience de la vitesse comparative de la vitesse de la lumière dans l'air et dans l'eau. *CR*, 30: 562–563.

1850b. Sur l'expérience relative à la vitesse comparative de la vitesse de la lumière dans l'air et dans l'eau. *CR*, 30: 771–774.

Flamm, Ludwig

1916. Beiträge zur Einsteinschen Gravitationstheorie. *PZ*, 17: 448–454.

Fok, Vladimir Aleksandrovič

1959. *The theory of space, time, and gravitation*. Oxford: Pergamon.

Fölsing, Albrecht

1993. *Albert Einstein: eine Biographie*. Frankfurt: Suhrkamp.

1997. *Albert Einstein: A biography*. New York: Viking.

Fontaine des Crutes, Pierre

1744. *Traité complet sur l'aberration*. Paris: Quillau.

Föppl, Ludwig

1894. *Einführung in die Maxwell'sche Theorie der Elektricität*. Leipzig: Teubner.

Foucault, Léon

1850. Méthode générale pour mesurer la vitesse de la lumière dans l'air et les milieux transparents. Vitesses relatives de la lumière dans l'air et dans l'eau. Projet d'expérience sur la vitesse de propagation du calorique rayonnant. *CR*, 30: 551–560.

1853. *Sur les vitesses relatives de la lumière dans l'air et dans l'eau. Thèse de doctorat*. Paris: Bachelier.

1854. Sur les vitesses relatives de la lumière dans l'air et dans l'eau. *ACP*, 41: 129–164.

Francœur, Louis Benjamin

1804. *Traité de mécanique élémentaire*, 3rd ed. Paris: Bachelier.

Frercks, Jan

2000. Creativity and technology in experimentation: Fizeau's terrestrial determination of the speed of light. *Centaurus*, 42: 249–287.

2001. *Die Forschungspraxis Hippolyte Fizeaus: eine Charakterisierung ausgehend von der Replikation seines Ätherwindexperiments von 1852*. Berlin: Wissenschaft und Technik Verlag.

2005. Fizeau's research program on ether drag: A long quest for a publishable experiment. *Physics in perspective*, 7: 35–65.

Fresnel, Augustin

1818. Lettre de M. Fresnel à M. Arago sur l'influence du mouvement terrestre dans quelques phénomènes de l'optique. *ACP*, 9: 57–66, 286.

1866–1870. *Œuvres complètes d'Augustin Fresnel, publiées par Henri de Sénarmont, Émile Verdet et Léonor Fresnel*. 3 vols. Paris: Imprimerie impériale.

Freundlich, Erwin

1916. *Die Grundlagen der Einsteinschen Gravitationstheorie*. Berlin: Springer.

Fric, Jacques
2013. *Painlevé et la relativité générale*. Thèse de doctorat, Université Denis Diderot. http://www.bibnum
.education.fr/sites/default/files/painleve_these.pdf

Friedländer, Benedict, and Immanuel Friedländer
1896. *Absolute oder relative Bewegung?* Berlin: Simions.

Friedman, Michael
1983. *Foundations of spacetime theories: Relativistic physics and philosophy of science*. Princeton: Princeton
University Press.
1995. Poincaré's conventionalism and logical positivism. *Foundations of science*, 2: 299–316.
2013. *Kant's construction of nature: A reading of the metaphysical foundations of natural science*. Cambridge:
Cambridge University Press.

Friedman, Robert Marc
2001. *The politics of excellence: Behind the Nobel prize in science*. New York: Times Books.

Gabbey, Alan
1980. Huygens and mechanics. In Henk Bos, M. Rudwick, H. Snelders, and R. Visser (eds.), *Studies on
Christiaan Huygens* (Lisse: Swets and Zeitlinger), pp. 166–199.

Galilei, Galileo
1623. *Il Saggiatore, nel quale con bilancia esquisita e giusta si ponderano le cose contenute nella Libra
astronomica e filosofica di Lotario Sarsi Sigensano*. Rome: Mascardi. Transl. in Stillman Drake,
Discoveries and opinions of Galileo (New York: Doubleday & Co, 1957).
1632. *Dialogo sopra i due massimi sistemi del mondo*. Florence: Landini. Transl. by Stillman Drake. 1964.
Berkeley: University of California Press.
1638. *Discorsi e dimostrazioni matematiche intorno a due nuove scienze*. Leiden: Elzevir.

Galison, Peter
1979. Minkowski's spacetime: From visual thinking to the absolute world. *HSPS*, 10: 85–121.
1984. Descartes's comparisons: From the invisible to the visible. *Isis*, 75: 311–326.
2003. *Einstein's clocks, Poincaré's maps: Empires of time*. New York: Norton.

Garber, Daniel
1992. *Descartes' metaphysical physics*. Chicago: University of Chicago Press.

Gauss, Karl Friedrich
[1825]. Neue allgemeine Untersuchungen über die krümmen Flächen. In Karl Friedrich Gauss, *Werke*,
Band 1–12 (Leipzig, Teubner), Vol. 8, pp. 408–443.
1828. Disquisitiones generales circa superficies curvas. *Commentationes Societatis Regiae scientiarum
Gottingensis recentiores classis mathematicae*, 6: 99–146.
1870–1927. *Werke*, Band 1–12. Leipzig: Teubner.

Gauss, Karl Friedrich, and Friedrich Wilhelm Bessel
1880. *Briefwechsel zwischen Gauss und Bessel*. Leipzig: Engelmann.

Gebhardt, Wolfgang
2016. Erich Kretschmann. The life of a theoretical physicist in difficult time. Max Planck Institut für
Wissenschaftsgeschichte, preprint 482.
Gerhardt, Carl Immanuel (ed.)
1899. *Der Briefwechsel von Gottfried Wilhelm Leibniz mit Mathematikern. Erster Band*. Berlin: Mayer &
Müller.

Giere, Ronald
1988. *Explaining science: A cognitive approach*. Chicago: The University of Chicago Press.

Giovanelli, Marco
2013. The forgotten tradition: How the logical empiricists missed the philosophical significance of the work of Riemann, Christoffel and Ricci. *Erkenntnis*, 78: 1219–1257.
2014. "But one must not legalize the mentioned sin": Phenomenological vs. dynamical treatments of rods and clocks in Einstein's thought. *SHPMP*, 48: 20–44.
2020. Like thermodynamics before Boltzmann. On the emergence of Einstein's distinction between constructive and principle theories. *SHPMP*, https://doi.org/10.1016/j.shpsb.2020.02.005
2021. Nothing but coincidences: The point-coincidence argument and Einstein's struggle with the meaning of coordinates. *European journal for philosophy of science*, 11: #45.

Giulini, Domenico
2008. What is (not) wrong with scalar gravity? *SHPMP*, 39: 154–180.
Glick, Thomas (ed.)
1987. *The comparative reception of relativity*. Dordrecht: Reidel.

Goenner, Hubert
2008. On the history of geometrization of space-time: From Minkowski to Finsler Geometry. arXiv:0811.4529
2010. Max Plancks Beiträge zur speziellen Relativitätstheorie. In Dieter Hofmann (ed.), *Max Planck und die moderne Physik* (Berlin: Springer), pp. 149–166.
2016. General relativity and the growth of a sub-discipline 'gravitation' in the German speaking physics community. arXiv:1607.03324

Goenner, Hubert, Jürgen Renn, Jim Ritter, and Tilman Sauer (eds.)
1999. *Einstein studies*, Vol. 7: *The expanding worlds of general relativity*. Boston: Birkhäuser.

Goldberg, Stanley
1967. Henri Poincaré and Einstein's theory of relativity. *AJP*, 35: 934–944.
1969. *Early response to Einstein's special theory of relativity, 1905–1912: A case study in national differences*. PhD dissertation, Harvard University.
1970a. Poincaré's silence and Einstein's theory of relativity: The role of theory and experiment in Poincaré's physics. *BJHS*, 5: 73–84.
1970b. The Abraham theory of the electron: The symbiosis of experiment and theory. *AHES*, 7: 7–25.
1970c. Bucherer, Alfred Heinrich. In *Dictionary of scientific biography* (New York, Scribner & Sons), Vol. 2, pp. 559–560.
1970d. In defense of ether: The British response to Einstein's special theory of relativity, 1905–1911. *HSPS*, 2: 89–124.
1976. Max Planck's philosophy of nature and his elaboration of the special theory of relativity. *HSPS*, 7: 125–160.
1984. *Understanding relativity: Origin and impact of a scientific revolution*. Boston: Birkhäuser.

Goldberg, Stanley, and Roger Stuewer (eds.)
1988. *The Michelson era in American science 1870–1930*. New York: AIP Conference Proceedings No. 179.

Gooding, David
1985. In Nature's school: Faraday as an experimentalist. In David Gooding and Frank James (eds.), *Faraday rediscovered: Essays on the life and work of Michael Faraday, 1791–1867* (Basingstoke: Macmillan Press), pp. 117–135.

's Gravesande, Willem Jacob
1720–1721. *Physices elementa mathematica, experimentis confirmata, sive introductio ad philosophiam Newtonianam*. 2 vols. Leiden: van der Aa.

1742. *Physices elementa mathematica, experimentis confirmata, sive introductio ad philosophiam Newtoni-anam*, 3rd ed. 2 vols. Leiden: Langerak.

Grattan-Guinness, Ivor
1984. Work of the workers: Advances in engineering mechanics and instruction in France, 1800–1830. *Annals of science*, 41: 1–33.
2005. "Exposition du système du monde" and "Traité de méchanique céleste." In I. Grattan-Guinness (ed.), *Landmark writings in Western mathematics 1640–1940* (Amsterdam: Elsevier), pp. 242–257.

Gray, Jeremy
1989. *Ideas of space: Euclidean, non-Euclidean, and relativistic*, 2nd ed. Oxford: Clarendon Press.
1999. (ed.). *The symbolic universe: Geometry and physics 1890–1930*. Oxford: Oxford University Press.
2006. Gauss and non-Euclidean geometry. In Andras Prékopav and Emil Molnár (eds.), *Non-Euclidean geometries. János Bolyai memorial volume* (Chicago: Springer), pp. 61–80.
2007. *Worlds out of nothing: A course in the history of geometry in the 19th century*. London: Springer.
2013. *Henri Poincaré: A scientific biography*. Princeton: Princeton University Press.

Gullstrand, Allvar
1921. Allgemeine Lösung des statischen Einkörperproblems in der Einsteinschen Gravitationstheorie. *AMAF*, 16 (8): 1–15 [printed May 31, 1921].
1922. Das statische Einkörperproblem in der Einstein'schen Theorie. Erwiderung zu Herrn E. Kretschmann. *AMAF*, 17 (3): 1–5.
1923. Antwort an Herrn E. Kretschmann. *AMAF*, 17 (25): 4.

Guth, Eugene
1970. Contribution to the history of Einstein's geometry as a branch of physics. In Moshe Carmeli, Stuart Fickler, and Louis Witten (eds.), *Relativity. Proceedings of the relativity conference in the Midwest, held at Cincinnati, Ohio, June 2–6, 1969* (New York: Springer), pp. 161–208.

Hacyan, Shahen
2015. Galileo and the equivalence principle: A faulty argument with the correct conclusion. *European journal of physics*, 36: 065044.

Halley, Edmond
1718. Considerations on the change of the latitude of some of the principle bright stars. *PT*, 30: 736–738.

Hamel, Georg 1921. Zur Einsteinschen Gravitationstheorie. Berliner mathematische Gesellschaft. *Sitzungs-berichte*, 19: 65–73.

Hankins, Thomas
1970. *Jean d'Alembert: Science and the enlightenment*. Oxford: Clarendon Press.

Harman, Peter
1998. *The natural philosophy of James Clerk Maxwell*. Cambridge: Cambridge University Press.

Haüy, René-Juste
1806. *Traité élémentaire de physique*, 2nd ed. 2 vols. Paris: Bachelier.

Havas, Peter
1964. Four-dimensional formulations of Newtonian mechanics and their relation to the special and the general theory of relativity. *Reviews of modern physics*, 36: 938–965.
1999. Einstein, relativity and gravitation research in Vienna before 1938. In Goenner et al. 1999, pp. 161–206.

Hawkins, Thomas
2000. *Emergence of the theory of Lie groups: An essay in the history of mathematics, 1869–1926*. New York: Springer.

Hearnshaw, John
2014. *The analysis of starlight*. Cambridge: Cambridge University Press.

Heaviside, Oliver
1885–1887. Electromagnetic induction and its propagation. *The electrician*. Also in *HEP*, 1, pp. 429–560 and *HEP*, 2, pp. 39–155.
1886–1887. On the self-induction of wires. *PM*. Also in *HEP*, 2, pp. 168–323.
1888–1889. Electromagnetic waves, the propagation of potential, and the electromagnetic effects of a moving charge. *The electrician*. Also in *HEP*, 2, pp. 490–499.
1889. On the electromagnetic effects due to the motion of electrification through a dielectric. *PM*. Also in *HEP*, 2, pp. 504–518.
1891–1892. On the forces, stresses, and fluxes of energy in the electromagnetic field. *PRS*. Also in *HEP*, 2, pp. 521–574.
1892. *Electrical papers*. 2 vols. London: Chelsea.
1893. *Electromagnetic theory*, Vol. 1. London: Ben.

Heilbron, John
2010. *Galileo*. Oxford: Oxford University Press.

Heinzmann, Gerhard, and David Stump
2017. Henri Poincaré. *SEP*, Winter 2017 edition. https://plato.stanford.edu/archives/win2017/entries/poincare/

Helmholtz, Hermann
1868a. Über die Thatsächlichen Grundlagen der Geometrie. *HWA*, 2, pp. 610–617.
1868b. Über die Thatsachen, die der Geometrie zum Grunde legen. *HWA*, 2, pp. 618–639.
1870a. Über die Theorie der Elektrodynamik. Erste Abhandlung: Über die Bewegungsgleichungen der Elektricität für ruhende Körper. *AP*. Also in *HWA*, 1, pp. 545–628.
[1870b]. Über den Ursprung und die Bedeutung der geometrischen Axiome. In Helmholtz 1884, Vol. 2, pp. 1–31.
1872. Über die Theorie der Elektrodynamik. Vorläufiger Bericht. Königlich Preussische Akademie der Wissenschaften zu Berlin, *Monatsberichte*. Also in *HWA*, 1, pp. 636–646.
1873a. Über die Theorie der Elektrodynamik. Zweite Abhandlung: Kritisches. *JRAM*. Also in *HWA*, 1, pp. 647–687.
1873b. Vergleich des Ampère'schen und des Neumann'schen Gesetzes für die Elektrodynamischen Kräfte. Königlich Preussische Akademie der Wissenschaften zu Berlin, *Monatsberichte*. Also in *HWA*, 1, pp. 688–701.
1874a. Über die Theorie der Elektrodynamik. Dritte Abhandlung: Die Elektrodynamischen Kräfte in bewegten Leitern. *JRAM*. Also in *HWA*, 1, pp. 702–762.
1874b. Kritisches zur Elektrodynamik. *AP*. Also in *HWA*, 1, pp. 763–773.
1875a. Versuche über die im ungeschlossenen Kreise durch Bewegung inducirten elektromotorischen Kräfte. *AP*. Also in *HWA*, 1, pp. 774–790.
1875b. Zur Theorie der anomalen Dispersion. *AP*, 154: 582–596.
1876. Bericht betreffend Versuche tiber die magnetische Wirkung elektrischer Convection, ausgeführt von Hm. Henry A. Rowland. *AP*. Also in *HWA*, 1, pp. 791–797.
1878. Über den Ursprung und Sinn der geometrischen Sätze; Antwort gegen Herrn Professor Lang. *HWA*, 2, pp. 640–660.
1881. On the modern development of Faraday's conception of electricity. *Journal of the Chemical Society*. Also in *HWA*, 3, pp. 53–87.
1884. *Vorträge und Reden*. 2 vols. Braunschweig: Vieweg.
1892. Das Prinzip der kleinsten Wirkung in der Elektrodynamik. *AP*. Also in *HWA*, 3, pp. 476–504.
1893a. Elektromagnetische Theorie der Farbenzerstreuung. *BB*, *AP*. Also in *HWA*, 3, pp. 505–525.
1893b. Folgerungen aus Maxwell 'scher Theorie über die Bewegung des reinen Aethers. *BB*, *AP*. Also in *HWA*, 3, pp. 526–535.

Henry, John
2016. Henry More. *SEP*, Winter 2016 edition. https://plato.stanford.edu/archives/win2016/entries/henry-more/

Hentschel, Klaus
1990. *Interpretationen und Fehlinterpretationen der speziellen und der allgemeinen Relativitätstheorie durch Zeitgenossen Albert Einsteins*. Basel: Birkhäuser.
1992. Einstein's attitude towards experiments. *SHPS*, 23: 593–624.
1994. Erwin Finlay Freundlich and testing Einstein's theory of relativity. *AHES*, 47: 143–201.
1998. *Zum Zusammenspiel von Instrument, Experiment und Theorie: Rotverschiebung im Sonnenspektrum und verwandte spektrale Verschiebungseffekte von 1880 bis 1960*. Hamburg: Kovac.
2013. Ernst Mach. In Arne Hessenbruch(ed.), *Reader's guide to the history of science* (London: Routledge), pp. 427–428.

Heras, Ricardo
2017. A review of Voigt's transformations in the framework of special relativity. arXiv:1411.2559

Herglotz, Gustav
1903. Zur Elektronentheorie. *GN* (1903), 357–382.
1910. Über den vom Standpunkt des Relativitätsprinzips aus als "starr" zu bezeichnenden Körper. *AP*, 31: 393–415.

Herivel, John
1965. *The background to Newton's Principia*. Oxford: Clarendon.

Hertz, Heinrich
1887a. Über sehr schnelle elektrische Schwingungen. *AP*. Also in Hertz 1892, pp. 32–58.
1887b. Über Induktionserscheinungen, hervorgerufen durch die elektrischen Vorgänge in Isolatoren. *BB*. Augmented *AP* version in Hertz 1892, pp. 102–114.
1888a. Über die Ausbreitungsgeschwindigkeit der elektrodynamischen Wirkungen. *BB, AP*. Also in Hertz 1892, pp. 115–132.
1888b. Über die Einwirkung einer geradlinigen elektrischen Schwingung auf eine benachbarte Strombahn. *AP*. Also in Hertz 1892, pp. 86–101.
1888c. Über elektrodynamische Wellen im Luftraume und deren Reflexion. *AP*. Also in Hertz 1892, pp. 133–146.
1889. Die Kräfte elektrischer Schwingungen, behandelt nach der Maxwell'schen Theorie. *AP*. Also in Hertz 1892, pp. 147–170.
1890a. Über die Grundgleichungen der Elektrodynamik für ruhende Körper. *AP*. Also in Hertz 1892, pp. 208–255.
1890b. Über die Grundgleichungen der Elektrodynamik für bewegte Körper. *AP*. Also in Hertz 1892, pp. 256–285.
1892. *Untersuchungen über die Ausbreitung der elektrischen Kraft*. Leipzig: Barth.
1894. *Die Prinzipien der Mechanik in neuem Zusammenhange dargestellt*. Leipzig: Barth.

Hertz, Paul
1904. *Untersuchungen über unstetige Bewegungen eines Elektrons*. PhD dissertation, Universität Göttingen.

Hilbert, David
1899. *Grundlagen der Geometrie*. Leipzig: Teubner.
[1915]. Proofs of Hilbert 1915, printed on December 6, 1915. Staats- und Universitätsbibliothek Göttingen (Handschriftenabteilung), Cod. Ms. D. Hilbert 634. Available at http://echo.mpiwg-berlin.mpg.de
1915. Die Grundlagen der Physik. (Erste Mitteilung). *GN* (1915), 395–407.
1917. Die Grundlagen der Physik. (Zweite Mitteilung). *GN* (1917), 53–76.

Hirosige, Tetu
1966. Electrodynamics before the theory of relativity. *Japanese studies in the history of science*, 5: 1–49.
1969. Origins of Lorentz's theory of electrons and the concept of the electromagnetic field. *HSPS*, 1: 151–209.
1976. The ether problem, the mechanistic world view, and the origins of the theory of relativity. *HSPS*, 7: 3–82.

Hoek, Martin
1868. Détermination de la vitesse avec laquelle est entraînée une onde lumineuse traversant un milieu en mouvement. *AN*, 2: 189–194.
1869. Sur la différence entre les constantes de l'aberration d'après Delambre et Struve. *Astronomische Nachrichten*, 73: 193–200.

Holton, Gerald
1960. On the origins of the special theory of relativity. *AJP*, 28: 627–636.
1969. Einstein, Michelson, and the "crucial" experiment. *Isis*, 60: 133–197.
1973. Influences on Einstein's early work. *The American scholar*, 37: 59–79.

Hon, Giora
1995. Is the identification of experimental error contextually dependent? The case of Kaufmann's experiment and its varied reception. In Jed Buchwald (ed.), *Scientific practice: Theories and stories of doing physics* (Chicago: University of Chicago Press), pp. 170–223.

Hon, Giora, and Bernard Goldstein
2005. How Einstein made asymmetry disappear: Symmetry and relativity in 1905. *AHES*, 59: 437–544.

Howard, Don
1999. Point coincidences and pointer coincidences: Einstein on the invariant content of space-time theories. In Goenner et al. 1999, pp. 463–500.

Howard, Don, and John Norton
1993. Out of the labyrinth: Einstein, Hertz and Göttingen answer to the hole argument. In Earman, Janssen, and Norton 1993, pp. 30–62.

Howard, Don, and John Stachel
1989. Einstein studies. Vol. 1: *Einstein and the history of general relativity*. Boston: Birkhäuser.
2000. Einstein studies. Vol. 8: *Einstein: The formative years, 1879–1909*. Boston: Birkhäuser.

Huggins, William
1868. Further observations of the spectra of some of the stars and nebulae, with an attempt to determine therefrom whether these bodies are moving toward or from the earth, also observations on the spectra of the sun and comet II, 1868. *PT*, 158: 528–564.

Hunger Parshall, Karen
1994. Toward a history of nineteenth-century invariant theory. In Knobloch and Rowe 1994, Vol. 1, pp. 157–208.

Hunt, Bruce
1986. Experimenting on the ether: Oliver Lodge and the great whirling machine. *HSPS*, 16: 111–134.
1991. *The Maxwellians*. Ithaca: Cornell University Press.

Husserl, Edmund
1913. Ideen zu einer reinen Phänomenologie und phänomenologischen Philosophie. Erstes Buch: Allgemeine Einführung in die reine Phänomenologie. *Jahrbuch für Philosophie und phänomenologische Forschung*, 1 (1): 1–323.

Huygens, Christiaan
[1659]. Untitled MS on pendulum motion. In Huygens 1888–1950, Vol. 17, pp. 126–137.
1669. Extrait d'une lettre de M. Hugens à l'auteur du journal. *Journal des sçavans*, March 18, 22–24.
1673. *Horologium oscillatorium, sive de motu pendulorum ad horologia aptato demonstrationes geometricae.* Paris: Muguet.
[c. 1690]. Pièces et fragments concernant la question de l'existence et de la perceptibilité du «mouvement absolu». In Huygens 1888–1950, Vol. 16, pp. 213–233.
1703. *De motu corporum ex percussione.* In *Christiani Hugenii Zelemii, dum viveret, toparchae opuscula postuma, quae continent dioptricam. Commentarios de vitris figurandis. Dissertationem de corona & parheliis. Tractatum de motu. Tractatum de vi centrifuga. Descriptionem automati planetarii* (Leiden: Boutesteyn), pp. 369–400. Translated by Michael Mahoney at http://www.princeton.edu/~hos/Mahoney/texts/huygens/impact/huyimpct.html (last accessed May 7, 2019).
1888–1950. *Œuvres complètes publiées par la Société hollandaise des sciences.* 22 vols. La Haye: Nijhof.

Hyder, David
2009. *The determinate world: Kant and Helmholtz on the physical meaning of geometry.* Berlin: De Gruyter.

Jamin, Jules
1858. Cours de physique de l'École polytechnique. Paris: Mallet-Blanchet.

Jammer, Max
1954. *Concepts of space: The history of theories of space in physics.* Cambridge: Harvard University Press.
1957. *Concepts of force: A study in the foundation of dynamics.* Cambridge: Harvard University Press.
2006. *Concepts of simultaneity from antiquity to Einstein and beyond.* Baltimore: The John Hopkins University Press.

Janssen, Michel
1993. H. A. Lorentz's attempt to give a coordinate-free formulation of the general theory of relativity. In Eisenstaedt and Kox 1993, pp. 344–363.
1995. *A comparison between Lorentz's ether theory and special relativity in the light of the experiments of Trouton and Noble.* PhD dissertation, University of Pittsburgh.
1999. Rotation as the nemesis of Einstein's *Entwurf* theory. In Goenner et al. 1999, pp. 127–157.
2003. The Trouton experiment, and a slice of Minkowski space-time. In Jürgen et al. (eds.), *Revisiting the foundations of relativistic physics: Festschrift in honor of John Stachel* (Dordrecht: Springer), pp. 27–54.
2004. Einstein's first systematic exposition of general relativity. PhilSci archive, December. http://philsci-archive.pitt.edu/id/eprint/212
2005. Of pots and holes: Einstein's bumpy road to general relativity. *AP*, 14 (supplement): 58–85.
2007. What did Einstein know and when did he know it. *GGR*, 2, pp. 785–837.
2012. The twins and the bucket: How Einstein made gravity rather than motion relative in general relativity. *SHPMP*, 43: 159–175.
2014. "No success like failure. . .": Einstein's quest for general relativity, 1907–1920. In Michel Janssen and Christoph Lehner (eds.), *The Cambridge companion to Einstein* (Cambridge: Cambridge University Press), pp. 167–227.
2019. How did Lorentz find his theorem of corresponding states? *SHPMP*, 67: 167–175.

Janssen, Michel, and Matthew Mecklenburg
2006. Electromagnetic models of the electron and the transition from classical to relativistic mechanics. In Vincent Hendricks et al. (eds.), *Interactions: Mathematics, physics and philosophy, 1860–1930* (Dordrecht: Springer), pp. 64–134.

Janssen, Michel, and Jürgen Renn
2007. Untying the knot: How Einstein found his way back to field equations discarded in the Zürich notebook. *GGR*, 2, pp. 839–926.
2015. Arch and scaffold: How Einstein found his field equations. *Physics today*, 68 (11): 30–36.

Janssen, Michel, and John Stachel
2004. The optics and electrodynamics of moving bodies. Max Planck Institute for History of Science, Berlin, preprint 265.

Johannesson, Paul
1896. *Das Beharrungsgesetz*. Berlin: Gärtners.

Jouguet, Émile
1908. *Lectures de mécanique*. 2 vols. Paris: Gauthier-Villars.

Jungnickel, Christa, and Russell McCormmach
1986. *Intellectual mastery of nature: Theoretical physics from Ohm to Einstein*. 2 vols. Chicago: The University of Chicago Press.

Kaiser, Walter
1987. Early theories of the electron gas. *HSPS*, 17: 270–297.

Kalckar, Jørgen, and Ole Ulfbeck
1982. Self-mass and equivalence in special relativity. Det Kongelige Danske Videnskabernes Selskab, *matematisk-fysiske Meddelelser*, 40 (11).

Kant, Immanuel
1746. *Gedanken von der wahren Schätzung der lebendigen Kräfte und Beurteilung der Beweise derer sich Herr von Leibniz und andere Mechaniker in dieser Streitsache bedienet haben, nebst einigen vorhergehenden Betrachtungen welche die Kraft der Körper überhaupt betreffen*. Königsberg: Dorn.
1758. *Neuer Lehrbegriff der Bewegung und Ruhe und der damit verknüpften Folgerungen in den ersten Gründen der Naturwissenschaft*. Königsberg: Driest.
1768. Von dem ersten Grunde des Unterschieds der Gegenden im Raume. *Wöchentliche Königsbergische Frag- und Anzeigungs-Nachrichten*, No. 6, 7, and 8.
1770. *De mundi sensibilis atque intelligibilis forma et principiis*. Königsberg: Kanter.
1781. *Kritik der reinen Vernunft*. Riga: Hartknoch.
1786. *Metaphysische Anfangsgründe der Naturwissenschaften*. Riga: Hartknoch.

Katzir, Shaul
2005. Poincaré's relativistic physics: Its origins and nature. *Physics in perspective*, 7: 268–292.

Kaufmann, Walter
1901. Die magnetische und die elektrische Ablenkbarkeit der Becquerelstrahlen und die scheinbare Masse des Elektrons. *GN* (1901), 143–155.
1902. Die elektromagnetische Masse des Elektrons. *PZ*, 4: 54–56.
1903. Über die elektromagnetische Masse der Elektronen. *GN* (1903), 326–330.
1905. Über die Konstitution des Elektrons. *BB* (1905), 949–956.
1906. Über die Konstitution des Elektrons. *AP*, 324: 487–553.

Kennefick, Daniel
2007a. *Traveling at the speed of thought: Einstein and the quest for gravitational waves*. Princeton: Princeton University Press.
2007b. Not only because of theory: Dyson, Eddington and the competing myths of the 1919 eclipse expedition. arXiv:0709.0685
2009. Testing relativity from the 1919 eclipse—a question of bias. *Physics today*, 62: 37–42.

Ketteler, Eduard
1873. *Astronomische Undulations-theorie, oder die Lehre von der Aberration des Lichtes*. Bonn: Neusser.

Killing, Wilhelm
1885. Die Mechanik in den nicht-Euklidischen Raumformen. *JRAM*, 98: 1–48.

Kirchhoff, Gustav
1857. Über die Bewegung der Elektrizität in Leitern. *AP*, 102: 529–544.
1876. *Vorlesungen über mathematische Physik*, Vol. 1: *Vorlesungen über Mechanik*. Leipzig: Teubner.

Klein, Christian Felix
1871. Über die sogenannte nicht-Euklidische Geometrie I. *Mathematische Annalen*, 4: 573–625.
1872. *Vergleichende Betrachtungen über neuere geometrische Forschungen*. Erlangen: Deichert.
1873. Über die sogenannte nicht-Euklidische Geometrie II. *Mathematische Annalen*, 6: 112–145.
1918a. Über die Differentialgesetze für die Erhaltung von Impuls und Energie in der Einsteinschen Gravitationstheorie. *GN* (1918), 171–189.
1918b. Über die Integralform der Erhaltungssätze und die Theorie der räumlich geschlossenen Welt. *GN* (1918), 394–423.
1926–1927. Vorlesungen über die Entwicklung der Mathematik im 19. Jahrhundert. 2 vols. Berlin: Springer.

Klein, Hermann
1872. *Die Prinzipien der Mechanik, historisch und kritisch dargestellt*. Leipzig: Teubner.

Kline, Morris
1972. *Mathematical thought from ancient to modern times*. 3 vols. Oxford: Oxford University Press.

Klinkerfues, Wilhelm
1867. *Die Aberration der Fixsterne nach der Wellentheorie*. Leipzig: Quandt & Händel.

Knobloch, Eberhard, and David Rowe (eds.)
1994. *The history of modern mathematics*. 3 vols. San Diego: Academic Press.

Knudsen, Ole
1980. 19th-century views on induction in moving conductors. *Centaurus*, 24: 346–360.

Kostro, Ludwik
1992. An outline of the history of Einstein's relativistic ether concept. In Eisenstaedt and Kox 1992, pp. 260–280.

Kottler, Friedrich
1916. Über Einsteins Äquivalenzhypothese und die Gravitation. *AP*, 4: 955–972.

Kox, Anne
1988. Hendrik Antoon Lorentz, the ether, and the general theory of relativity. *AHES*, 38: 67–78.
1992. General relativity in the Netherlands, 1915–1920. In Eisenstaedt and Kox 1992, pp. 39–56.
1997. The discovery of the electron: II. The Zeeman effect. *European journal of physics*, 18: 139–144.
2008. (ed.) *The scientific correspondence of H. A. Lorentz*, Vol. 1. New York: Springer.

Kox, Anne, and Jean Eisenstaedt (eds.)
2005. *Einstein studies*, Vol. 11: *The universe of general relativity*. Boston: Birkhäuser.

Kragh, Helge
1985. The fine structure of hydrogen and the gross structure of the physics community. *HSPS*, 15: 67–125.
2012. Geometry and astronomy: Pre-Einstein speculations of non-Euclidean space. arXiv:1205.4909

Kretschmann, Erich
1917. Über den physikalischen Sinn der Relativitätspostulate. A. Einsteins neue und seine ursprüngliche Relativitätstheorie. *AP*, 53: 575–614.
1922. Eine Bemerkung zu Hrn. A. Gullstrands Abhandlung: "Allgemeine Lösung des statischen Einkörperproblems in der Einsteinschen Gravitationstheorie." *AMAF*, 17 (2): 1–4.
1923. Das statische Einkörperproblem in der Einstein'schen Theorie. Antwort an Hrn. A. Gullstrand. *AMAF*, 17 (25): 1–4

Kristensen, Leif Kahl, and Kurt Møller Pedersen
2012. Roemer, Jupiter's satellites and the velocity of light. *Centaurus*, 54: 4–38.

Kuhn, Thomas
1962. *The structure of scientific revolutions*. Chicago: The University of Chicago Press.

Lagrange, Joseph Louis
1797. *Théorie des fonctions analytiques*. Paris: Imprimerie de la République.
1811–1815. *Mécanique analytique*, 2nd ed. 2 vols. Paris: Courcier.

Laguens, Florian
2018. *Eddington philosophe : la nature et la portée de la science physique d'après Arthur S. Eddington*. Thèse de doctorat, Université Paris 1.

Lalande, Jérôme le Français de
1771. *Astronomie*, 2nd ed., Vol. 3. Paris: Veuve Desaint.
1781. *Astronomie*, 2nd ed., Vol. 4. Paris: Veuve Desaint.

Lambert, Johann Heinrich
1786. [written in 1766] Theorie der Parallellinien. *Magazin für die reine und angewandte Mathematik*, 2: 137–164, 3: 325–258.

Lanczos, Cornelius
1922. Ein vereinfachendes Koordinatensystem für die Einsteinschen Gravitationsgleichungen. *PZ*, 23: 537–539.
1923. Zur Theorie der Einsteinschen Gravitationsgleichungen. *Zeitschrift für Physik*, 13: 7–16.

Landau, Lev Davidovich, and Evgeny Mikhailovich Lifshitz
1951. *The classical theory of fields*. Reading: Addison-Westley.
1960. *Mechanics*. Oxford: Pergamon Press.

Lange, Ludwig
1884. Über die wissenschaftliche Fassung des Galilei'schen Beharrungsgesetz. *Philosophische Studien*, 2: 266–297.
1885a. Nochmals über das Beharrungsgesetz. *Philosophische Studien*, 2: 539–545.
1885b. Über das Beharrungsgesetz. *Berichte über die Verhandlungen der Sächsischen Gesellschaft der Wissenschaften zu Leipzig, mathematisch-physische Klasse*, 37: 333–351.
1886. *Die geschichtliche Entwickelung des Bewegungsbegriffs und ihr voraussichtliches Endergebniss: Ein Beitrag zur historischen Kritik der mechanischen Prinzipien*. Leipzig: Engelmann.

Langevin, Paul
1905. La physique des électrons. *RGSPA*, 16: 257–276.
[1910–1911]. Lecture notes taken by Léon Brillouin: "Cours de Relativité au Collège de France 1910–1911," Léon Brillouin Papers, Box 7, folder 8, American Institute of Physics, Niels Bohr Library.
1911. L'évolution de l'espace et du temps. *Scientia*, 10: 31–54.
1912. Le temps, l'espace et la causalité dans la physique moderne. *Bulletin de la Société française de philosophie*, 12: 1–28.

Laplace, Pierre
1796. *Exposition du système du monde*, Vol. 1. Paris: Cercle Social.

Laplace, Pierre Simon de
1799. *Traité de mécanique céleste*, Vol. 1. Paris: Duprat.
1805. *Traité de mécanique céleste*, Vol. 4. Paris: Veuve Courcier.
1809. *The system of the world*, transl. by J. Pond. Vol. 1. London: Phillips.
1813. *Exposition du système du monde*, 4th ed. 2 vols. Paris: Veuve Courcier.

Larmor, Joseph
1894. A dynamical theory of the electric and luminiferous medium. Part I. *PT*. Also in *LMPP*, 1, pp. 414–535.
1895. A dynamical theory of the electric and luminiferous medium. Part II: Theory of electrons. *PT*. Also in *LMPP*, 1, pp. 543–597.
1896. On the theory of moving electrons and electric charges. *PM*. Also in *LMPP*, 1, pp. 615–618.
1897a. On the theory of the magnetic influence on spectra; and on the radiation from moving ions. *PM*, 44: 503–512.
1897b. A dynamical theory of the electric an luminiferous medium. Part III: Relations with material media. *PT*. Also in *LMPP*, 2, pp. 11–132.
1900. *Aether and matter*. Cambridge: Cambridge University Press.
1919. Gravitation and light. *Nature*, 104: 412.
1920. Gravitation and light. *Nature*, 104: 530.

Laub, Jakob
1907. Zur Optik der bewegten Körper I. *AP*, 22: 738–744.
1910. Über die experimentellen Grundlagen des Relativitätsprinzips. *Jahrbuch der Radioaktivität und Elektronik*, 7: 405–463.

Laue, Max
1907. Die Mitführung des Lichtes durch bewegte Körper nach dem Relativitätsprinzip. *AP*, 22: 538–547.
1911a. Zur Diskussion über den starren Körper in der Relativitätstheorie. *PZ*, 12: 85–87.
1911b. *Das Relativitätsprinzip*. Braunschweig: Vieweg.
1911c. Zur Dynamik der Relativitätstheorie. *AP*, 340: 524–542.
1913. *Das Relativitätsprinzip*, 2nd ed. Braunschweig: Vieweg.
1920a. Theoretisches über neuere optische Beobachtungen zur Relativitätstheorie. *PZ*, 21: 659–662.
1920b. Zur Theorie der Rotverschiebung der Spektrallinien an der Sonne. *Zeitschrift für Physik*, 3: 389–395.
1921. *Die Relativitätstheorie*, Vol. 2: *Die allgemeine Relativitätstheorie und Einsteins Lehre der Schwerkraft*. Braunschweig: Vieweg.

Laugwitz, Detlef
1996. *Bernhard Riemann, 1826–1866: Wendepunkte in der Auffassung der Mathematik*. Basel: Birkhäuser.

Lebedev, Pyotr Nikolaevich
1901. Untersuchungen über die Druckkräfte des Lichts. *AP*, 6: 433–458.

Le Bellac, Michel
2010. The Poincaré group. In E. Charpentier, E. Ghys, and A. Lesne (eds.), *The scientific legacy of Poincaré* (London: London Mathematical Society), pp. 329–350.

Lecornu, Léon
1914. *Cours de mécanique professé à l'École Polytechnique*. Paris: Gauthier-Villars.

Legendre, Adrien Marie
1787. Mémoire sur les opérations trigonométriques, dont le résultat dépend de la figure de la terre. *MAS* (1787), 352–383.
1794. *Éléments de géométrie*. Paris: Firmin Didot.

Lehmkuhl, Dennis
2014. Why Einstein did not believe that general relativity geometrizes gravity. *SHPMP*, 46: 316–326.

Lehner, Christoph
2005. Einstein and the principle of general relativity, 1916–1921. In Kox and Eisenstaedt 2005, pp. 103–108.

Leibniz, Gottfried Wilhelm
[1692]. Essay de dynamique. In Costabel 1960, pp. 97–106.
1695. Specimen dynamicum, pro admirandis naturae legibus circa corporum vires et mutuas actiones detegendis, et ad suas causas revocandis [part 1]. *Acta eruditorum* (1695), 145–157. Also in Leibniz 1849–1863, Vol. 6, pp. 234–246 (part 1), pp. 246–254 (part 2).
[1698]. Essay de dynamique sur les loix du mouvement, où il est monstré, qu'il ne se conserve pas la même quantité de mouvement, mais la même force absolue, ou bien la même quantité de l'action motrice. In Leibniz 1849–1860, Vol. 6, pp. 215–231.
1849–1863. *Mathematische Schriften*, ed. C. I. Gerhardt, 7 vols. Berlin: Asher.
1875–1890. *Die philosophischen Schriften*, ed. C. I. Gerhardt, 7 vols. Berlin: Weidmann.

Lense, Josef
1918. Über Relativitätseinflüsse in den Mondsystemen. *Astronomische Nachrichten*, 206: 117–120.

Lense, Josef, and Hans Thirring
1918. Über den Einfluss der Eigenrotation der Zentralkörper auf die Bewegung der Planeten und Monde nach der Einsteinschen Gravitationstheorie. *PZ*, 19: 156–163.

Le Roux, Jean
1921. La loi de gravitation et ses conséquences. *CR*, 172: 1467–1469.

Le Verrier, Urbain
1859. Théorie du mouvement de Mercure. *Annales de l'observatoire de Paris*, 5: 51–103.

Levi-Civita, Tullio
1917a. Nozione di parallelismo in una varietà qualunque e conseguente specificazione geometrica della curvatura Riemanniana. *RCMP*, 42: 173–205.
1917b. Statica Einsteiniana. *RAL*, 26: 458–470.
1918. La teoria di Einstein e il principio di Fermat. *Il nuovo cimento*, 16: 105–114.
1927. Sur l'écart géodésique. *Mathematische Annalen*, 97: 291–320.

Lie, Sophus
1886. Bemerkungen zu v. Helmholtzs Arbeit: Über die Thatsachen, die der Geometrie zugrunde legen. *Berichte über die Verhandlungen der Sächsischen Akademie der Wissenschaften zu Leipzig, mathematisch-physische Klasse*. Also in *Gesammelte Abhandlungen* (Leipzig: Teubner), Vol. 2 (1935), pp. 374–379.
1890. Über die Grundlagen der Geometrie. In *Berichte über die Verhandlungen der Sächsischen Akademie der Wissenschaften zu Leipzig, mathematisch-physische Klasse*. Also in *Gesammelte Abhandlungen* (Leipzig: Teubner), Vol. 2 (1935), pp. 380–413 (Abh. I), pp. 414–468 (Abh. II).

Liebmann, Heinrich
1902. Die Kegelschnitte und die Planetenbewegung im nichteuklidischen Raum. *Berichte über die Verhandlungen der Königlich-Sächsischen Gesellschaft der Wissenschaften zu Leipzig, mathematisch-physische Klasse*, 54: 493–423.

Liénard, Alfred
1896. La théorie de Lorentz. *EE*, 14: 417–424, 456–461.

1898a. Champ électrique et magnétique produit par une charge électrique concentrée en un point et animée d'un mouvement quelconque. *EE*, 16: 5–14, 53–59, 106–112.
1898b. La théorie de Lorentz et celle de Larmor. *EE*, 16: 320–334, 360–365.

Lipschitz, Rudolf
1869. Untersuchungen in Betreff der ganzen homogenen Functionen von *n* Differentialen. *JRAM*, 70: 71–102.
1870a. Entwicklung einiger Eigenschaften der quadratischen Formen von *n* Differentialen. *JRAM*, 71: 274–287, 288–295.
1870b. Fortgesetzte Untersuchungen in Betreff der ganzen homogenen Functionen von *n* Differentialen. *JRAM*, 72: 1–56.
1877. Bemerkungen zu dem Princip des kleinsten Zwanges. *JRAM*, 82: 316–342.

Liu, Chuang
1991. *Relativistic thermodynamics: Its history and foundations.* PhD dissertation, University of Pittsburgh.
1997. Planck and the special theory of relativity. In Roger Stuewer, Fabio Bevilacqua, and Dieter Hofmann (eds.), *The emergence of modern physics* (Pavia: La Goliardica Pavese), pp. 287–296.

Lloyd, Geoffrey
1970. *Early Greek science: Thales to Aristotle.* London: Chatto & Windus.

Lobachevsky, Nikolai Ivanovich
1835–1838. New elements of geometry, with a complete theory of parallels [In Russian]. *Scientific writings of the Imperial university in Kasan.* German transl. in Friedrich Engel and Paul Stäckel (eds.), *Urkunden zur Geschichte der nichteuklidischen Geometrie,* Vol. 1 (Leipzig: Teubner, 1898), pp. 67–236.
1837. Géométrie imaginaire. *JRAM*, 17: 295–320.
1840. *Geometrische Untersuchungen.* Berlin: Fincke.

Lodge, Oliver
1892. On the present state of our knowledge of the connection between ether and matter: An historical summary. *Nature*, 46: 164–165.
1893. Aberration problems. *PT*, 184A: 727–804.

Lodge, Paul
2003. Leibniz on relativity and the motion of bodies. *Philosophical topics*, 31: 277–308.

Lorentz, Hendrik Antoon
1875. *Over de theorie der terugkaatsing en breking van het licht.* Arnhem: Van der Zande. Transl. as "Sur la théorie de la réflexion et de la réfraction de la lumière" in *LCP*, 1, pp. 193–383.
1878. Over het verband tusschen de voortplantings sneldheid en samestelling der midden stofen. *VKA*. Transl. as "Concerning the relation between the velocity of propagation of light and the density and composition of media" in *LCP*, 2, pp. 3–119.
1886. Over den invloed, dien de beweging der aarde op de lichtverschijnselen uitoefent. *VKA*, 2: 297–372.
1887. De l'influence du mouvement de la terre sur les phénomènes lumineux. *AN*, 21: 103–176.
1892a. La théorie électromagnétique de Maxwell et son application aux corps mouvants. *AN*. Also in *LCP*, 2, pp. 164–321.
1892b. De relative beweging van der aarde en den aether. *VKA*. Transl. as "The relative motion of the earth and the ether" in *LCP*, 4, pp. 220–223.
1895. Versuch einer Theorie der elektrischen und optischen Erscheinungen in bewegten *Körpern*. Leiden. Also in *LCP*, 5, pp. 1–139.
1897a. Over de gedeeltelijke polarisatie van het licht dat door eene lichtbron in een magnetisch veld wordt uitgestraald. *VKA*. French transl. in *LCP*, 3, pp. 47–66.
1897b. Über den Einfluss magnetischer Kräfte auf die Emission des Lichtes. *AP*, 63: 278–284. Also in *LCP*, 3, pp. 40–46.

1898. Die Fragen welche die translatorische Bewegung des Lichtäthers betreffen [Düsseldorf Meeting]. *VDNA*. Also in *LCP*, 7, pp. 101–115.

1899a. Vereenvoudigde theorie der electrische en optische verschijnselen in Iichamen die zich bewegen. *VKA*. Transl. as "Théorie simplifiée des phénomènes électriques et optiques dans les corps en mouvement" in *AN* (1902) and in *LCP*, 5, pp. 139–155.

1899b. Stokes's theory of aberration in the supposition of a variable density of the aether. *PRA*, 1: 443–448.

1900. Beschouwingen over de zwaartekracht. *VKA*, 8: 603–620. French transl. in *LCP*, 5, pp. 198–215.

1902. The intensity of radiation and the motion of the earth. *PRA*, 4: 678–681.

1904. Electromagnetische verschijnselen in een stelsel dat zich met willekeurige snelheid, kleiner dan die van het licht, beweegt. *VKA*, 12: 986–1009. Transl. as "Electromagnetic phenomena in a system moving with any velocity smaller than light" in *PRA*. Also in *LCP*, 5, pp. 172–197.

1905. *Ergebnisse und Probleme der Elektronentheorie. Vortrag, gehalten am 20. Dezember 1904. im Elektrotechnischen Verein zu Berlin*. Berlin: Springer.

1909. *The theory of electrons and its applications to the phenomena of light and radiant heat. A course of lectures delivered in Columbia University, New York, in March and April 1906*. Leipzig: Teubner.

1910. Alte und neue Fragen der Physik. *PZ*, 11: 1234–1257.

1916. On Einstein's theory of gravitation. *PRA*. Also in *LCP*, 5, pp. 246–313.

1934–1939. *Collected papers*. 9 vols. The Hague: Nijhoff.

Love, Augustus

1897. *Theoretical mechanics: An introductory treatise on the principles of dynamics with applications and numerous examples*. Cambridge: Cambridge University Press.

Lützen, Jesper

1994. The geometrization of analytical mechanics: A pioneering contribution by J. Liouville (ca. 1850). In Knobloch and Rowe 1994, Vol. 2, pp. 77–98.

1999. Geometrizing configurations: Heinrich Hertz and his mathematical precursors. In Gray 1999, pp. 25–46.

2005. *Mechanistic images in geometric form: Heinrich Hertz's Principles of Mechanics*. Oxford: Oxford University Press.

MacCullagh, James

1848. [read December 9, 1839] An essay towards the dynamical theory of crystalline reflexion and refraction. Royal Irish Academy, *transactions*, 21: 17–50.

MacGregor, James Gordon

1893. On the hypotheses of dynamics. *PM*, 36: 233–264.

Mach, Ernst

1872. *Die Geschichte und die Wurzel des Satzes von der Erhaltung der Arbeit*. Prague: Calve.

1883. *Die Mechanik in ihrer Entwicklung. Historisch-kritisch dargestellt*. Leipzig: Brockhaus.

1889. *Die Mechanik in ihrer Entwicklung. Historisch-kritisch dargestellt*, 2nd ed. Leipzig: Brockhaus.

1897. *Die Mechanik in ihrer Entwicklung. Historisch-kritisch dargestellt*, 3rd ed. Leipzig: Brockhaus.

Machamer, Peter

1978. Aristotle on natural place and motion. *Isis*, 69: 377–387.

Maltese, Giulio

2000. On the relativity of motion in Leonhard Euler's science. *AHES*, 54: 319–348.

Maltese, Giulio, and Lucia Orlando

1995. The definition of rigidity in the special theory of relativity and the genesis of the general theory of relativity. *SHPMP*, 26: 263–306.

Martínez, Alberto
2004. Ritz, Einstein, and the emission hypothesis. *Physics in perspective*, 6: 4–28.
2009. *Kinematics: The lost origins of Einstein's relativity*. Baltimore: Johns Hopkins University Press.

Mascart, Éleuthère
1866. *Éléments de mécanique*. Paris: Hachette.
1872. Sur les modifications qu'éprouve la lumière par suite du mouvement de la source et du mouvement de l'observateur. *Annales scientifiques de l'Ecole Normale*, 1: 157–214.
1874. Sur les modifications qu'éprouve la lumière par suite du mouvement de la source et du mouvement de l'observateur. *Annales scientifiques de l'Ecole Normale*, 3: 363–420.
1889–1893. *Traité d'optique*. 3 vols. Paris: Gauthier-Villars.

Maxwell, James Clerk
1856. On Faraday's lines of force. *Transactions of the Cambridge Philosophical Society*, 10: 155–229.
1861–1862. On physical lines of force. *PM*, 21: 161–175, 281–291, 338–348; 23: 12–24, 85–95.
1865. A dynamical theory of the electromagnetic field. *PT*, 155: 459–512.
[1867]. On the influence of the motion of the heavenly bodies on the index of refraction of light [Letter to Airy of 12 June 1867]. In William Huggins, "Further observations of the spectra of some of the starts and nebulae, with an attempt to determine therefrom whether these bodies are moving toward or from the earth, also observations on the spectra of the sun and comet II." *PT*, 158 (1868): 532–535.
1869. Question IX [for the Mathematical Tripos examination]. *Cambridge University calendar*. Cambridge: Cambridge University Press.
1873. *A treatise on electricity and magnetism*. 2 vols. Oxford: The Clarendon Press.
1876. *Matter and motion*. London: Society for Promoting Christian Knowledge.
1878. Ether. *Encyclopaedia Britannica*. 9th ed. Vol. 8, pp. 568–572.
1880. On a possible mode of detecting a motion of the solar system through the luminiferous ether [Maxwell to Todd, March 19, 1879]. *Nature*, 21: 314–315.

Mayrargue, Arnaud
1991. *L'aberration des étoiles et l'éther de Fresnel*. Thèse de doctorat, Université de Paris 7.

McCormmach, Russell
1970. H. A. Lorentz and the electromagnetic view of nature. *Isis*, 61: 459–497.

Michelson, Albert
1881. The relative motion of the earth and the luminiferous ether. *American journal of science*, 22: 120–129.
1882. Sur le mouvement relatif de la terre et de l'éther. *CR*, 94: 520–523.

Michelson, Albert, and Edward Morley
1886. Influence of the motion of the medium on the velocity of light. *American journal of science*, 31: 377–386.
1887. On the relative motion of the earth and the luminiferous ether. *American journal of science*, 34: 333–345.

Mie, Gustav
1912a. Grundlagen einer Theorie der Materie. Erste Mitteilung. *AP*, 37: 511–534.
1912b. Grundlagen einer Theorie der Materie. Zweite Mitteilung. *AP*, 39: 1–40.
1913. Grundlagen einer Theorie der Materie. Dritte Mitteilung. *AP*, 40: 1–66.
1920. Die Einführung eines vernunftgemässen Koordinatensystems in die Einsteinsche Gravitationstheorie und das Gravitationsfeld einer schweren Kugel. *AP*, 62: 46–74.

Miller, Arthur
1973. A study of Henri Poincaré's "Sur la dynamique de l'électron." *AHES*, 10: 207–328.
1980. On some other approaches to electrodynamics in 1900. In Harry Woolf (ed.), *Some strangeness in the proportion. A centennial symposium to celebrate the achievements of Albert Einstein* (Reading: Addison-Westley), pp. 66–91.

1981a. *Albert Einstein's special relativity: Emergence and early interpretation (1905–1911)*. Reading: Addison-Westley.
1981b. Unipolar induction: A case study of the interaction between science and technology. *Annals of science*, 3: 155–189.

Minkowski, Hermann
[1907]. Das Relativitätsprinzip. *AP*, 47 (1915), 927–938.
1908. Die Grundgleichungen für die elektromagnetischen Vorgänge in bewegten Körpern. *GN* (1908), 53–111.
1909. Raum und Zeit. *PZ*, 10: 104–111.

Minkowski, Hermann, and Max Born
1910. Eine Ableitung der Grundgleichungen vom Standpunkte der Elektronentheorie. Aus dem Nachlasse bearbeitet von Max Born. *Mathematische Annalen*, 68: 552–564.

Moigno, François
1847. *Répertoire d'optique moderne ou analyse complète des travaux modernes relatifs aux phénomènes de la lumière. Parts 1–2*. Paris: Franck.
1850. *Répertoire d'optique moderne ou analyse complète des travaux modernes relatifs aux phénomènes de la lumière. Parts 3–4*. Paris: Franck.

More, Henry
1659. *The immortality of the soul, so farre forth as it is demonstrable from the knowledge of Nature and the Light of Reason*. London: Flescher.
1712. *A collection of several philosophical writings of Henry More*, 4th ed. London: Downing.

Mormorino, Gianfranco
1994. *Penetralia motus: La fondazione relativistica della meccanica in Christiaan Huygens, con l'edizione del Codex Hugeniorum 7A*. Florence: La Nova Italia Editrice.

Mosengeil, Kurd von
1906. Theorie der stationären Strahlung in einem gleichförmig bewegten Hohlraum. PhD thesis, Friedrich-Wilhelms-Universität zu Berlin.

Muñoz, Gerardo, and Preston Jones
2010. The equivalence principle, uniformly accelerated reference frames, and the uniform gravitational field. *AJP*, 78: 377–383.

Nabonnand, Philippe
2010. La théorie de l'espace de Poincaré. In Pierre Édouard Bour and Sophie Roux (eds.), *Recherches sur la philosophie et le langage. Lambertiana* (Paris: Vrin), pp. 373–391.
Navarro, Jaume (ed.)
2018. *Ether and modernity: The recalcitrance of an epistemic object in the early twentieth century*. Oxford: Oxford University Press.

Narr, Friedrich
1875. *Einleitung in die theoretische Mechanik*. Leipzig: Teubner.

Nersessian, Nancy
1986. Why wasn't Lorentz Einstein? *Centaurus*, 29: 205–242.
1988. "Ad hoc" is not a four-letter word: H. A. Lorentz and the Michelson-Morley experiment. In Goldberg and Stuewer 1988, pp. 71–77.

Neumann, Carl
1870. *Ueber die Principien der Galilei-Newtonschen Theorie.* Leipzig: Teubner.
1886. Ausdehnung der Kepler'schen Gesetze auf den Fall, dass die Bewegung auf einer Kugelfläche statt-findet. *Berichte über die Verhandlungen der Königlich-Sächsischen Gesellschaft der Wissenschaften zu Leipzig, mathematisch-physische Klasse,* 38: 1–2.

Neumann, Franz
1846. *Die mathematische Gesetze der inducirten elektrischen Ströme* [read on October 27, 1845]. Berlin: Reimer.

Newton, Isaac
[c. 1665]. Waste Book, MS Add. 4004, Cambridge University Library, Cambridge, UK. In *The Newton Project,* http://www.newtonproject.ox.ac.uk/view/texts/normalized/NATP00220
[c. 1668?]. De gravitatione et aequipondio fluidorum et solidorum in fluidis. MS Add. 4003, Cambridge University library.
1687. *Philosophiae naturalis principia mathematica.* London: Streater & Smith.
1706. *Optice, sive de reflexionibus, refractionibus et coloribus lucis.* London: Smith & Walford.
1713. *Philosophiae naturalis principia mathematica,* 2nd ed. Cambridge: Crownfield.
1729. *The mathematical principles of natural philosophy.* Transl. by Andrew Motte. 2 vols. London: Motte.

Nichols, Ernest Fox, and Gordon Ferrie Hull
1903. The pressure due to radiation. *The astrophysical journal,* 17: 315–351.

Noether, Fritz
1910. Zur Kinematik des starren Körpers in der Relativitätstheorie. *AP,* 31: 919–944.

Nolte, David
2020. The fall and rise of the Doppler effect. *Physics today,* March 2020: 31–35.

Nordmann, Charles
1922. Einstein expose et discute sa théorie. *Revue des deux mondes,* 9: 129–166.

Nordmeyer, Paul
1903. Über den Einfluss der Erdbewegung auf die Verteilung der Intensität der Licht- und Wärmestrahlung. *AP,* 11: 284–302.

Nordström, Gunnar
1909. Zur Elektrodynamik Minkowskis. *PZ,* 10: 681–687.
1910. Zur elektromagnetischen Mechanik. *PZ,* 11: 440–445.
1911. Zur Relativitätsmechanik deformierbar Körper. *PZ,* 12: 854–857.
1912. Relativitätstheorie und Gravitation. *PZ,* 13: 1126–1129.
1913. Zur Theorie der Gravitation vom Standpunkt des Relativitätsprinzips. *AP,* 42: 533–554.

North, John
1983. The satellites of Jupiter, from Galileo to Bradley. In Alwyn van der Merwe (ed.), *Old and new questions in physics, cosmology, philosophy, and theoretical biology* (New York: Plenum Press), pp. 689–717.
1984. How Einstein found his field equations: 1912–1915. *HSPS,* 14: 253–316.
1985. What was Einstein's principle of equivalence? *SHPS,* 16: 203–246.
1987. Einstein, the hole argument and the reality of space. In John Forge (ed.), *Measurement, realism, and objectivity* (Dordrecht: Reidel), pp. 153–188.
1992a. Einstein, Nordström and the early demise of scalar, Lorentz-covariant theories of gravitation. *AHES,* 45: 17–94.
1992b. The physical content of general covariance. In Eisenstaedt and Kox 1992, pp. 281–315.
1993a. Einstein and Nordström: Some lesser known thought experiments in gravitation. In Earman, Janssen, and Norton 1993, pp. 3–29.

1993b. General covariance and the foundations of general relativity: Eight decades of dispute. *Reports on progress in physics*, 56: 791–858.

2004. Einstein's investigations of Galilean covariant electrodynamics prior to 1905. *AHES*, 59: 45–105.

2005. A conjecture on Einstein, the independent reality of spacetime coordinate systems and the disaster of 1913. In Kox and Einsenstaedt 2005, pp. 67–102.

2007. What was Einstein's "fateful prejudice"? *GGR*, 2, pp. 715–784.

2013. Chasing the light: Einstein's most famous thought experiment. In James Robert Brown, Mélanie Frappier, and Letitia Meynell (eds.), *Thought experiments in philosophy, science and the arts* (New York: Routledge), pp. 123–140.

2014. Einstein's special theory of relativity and the problems in the electrodynamics of moving bodies that led him to it. In Michel Janssen and Christoph Lehner (eds.), *The Cambridge companion to Einstein* (Cambridge: Cambridge University Press), pp. 72–102.

2020. Einstein's conflicting heuristics: The discovery of general relativity. In C. Beisbart, T. Sauer, and C. Wüthrich (eds.), *Thinking about space and time: 100 years of applying and interpreting general relativity* (Cham: Springer), pp. 17–48.

O'Hara, James, and Willibald Pricha
1987. *Hertz and the Maxwellians*. London: Peregrinus.

Olesko, Kathryn
1991. *Physics as a calling: Discipline and practice in the Königsberg seminar of physics*. Ithaca: Cornell University Press.

Ørsted, Hans Christian
1820. *Experimenta circa effectum conflictus electrici in acum magneticam*. Copenhagen: Schultz.

Oseen, Carl
1921. Über das allgemeine statische, kugelsymmetrische Gravitationsfeld nach der Einsteinschen Theorie. *AMAF*, 16 (9): 1–6.

Painlevé, Paul
1921a. La mécanique classique et la théorie de la relativité. *CR*, 173: 677–680.
1921b. La gravitation dans la mécanique de Newton et dans la mécanique d'Einstein. *CR*, 173: 873–887.
1922. La théorie classique et la théorie einsteinienne de la gravitation. *CR*, 174: 1137–1143.

Pais, Abraham
1982. *"Subtle is the Lord . . .": The science and life of Albert Einstein*. Oxford: Oxford University Press.

Paty, Michel
1987. The scientific reception of relativity in France. In Glick 1987, pp. 113–167.
1993. *Einstein philosophe : la physique comme pratique philosophique*. Paris: Presses Universitaires de France.
2002. Poincaré, Langevin et Einstein. *Épistémologiques*, 2: 33–73.

Pauli, Wolfgang
1921. *Relativitätstheorie*. Leipzig: Teubner.

Pav, Peter Anton
1966. Gassendi's statement of the principle of inertia. *Isis*, 57: 24–34.

Péclet, Eugène
1823. *Cours de physique*. Marseille: Ricard.
1838. *Traité élémentaire de physique*. Paris: Hachette.

Pedersen, Kurt Møller
1978. La vie et l'œuvre de Roemer. In René Taton (ed.), *Roemer et la vitesse de la lumière : table ronde du Centre national de la recherche scientifique, Paris, 16 et 17 juin 1976* (Paris, Vrin), pp. 113–128.
1980. Roger Joseph Boscovich and John Robison on terrestrial aberration. *Centaurus*, 24: 335–345.
2000. Water-filled telescopes and the prehistory of Fresnel ether dragging. *AHES*, 54: 499–564.
2008. Leonhard Euler's wave theory of light. *Perspectives on science*, 16: 392–416.

Pestre, Dominique
1984. *Physique et physiciens en France, 1918–1940*. Paris: Archives contemporaines.

Pietrocola Pinto de Oliveira, Mauricio
1992. *Élie Mascart et l'optique des corps en mouvement*. Thèse de doctorat, Université de Paris 7.

Pinaud, Auguste
1846. *Programme d'un cours élémentaire de physique*, 4th ed. Paris: Hachette.

Planck, Max
1906a. Das Prinzip der Relativität und die Grundgleichungen der Mechanik. *Verhandlungen der Deutschen physikalischen Gesellschaft*, 8: 136–141.
1906b. Die Kaufmannschen Messungen der Ablenkbarkeit der β-Strahlen in ihrer Bedeutung für die Dynamik der Elektronen. *PZ*, 7: 753–761.
1907. Zur Dynamik bewegter Systeme. *BB* (1907), 542–570.
1908a. Bemerkungen zum Prinzip der Aktion und Reaktion in der allgemeinen Dynamik. *PZ*, 9: 828–830.
1908b. Zur Dynamik bewegter Systeme [reprint of Planck 1907]. *AP*, 25: 1–34.
1910. Gleichförmige Rotation und Lorentz-Kontraktion. *PZ*, 11: 294.

Poincaré, Henri
1881. Sur les fonctions fuchsiennes. *CR*, 92: 333–335, 395–398, 957, 1198–1200, 1274–1276, 1484–1487.
1887. Sur les hypothèses fondamentales de la géométrie. *Bulletin de la Société mathématique de France*, 15: 203–216.
1889. *Leçons sur la théorie mathématique de la lumière* [1st semester 1887–1888]. Ed. J. Blondin. Paris: Carré.
1890. *Electricité et optique. I. Les théories de Maxwell et la théorie électromagnétique de la lumière* [2nd semester 1887–1888]. Paris: Carré.
1891. Les géométries non euclidiennes. *RGSPA*, 2: 761–774.
1892. Sur les géométries non euclidiennes. *RGSPA*, 3: 74–75.
1893. *Théorie mathématique de la lumière. II. Nouvelles études sur la diffraction. Théorie de la dispersion de Helmholtz* [1st semester, 1891–1892]. Eds. M. Lamotte and D. Hurmuzescu. Paris: Carré.
1895a. A propos de la théorie de Larmor. *EE*, 3: 5–13, 285–295.
1895b. A propos de la théorie de Larmor. *EE*, 5: 5–14, 385–392.
1897. Les idées de Hertz sur la mécanique. *RGSPA*, 8: 734–743.
1898. La mesure du temps. *Revue de métaphysique et de morale*, 6: 1–13.
1899. Des fondements de la géométrie : à propos d'un livre de M. Russell. *Revue de métaphysique et de morale*, 7: 251–279.
1900a. La théorie de Lorentz et le principe de la réaction. In *Recueil de travaux offerts par les auteurs à H. A. Lorentz à l'occasion du 25ème anniversaire de son doctorat le 11 décembre 1900. AN*, 5: 252–278.
1900b. Les relations entre la physique expérimentale et la physique mathématique. *RGSPA*, 11: 1163–1175.
1900c. Über die Beziehungen zwischen der experimentellen und der mathematischen Physik. *PZ*, 2: 166–171, 182–186, 196–201.
1901a. Sur les principes de la mécanique. *Bibliothèque du congrès international de philosophie* (Paris: Armand Collin), Vol. 3 pp. 457–494.
1901b. *Électricité et optique. La lumière et les théories électrodynamiques [Sorbonne lectures of 1888, 1890, and 1899]*. Paris: Gauthier-Villars.
1902a. *La science et l'hypothèse*. Paris: Flammarion.
1902b. Sur la valeur objective de la science. *Revue de métaphysique et de morale*, 10: 263–293.

1904a. L'état actuel et l'avenir de la physique mathématique [Saint-Louis lecture]. *Bulletin des sciences mathématiques*, 28: 302–324.

1904b. *Wissenschaft und Hypothese*. Leipzig: Teubner.

1905a. Sur la dynamique de l'électron. *CR*, 140: 1504–1508.

1905b. La vie et les œuvres d'Alfred Cornu. *Journal de l'Ecole Polytechnique*, 10: 143–155.

1906a. Sur la dynamique de l'électron. *RCMP*, 21: 129–175.

1906b. La fin de la matière. *Athenaeum*, 4086: 201–202.

[1906–1907]. Les limites de la loi de Newton [Sorbonne lectures]. *Bulletin astronomique publié par l'Observatoire de Paris*, 17 (1953): 121–365.

1908. La dynamique de l'électron. *RGSPA*, 19: 386–402.

1909. La mécanique nouvelle. *Revue scientifique*, 12: 170–177.

1910a. La mécanique nouvelle. In Henri Poincaré, *Sechs Vorträge über ausgewählte Gegenstände aus der reinen Mathematik und mathematischen Physik* (Leipzig: Teubner), pp. 49–58.

1910b. Die neue mechanik. *Himmel und Erde*, 23: 97–116.

1912. L'espace et le temps. *Scientia*, 12: 159–170.

1913a. *La dynamique de l'électron* [July 1912 lectures at the Ecole Supérieure des Postes et des Télégraphes, ed. by Viard and Pomey]. Paris: Dumas.

1913b. *The foundations of science*. Transl. of *La science et l'hypothèse* (Paris, 1902), *La valeur de la science* (Paris, 1905), and *Science et méthode* (Paris, 1908). New York: The Science Press.

1913c. Les conceptions nouvelles de la matière. In Paul Doumergue (ed.), *Le matérialisme actuel* (Paris), pp. 49–67.

Poisson, Siméon Denis

1811. *Traité de mécanique*, Vol. 1. Paris: veuve Courcier.

1833. *Traité de mécanique*, Vol. 1, 2nd ed. Paris: Bachelier.

Pojman, Paul

2020. Ernst Mach. *SEP*, Winter 2020 edition. https://plato.stanford.edu/archives/win2020/entries/ernst-mach/

Portnoy, Esther

1982. Riemann's contribution to differential geometry. *Historia mathematica*, 9: 1–18.

Potier, Alfred

1874. Conséquences de la formule de Fresnel relative à l'entrainement de l'éther par les milieux transparents. *JP*, 3: 201–204.

1876. De l'entraînement des ondes lumineuses par la matière pondérable en mouvement. *JP*, 5: 105–108.

Potters, Jan

2019. Heuristics versus norms: On the relativistic responses to the Kaufmann experiments. *SHPMP*, 66: 69–89.

Poynting, John Henry

1884. On the transfer of energy in the electromagnetic field. *PT*, 175: 343–361.

Powell, Baden

1846. Remarks on some points of the reasoning in the recent discussions on the theory of the aberration of light. *PM*, 29: 425–440.

Preston, Samuel Tolver

1885. On some electromagnetic experiments of Faraday and Plücker. *PM*, 19: 131–140.

Príncipe Silva, João Paulo

2008. *La réception française de la mécanique statistique*. Thèse de doctorat, Université Denis Diderot.

2012. Sources et nature de la philosophie de la physique d'Henri Poincaré. *Philosophia scientiae*, 16: 197–222.

Privat-Deschanel, Augustin
1869. *Traité élémentaire de physique*. Paris: Hachette.

Provost, Jean-Pierre, and Christain Bracco
2006. La théorie de la relativité de Poincaré de 1905 et les transformations actives. *AHES*, 60: 337–351

Pulte, Helmut
1989. *Das Prinzip der kleinsten Wirkung und die Kraftkonzeptionen der rationalen Mechanik: Eine Unter- suchung zur Grundproblematik bei Leonhard Euler, Pierre Louis Moreau de Maupertuis und Joseph Louis Lagrange*. Stuttgart: Steiner.

Pyenson, Lewis
1979. Physics in the shadow of mathematics: The Göttingen electron-theory seminar of 1905. *AHES*, 21: 55–89.
1980. Einstein's education: Mathematics and the laws of nature. *Isis*, 71: 399–425.
1982. Audacious enterprise: The Einsteins and electrotechnology in late nineteenth-century Munich. *HSPS*, 12: 373–392.
1985. *The young Einstein: The advent of relativity*. Bristol: Adam Hilger.

Ray, Christopher
1987. *The evolution of relativity*. Bristol: Adam Hilger.

Rayleigh, Lord
1892. [1887] Aberration. *Nature*, 45: 499–502.
1902a. Is rotatory polarization influenced by the earth's motion? *PM*, 4: 215–220.
1902b. Does motion through the aether cause double refraction? *PM*, 4: 678–683.

Reich, Karin
1973. Die Geschichte der Differentialgeometrie von Gauss bis Riemann (1828–1868). *AHES*, 11: 273–382.
1992. Levi-Cititasche Parallelverschiebung, affiner Zusammenhang, Übertragungsprinzip: 1916/17– 1922/23. *AHES*, 44: 77–105.
1994. *Die Entwicklung des Tensorkalküls. Vom absoluten Differentialkalkül zur Relativitätstheorie*. Basel: Birkhäuser.

Reiff, Richard
1893. Die Fortpflanzung des Lichtes in bewegten Medien nach der elektrischen Lichttheorie. *AP*, 1: 361–367.

Renn, Jürgen
1993. Einstein as a disciple of Galileo: A comparative study of concept development in physics. *Science in context*, 6: 311–341.
2007a. The summit almost scaled: Max Abraham as a pioneer of relativistic theory of gravitation. *GGR*, 3, pp. 305–330.
2007b. The third way to general relativity: Einstein and Mach in context. *GGR*, 4, pp. 21–76.

Renn, Jürgen, and Tilman Sauer
1999. Heuristics and mathematical representation in Einstein's search for a gravitational field equation. In Goenner et al. 1999, pp. 87–125.
2007. Pathways out of classical physics: Einstein's double strategy in his search for the gravitational field equation. *GGR*, 1, pp. 113–312.

Renn, Jürgen, and Matthias Schemmel (eds.)
2007. Gravitation in the twilight of classical physics: Between mechanics, field theory, and astronomy. *GGR*, 3, pp. 3–22.

Renn, Jürgen, and John Stachel
2007. Hilbert's foundation of physics: From a theory of everything to a constituent of general relativity. *GGR*, 4, pp. 1778–1895.

Resal, Henry
1862. *Éléments de mécanique*. Paris: Mallet-Bachelier.

Rice, James
1920. The predicted shift of the Fraunhofer lines. *Nature*, 104: 598.

Ricci-Curbastro, Gregorio
1884. Principii di una teoria delle forme differenziali quadratiche. *Annali di matematica pure ed applicate*, 12: 135–167. Also in Ricci 1956–1957, pp. 138–171.
1886. Sui parametri e gli invarianti delle forme quadratiche differenziali. *Annali di matematica pure ed applicate*, 14: 1–11. Also in Ricci 1956–1957, pp. 177–188.
1887. Sulla derivazione covariante ad una forma quadratica differenziale. *RAL*, 3: 15–18. Also in: Ricci 1956–1957, pp. 198–203.
1888. Delle derivazioni covarianti e contravarianti e del loro uso nella analisi applicata. *Studi editi dalla Università di Padova a commemorare l'ottavo centenario della Università de Bologna*, 3: 3–23. Also in Ricci 1956–1957, pp. 244–267.
1892. Résumé de quelques travaux sur les systèmes variables de fonctions associés à une forme différentielle quadratique. *Bulletin des sciences mathématiques*, 16: 167–189. Also in Ricci 1956–1957, pp. 288–310.
1956–1957. *Opere*. 2 vols. Roma: Edizioni cremonese.

Ricci-Curbastro, Gregorio, and Tullio Levi-Civita
1901. Méthodes de calcul différentiel absolu et leurs applications. *Mathematische Annalen*, 54: 125–201.

Riemann, Bernhard
[1861]. Commentatio mathematica, qua respondere tentatur quaestioni ab Illma Academia Parisiensi proposidae: 'Trouver quel doit être l'état calorifique d'un corps solide homogène indéfini pour qu'un système de courbes isothermes, à un instant donné, restent isothermes après un temps quelconque, de telle sorte que la température d'un point puisse s'exprimer en fonction du temps et de deux autres variables indépendantes.' In Riemann 1876, pp. 370–383.
1867. Über die Hypothesen, welche der Geometrie zugrunde legen [Habilitation lecture, 1854]. *Abhandlungen der Königlichen Gesellschaft der Wissenschaften zu Göttingen*, 13: 132–152.
1876. *Werke und Wissenschaflicher Nachlass*, ed. by Heinrich Weber with Richard Dedekind's collaboration. Leipzig: Teubner.

Ritz, Walther
1908a. Recherches critiques sur l'électrodynamique générale. *ACP*, 13: 145–275.
1908b. Recherches critiques sur les théories électrodynamiques de Cl. Maxwell et de H.-A. Lorentz. *Archives des sciences physiques et naturelles*, 26: 209–239.
1908c. Du rôle de l'éther en physique. *Rivista di scienza: Scientia*, 3: 260–274.
[1909]. Das Prinzip der Relativität in der Optik (Antrittsrede zur Habilitation, MS ed. by Pierre Weiss). In Ritz 1911, pp. 509–518.
1911. *Gesammelte Werke. Oeuvres* (ed. Pierre Weiss). Paris: Gauthier-Villars.

Roberts, John
2003. Leibniz on force and absolute motion. *Philosophy of science*, 70: 553–573.

Robison, John
1790. [read April 7, 1788] On the motion of light, as affected by refracting and reflecting substances, which are also in motion. Royal Society of Edinburgh, *transactions*, 2: 83–111.

Robb, Alfred
1911. *Optical geometry of motion, a new view of the theory of relativity*. Cambridge: Heffner & Sons.

Rohrlich, Fritz
1960. Self-energy and the stability of the classical electron. *AJP*, 28: 639–643.
1965. *Classical charged particles: Foundations of their theory*. Reading, MA: Addison-Wesley.

Rømer, Ole Christensen
1676. Démonstration touchant le mouvement de la lumiere trouvé par M. Römer de l'Académie Royale des Sciences [anonymous report on Rømer's argument]. *Journal des sçavans*, 1676: 233–236.

Roseveare, Nicholas
1982. *Mercury's perihelion. From Le Verrier to Einstein*. Oxford: Clarendon Press.

Rowe, David
2008. Max von Laue's role in the relativity revolution. *The mathematical intelligencer*, 30: 54–60.

Rowland, Henry
1878. On the magnetic effect of electric convection. *American journal of science*, 15: 30–38.

Ryckman, Thomas
2005. *The reign of relativity: Philosophy in physics 1915–1925*. Oxford: Oxford University Press.
2017. *Einstein*. New York: Routledge.

Rynasiewicz, Robert
1988. Lorentz's local time and the theorem of corresponding states. In Arthur Fine and Jarrett Leplin (eds.), *PSA 1988* (East Lansing: Philosophy of Science Association), Vol. 1, pp. 67–74.
1999. Kretschmann's analysis of covariance and relativity principles. In Goenner, Renn, Ritter, and Sauer 1999, pp. 431–462.
2000. The construction of the special theory: Some queries and considerations. In Howard and Stachel 2000, pp. 159–201.
2014. Newton's views on space, time, and motion. *SEP*, Summer 2014 edition, https://plato.stanford.edu/archives/sum2014/entries/newton-stm/

Saccheri, Girolamo
1733. *Euclides ab omni naevo vindicatus, sive, conatus geometricus quo stabiliuntur prima ipsa universae geometriae principia*. Milan: Montanus.

Saint-Venant, Adhémar Barré de
1851. *Principes de mécanique fondés sur la cinématique*. Paris: Bachelier.

Sánchez-Ron, José
1987. The reception of Einstein's relativity in Great Britain. In Glick 1987, pp. 27–58.
2021. *Einstein's relativity in Great Britain: From Eddington to Hawking and Penrose; a tale of physicists, astronomers, mathematicians and philosophers*. Singapore: World Scientific.

Sauer, Tilman
1999. The relativity of discovery: Hilbert's first note on the foundations of physics. *AHES*, 53: 529–175.
2005a. Einstein equations and Hilbert Action: What is missing on page 8 of the proofs for Hilbert's first communication on the foundations of physics. *AHES*, 59: 577–590.

2005b. Einstein's review paper on general relativity theory. In Ivor Grattan-Guiness (ed.), *Landmark writings in western mathematics, 1640–1940* (Amsterdam: Elsevier), pp. 802–822.
2013. Marcel Grossmann's contribution to the general theory of relativity. arXiv:1312.4068

Saunders, Simon
2013. Rethinking Newton's *Principia*. *Philosophy of science*, 80: 22–48.

Schell, Wilhelm
1870. *Theorie der Bewegung und der Kräfte: Ein Lehrbuch der theoretischen Mechanik, mit besonderer Rücksicht auf die Bedürfnisse technischer Hochschulen.* Leipzig: Teubner.
1879. *Theorie der Bewegung und der Kräfte: Ein Lehrbuch der theoretischen Mechanik, mit besonderer Rücksicht auf die Bedürfnisse technischer Hochschulen,* 2nd ed. 2 vols. Leipzig: Teubner.

Schilpp, Paul Arthur
1949. *Albert Einstein, philosopher-scientist.* Evanston: Library of Living Philosophers.

Scholz, Erhard
1980. *Geschichte des Mannigfaltigkeitsbegriffs von Riemann bis Poincaré.* Basel: Birkhäuser.
1992. Riemann's vision of a new approach to geometry. In Luciano Boi, Dominique Flament, and Jean-Michel Salenskis (eds.), *1830–1930: A century of geometry* (Berlin: Springer), pp. 22–34.

Schrödinger, Erwin
1918a. Die Energiekomponenten des Gravitationsfeldes. *PZ*, 19: 4–7.
1918b. Über ein Lösungssysteme der allgemein kovarianten Gravitationsgleichungen. *PZ*, 19: 20–22.

Schuster, Peter Maria
2005. *Moving the stars: Christian Doppler, his life, his works and principle, and the world after.* Pöllauberg: Living Edition.

Schütz, Ignaz
1897. Prinzip der absoluten Erhaltung der Energie. *GN* (1897), 110–123.

Schwarzschild, Karl
1900. Über das zulässige Krümmungsmass des Raumes. *Vierteljahrschrift der Astronomischen Gesellschaft*, 35: 337–347.
1903a. Zur Elektrodynamik. I: Zwei Formen des Princips der Action in der Elektronentheorie. *GN* (1903), 126–131.
1903b. Zur Elektrodynamik. II: Die elementare elektrodynamische Kraft. *GN* (1903), 132–141.
1903c. Zur Elektrodynamik. III: Über die Bewegung des Elektrons. *GN* (1903), 245–278.
1916. Über das Gravitationsfeld eines Massenpunktes nach der Einsteinschen Theorie. *BB* (1916), 189–196.

Scott, Robert B.
2015. Teaching the gravitational redshift: Lessons from the history and philosophy of physics. Spanish relativity meeting (ERE 2014): Almost 100 years after Einstein's revolution. *Journal of physics: Conference series*, 600 (2015): 012055

Searle, George Frederick Charles
1897. On the steady motion of an electrified ellipsoid. *PM*, 5: 329–341.

Seelig, Carl
1954. *Albert Einstein: eine dokumentarische Biographie, 1894–1962.* Zürich: Europa Verlag.

Sellmeier, Wolfgang
1871. Zur Erklärung der anomalen Farbenfolge im Spectrum einiger Substanzen. *AP*, 143: 272–282.

Siegel, Daniel
1991. *Innovation in Maxwell's electromagnetic theory: Molecular vortices, displacement current, and light.* Cambridge: Cambrige University Press.

Silberstein, Ludwik
1922. *The theory of general relativity and gravitation* [Toronto lectures of January 1921]. New York: Van Nostrand.

Simon, Josep
2013. Physics textbooks and textbook physics in the nineteenth and twentieth centuries. In Jed Buchwald and Robert Fox (eds.), *The Oxford handbook in the history of physics* (Oxford: Oxford University Press), pp. 651–678.

Simpson, Thomas
1740. *Essays on several curious and useful subjects, in speculative and mix'd mathematicks.* London: Woodfall.

Sinaceur, Mohammed Allai
1990. Dedekind et le programme de Riemann. *RHS*, 43: 221–296.

Sklar, Lawrence
1974. *Space, time, and spacetime.* Berkeley: University of California Press.
1985. *Philosophy and spacetime physics.* Berkeley: University of California Press.
2013. *Philosophy and the foundations of dynamics.* Cambridge: Cambridge University Press.

Slowik, Edward
2006. The "dynamics" of Leibnizian relationism: Reference frames and force in Leibniz's continuum. *SHPMP*, 37: 617–634.

Smeenk, Chris, and Eric Schliesser
2013. Newton's *Principia*. In Jed Buchwald and Robert Fox (eds.), *The Oxford handbook in the history of physics* (Oxford: Oxford University Press), pp. 109–165.
Solovine, Maurice (ed.)
1956. *Albert Einstein: Lettres à Maurice Solovine.* Paris: Gauthier-Villars.

Sommerfeld, Arnold
1904a. Zur Elektronentheorie. I. Allgemeine Untersuchung des Feldes eines beliebig bewegten Elektrons. *GN* (1904), 99–130.
1904b. Zur Elektronentheorie. II. Grundlagen für eine allgemeine Dynamik des Elektrons. *GN* (1904), 363–439.
1905. Zur Elektronentheorie. III. Über Lichtgeschwindigkeits- und Überlichtgeschwindigkeits-Elektronen. *GN* (1905), 201–235.
1910a. Zur Relativitätstheorie. I. Vierdimensionale Vektoralgebra. *AP*, 28: 749–776.
1910b. Zur Relativitätstheorie. II. Vierdimensionale Vektoranalysis. *AP*, 33: 649–689.
1949. To Albert Einstein's seventieth birthday. In Schilpp 1949, pp. 99–105.

Spivak, Michael
1970–1975. *A comprehensive introduction to differential geometry.* Boston: Publish or Perish.

Stachel, John
1980. Einstein and the rigidly rotating disk. In Alan Held (ed.), *General relativity and gravitation: A hundred years after the birth of Einstein* (New York: Plenum), pp. 1–15.
1989a. The rigidly rotating body as the 'missing link' in the history of general relativity. In Howard and Stachel 1989, pp. 48–62.

1989b. Einstein's search for general covariance [from a talk given in July 1980]. In Howard and Stachel 1989, pp. 63–100.

1993. The meaning of general covariance: The hole story. In John Earman, Allen I. Janis, Gerald J. Massey, and Nicholas Rescher (eds.), *Philosophical problems of the internal and external world: Essays on the philosophy of Adolf Grünbaum* (Konstanz: Universitätsverlag/Pittsburgh: University of Pittsburgh Press), pp. 129–160.

1995. History of relativity. In L. Brown, A. Pais, B. Pippard (eds.), *Twentieth century physics* (New York: American Institute of Physics Press), pp. 249–356.

2005. Fresnel's (dragging) coefficient as a challenge to 19th century optics of moving bodies. In Kox and Stachel 2005, pp. 1–13.

2007. The first two acts. *GGR*, 1, pp. 81–111.

2014. The hole argument and some physical and philosophical implications. *Living reviews in relativity*, 17: 1–64.

Stachel, John, et al. (eds.)
1989. *The collected papers of Albert Einstein*. Vol. 2. *The Swiss years: writings, 1900–1909*. Princeton: Princeton University Press.

Staley, Richard
2008. *Einstein's generation: The origins of the relativity revolution*. Chicago: University of Chicago Press.
2011. Culture and mechanics in Germany, 1869–1918: A sketch. In Cathryn Carson, Alexei Kojevnikov, and Helmuth Trischler (eds.), *Weimar culture and quantum mechanics: Selected papers by Paul Forman and contemporary perspectives on the Forman thesis* (London: Imperial College Press), pp. 277–292.

Stallo, John Bernhard
1882. *The concepts and theories of modern physics*. New York: Appleton.

Stan, Marius
2009. Kant's early theory of motion: Metaphysical dynamics and relativity. *The Leibniz review*, 19: 29–61.
2016. Huygens on inertial structure and relativity. *Philosophy of science*, 83: 277–298.

Stanley, Matthew
2003. "An expedition to heal the wounds of war": The 1919 eclipse and Eddington as Quaker adventurer. *Isis*, 94: 57–89.

Stein, Howard
1977. Some philosophical prehistory of Einstein's general relativity. In John Earman, Clark Glymour, and John Stachel (eds.), *Foundations of space-time theories: Minnesota studies in the philosophy of science* (Minneapolis: University of Minnesota Press), pp. 3–49.

Steinle, Friedrich
2005. *Explorative Experimente: Ampère, Faraday und die Ursprünge der Elektrodynamik*. Stuttgart: Steiner.

Stewart, Albert
1964. The discovery of stellar aberration. *Scientific American*, 210 (3): 100–109.

Stokes, George Gabriel
1845a. [read on 14 April 1845] On the theories of the internal friction of fluids in motion, and of the equilibrium and motion of elastic fluids. *TCPS*, 8 (1849). Also in Stokes 1880–1905, Vol. 1, pp. 75–129.

1845b. On the aberration of light. *BAR*, 1845: 9.

1845c. On the aberration of light. *PM*, 27: 9–15.

1846a. Remarks on Professor Challis's theoretical explanation of the aberration of light. *PM*, 28: 15–17.

1846b. On Fresnel's theory of the aberration of light. *PM*, 28: 76–81.

1846c. On the aberration of light. *PM*, 28, 335–336.

1846d. On the constitution of the luminiferous aether, viewed with reference to the phenomenon of the aberration of light. *PM*, 29: 6–10.

1848. On the constitution of the luminiferous aether. *PM*, 22. Also in Stokes 1880–1905, Vol. 2, pp. 8–13.

1880–1905. *Mathematical and physical papers*. 5 vols. Cambridge: Cambridge University Press.

Stoney, George Johnstone

1891. On the cause of the double lines and of equidistant satellites in the spectra of gases. Royal Dublin Society, *scientific transactions*, 4: 563–608.

Streintz, Heinrich

1883. *Die physikalischen Grundlagen der Mechanik*. Leipzig: Teubner.

Struik, Dirk Jan

1989. Schouten, Levi-Civita, and the emergence of tensor calculus. In Eberhard Knobloch and David Rowe (eds.), *The history of modern mathematics* (San Diego: Academic Press), Vol. 2, pp. 99–108.

1993. From Riemann to Ricci: The origins of the tensor calculus. In Hari Srivastava and Themistocles Rassias (eds.), *Analysis, geometry, and groups: A Riemann legacy volume* (Palm Harbor: Hadronic Press), pp. 654–674.

Stump, David

1989. Henri Poincaré's philosophy of science. *SHPS*, 20: 335–363.

Sturm, Charles

1861. *Cours de mécanique de l'École Polytechnique*. Paris: Mallet-Bachelier.

Suppe, Frederick

1974. *The structure of scientific theories*. Urbana: The University of Illinois Press.

Swenson, Loyd

1972. *The ethereal aether: A history of the Michelson-Morley-Miller aether-drift experiments, 1880–1930*. Austin: University of Texas Press.

Swerdlow, Noel Mark

2013. Galileo's mechanics of natural motion and projectiles. In Jed Buchwald and Robert Fox (eds.), *The Oxford handbook in the history of physics* (Oxford: Oxford University Press), pp. 25–55.

Synge, John Lighton

1921. A system of space-time co-ordinates. *Nature*, 108: 275.

1960. *Relativity: The general theory*. Amsterdam: North Holland.

Tait, Peter Guthrie

1884. Note on reference frames. *Proceedings of the Royal Society of Edinburgh*, 22: 743–745.

Taltavull, Marta Jordi

2013. Sorting things out: Drude and the foundations of classical optics. In Massimiliano Badino and Jaume Navarro (eds.), *Research and pedagogy: A history of quantum physics through its textbooks* (Berlin: Edition Open Access), pp. 23–64.

Tazzioli, Rossana

1994. Rudolf Lipschitz's work on differential geometry and mechanics. In Eberhard Knobloch and David Rowe (eds.), *The history of modern mathematics* (San Diego: Academic Press), Vol. 3, pp. 113–138.

Thirring, Hans
1918. Über die Wirkung rotierender ferner Massen in der Einsteinschen Gravitationstheorie. *PZ*, 19: 13–39.

Thomson, James
1884a. On the law of inertia; the principle of chronometry, and the principle of absolute clinural rest, and of absolute rotation. *Proceedings of the Royal Society of Edinburgh*, 12: 568–578.
1884b. A problem on point-motions for which a reference-frame can so exist as to have the motions of the points relative to it, rectilinear and mutually proportional. *Proceedings of the Royal Society of Edinburgh*, 12: 730–743.

Thomson, Joseph John
1881. On the electric and magnetic effects produced by the motion of electrified bodies. *PM*, 11: 229–249.

Thomson, William, and Peter Guthrie Tait
1867. *Treatise of natural philosophy*. Vol. 1. Oxford: Clarendon Press.

Toncelli, Raffaella
2010. *Le rôle des principes dans la construction des théories relativistes de Poincaré et Einstein*. Bruxelles: Université libre de Bruxelles.

Tobin, William
2003. *The life and science of Léon Foucault: The man who proved the earth rotates*. Cambridge: Cambridge University Press.

Tonnelat, Marie-Antoinette
1971. *Histoire du principe de relativité*. Paris: Flammarion.

Tonolo, Angelo
1961. Sulle origini del calcolo di Ricci. *Annali di matematica pura ed applicata*, 53: 189–207.

Torretti, Roberto
1978. *Philosophy of geometry from Riemann to Poincaré*. Dordrecht: Reidel.
1983. *Relativity and geometry*. Oxford: Pergamon Press.

Trautmann, Andrzej
1962. Conservation laws in general relativity. In L. Witten (ed.), *Gravitation: An introduction to current research* (New York: Wiley), pp. 169–198.

Trouton, Frederick
1902. The results of an electrical experiment, involving the relative motion of the earth and ether, suggested by the late Professor FitzGerald. *Transactions of the Royal Dublin Society*, 7: 379–384.

Trouton, Frederick, and Henry Noble
1904. The mechanical forces acting on a charged condenser moving through space. *PT*, A202: 165–181.

Truesdell, Clifford
1960. The rational mechanics of flexible or elastic bodies, 1638–1788. In *Leonhardi Euleri opera omnia*, Series II, Vol. 11, Part 2.

van Berkel, Klaas, and Maarten Ultee
2013. *Isaac Beeckman on matter and motion: Mechanical philosophy in the making*. Baltimore: Johns Hopkins University Press.

van Dongen, Jeroen
2012. Mistaken identity and mirror images: Albert and Carl Einstein, Leiden and Berlin, relativity and revolution. *Physics in perspective*, 14: 126–177.

van Fraassen, Bas
1980. *The scientific image*. Oxford: Clarendon Press.

Veltmann, Wilhelm
1870a. Fresnel's Hypothese zur Erklärung der Aberrationserscheinungen. *Astronomische Nachrichten*, 75: 145–160.
1870b. Über die Fortpflanzung des Lichtes in bewegten Medien. *Astronomische Nachrichten*, 76: 129–144.
1873. Über die Fortpflanzung des Lichtes in bewegten Medien. *AP*, 30: 493–535.

Verbunt, Frank, and Marc van der Sluys
2019. Why Halley did not discover proper motion and why Cassini did. *Journal for the history of astronomy*, 50: 383–397.

Vilain, Christiane
1992. Spherical coordinates in general relativity from 1915 to 1960: A physical interpretation. In Eisenstaedt and Kox 1992, pp. 419–436.
1996. *La mécanique de Christian Huygens : la relativité du mouvement au XVIIᵉ siècle*. Paris: Blanchard.

Violle, Jules
1883, 1892. *Cours de physique*. 2 vols. Paris: Masson.
1892, 1897. *Lehrbuch der Physik*. 2 vols. Berlin: Springer.

Vizgin, Vladimir Pavlovich
2001. On the discovery of the gravitational field equations by Einstein and Hilbert: New materials. *Uspekhi Fizicheskikh Nauk*, 44: 1283–1298.

Voigt, Woldemar
1887. Über das Doppler'sche Princip. *GN* (1887), 41–51.

Volgraff, Johan Adriaan
1929a. Avertissement. In Huygens 1888–1950, pp. 171–178.
1929b. Avertissement. In Huygens 1888–1950, pp. 189–201.

Wallace, William
1981. Galileo and scholastic theories of impetus. In Alfonso Maier and Agostino Paravicini Bagliani (eds.), *Studi sul XIV secolo in memoria di Anneliese Maier* (Roma: Edizioni di storia e letteratura), pp. 275–297.

Wallis, John
1668. A summary account given by Dr. John Wallis of the general laws of motion. *PT*, 3: 864–866.

Walter, Scott
1996a. *Hermann Minkowski et la mathématisation de la théorie de la relativité restreinte. 1905–1915*. PhD thesis, Université de Paris 7.
1996b. Henri Poincaré's student notebooks, 1870–1878. *Philosophia scientiae*, 1: 1–17.
1997. La vérité en géométrie : sur le rejet de la doctrine conventionnaliste. *Philosophia scientiae*, 2: 103–135.
1999a. Minkowski, mathematicians, and the mathematical theory of relativity. In Goenner, Renn, Ritter, and Sauer 1999, pp. 45–86.
1999b. The non-Euclidean style of Minkowskian relativity. In Gray 1999, pp. 91–127.
2007a. (ed.) La correspondance entre Henri Poincaré, les physiciens, chimistes et ingénieurs. Basel: Birkhäuser.
2007b. Breaking in the 4-vectors: The four-dimensional movement in gravitation, 1905–1910. *GGR*, 3, pp. 193–252.
2008. Hermann Minkowski's approach to physics. *Mathematische Semesterberichte*, 55: 213–235.

2010. Minkowski's modern world. In Vesselin Petkov (ed.), *Minkowski spacetime: A hundred years later* (Dordrecht: Springer), pp. 43–61.
2011. Henri Poincaré, theoretical physics, and relativity theory in Paris. In Karl-Heinz Schlote and Martina Schneider (eds.), *Mathematics meets physics* (Frankfurt: Harri Deutsch), pp. 213–239.
2018. Ether and electrons in relativity theory (1900–11). In Jaume Navarro (ed.), *Ether and modernity: The recalcitrance of an epistemic object in the early twentieth century* (Oxford: Oxford University Press), pp. 67–87.

Warwick, Andrew
2003. *Masters of theory: Cambridge and the rise of mathematical physics.* Chicago: University of Chicago Press.

Weber, Heinrich
1876. Vorrede. In Riemann 1876, pp. III–V.
1892. Anmerkungen. In Bernhard Riemann, *Gesammelte mathematische Werke und wissenschaftlicher Nachlass,* 2nd ed. (Leipzig: Teubner), pp. 405–423.

Weber, Wilhelm
1839. Unipolare Induktion. Resultate aus den Beobachtungen des magnetischen Vereins, 3: 63–90. Also in *AP,* 52 (1841): 353–386.
1846. *Elektrodynamische Maassbestimmungen.* Leipzig: Weidmann.

Weber, Wilhelm, and Rudolf Kohlrausch
1857. Elektrodynamische Maasbestimmungen insbesondere Zurückführung der Stromintensitäts-Messungen auf mechanisches Maass. Königlich Sächsische Gesellschaft der Wissenschaften zu Leipzig, *Berichte.* Also in Weber, *Werke,* 6 vols. (Berlin: Springer, 1892–94), Vol. 3: 609–676.

Weinstein, Galina
2000. Poincaré's contributions to relativistic dynamics. *SHPMP,* 31: 15–48.
2015a. *Einstein's pathway to the special theory of relativity.* Cambridge: Cambridge Scholars Publishing.
2015b. *General relativity conflict and rivalries: Einstein's polemics with physicists.* Cambridge: Cambridge Scholars Publishing.
2018. Why did Einstein reject the November tensor in 1912–1913, only to come back to it in November 1915? *SHPMP,* 62: 98–122.

Westfall, Richard
1962. The foundation of Newton's philosophy of nature. *BJHS,* 1: 171–182.
1980. *Never at rest: A biography of Isaac Newton.* Cambridge: Cambridge University Press.

Weyl, Hermann
1917. Zur Gravitationstheorie. *AP,* 54: 117–145.
1918. *Raum · Zeit · Materie: Vorlesungen über allgemeine Relativitätstheorie.* Berlin: Springer.
1919. *Raum · Zeit · Materie: Vorlesungen über allgemeine Relativitätstheorie,* 3rd ed. Berlin: Springer.
1921. *Raum · Zeit · Materie: Vorlesungen über allgemeine Relativitätstheorie,* 4th ed. Berlin: Springer.

Whewell, William
1819. *An elementary treatise on dynamics.* Cambridge: Whittaker.

Whittaker, Edmund
1904. *A treatise on the analytical dynamics of particles and rigid bodies.* Cambridge: Cambridge University Press.
1951. *A history of the theories of aether and electricity.* Vol. 1: *The classical theories.* London: Nelson.

Wiechert, Emil

1894. Die Bedeutung des Weltäthers. Physikalisch-ökonomische Gesellschaft zu Königsberg, *Schriften*, 35: [4]–[11].

1896a. Über die Grundlagen der Elektrodynamik. *AP*, 59: 283–323.

1896b. Die Theorie der Elektrodynamik und die Röntgen 'sche Entdeckung. Physikalisch- ökonomische Gesellschaft zu Königsberg, *Schriften*, 37: 1–48.

1897. Über das Wesen der Elektrizität [Königsberg, January 7, 1897]. Physikalisch- ökonomische Gesellschaft zu Königsberg, *Schriften*, 38: 3–16.

1898a. Experimentelle Untersuchungen über die Geschwindigkeit und die magnetische Ablenkbarkeit der Kathodenstrahlen. *GN* (1898), 260–293. Variant in *AP*, 69, 1899: 739–766.

1898b. Hypothesen für eine Theorie der elektrischen und magnetischen Erscheinungen. *GN* (1898), 87–106.

1899. Grundlagen der Elektrodynamik. In *Festschrift zur Enthüllung des Gauss-Weber Denkmals in Göttingen* (Leipzig: Teubner), pp. 1–112.

Wien, Wilhelm

1898. Über die Fragen, welche die translatorische Bewegung des Lichtäthers betreffen. *VDNA*, 70: 49–56. More detailed in *AP*, 65: I–XVIII.

1900. Über die Möglichkeit einer elektromagnetischen Begründung der Mechanik. In *Recueil de travaux offerts par les auteurs à H. A. Lorentz à l'occasion du 25ème anniversaire de son doctorat le 11 décembre 1900. AN*, 5: 96–107.

1901. Über mögliche Aetherbewegungen. *PZ*, 2: 148–150.

1904a. Über einen Versuch zur Entscheidung der Frage, ob sich der Lichtäther mit der Erde bewegt oder nicht. *PZ*, 5: 585–586.

1904b. Über die Differentialgleichungen der Elektrodynamik für bewegte Körper. *AP*, 13: 641–668.

1904c. Erwiderung auf die Kritik des Herrn M. Abraham. *AP*, 14: 635–637.

1904d. Zur Elektronentheorie. *PZ*, 5: 576–579.

1905. *Über Elektronen*. Leipzig: Teubner.

Wilson, David

1972. George Gabriel Stokes on stellar aberration and the luminiferous ether. *BJHS*, 6: 57–72.

1987. *Kelvin and Stokes: A comparative study in Victorian physics*. Bristol: Adam Hilger.

Wilson, Patrick

1782. An experiment proposed for determining, by the aberration of the fixed stars, whether the rays of light, in pervading different media, change their velocity according to the law which results from Sir Isaac Newton's ideas concerning the cause of refraction; and for ascertaining their velocity in every medium whose refractive density is known. *PT*, 72: 58–70.

Winterberg, Friedwart

2004. On "Belated Decision in the Hilbert-Einstein Priority Dispute," published by L. Corry, J. Renn, and J. Stachel. *Zeitschrift für Naturforschung*, A59: 715–719.

Wise, Norton

1979. The mutual embrace of electricity and magnetism. *Science*, 203: 1310–1318.

1981. The flow analogy to electricity and magnetism—Part I: William Thomson's reformulation of action at a distance. *AHES*, 25: 19–70.

Wolfschmidt, Gudrun

2005. Christian Doppler (1803–1853) and the impact of the Doppler effect in astronomy. *Acta Universitatis Carolinae, Mathematica et physica*, 46, suppl.: 199–211.

Wren, Christoffer

1668. Lex naturae de collisione corporum. *PT*, 3: 867–868.

Yavetz, Ido
1995. *From obscurity to enigma: The work of Oliver Heaviside, 1872–1889*. Basel: Birkhäuser.

Young, Thomas
1804. Experiments and calculations relative to physical optics. *PT*, 94: 1–16.
1807. *A course of lectures on natural philosophy*, 2 vols. London: Johnson.

Zeeman, Pieter
1896. On the influence of magnetism on the nature of the light emitted by a substance. Physical laboratory
      at the University of Leiden. *Communications*, 23: 1–19.
1897a. On the influence of magnetism on the nature of the light emitted by a substance. *PM*, 43: 226–239.
1897b. Doublets and triplets in the spectrum produced by external magnetic forces. *PM*, 44: 55–60.

Zöllner, Friedrich
1872. *Über die Natur der Cometen. Beiträge zur Geschichte und Theorie der Erkenntnis*. Leipzig: Engelmann.
1894. *Beiträge zur deutschen Judenfrage, mit akademischen Arabesken als Unterlagen zu einer Reform der
      deutschen Universitäten*. Leipzig: Mutze.

Zund, Joseph David
1983. Some comments on Riemann's contributions to differential geometry. *Historia mathematica*, 1: 84–89.

INDEX